Leif Haugen

TECHNICAL COMMUNICATION

a reader-centered approach

TECHNICAL COMMUNICATION

a reader-centered approach

SEVENTH EDITION

PAUL V. ANDERSON

Miami University (Ohio)

WADSWORTH
CENGAGE Learning

Australia • Brazil • Japan • Korea • Mexico • Singapore • Spain • United Kingdom • United States

WADSWORTH
CENGAGE Learning

**Technical Communication:
A Reader-Centered Approach,
Seventh Edition**
Paul V. Anderson

Senior Publisher: Lyn Uhl

Publisher: Michael Rosenberg

Development Editor: Ed Dodd

Assistant Editor: Jillian D'Urso

Editorial Assistant: Erin Pass

Media Editor: Amy Gibbons

Marketing Manager: Christina Shea

Marketing Coordinator: Ryan Ahern

Marketing Communications Manager:
Laura Localio

Senior Content Project Manager:
Michael Lepera

Art Director: Jill Ort

Senior Print Buyer: Betsy Donaghey

Permissions Editor:
Mardell Glinski Schultz

Production Service: MPS Limited,
A Macmillan Company

Text Designer: Riezebos Holzbaur/
Tim Heraldo

Photo Manager: Deanna Ettinger

Cover Designer: Riezebos Holzbaur/
Brienna Hattey

Compositor: MPS Limited, A Macmillan
Company

For product information and technology assistance, contact us at
Cengage Learning Customer & Sales Support, 1-800-354-9706

For permission to use material from this text or product,
submit all requests online at **www.cengage.com/permissions**.
Further permissions questions can be emailed to
permissionrequest@cengage.com.

Library of Congress Control Number: 2009929260

ISBN-13: 978-1-4282-6393-2

ISBN-10: 1-4282-6393-4

Wadsworth
20 Channel Center Street
Boston, MA 02210
USA

Cengage Learning is a leading provider of customized learning solutions with office locations around the globe, including Singapore, the United Kingdom, Australia, Mexico, Brazil and Japan. Locate your local office at **international.cengage.com/region**

Cengage Learning products are represented in Canada by Nelson Education, Ltd.

For your course and learning solutions, visit **www.cengage.com**.

Purchase any of our products at your local college store or at our preferred online store **www.CengageBrain.com**.

Printed in Canada
1 2 3 4 5 6 7 14 13 12 11 10

BRIEF CONTENTS

APPENDICES

CONTENTS

PREFACE

Welcome to the seventh edition of *Technical Communication: A Reader-Centered Approach*. Technical communication continues to change in substantial ways, and research has produced new insights into effective communication practices and pedagogy. This edition, building on the many strengths of previous editions, offers updated treatment of core topics, coverage of additional topics, and new pedagogical features that make it even more effective for teaching and learning.

Most importantly, this edition retains the book's distinctive reader-centered approach. Its premise is that in technical communication, success is measured by reader response. If the communication helps the reader use the information provided quickly and easily and if it influences the reader's attitudes and actions in the way the writer intends, it is effective. As it addresses each aspect of writing, from the largest considerations of content and organization, to the smallest details of sentence construction and table design, the book teaches students how to elicit the desired responses from their readers.

The hallmarks of the reader-centered approach are that it

- **Teaches highly transferable strategies.** Through its uniquely strong emphasis on teaching a flexible set of research-based strategies, the reader-centered approach teaches students how to craft all the communications they will prepare in their classes and careers in a thoughtful, resourceful, and creative manner.

- **Provides in-depth coverage in an easy-to-learn manner.** In most chapters, the major points are distilled into easy-to-remember guidelines whose implications and applications are then elaborated. The guidelines themselves reinforce one another because they all flow from a common set of reader-centered principles and processes.

- **Benefits students in many fields.** The strategic emphasis and wide applicability of the reader-centered approach means that the book is well suited to students majoring in a broad range of technical, engineering, scientific, business, and other specialized fields.

- **Supports the specific course you choose to teach.** The reader-centered approach is very flexible, enabling you to select any array of chapters and projects while still preparing your students with the sophisticated, transferable skills they will need wherever they choose to work after graduation.

MAJOR CHANGES TO THIS EDITION

For this edition, I created many new features and refined others. These changes enriched my courses by adding novel teaching and learning opportunities, and they increased the fun my students and I had while working together. I believe these additions will have similar effects for you and your students.

- **A new chapter, Analyzing Information and Thinking Critically, fills a major gap in technical communication instruction.** We all tell students that brainstorming, library research, and other methods of gathering information are essential parts of the technical communication process. But textbooks (including earlier editions of this one) skip directly from gathering information to drafting, as if writers intuitively know what to tell their readers about the information and data they've gathered. The new Chapter 7 fills this gap by providing specific, practical suggestions for finding meaningful patterns in research results, examining these results critically, and determining how to present them usefully and persuasively to readers.

- **New discussions help students see how their technical communication course connects with their other writing classes.** New discussions in Chapters 1 (Communication, Your Career, and This Book) and 3 (Defining Your Communication's Objectives: Purpose, Reader, Context) describe the relationship between this book's advice and what students learn in other writing courses, including first-year composition. The new Chapter 7 (Analyzing Information and Thinking Critically) tells what the familiar term *critical thinking* means in technical fields and technical communication.

- **A new chapter, Communicating and Collaborating in the Globally Networked World, provides advice about the new technologies that are finding a place in the working world.** New Internet-supported technologies are now changing the nature of work and communication in small companies as well as international corporations. Several of these technologies, such as IM, blogs, and social networks, are familiar to students. Chapter 17 provides advice about using them in the ways they are employed *in the workplace*.

- **New Try This features that appear in the margins invite students to explore beyond the book even as they are reading it.** For instance, while reading about strategies of persuasion in Chapter 5 (Planning Your Persuasive Strategies), students are invited to visit the Scambusters website to see whether these same strategies are used by criminals. While reading Chapter 9 (Developing an Effective Style), students are invited to characterize the style of a letter or notice from an office at their university.

- **New Guiding You through the Process sections in Chapters 23 through 27 help students see how to follow advice from the first sixteen chapters when preparing proposals, reports, and instructions.** Each Guiding section is tailored to the kind of communication the student is preparing. Each describes ways to apply the advice from the earlier chapters and also supplements it.

- **New discussions describe—and give urls for—free Internet programs used in the workplace for collaborative writing.** These additions to Chapter 17 (Communicating and Collaborating in the Globally Networked World) support instructors who wish to globalize their courses by establishing intercultural projects with instructors and students in other countries. My students regularly collaborate with students in Sweden. The advice can also be used for collaborative work within a single class.
- **Revised Chapter 20 shows students how to create websites with Microsoft Word—and nothing more.** This revision to Chapter 20 (Creating Reader-Centered Websites) means that you can, if you wish, include website construction in your class without additional expense to you or your students and without needing to teach another computer program.
- **New chapter-long examples help students apply the chapters' guidelines.** By tracing a single example throughout, several chapters demonstrate the ways that their guidelines can be applied in a single project—and what each guideline contributes to its overall success. Chapter-long examples appear in Chapter 3 (Defining Your Communication's Objectives: Purpose, Reader, Context), Chapter 6 (Gathering Reader-Centered Information), Chapter 7 (Analyzing Information and Thinking Critically), and Chapter 20 (Creating Reader-Centered Websites). In fact, the same example continues through both chapters on research (Chapters 6 and 7).
- **Updated tutorials and examples use the latest releases of Microsoft Word.** Versions of the tutorials for older releases of Word are available at the book's website.
- **Updated Appendix A explains the latest APA and MLA styles.**

ADDITIONAL KEY FEATURES OF THIS EDITION

Among the many features retained from the previous edition, I have paid special attention to the following because of the special ways I believe they increase the book's breadth and effectiveness for teaching and learning.

- **Easy-to-remember guidelines** in the first twenty chapters help students retain the most central strategies, thereby increasing their resourcefulness and confidence as writers.
- **Two-page spreads** use visual design to present information and advice in an easy-to-apply manner. (See pages 176, 248, and 358 for examples.)
- **Writer's Tutorials**, many using a two-page spread design, guide students step-by-step through certain processes. (See pages 40 and 350 for examples.)
- **Writer's Reference Guides** enable students to access quickly the advice that pertains to the particular research methods, organizational patterns, and graphics they are using in the specific communication they are creating.

- **Ethics Guidelines** are integrated into the chapters, so ethics becomes a continuous topic throughout a course rather than the topic for one day's reading. In addition, most chapters include special exercises that focus on ethical issues particular to the topic of those chapters.
- **Marginal notes** highlight key points in the text and bookmark key discussions.
- **WWW guideposts** direct students to supplementary resources and information at the book's companion website.
- **Learn More guideposts** direct students to related discussions in other chapters.
- **Planning Guides and Revision Checklists** assist students as they work on their course projects and professional communications. Downloadable versions of these are available at the text's companion website.
- **Global Guidelines,** which are also integrated into the chapters, help students learn the many ways that cultural difference affects communication and provide concrete suggestions for increasing their effectiveness in cross-cultural communications.
- **Exercises** at the ends of chapters promote students' ability to apply the book's advice by asking them to Use What You've Learned. The exercises are grouped in four categories: Apply Your Expertise, Explore Online, Collaborate with Your Classmates, and Apply Your Ethics.

ORGANIZATION AND COVERAGE OF THIS EDITION

I have also retained the book's overall organization into four major parts. This structure balances attention to communication processes and products, and it provides you and other instructors with the flexibility to take up the chapters from each section in the way that best advances your courses.

- **Introduction.** Chapter 1 helps students understand the nature of writing in the workplace, the way this book builds on what they've learned in other courses, and the kinds of expertise this book will help them develop. Chapter 2 provides a detailed overview of the reader-centered approach by leading students through the process of creating highly effective résumés and job application letters.
- **Communication Process.** Chapters 3 through 16 guide students through the activities in the writing process, helping them become confident, resourceful writers. Three reference guides help students use a variety of research methods, employ seven organizational patterns that are often useful in workplace communications, and create eleven types of graphics. Richly annotated examples enable students to see the book's advice in action.

- **Applications.** Chapters 17 through 21 provide detailed advice for applying the reader-centered approach when communicating and collaborating through the Internet, creating communications with a team, creating oral presentations, making websites, and working on client and service-learning projects.
- **Superstructures.** Chapters 22 through 27 take a reader-centered approach to six of the most common types of career-related communications: correspondence, proposals, empirical research reports, feasibility reports, progress reports, and instructions.

In addition, Appendix A explains the APA and MLA documentation styles. Appendix B includes a variety of effective projects for student assignments. Downloadable and editable versions of projects are available at the book's website.

SUPPORTING MATERIALS FOR STUDENTS AND INSTRUCTORS

Did you know?? Many Cengage Learning texts are available in our textbook rental program or are also available as an eBook where you can buy by the chapter. Keep CengageBrain in mind for your next Cengage Learning purchase.

Visit CengageBrain.com. Good Grades Will Follow.

- **Instructor's Manual.** Accompanying this edition of *Technical Communication* is a new instructor's manual prepared by Professor Lisa McClure of Southern Illinois University. It includes a thorough introduction to the course, information on how to integrate supplemental materials into the class, advice on teaching the exercises and cases in the textbook, alternative cases, projects and exercises, and more. Instructors may request printed copies by calling 1-800-423-0563 and requesting ISBN 0-495-80298-0, by contacting their local Cengage sales representative, or by downloading a PDF version from the book's website: **www.cengage.com/english/anderson7e.**
- **English CourseMate.** Cengage Learning's **CourseMate** brings course concepts to life with interactive learning, study, and exam preparation tools that support the printed textbook. Watch student comprehension soar as your class works with the printed textbook and the textbook-specific website. **CourseMate** goes beyond the book to deliver what you need! Features of **English CourseMate for Technical Communication** include an **easy-to-navigate presentation; over 60 annotated sample documents** that realistically model reader-centered communication; extensive **tutorials,** many in video form, on key technical communication and grammar topics; **cases** that can serve as a basis for class discussion and projects; an **interactive eBook; quick-reference guides** that aid students in creating many specific types of reader-centered communication; and **chapter quizzes** and a **final exam** that allow students to assess their understanding.

- **Premium Website.** Completely redone for the seventh edition, the new and improved premium website for *Technical Communication*, Seventh Edition, offers an easy-to-navigate presentation coupled with extensive resources that promote broader, deeper teaching and learning, including the following:

  - **Annotated sample documents,** over sixty in total, model reader-centered communication in a realistic format.
  - **Tutorials,** many in video form, offer thorough guidance on key technical communication, grammar, and writing topics.
  - **Cases**—some suitable for homework or class discussion, others appropriate for course projects—help students hone their reader-centered communication skills.
  - **Exercises** supplementing those in the text, **interactive chapter quizzes,** and a **final exam** provide students with additional opportunities to work with core concepts from the text.
  - An **interactive eBook** incorporates user-friendly navigation, search, and note-taking and highlighting tools with the full text of the print version.
  - Downloadable and customizable **Planning Guides** help students navigate the process of creating many kinds of communication.
  - **Superstructures** serve as quick reference guides to crafting specific types of documents, from reports to proposals.
  - **PowerPoint Presentations** for every chapter provide handy tools for concept review and class discussion.
  - A **Style Guide** provides brief, user-friendly guidance on issues of grammar, punctuation, style, and usage.
  - **Web Resources** direct students to additional online tools and technical communication sites of interest.
  - **Instructor Resources** include a downloadable version of the Instructor's Manual plus teaching tips and guidance on adapting the text to common teaching approaches in technical communication.
  - **Projects.** All projects from Appendix B appear on the website so students can consult their assignments even when their book isn't handy. Additional projects provide instructors with a wider selection from which to choose assignments that are most appropriate for their students.

- With **Enhanced InSite™ for Technical Communication,** instructors and students gain access to the proven, class-tested capabilities of **InSite**—such as peer reviewing and originality checking—*plus* resources designed to help students become more successful and confident writers, including access to **Personal Tutor's paper assistance and live tutoring services,** an interactive handbook, tutorials, and more. Also included are book-specific resources such as an interactive eBook version of the text, QuickMarks customized to the text for seamless online grade marking, and extensive preloaded assignments written specifically for the Technical Communication course. Other features include fully integrated discussion boards, streamlined assignment creation, and access to **InfoTrac College Edition.** To learn more, visit us online at www.cengage.com/insite.

- **Infotrac College Edition with InfoMarks.** This online research and learning center offers over 20 million full-text articles from nearly 6,000 scholarly and popular periodicals. The articles cover a broad spectrum of disciplines and topics—ideal for every type of researcher.

- **Turnitin.** This proven online plagiarism-prevention software promotes fairness in the classroom by helping students learn to correctly cite sources and allowing instructors to check for originality before reading and grading papers. Visit www.cengage.com/turnitin to view a demonstration.

AUTHOR'S ACKNOWLEDGMENTS

Writing a textbook is truly a collaborative effort to which numerous people make substantial contributions. I take great pleasure in this opportunity to thank the many people who generously furnished advice and assistance while I was working on this seventh edition of *Technical Communication: A Reader-Centered Approach.* I am grateful to the following individuals, who prepared extensive and thoughtful reviews of the sixth edition and my preliminary plans for the seventh edition: Gertrude L. Burge, *University of Nebraska*; Diljit K. Chatha, *Prairie View A&M University*; Zana Katherine Combiths, *Virginia Polytechnic University*; Janice Cooke, *University of New Orleans*; Nancy Coppola, *New Jersey Institute of Technology*; Dawnelle A. Jager, *Syracuse University*; Linda G. Johnson, *Southeast Technical Institute*; Matthew S. S. Johnson, *Southern Illinois University Edwardsville*; L. Renee Riess, *Hill College*; Barbara Schneider, *University of Toledo*; William West, *University of Minnesota-Minneapolis*.

I would also like to thank the following persons who reviewed the previous edition: Anne Bliss, *University of Colorado-Boulder*; Adam Collins, *Grambling State University*; Suzanne Karberg, *Purdue University*; Karen R. Schnakenberg, *Carnegie Mellon University*; Elizabeth Wardle, *University of Dayton*.

While developing this edition, I have benefited from the thoughtful and energetic assistance of an extraordinary group at Wadsworth/Cengage Learning. Ed Dodd skillfully shepherded the book through the many phases of development and production. I am very grateful to Michael Lepera for his exceptional support during production. Megan Lessard contributed abundant resourcefulness and tenacity while conducting photo research. Michael Rosenberg's confidence that this edition would turn out well has been indispensable.

I am indebted to all my students for their keen critiques of the book and for the originality of their work, which provides a continuous supply of new ideas. In particular, I thank Jessica Bayles, Erin Flinn, Billy O'Brien, Joseph Terbrueggen, and

Tricia M. Wellspring, whose thorough, thoughtful work provided examples and helped me develop several of the new discussions in this edition. I am also indebted to John Folk for helping me correct one of the figures. For this edition, Betty Marak, who has read this book more than anyone else over the years, assisted once more in preparing the manuscript. Others who supplied indispensible assistance include Becky Bergman, Linda Bradley, and Magnus Gustafsson (Chalmers University of Technology, Gothenburg, Sweden), who collaborated with me on teaching and research projects through which I developed much of the new material concerning Internet communication and collaboration. Melissa Andreychek helped me think things through while we co-taught a class in which I developed and tested various innovations in this edition. Kirsten Anderson of Turnstone Design assisted me in several ways again for this edition. Tom Collins and Steve Oberjohn have provided enduring assistance over the years. Tom graciously offered a hand again this time. All my work in technical communication benefits from many conversations with and numerous examples of excellent teaching provided by Jean Lutz. I am deeply grateful to all of these individuals.

Finally, I thank my family. Their encouragement, kindness, and good humor have made yet another edition possible.

PAUL V. ANDERSON
Oxford, Ohio

■ Preface

PART I

INTRODUCTION

1

1 | Communication, Your Career, and This Book

CHAPTER OVERVIEW

From the perspective of your professional career, communication is one of the most valuable subjects you will study in college.

Why? Imagine what your days at work will be like. If you are majoring in an engineering, technical, or other specialized field, you will spend much of your time using the special knowledge and skills you learned in college to answer questions asked by coworkers and complete projects assigned by managers. Furthermore, you will generate many good ideas on your own. Looking around, you'll discover ways to make things work better or do them less expensively, overcome problems that have stumped others, or make improvements others haven't begun to dream about.

Yet all your knowledge and ideas will be useless unless you communicate them to someone else. Consider the examples of Sarah Berlou and Larry Thayer. A recent college graduate who majored in metallurgy, Sarah has spent three weeks analyzing pistons that broke when her employer tested a fuel-saving automobile engine. Her analysis has been skillful. Her conclusions are valid. However, the insights she gained about why the pistons failed will be useless to her employer unless she communicates them clearly and usefully to the engineers who must redesign the pistons. Similarly, Larry, a nutritionist newly hired by a hospital, has several ideas for improving the efficiency of the hospital's kitchen. However, his ideas will reduce costs and improve service to patients only if he presents his recommendations persuasively to the people who have the power to implement them.

WWW

For additional information, examples, and exercises related to this chapter, visit this book's website **www.cengage.com/english/anderson7e** and click on Chapter 1.

COMMUNICATION EXPERTISE WILL BE CRITICAL TO YOUR SUCCESS

College graduates typically spend one day a week—or more—writing.

Like Sarah and Larry, you will be able to make your work valuable to others only if you communicate it effectively. Numerous studies indicate that the typical college graduate spends about 20 percent of his or her on-the-job time writing (Beer & McMurrey, 1997). That's one day out of every five-day work week! And it doesn't include the additional time spent talking—whether on the phone or in person, whether in groups and meetings or one-on-one. Writing is so important to employers that they spend an estimated $3.1 billion annually on writing instruction for their employees (National Commission on Writing, 2004).

The ability to communicate effectively increases your promotability.

Moreover, the ability to communicate effectively can significantly increase the success you enjoy in your career. Researcher Stephen Reder discovered that college graduates judged to be in the top 20 percent of writing ability earn, on average, more than three times as much as workers rated in the bottom 20 percent of writing ability (Fisher, 1998). In a survey of leading employers, the National Commission on Writing (2004) found that writing skills is a major consideration for promotion.

This book's goal is to help you develop the communication expertise you need in order to realize the full potential of your expertise in your specialized field.

Developing your communication expertise will be even more important if you have chosen a career as a technical, scientific, medical, or professional communicator. Work in these fields is ideal for persons who love to learn and communicate about technical and scientific advances. Employed by private corporations, nonprofit organizations, and government agencies, these professional communicators typically create print and online communications on topics that match their personal interests, such as computers, health, and environmental science.

In addition to being essential to your career, communication expertise will enable you to make valuable contributions to your campus or community. Volunteer organizations, service clubs, and other organizations will welcome your assistance in writing clear and compelling reports, proposals, and other documents. When confronting complex decisions about environmental standards, economic policy, and other issues, the citizens of your community will be grateful for your ability to explain technical, scientific, and other specialized subjects in ways they can understand.

WWW

For more information on careers in technical communication and related fields, go to **www.cengage. com/english/anderson7e** and click on Chapter 1.

CHARACTERISTICS OF WORKPLACE WRITING

This book assumes that you already know a great deal about effective communication that will be invaluable to you in your career. It also assumes that you must learn new skills—and even new ways of thinking about communication—in order to develop the expertise required on the job. A first step in building on and expanding your current communication abilities is to learn the key characteristics of workplace writing so you can see how it resembles and differs from the writing you are accustomed to doing in school and other settings.

Serves Practical Purposes

On the job, you will write for practical purposes, such as helping your employer improve a product or increase efficiency. Your readers will be coworkers, customers, or other individuals who need information and ideas from you in order to pursue their own practical goals. You may have prepared this type of communication for instructors who asked you to write to real or imagined readers who need your information to make a decision or take an action. In contrast, term papers, essay exams, and similar school assignments have an equally valid, but much different purpose. They are intended to help you learn and demonstrate your mastery of course material. Although your instructors will read them in order to assess your knowledge and decide what grade to assign, they usually don't need the information and ideas you provide in order to guide their decisions and actions as they pursue their own goals.

Different purposes profoundly affect the kinds of communication you need to produce. For example, when you write an essay exam or term paper whose purpose is to show how much you know, you succeed by saying as much as you can about your subject. At work, where your goal is to help your readers make a decision or perform a task, you succeed by telling your readers only what they need, no matter how much more you know. Extra information will only clog your readers' paths to what they require. In your career, you will need to continue to develop your ability to determine what your readers want and need.

Must Satisfy Many Different Readers in a Single Communication

At work, writers often create a single communication that addresses a group of people who differ from one another in many important ways, such as their familiarity with your specialty, the ways they will use your information, and their professional and personal concerns. The audience for Larry's report recommending changes to the hospital kitchen is complex in this way. His readers include his supervisor, who will want to know how operations in her area would have to change if Larry's recommendations were adopted; the vice president for finance, who will want to analyze Larry's cost estimates; the director of personnel, who will want to know how job descriptions will need to be rewritten; and members of the labor union, who will want assurances that the new work assignments will treat them fairly. On the job, you will need expertise at constructing one communication that, like Larry's, must simultaneously satisfy an array of persons who will each read it with his or her own set of concerns and goals in mind.

Addresses International and Multicultural Audiences

Learn More

To learn strategies for addressing international and intercultural communication audiences, read the Global Guidelines included in most of the other chapters.

Also, when writing at work, you may often address readers from other nations and cultural backgrounds. Many organizations have clients, customers, and suppliers in other parts of the world. Thirty-three percent of U.S. corporate profits are generated by international trade (Lustig & Koester, 1993), and the economies of many other nations are similarly linked to distant parts of the globe. Corporate and other websites are accessed by people around the planet. Even when communicating to coworkers at your own location, you may address a multicultural audience—persons of diverse national and ethnic origins.

Uses Distinctive Types of Communication

At work, people create a wide variety of job-related communications that aren't usually prepared at school, including memos, business letters, instructions, project proposals, and progress reports. Each of these types of communication has its own conventions, which you must follow to write successfully.

Employs Graphics and Visual Design to Increase Effectiveness

In communications written at work, tables, charts, drawings, photographs, and other graphics are often as important as written text. To write effectively, you will need expertise at creating graphics and at arranging your graphics and text on a page or computer screen in ways that make your communications visually appealing, easy to understand, and easy to navigate. Figure 1.1 shows a page from an instruction manual that illustrates the importance of graphics and visual design.

Requires Collaboration

Whatever experience you gain at school in writing collaboratively will benefit you on the job. Eighty-seven percent of the college graduates studied by researchers Lisa Ede and Andrea Lunsford (1990) reported that they write with cowriters at least some of the time. For long documents, the number of cowriters is sometimes astonishingly large. Martin Marietta Corporation's multivolume proposal to build the international space station contained text and drawings by more than 300 engineers (Mathes & Stevenson, 1991). Even when you prepare communications alone, you may consult your coworkers, your boss, and even members of your intended audience as part of your writing process.

In one common form of collaboration, you will need to submit drafts of some of your communications for review by managers and others who have the power to demand changes. The number of reviewers may range from one to a dozen or more, and some drafts go through many cycles of review and revision before obtaining final approval.

© Anton Vengo/SuperStock

On the job, groups of employees often work together to plan, draft, and revise proposals, reports, and other printed, online, and oral communications. To learn strategies for using graphics and visual design to achieve your communication objectives on the job, see Chapters 13 and 14.

Created in a Globally Networked Environment

At work, increasing numbers of employees are doing their collaborative writing in a globally networked environment. Using the Internet, project team members write reports, proposals, and other documents even though these people may be located in different countries or on different continents. The teams may hold their work sessions entirely online (see Figure 1.2), never meeting in person. Their members may come from different cultures and have different languages as their native languages.

Shaped by Social and Political Factors

Every communication situation has social dimensions. In your writing at school, the key social relationship usually is that of a student to a teacher. At work, you will have a much wider variety of relationships with your readers, such as manager to subordinate, customer to supplier, or coworker to coworker. Sometimes these

Learn More

To learn strategies for writing collaboratively in the globally networked workplace, see Chapter 18, especially Guidelines 7, 8, and 9. See also the Global Guidelines included in many chapters.

FIGURE 1.1

Page That Illustrates the Way Graphics and Visual Design Work Together with Words in Technical Communications

The visually prominent heading explains what readers will learn from this page.

The drawings show exactly what a reader needs to do to clean the printer; they even show a hand performing these tasks.

The drawings include arrows to indicate the direction of movement.

Each numbered drawing corresponds to the step with the same number.

The numbers for the drawings and steps are visually prominent to help readers match each drawing with its corresponding step.
- In the steps, the numbers are bold and placed in a column of their own.
- In the drawings, the numbers are large and bold.

The cautions are highlighted visually.
- Horizontal lines (called rules) set the cautions off from the text.
- The word Caution is printed in bold and blue.

To Clean the Printer

1   Turn the printer off and unplug the power cable, and then open the printer's top cover by pressing the top cover release on the side of the printer.

2   Remove toner cartridge.

Caution
Because light damages the cartridge's photosensitive drum, do not expose the cartridge to light for more than a few minutes.

3   With a dry lint-free cloth, wipe any residue from the paper path area and the toner cartridge cavity as shown.

4   Remove the cleaning brush from the shoulder above the toner cartridge area. Place the flat part of the brush on the shoulder while allowing the brush to be inserted below the shoulder where the mirror is located. Move the brush from side to side several times to clean the mirror.

5   Replace the brush and toner cartridge, close the top cover, plug in the power cable, and then turn the printer on.

Caution
Do not touch the transfer roller (shown in the illustration) with your fingers. Skin oils on the roller can cause print quality problems.

Caution
If toner gets on your clothes, wipe it off with a dry cloth and wash your clothes in **cold** water. Hot water sets toner into fabric.

Troubleshooting and Maintenance 4-19

4 Troubleshooting and Maintenance

At the side of this and every page in the manual is a colored rectangle that gives the section number and title. These rectangles help readers flip quickly to the specific information they need.

relationships will be characterized by cooperation and goodwill. At others, they will be fraught with competitiveness as people strive for recognition, power, or money for themselves and their departments. To write effectively, you will need to attune the style, tone, and overall approach of each communication to these social and political considerations.

FIGURE 1.2
Writing Team Members in Australia and Germany Conduct a Work Session Online

Shaped by Organizational Conventions and Culture

In addition, each organization has a certain style that reflects the way it perceives itself and presents itself to outsiders. For example, an organization might be formal and conservative or informal and innovative. Individual departments within organizations may also have their own styles. On the job, you will be expected to understand the style of your organization and employ it in your writing.

Must Meet Deadlines

In many jobs, deadlines are much more significant—and changeable—than most deadlines at school. A deadline may be pushed back or advanced several times during on a project. No matter what the deadline, however, the work must be completed on time. For example, when a company prepares a proposal or sales document, it must reach the client on time. Otherwise, it may not be considered at all—no matter how good it is. Employers sometimes advise that "it's better to be 80 percent complete than 100 percent late."

Sensitive to Legal and Ethical Issues

Under the law, most documents written by employees represent the position and commitments of the organization itself. Company documents can even be subpoenaed as evidence in disputes over contracts and in product liability lawsuits.

WWW

For more on the legal and ethical dimensions of workplace communication, go to **www.cengage.com/english/anderson7e** and click on Chapter 1.

Ethics are discussed further on page 20.

Even when the law does not come into play, many communications written at work have moral and ethical dimensions. The decisions and actions they advocate can affect many people for better or worse. Because of the importance of the ethical dimension of workplace writing, this book incorporates in most chapters a discussion of ethical issues that may arise in your on-the-job communications.

AT WORK, WRITING IS AN ACTION

As you can infer from the preceding section, there is tremendous variety among communications written on the job, depending on such variables as their purposes and readers; organizational conventions and cultures; and the political, social, legal, and ethical contexts in which they are prepared. Some people diminish their ability to write effectively in these multifaceted, shifting situations because they mistakenly think of writing as an afterthought, as merely recording or transporting information they developed while acting as specialists in their chosen fields.

Nothing could be further from the truth.

When you write at work, you act. You exert your power to achieve a specific result, to change things from the way they are now to the way you want them to be. Consider, again, the examples of Sarah and Larry. Sarah wants to help her team develop a successful engine. Acting as a metallurgical specialist, she has tested the faulty pistons to determine why they failed. To contribute to the success of the engine, she must now perform a writing act. She must now compose sentences, construct tables of data, and perform other writing activities to present her results in a way the engineers will find useful. Similarly, Larry believes that the hospital kitchen is run inefficiently. Acting as a nutrition specialist, he has devised a plan for improving its operation. For his plan to be put into effect, Larry must perform an act of writing. He must write a proposal that persuades the hospital's decision makers to implement his plan.

The most important thing to remember about the "writing acts" you will perform at work is that they are social actions. Every communication you write will be an interchange between particular, individual people: you and your readers. Perhaps you will be a supervisor telling a coworker what you want done, an adviser trying to persuade your boss to make a certain decision, or an expert helping another person operate a certain piece of equipment. Your reader may be an experienced employee who is uncertain of the purpose of your request, a manager who has been educated to ask particular questions when making a decision, or a person who has operated a similar piece of equipment for years.

Even when writing to a large group of people, your communication will establish an individual relationship between you and each person in the group. Each person will read with his or her own eyes, react with his or her own thoughts and feelings.

THE MAIN ADVICE OF THIS BOOK: THINK CONSTANTLY ABOUT YOUR READERS

The observation that writing is a social action leads to the main advice of this book: When writing, think constantly about your readers. Think about what they want from you—and why. Think about how you want to help or influence them and how they will react to what you have to say. Think about them as if they were standing right there in front of you while you talked together.

You may be surprised that this book emphasizes the personal dimension of writing more than such important characteristics as clarity and correctness. Although clarity and correctness are important, they cannot, by themselves, ensure that something you write at work will be successful.

For example, if Larry's proposal for modifying the hospital kitchen is to succeed, he will have to explain the problems created by the present operation in a way that his readers find compelling, address the kinds of objections his readers will raise to his recommendations, and deal sensitively with the possibility that his readers may feel threatened by having a new employee suggest improvements to a system they themselves set up. If his proposal fails to do these things, it will not succeed, no matter how "clear and correct" the writing is. A communication may be perfectly clear, perfectly correct, and yet be utterly unpersuasive, utterly ineffective.

The same is true for all the writing you will do at work: What matters is how your readers respond. That's the reason for taking the reader-centered approach described in this book. This approach focuses your attention on the ways you want to help and influence your readers, and teaches specific strategies that you can use to achieve these goals.

The importance of learning to think constantly about your readers is highlighted by research that compared the ways students and successful workplace communicators approach their writing tasks (Beaufort, 1999; Bereiter & Scardamalia, 1993; Dias & Paré, 2000; Spilka, 1993). These studies reveal that, unlike most students, successful workplace communicators think about readers while deciding about nearly every detail of their communications. When they aren't thinking specifically about their readers, successful workplace writers are thinking about the conventions, expectations, and situations that could influence their readers' responses to their writing. This book will assist you in developing the same reader-centered approach that is used by these expert communicators.

QUALITIES OF EFFECTIVE ON-THE-JOB COMMUNICATION: USABILITY AND PERSUASIVENESS

A first step in developing your expertise in on-the-job writing is to focus your attention on the two qualities that a workplace communication must have to be successful: usability and persuasiveness. Both qualities, of course, must be defined from the readers' perspective.

Usability refers to a communication's ability to help its readers. As explained above, people read at work to gain information they need in order to *do something*. Their tasks may be physical, such as installing a new memory card in a computer. Or their tasks may be mental, as when Sarah's readers use her report to redesign the pistons and Larry's readers compare his recommended procedures with those currently used in the hospital kitchen. No matter what the readers' task, a workplace communication is highly usable if it enables them to:

- Locate quickly the information they need in order to accomplish their goal.
- Understand the needed information easily and accurately.
- Use the information to complete their task with minimum effort.

If readers have difficulty locating, understanding, or using the information they need, the communication is not as usable, not as helpful, as it should be.

A communication's persuasiveness is its ability to influence its readers' attitudes and actions. It's easy to imagine why persuasiveness is important in proposals and recommendation reports. However, persuasiveness is also an indispensable quality in *all* on-the-job communications, including ones we often think of as purely informational. For example, experience shows that many people are impatient with instructions. They read them carelessly, if at all. Thus, an instruction manual must persuade potential readers to consult it rather than try things out on their own, possibly damaging products, fouling the equipment, or harming themselves.

As the preceding paragraphs suggest, every communication written at work must have *both* usability and persuasiveness to succeed. In most communications, one of these qualities dominates: usability in instructions, for example, and persuasiveness in proposals. However, the two are always inextricably intertwined. Instructions are effective only if the intended readers are persuaded to use them. A proposal can persuade only if its readers can easily find, understand, and analyze its content. To see how you can combine usability and persuasiveness in a single communication, look at the web page and memo shown in Figures 1.3 and 1.4.

WWW

To view other websites and print documents that combine usability and persuasiveness, go to **www.cengage.com/ english/anderson7e** and click on Chapter 1.

THE DYNAMIC INTERACTION BETWEEN YOUR COMMUNICATION AND YOUR READERS

Of course, the people who actually decide whether your on-the-job communications are usable and persuasive will be your readers. Consequently, the detailed suggestions in this book are based, in large part, on what researchers have learned about how people read. The following paragraphs describe the three research findings that will be the most useful to you.

- Readers construct meaning.
- Readers' responses are shaped by the situation.
- Readers react moment by moment.

In this home page, writers at Dell met the challenge of creating a site that would be both usable and persuasive to a wide variety of persons, including consumers and business people, novice and expert computer users, new visitors and longtime customers, as well as persons in the United States and other countries around the world.

FIGURE 1.3
Web Page that Combines Usability and Persuasiveness

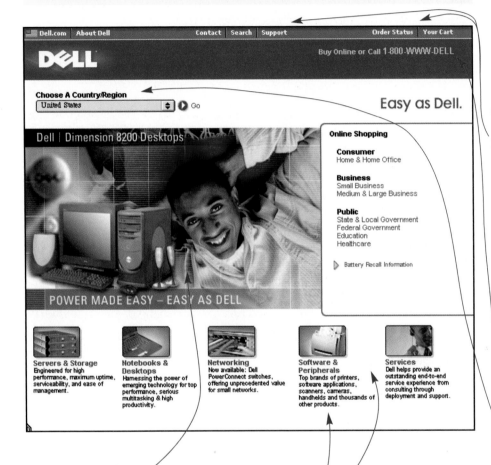

Usability strategies

At the top of the page, the writers provide several links that enable people to find specific information quickly.

- A link to the mailing address and phone number for various company offices.
- A link to a search engine that enables people to locate specific information without navigating through multiple screens.
- A link to support.
- A link that enables people to check the status of orders they've made.

To help people who prefer to buy on the phone rather than online, the writers include the number for Dell's sales agents.

For users around the world, the writers provide links to Dell sites in many countries.

The writers organize their links in ways that help users quickly find the information they want.

- Links on the right-hand side are organized around types of customers.
- Links across the bottom are organized around the products and services people might want.

Persuasive Strategies

To emphasize customer satisfaction with Dell, the writers included a large photograph that shows a smiling, relaxed, and unstressed person.

At the bottom of the page, the writers describe Dell's products in highly favorable ways.

They use bright colors against a white background to create an upbeat, appealing design.

By using large amounts of white space, the writers create an uncrowded, orderly design that reinforces the slogan, "Easy as Dell."

FIGURE 1.4
Memo That Combines
Usability and
Persuasiveness

In this memo, Frank, a chemist, uses many usability and persuasive strategies while reporting test results to his boss, Alyssa. His report will help her decide whether to use a newly developed plastic to bottle an oven cleaner manufactured by their employer. This product is sold on supermarket shelves.

Usability strategies

In the subject line, Frank tells Alyssa exactly what the memo is about.

Frank uses the first sentence to tell Alyssa how the memo relates to her: It reports on tests she requested.

In the next two sentences, Frank presents the information that Alyssa most needs: The new plastic won't work. She doesn't have to read the entire memo to find this key information.

Using headings and topic sentences, Frank helps Alyssa locate specific details quickly.

He uses bold type to make the headings stand out visually.

By presenting the test results in a list, he enables Alyssa to read through them quickly.

Frank states clearly the significance of each test result.

Persuasive strategies

To build credibility for his results, Frank describes the test method in detail.

To establish the credibility of the conclusion he draws, Frank presents specific details concerning each of the three major problems discovered in the storage tests.

By using lively active verbs (highlighted in color), he creates a vivid, action-oriented writing style.

By preparing a neat and carefully proofread memo, Frank bolsters confidence in the care he takes with his work.

PIAGETT HOUSEHOLD PRODUCTS
Intracompany Correspondence

October 12, 2007

To Alyssa Wyatt
From Frank Thurmond
Subject Test of Salett 321 Bottles for Use with StripIt

We have completed the tests you requested to find out whether we can package StripIt Oven Cleaner in bottles made of the new plastic, Salett 321. We conclude that we cannot, chiefly because StripIt attacks and begins to destroy the plastic at 100°F. We also found other significant problems.

Test Methods

To test Salett 321, we used two procedures that are standard in the container industry. First, we evaluated the storage performance of filled bottles by placing them in a chamber for 28 days at 73°F. We stored other sets of 24 bottles at 100°F and 125° for the same period. Second, we tested the response of filled bottles to environmental stress by exposing 24 of them for 7 days to varying humidities and varying tremperatures up to 140°F.

We also subjected glass bottles containing StripIt to the same test conditions.

Results and Discussion

In the 28 day storage tests, we discovered three major problems:

- StripIt attacked the bottles made from Salett 321 at 100°F and 125°F. At 125°F, the damage was particularly serious, causing localized but severe deformation. Most likely, StripIt's ketone solvents weakened the plastic. The deformed bottles leaned enough to fall off shelves in retail stores.

- The sidewalls sagged slightly at all temperatures, making the bottles unattractive.

- StripIt yellowed in plastic bottles stored at 125°F. No discoloration occured in glass bottles at this temperature. We speculate that StripIt interacted with the resin used in Salett 321, absorbing impurities from it.

In the environmental test, StripIt attacked the bottles at 140°F.

Conclusion

Salett 321 is not a suitable container for StripIt. Please call me if you want additional information about these tests.

All three research findings stress that writing is not a passive activity for readers, but a dynamic interaction between readers and texts.

Readers Construct Meaning

When researchers say that readers *construct* meaning, they are emphasizing the fact that the meaning of a written message doesn't leap into our minds solely from the words we see. Instead, to derive meaning from the message, we actively interact with it. In this interaction, we employ a great deal of knowledge that's not on the page, but in our heads. Consider, for example, the knowledge we must possess and apply to understand this simple sentence: "It's a dog." To begin, we must know enough about letters and language to decipher the three printed words (*it's, a,* and *dog*) and to understand their grammatical relationships. Because they do not possess this knowledge, young children and people from some other cultures cannot read the sentence. Often, we must also bring other kinds of knowledge to a statement in order to understand it. For instance, to understand the meaning of the sentence "It's a dog," we must know whether it was made during a discussion of Jim's new pet or during an evaluation of ABC Corporation's new computer.

In addition to constructing the meanings from individual words and sentences, we build these smaller meanings into larger structures of knowledge. These structures are not merely memories of words we have read. They are our own creations. To demonstrate this point, write a sentence that explains the following heading, which you read earlier in this chapter: "At work, writing is an action." Next, look for a sentence in the book that exactly matches yours. Most likely, you won't find one. The sentence you wrote is not one you remembered. Rather, it is the meaning you constructed through your interaction with the text.

The fact that readers construct the meaning they derive from a communication has many implications for writers that are explored later in this book. An especially important one is this: You should learn as much as possible about the knowledge your readers will bring to your communication so you can create a communication that helps them construct the meanings you want them to build.

Readers' Responses Are Shaped by the Situation

A second important fact about reading is that people's responses to a communication are shaped by the context in which they read—including such things as their purpose for reading, their perception of the writer's purpose, their personal stake in the subject discussed, and their past relations with the writer.

For example, Kate has finished investigating several brands of laptop computers to determine which one she should recommend her company purchase for each of its fifty field engineers. Opening her e-mail, she finds a message asserting that the ABC computer is "a dog." Her response to this statement will depend on many factors. Did the information she gathered support this assessment?

Was the statement made by a computer specialist, a salesperson for one of ABC's competitors, or the president of Kate's company? Has she already announced publicly her own assessment of the computer, or is she still undecided about it? Depending on the answers to these questions, Kate's response might range anywhere from pleasure that her own judgments have been supported by a well-respected person to embarrassment that her publicly announced judgment has been called into question.

The range of situational factors that can affect a reader's response is obviously unlimited. The key point is that in order to predict how your readers might respond, you must understand thoroughly the situation in which they will read your message.

Readers React Moment by Moment

The third important fact about reading is that readers react to communications moment by moment. When we read a humorous novel, we chuckle as we read a funny sentence. We don't wait until we finish the entire book. Similarly, people react to each part of a memo, report, or proposal as soon as they come to it. Consider the following scenario.

Imagine that you manage a factory's personnel department. A few days ago, you discussed a problem with Donald Pryzblo, who manages the data processing department. Recently, the company's computer began issuing some payroll checks for the wrong amount. Your department and Pryzblo's work together to prepare each week's payroll in a somewhat antiquated way. First, your clerks collect a time sheet for each employee, review the information, and transfer it to time tickets, which they forward to Pryzblo's department. His clerks enter the information into a computer program that calculates each employee's pay and prints the checks. The whole procedure is summarized in the diagram shown here.

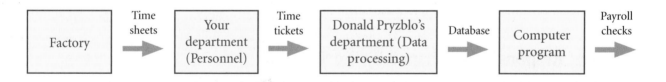

In your discussion with Pryzblo, you proposed a solution he did not like. Because you two are at the same level in the company, neither of you can tell the other what to do. When you turn on your computer this morning, you find an e-mail message from Pryzblo.

For this demonstration, your task is to read the memo *very slowly*—so slowly that you can focus on the way you react, moment by moment, to each statement. First, get a sheet of paper. Then, turn to the e-mail message shown in Figure 1.5 on page 17, covering

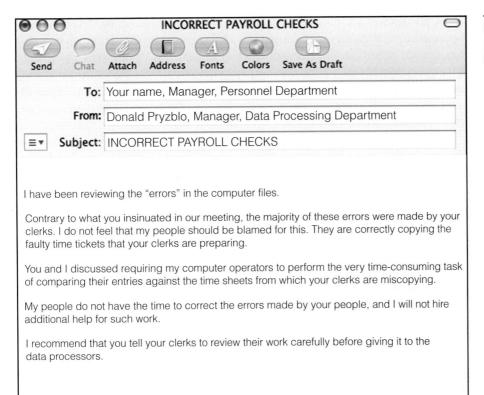

FIGURE 1.5
E-mail Message for
Demonstration

INCORRECT PAYROLL CHECKS

Send Chat Attach Address Fonts Colors Save As Draft

To: Your name, Manager, Personnel Department

From: Donald Pryzblo, Manager, Data Processing Department

Subject: INCORRECT PAYROLL CHECKS

I have been reviewing the "errors" in the computer files.

Contrary to what you insinuated in our meeting, the majority of these errors were made by your clerks. I do not feel that my people should be blamed for this. They are correctly copying the faulty time tickets that your clerks are preparing.

You and I discussed requiring my computer operators to perform the very time-consuming task of comparing their entries against the time sheets from which your clerks are miscopying.

My people do not have the time to correct the errors made by your people, and I will not hire additional help for such work.

I recommend that you tell your clerks to review their work carefully before giving it to the data processors.

it with the sheet of paper. Next, slide the paper down the page, stopping after you read the first sentence. Immediately record your reactions (in your role as manager of the personnel department). Proceed in this way through the rest of the message.

Finished? Now look over your notes. Most people who participate in this demonstration find themselves responding strongly to almost every sentence. For example, they react to the quotation marks that surround the word errors in the first sentence. The word *insinuated* in the second sentence also draws an immediate response from most readers. (They laugh if they forget to play the role of personnel manager; they cringe if they remember to play the role.)

The fact that readers respond to a communication moment by moment is important to you as a writer because your readers' reaction to any one sentence in a communication you write will influence their reaction to everything they read from that point forward. For example, most people who read Donald Pryzblo's memo while playing the role of personnel manager grow defensive the moment they see the quotation marks around the word *errors*, and they become even more so when they read the word *insinuated*. After they read the third paragraph, their defensiveness hardens into a grim determination to resist any recommendation Pryzblo may make.

Readers respond on a moment-by-moment basis.

Readers' reactions in one moment shape their subsequent reactions.

When writing, keep your
readers foremost in mind.

A few readers are more even tempered. Instead of becoming defensive, they become skeptical. As they read the first two sentences, they realize Pryzblo is behaving emotionally rather than intellectually, so they evaluate his statements very carefully. When they read his accusation that the personnel department clerks are miscopying the time sheets, they want to know what evidence supports that claim. When the next sentence fails to provide any evidence, they feel disinclined to go along with any recommendations from Pryzblo.

Thus, even though individual readers react to the first few sentences of this memo differently, their early reactions shape their responses to the sentences that follow. Consequently, even though Pryzblo's recommendation seems sensible enough, no reader I've met feels inclined to accept it.

Of course, Pryzblo may have had some other purpose in writing his memo. For example, he may have wanted to inflame the personnel manager into responding rashly so he or she would get in trouble with the boss. For that purpose, his memo might have worked very well. But even in that case, the basic points of this demonstration would remain unchanged: People react to what they read on a moment-by-moment basis, and their reactions at each moment shape their reactions to what follows.

The preceding discussion underscores the importance of taking a reader-centered approach to writing, one in which you think constantly about your readers. Each reader creates his or her own response to your communications. To write effectively, you must predict these responses and design your messages accordingly. You will be best able to do this if you keep your readers—their needs and goals, feelings and situations, preferences and responsibilities—foremost in mind throughout your work on each communication.

SOME READER-CENTERED STRATEGIES YOU CAN BEGIN USING NOW

Despite the many ways readers' goals, concerns, feelings, and likely responses can vary from situation to situation, readers approach almost all on-the-job communications with several widely shared aims and preferences. The following list briefly introduces several reader-centered strategies you can begin using immediately that address these common aims and preferences. All are discussed more fully later in this book. Many are illustrated in the web page and memo shown in Figures 1.3 and 1.4.

Try This

At the website of an employer you might like to work for, look for information for college students seeking jobs. Which of the first three strategies does it use well? Which could it use more effectively?

- **Help your readers find the key information quickly.** State your main points at the beginning rather than middle or end of your communications. Use headings, topic sentences, and lists to guide your readers to the specific information they want to locate. Eliminate irrelevant information that can clog the path to what your readers want.

- **Use an easy-to-read writing style**. Trim away unnecessary words. Use the active voice rather than the passive. Put action in verbs rather than in other parts of speech. To test the effectiveness of these three techniques, researchers James Suchan and Robert Colucci (1989) created two versions of the same report. The high-impact version used the techniques. The low-impact one did not. The high-impact version reduced reading time by 22 percent, and tests showed that readers understood it better.
- **Highlight the points your readers will find to be persuasive.** Present the information your readers will find more persuasive before you present the information they will find to be less persuasive. Show how taking the actions you advocate will enable them to achieve their own goals. When selecting evidence to support your arguments, look specifically for items you know your readers will find to be credible and compelling.
- **Talk with your readers.** Before you begin work on a report or set of instructions, ask your readers, "What do you want in this communication? How will you use the information it presents?" When planning the communication, share your thoughts or outline, asking for their reactions. After you've completed a first draft, ask for their feedback.

If you cannot speak directly with your readers, talk with them imaginatively by creating an imaginary movie of your reader in the act of reading your communication (Elbow, 1998). In your movie, it's crucial that you talk *with* your readers, not *to* them. When you talk *to* other people, you are like an actor reciting a speech: You stick to your script without regard to the way your audience is responding. When you talk *with* other people, you adjust your statements to fit their reactions. Does someone furrow his brow in puzzlement? You explain the point more fully. Does someone twist her hands impatiently? You abbreviate your message.

By creating a mental movie of her readers, Marti, a mechanical engineer, gradually refined one step in a set of instructions so that it met her readers' needs. Here is her first draft.

15. Check the reading on Gauge E.

While watching her readers' reactions in her mental movie, Marti saw them ask, "What should the reading be?" So she answered their question.

15. Determine whether the reading on Gauge E matches the appropriate value listed in the Table of Values.

In her mental movie, Marti then imagined her readers asking, "Where is the Table of Values?" She revised again.

15. Determine whether the reading on Gauge E matches the appropriate value listed in the Table of Values (page 38).

Finally, Marti imagined her reader discovering in the Table of Values that the reading on Gauge E was wrong. Her imaginary readers asked, "What should we do now?" Marti added another sentence.

| If the value does not match, follow the procedure for correcting imbalances (page 27). |

COMMUNICATING ETHICALLY

So far, this chapter has introduced concepts and strategies that will serve as the springboard from which you can understand and apply the rest of this book's advice for creating effective communications at work. Because effective communications create change—they make things happen—they also have an important ethical dimension. When things happen, people are affected. Their happiness and even their health and well-being may be impacted. For example, you may write a proposal for a new product that can cause physical harm—at least if not handled properly. You may prepare a report that managers will use to make other people's jobs significantly more—or less—desirable, or even determine whether these people will continue to be employed. Ethical considerations can also apply to impacts on people outside your employer's organization, even future generations, and the environment.

Because technical communications have an important ethical dimension, in addition to providing advice for creating usable, persuasive communications, this book also provides advice for creating ethical ones.

What Is Ethical?

To act ethically, you must first have a sense of what is ethical. At work, you have three major sources for guidance.

WWW

To view professional and corporate codes of ethics, go to **www.cengage.com/english/anderson7e** and click on Chapter 1.

- Professionals in your specialty have probably developed a code of ethics.
- Your employer may also have developed an ethics code. Some companies have even hired professional ethics specialists whom employees may consult.
- Your own sense of values, the ones you developed in your home, community, and studies.

Some students wonder whether it will really be necessary for them to think about their values at work. In your career, will you actually be asked to write unethically? The answer depends largely on the company you work for and the managers to whom you report. Based on interviews with top executives, Charles E. Watson (1991) reports that 125 of the largest companies in the United States have a strong commitment to ethical behavior. Employees of these companies, one assumes, receive support and guidance when facing an ethical conflict. Similarly, in *Companies with a Conscience*, Mary Scott and Howard Rothman (1992) describe many instances of values-minded management. And the Center for Business Ethics has a file with hundreds of ethics codes that corporations throughout the United States have adopted as their official policies (see page 123 for an example). Still, we all have read stories about companies that engage in bribery, price gouging, dumping of toxic wastes in public waterways, and other unethical practices. Further, even a company that has adopted an ethics code may have employees who act unethically. A nationwide survey indicates that 50 percent of U.S. employees have felt pressure on the job to act in ways they consider to be unethical (Golen, Powers, & Titkemeyer, 1985).

Because they want to avoid conflicts over values in their careers, some employees have decided that their personal values have no place on the job. But that is a dangerous course. It can lead you into going along with actions at work that you would condemn at home. Furthermore, as companies decide what to do in certain situations, they sometimes discuss quite explicitly the ethical dimensions of the actions they might take. In these discussions, you can influence your employer's organization to act in accordance with your own ethical views, but only if you have brought your values with you to work—and only if you have the communication expertise required to present your view in ways that others find persuasive.

This Book's Approach to Ethics

A basic challenge facing anyone addressing ethical issues in workplace settings is that different people have different values and, consequently, different views of the right actions to take in various situations. This fact should not surprise you. For thousands of years, philosophers have offered various incompatible ethical systems. They have yet to reach agreement. Moreover, people from different cultural backgrounds and different nations adhere to different values. On your own campus, you surely know other students with whom you disagree on ethical issues.

The same thing happens in the workplace. Consequently, this book won't tell you what your values ought to be. Instead, it seeks to help you act in accordance with your own values. Toward that end, it seeks to enhance your sensitivity to often subtle and difficult-to-detect ethical implications so that you don't inadvertently end up preparing a communication that affects people in ways you would wish it hadn't. This book also presents some ways of looking at the ethical aspects of various writing decisions—such as the way you use colors in graphs—that you may not have considered before. Ultimately, however, the book's goals are to help you communicate in ways that, after careful consideration, you believe to be ethical and to enable you to build the communication expertise needed to influence others when you want to raise ethical questions.

WHAT LIES AHEAD IN THIS BOOK

Throughout this book, you'll find detailed advice and information to help you create highly usable, highly persuasive, and thoroughly ethical communications on the job. Its contents are divided into three major parts.

These corporations are among the thousands worldwide that have adopted ethics codes to guide employee and company conduct.

Learn More

To learn strategies for recognizing and addressing the ethical issues that arise on the job, read the Ethics Guidelines included in most of the other chapters.

Major Parts of This Book

- The Reader-Centered Writing Process. Chapters 1 through 16 provide reader-centered advice for performing each of the five major activities of writing:
  - Defining your communication's objectives.
  - Conducting research.
  - Planning your communication.
  - Drafting its text and visual elements.
  - Revising your draft.
- Applications of the Reader-Centered Writing Process. Chapters 17 through 21 explain how to apply the reader-centered writing process in order to do the following:
  - Create a communication collaboratively.
  - Prepare and deliver oral presentations.
  - Create web pages and websites.
  - Prepare a communication for a client, whether on the job or as a course project.
- Superstructures. Chapters 22 though 27 introduce you to the general frameworks, called *superstructures*, and conventions used in the workplace for correspondence (including e-mail), reports, proposals, and instructions.

GUIDELINES, YOUR CREATIVITY, AND YOUR GOOD JUDGMENT

This book's guidelines are not rules.

Many of the chapters offer guidelines—brief summary statements designed to make the book's suggestions easy to remember and use. As you read the guidelines, remember that they are merely that: guidelines. They are not rules. Each one has exceptions. They may, at times, conflict with one another. You may even work for a boss or client who insists that you write in ways that contradict them. To write successfully in your career, you will need to use your creativity and good judgment to decide when to follow, adapt, or ignore the guidelines.

To help you learn how to apply the guidelines, the chapters describe many sample situations and show sample communications. These samples reflect typical workplace communication practices. However, what's *typical* is not what's universal. The readers and circumstances you encounter in your job will certainly differ to some extent from those described here. In fact, you may work for a boss or client whose regulations, values, or preferences are very untypical. To write successfully, then, you may need to ignore one or more of this book's guidelines.

One set of principles will serve you well in all situations: Focus on your readers and the way you want to affect them, use your knowledge of your readers to shape your message in a way they will find highly usable and highly persuasive, and treat ethically all persons who might be affected by your message.

In your effort to develop expertise at communicating in your career, I wish you good luck and great success.

USE WHAT YOU'VE LEARNED

For additional ways to use what you've learned, go to **www.cengage.com/english/anderson7e.** *Instructors:* The book's website includes suggestions for teaching the exercises.

EXERCISE YOUR EXPERTISE

1. Interview someone who holds the kind of job you might like to have. Ask about the types of communications the person writes, the readers he or she addresses, the writing process and technology the person uses, and the amount of time the person spends writing. Supplement these questions with any others that will help you understand how writing fits into this person's work. According to your instructor's directions, bring either notes or a one-page report to class.

2. Find a communication written by someone who has the kind of job you want, perhaps by asking a friend, family member, or your own employer. Explain the communication's purposes from the point of view of both writer and readers. Describe some of the writing strategies the writer has used to achieve these purposes.

EXPLORE ONLINE

Explore websites created by two organizations in the same business (airlines, computers, museums, etc.) or two employers for whom you might like to work. Compare the strategies used to make the sites usable and persuasive.

Note ways their usability and persuasiveness might be increased.

COLLABORATE WITH YOUR CLASSMATES

Working with another student, rewrite the e-mail message by Donald Pryzblo (Figure 1.5, page 17) so that it will be more likely to persuade the personnel manager to follow Pryzblo's recommendation. Assume that Pryzblo knows that the manager's clerks are miscopying because he has examined the time sheets, time tickets, and computer files associated with 37 incorrect payroll checks; in 35 cases, the clerks made the errors. Take into account the way you expect the personnel manager to react upon finding an e-mail from Pryzblo in his or her in-box. Make sure that the first sentence of your revision addresses a person in that frame of mind and that your other sentences lead effectively from there to the last sentence, which you should leave unchanged.

APPLY YOUR ETHICS

As a first step in bringing your personal values to your on-the-job communication, list those values that you think will be especially important in your career. Explain situations in which you think it may be especially important for you to be guided by them.

CASE	HELP MICKEY CHELINI SELECT THE RIGHT FORKLIFT TRUCK

For additional cases, go to www.cengage.com/english/anderson7e. *Instructors:* The book's website includes suggestions for teaching the cases.

It has been two weeks since you received this assignment from your boss, Mickey Chelini, who is the Production Engineer at the manufacturing plant that employs you. "We've been having more trouble with one of our forklift trucks," he explained. You are not surprised. Some of those jalopies have been breaking down regularly for years. "And

Ballinger's finally decided to replace one of them," Mickey continued. Ballinger is Mickey's boss and the top executive at the plant. His title is Plant Manager.

"What finally happened to make him decide that?" you asked. "I thought he was going to keep trying to repair those wrecks forever."

"Actually, he wants to replace one of the newer ones that we bought just two years ago," Mickey replied. "That particular forklift was manufactured by a company that has since gone bankrupt. Ballinger's afraid we won't be able to get replacement parts. I think he's right."

"Hmm," you commented.

"Anyway," Mickey said, "Ballinger wants to be sure he spends the company's money more wisely this time. He's done a little investigation himself and has narrowed the choice to two machines. He's asked me to figure out which one is the best choice."

You could see what was coming. You've had a hundred assignments like this before from Mickey.

"I'd like you to pull together all the relevant information for me. Don't make any recommendation yourself; just give me all the information I need to make my recommendation. Have it to me in two weeks."

"Will do," you said, as you started to think about how you could squeeze this assignment into your already tight schedule.

Your Assignment

It's now two weeks later. You've gathered the information given below in the "Notes on Forklifts." First, plan your final report by performing the following activities:

- List the specific questions Mickey will want your report to answer. Note: Your boss does not want you to include a recommendation in your report.
- Underline the facts in your notes that you would include in your report; put an asterisk by those you would emphasize.
- Decide how you would organize the report.
- Explain which techniques from the list on pages 18–19 you would use when writing this report. What other things would you do to assist your readers?

Second, imagine that you have been promoted to Mickey's job (Production Engineer). Tell how you would write the report that you would send to Ballinger about the purchase of a new forklift. It must contain your recommendation.

Notes On Forklifts

Present Forklift. The present forklift, which is red, moves raw material from the loading dock to the beginning of the production line and takes finished products from the Packaging Department back to the loading dock. When it moves raw materials, the forklift hoists pallets weighing 600 pounds onto a platform 8 feet high, so that the raw materials can be emptied into a hopper. When transporting finished products, the forklift picks up and delivers pallets weighing 200 pounds at ground level. The forklift moves between stations at 10 mph, although some improvements in the production line will increase that rate to 15 mph in the next two months. The present forklift is easy to operate. No injuries and very little damage have been associated with its use.

Electric Forklift. The electric forklift carries loads of up to 1,000 pounds at speeds up to 30 mph. Although the electric forklift can hoist materials only 6 feet high, a 2-foot ramp could be built beneath the hopper platform in three days (perhaps over a long weekend, when the plant is closed). During construction of the ramp, production would have to stop. The ramp would cost $1,600. The electric forklift costs $37,250, and a special battery charger costs $1,500 more. The forklift would use about $2,000 worth of electricity each year. Preventive maintenance costs would be about $700 per year, and repair costs would be about $800 per year. While operating, the electric forklift emits no harmful fumes. Parts are available from a warehouse 500 miles away. They are ordered by phone and delivered the next day. The electric forklift has a good operating record, with very little damage to goods and with no injuries at all. It comes in blue and red.

Gasoline Forklift. The gasoline forklift is green and carries loads of up to 1 ton as rapidly as 40 mph. It can hoist materials 12 feet high. Because this forklift is larger than the one presently being used, the company would have to widen a doorway in the cement wall separating the Packaging Department from the loading dock. This alteration would cost $800 and would stop production for two days. The gasoline forklift costs $49,000 but needs no auxiliary equipment. However, under regulations established by the Occupational Safety and Health Administration, the company would have to install a ventilation fan to carry the exhaust fumes away from the hopper area. The fan costs $870. The gasoline forklift would require about $1,800 of fuel per year. Preventive maintenance would run an additional $400 per year, and repairs would cost about $600 per year. Repair parts are available from the factory, which is 17 miles from our plant. Other owners of this forklift have incurred no damage or injuries during its operation.

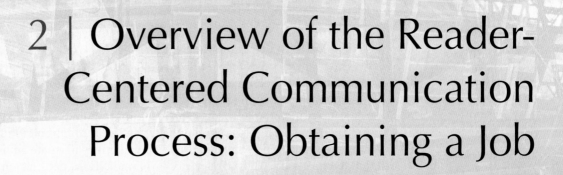

2 | Overview of the Reader-Centered Communication Process: Obtaining a Job

As explained in Chapter 1, this book aims to help you develop your communication expertise by teaching you the strategies used by successful workplace communicators. At the heart of these communicators' success is the reader-centered way they approach their communications.

This chapter's purpose is to provide you with an overview of the reader-centered approach, which is described in detail throughout the rest of this book. The chapter shows how to take a reader-centered approach in preparing two communications that will play important roles in your career: your résumé and job application letter.

The résumé and job application letter are ideal communications to use in an overview of the reader-centered process because they illustrate with special clarity that the success of your work-related communications depends on your readers' response to them. If your résumé and letter persuade employers to invite you for an interview or offer you a job, they have succeeded. If they don't produce such a response in their readers, they haven't been effective. Thus, in addition to helping you employ the reader-centered approach to your own on-the-job writing, this chapter will help you see *why* it is important for you to do so.

WWW

To read additional information, see more examples, and access links related to this chapter, go to **www.cengage.com/ english/anderson7e** and click on Chapter 2.

CENTRAL PRINCIPLES OF THE READER-CENTERED APPROACH

To gain the maximum value from reading this chapter, take a moment to review the central principles of the reader-centered approach. As explained in Chapter 1, this approach's key strategy is simple: Throughout all your work on a communication, think constantly about your readers.

- **As you make each writing decision, consider your readers' characteristics, goals, expectations, situations, and other factors that will shape their response to what you say.**
- **Concentrate on crafting a communication that will be persuasive and usable in your readers' eyes.** Persuasiveness and usability are indispensable qualities of all successful work-related communications.
- **Focus specifically on the ways your readers will respond, moment by moment, while they are reading your communication.** These moments are the only opportunity your communication has to influence your readers directly.

A READER-CENTERED APPROACH TO WRITING YOUR RÉSUMÉ

The following sections explain how to apply the central principles of the reader-centered approach to all four activities of the résumé-writing—and communication—process:

- **Defining objectives** for your résumé
- **Planning** what you will say in your résumé and how you will organize it
- **Drafting** your résumé's text, graphics, and visual design
- **Revising** your draft by reviewing it from your readers' perspective and then making the improvements your review suggests

A special section provides advice for preparing web and scannable résumés, which many employers request.

Define Your Résumé's Objectives

When writing your résumé—or any other workplace communication—you must make many decisions about such things as what to say, how to organize, how to design pages, and so on. The most effective basis for making these decisions is to think about your readers in the act of reading. What will they be looking for? How will they look for this information? How will they use it when they find it? What are their attitudes about your subject and what do you want their attitudes to be when they have finished reading? The answers to these questions will help you create a communication that, during each reading moment, leads readers closer to taking the action you desire. With your résumé, this action is to invite you for an interview or offer you a job.

Identify Employers You'd Like To Work For

To understand your readers, you must know who they are. When writing a résumé, start by naming the employers to whom you will apply. If you haven't identified the jobs, internships, or co-op positions you'd like, you might talk with instructors in your major and with staff at the college placement office. You might also network by asking family members, friends, former employers, and others to help you spot opportunities. As you speak with each person, ask him or her to suggest others who can assist. Other good sources for leads are newspapers, professional organizations, and professional publications directed to people who have graduated in your major. You can also use the web, consulting such sites as CareerOneStop (www.careeronestop.org), commercial boards such as Monster (www.monster.com), and state and local job listings. If you don't know what kind of job you want, there are resources on the web and, most likely, at your college that can help.

Define Your Résumé's Persuasive Objectives

Having identified employers to whom you will apply, create as accurate a mental portrait as you can of the way they will read your résumé. Because your résumé's primary objective is to persuade employers that you possess the qualifications they are looking for, begin by determining what these qualifications are.

Try This

Go to a website that offers a free service or test for people trying to find careers they would like. See if it suggests any careers you hadn't thought of. For links to such websites, click on Chapter 2 at **www.cengage.com/ english/anderson7e**.

Learn More

To learn more about creating a portrait of your reader in the act of reading, turn to page 76.

At a general level, these qualifications fall into the following three categories:

WHAT EMPLOYERS SEEK IN NEW EMPLOYEES

- **Technical expertise.** Employers want to hire people who can perform their jobs adeptly with a minimum of on-the-job training.
- **Supporting abilities.** Most jobs require a wide range of abilities beyond the purely technical ones of your degree or specialty. Among others, these often include communication, interpersonal, time management, and project management skills.
- **Favorable personal qualities.** Employers want to hire motivated, self-directed, and responsible individuals. They also want people who will work well with their other employees as well as with customers, vendors, and others outside the organization.

Of course, for any particular job, employers seek specific technical know-how, supporting abilities, and personal attributes. Instead of asking, "Does this applicant have technical expertise?" an employer will ask, "Can this person analyze geological samples?" "Write computer programs in Linux?" "Manage a consumer electronics store?" Your task is to find out the precise set of technical know-how, supporting abilities, and personal attributes that is sought by the employers to whom you are applying.

Learn More

For detailed advice about conducting interview and library research, turn to the Writer's Reference Guide: Using Five Reader-Centered Research Methods, page 165.

To identify these specific qualifications, learn as much as you can about the daily work of people in the position you would like to hold. Interview someone who holds such a position, talk with instructors in your major, and read materials about careers at your library or campus employment center. If you have an advertisement or job description from the employer to whom you are applying, study it carefully. As you identify qualifications the employer wants, put them in a list. The longer and more detailed your list, the better prepared you will be to write a résumé that achieves its persuasive objective, which is to demonstrate you have the knowledge, skills, and personal qualities that employers are seeking.

Define Your Résumé's Usability Objectives

In order to succeed, your résumé must not only highlight your qualifications, but also present these facts in a manner that employers find to be highly usable. For résumé readers, the key factor in usability is the speed with which they can find the information they want about you.

Résumés are read in a three-stage process by most employers. At all three stages, readers perform the same moment-by-moment reading activity: They compare the information each applicant supplies against the requirements of the job and against the qualifications of the other applicants. However, at each stage they perform this activity in somewhat different ways:

- **Initial screening.** Many employers receive hundreds, even thousands, of applications a week. One person may be charged with scanning quite quickly through this deluge, hoping to spot promising candidates. Often such initial readers are employees in a personnel office, not specialists in the area where you are seeking a position. They may spend less than a minute on each résumé. Unless they can rapidly spot clear, relevant information about an applicant's qualifications, they set the résumé aside and pick up the next one. To reduce this screening workload, an increasing number of employers submit all résumés to a computer first; only those that meet predefined criteria are read by a human.

Three ways employers use résumés

- **Detailed examination of the most promising applications.** Résumés that pass the initial screening are forwarded to managers and others in the department that wishes to hire a new employee. These readers know exactly what qualifications they desire, and they look specifically for them.
- **Preparation for in-depth interviewing.** Applicants whose résumés are persuasive to second-stage readers are invited for interviews. Their résumés are usually read with great care at this stage. Often managers and others will use the résumé as the basis for questions to ask during the interview.

The differing needs of the readers at these three stages mean that you must address the challenge of preparing a résumé that satisfies a first-stage reader's need to find information quickly while also meeting the second- and third-stage readers' need for a substantial amount of specific details about your qualifications.

Plan Your Overall Writing Strategies

In the reader-centered writing process, you first list your objectives as you plan the overall strategies you will use in your communication. In this planning, you decide what to include and how you will organize and present this content.

Think Creatively about Your Qualifications

Many people don't realize how many qualifications they have for the jobs they want. To develop a rich pool of possible contents for your résumé, try beginning with the list you created when detailing the qualifications desired by the employer. Next to each item, put down the course, activity, previous job, or other experience through which you gained or demonstrated the desired knowledge or ability. You might also create a list of your accomplishments and areas of knowledge, noting the connection each item may have with a qualification for the job you want. Talking with someone who holds the job now could spark additional ideas. Later sections in this chapter suggest some other possibilities. You may later decide not to include some of the ideas you have generated, but don't hold back on possibilities now.

Decide How Long Your Résumé Should Be

In planning, people often ask, "How long should my résumé be?" The answer is that it should be as short as possible while still presenting the facts employers will find most persuasive. For most undergraduates, this is one page. However, some undergraduates have extensive qualifications that justify a second page. Many experienced workers do as well. If your résumé runs past one page by a small amount, redesign your layout, edit your content, or delete something to make it fit it on a single page.

Choose the Type of Résumé You Will Prepare

There are two types of résumés you can choose:

- **Experiential résumé.** In an experiential résumé, you organize information about yourself around your experiences, using headings such as "Education," "Employment," and "Activities." Under these headings, you describe your experiences in ways that demonstrate that you possess the qualifications the employers want. An experiential résumé is the best choice for most college students and persons new in their careers. Figures 2.1 and 2.2 (pages 34 and 35) show examples of an experiential résumé.

- **Skills résumé.** In a skills résumé, you create key sections around your abilities and accomplishments, using such headings as "Technical Abilities," "Management Experience," and "Communication Skills." Later, you list the college you attended and jobs you've held. Skills résumés work best for people with enough professional experience to be able to list several on-the-job responsibilities and accomplishments in each of several categories. Figure 2.3 (page 39) shows a skills résumé.

Draft Your Résumé

For many people, drafting the text is the most challenging part of writing a résumé. Because employers must find information quickly and they want specific, relevant details, every word counts. The following sections provide advice about your general style and about presenting your qualifications effectively in each section of your résumé.

Writing Style for Résumés

Your writing style, like every other feature of your résumé, should help employers quickly find detailed information about your qualifications that match the qualifications for the job the employer has open.

- **Write concisely.** Typically, résumés contain lists of sentence fragments and phrases, not full sentences. Eliminate every word that you can without losing essential information.

Conducted an analysis of the strength of the load-bearing parts	Wordy
Analyzed strength of load-bearing parts	Concise

Member, PreMed Club, 2010–2011	Wordy
Vice President, PreMed Club, 2010–2011	Concise
PreMed Club, Vice President, 2010–2011	

- **Be specific.** Employers find generalities to be ambiguous and unimpressive. Give precise details.

Proficient on the computer	General
Program in Linux, SQL, CAD, and php	Specific

Reduced breakage on the assembly line	General
Reduced breakage on the assembly line by 17%	Specific

- **Watch your verb tenses.** Use the past tense for activities completed. Use the present tense for those you continue to perform.

Designed automobile suspensions	Completed activity
Design aircraft landing gear	Continuing activity

By convention, the word "I" is used in only one place in résumés: the professional objective.

Name and Contact Information

Help your readers locate your résumé in a stack of applications by placing your name prominently at the top of the page. Enable them to contact you quickly for an interview or with a job offer by including your postal address, e-mail address, and phone number. If you will live at another address during the summer or other part of the year, give the relevant information.

Professional Objective

The statement of your objective serves two important purposes. It provides a focus for your résumé and tells employers what job you want. Your goal in the rest of your résumé is to make a compelling case that you have the qualifications necessary to do an outstanding job in the position you describe in your objective.

When drafting your objective, take the reader-centered approach of telling what you will *give* to your future employer rather than the writer-centered approach of telling what you want to gain from the employer. Here's a simple, two-step procedure for doing this:

1. **Identify the results sought by the department or unit in which you wish to work.** If you don't know what they are, ask a professor, talk with someone

in the kind of job you want, or contact your campus employment center for assistance.

2. **State that your objective is to help achieve those results.**

Howard, a senior in sports studies, and Jena, a computer science major, used this process to improve the first drafts of their objectives. Their initial drafts read as follows:

Writer-centered vague objectives

Howard: A position where I can extend my knowledge of nutrition and sports studies

Jena: A position as a computer systems analyst in the field of software development with an innovative and growing company

When Howard and Jena considered their statements from their readers' perspective, they realized that they were merely stating what they wanted to gain from their jobs. They said nothing about how they would use their knowledge to benefit their employers. Howard and Jena revised their objectives to read as follows:

Reader-centered, results-oriented revisions

Howard: To develop nutrition and exercise programs that help athletes achieve peak performance

Jena: To create software systems that control inventory, ordering, and billing

If you are applying for an internship, co-op, or summer job, tell what you will "help" or "assist" the employer to do.

Objective for internship

An internship helping design high-quality components for the electronics industry

Some job seekers ask whether they really need to include an objective statement. In fact, an objective may be unnecessary for people advanced in their careers who begin their résumés with a summary of their qualifications and accomplishments that indicates clearly the specific kind of job they are seeking. For people seeking an entry-level job, the objective is essential. As résumé expert Susan Ireland (2003) says, for a new college graduate to omit an objective is like saying, "Here's what I've done. Now tell me what I should do next." Employers are looking for applicants whose qualifications meet their needs. They are not trying to find jobs for applicants.

Education

Place your education section immediately after your identifying information, unless you have had enough work experience to make the knowledge you gained at work more impressive to employers than the knowledge you gained in classes. Name your college, degree, and graduation date. If your grade point average is good, include it. If your average in your major is higher, give it. But don't stop there. Provide additional details about your education that employers will see as relevant to the job you want. Here are some examples.

FACTS TO HIGHLIGHT ABOUT YOUR EDUCATION

- Advanced courses directly relevant to the job you want (give titles, not course numbers)
- Courses outside your major that broaden the range of abilities you would bring to an employer
- Internships, co-op assignments, or other on-the-job academic experiences
- Special projects, such as a thesis or a design project in an advanced course
- Academic honors and scholarships
- Study abroad
- Training programs provided by employers

You can highlight any of these credentials by giving it a separate heading (for example, "Honors" or "Internship").

The résumés of Jeannie Ryan (Figure 2.1) and Ramón Perez (Figure 2.2) show how two very different college seniors elaborated on the ways their education qualifies them for the jobs they are seeking.

Work Experience

When drafting your work experience section, list the employers, their cities, your job titles, and your employment dates. Then present facts about your work that will impress employers.

FACTS TO HIGHLIGHT ABOUT YOUR WORK EXPERIENCE

- **Your accomplishments.** Describe projects you worked on, problems you addressed, goals you pursued, products you designed, and reports you helped write. Where possible, emphasize specific results—number of dollars saved, additional units produced, or extra customers served.
- **Knowledge gained.** Be resourceful in highlighting things you learned that increased your ability to contribute to your future employer. Realizing that her duties of stacking ice cream packages neatly in supermarket freezers might not seem relevant to a job in marketing, Jeannie describes the insight she gained in that job: "Learned to see consumer marketing from the perspectives of both the manufacturer and the retailer" (Figure 2.1).
- **Responsibilities given.** If you supervised others, say how many. If you controlled a budget, say how large it was. Employers will be impressed that others have entrusted you with significant responsibility.

FIGURE 2.1
Experiential Résumé

Jeannie tells where she can be reached before and after graduation.

She includes her e-mail address.

Jeannie states her specific career objective.

Jeannie emphasizes her thorough preparation by listing many relevant courses.

Jeannie describes a special course related to her career objective.

She provides specific details when highlighting her accomplishment.

She uses bullet lists throughout her résumé to enable her readers to scan her qualifications quickly.

Jeannie tells what she learned that would help her succeed in the job she desires.

Jeannie lists references who can verify her knowledge in each of her areas of expertise related to the job she wants.

WWW

To see how Jeannie designed her résumé in a word-processing program, go to **www.cengage.com/english/anderson7e**.

Jeannie Ryan

Present Address		**After May 29, 2010**
325 Foxfire Drive, Apt 214		85 Deitrich Court
Denver, Colorado 70962		Flint, Colorado 73055
(303) 532-1401		(303) 344-7329
	ryanja@acs.udenver.edu	

Objective To conduct market research in a full-service marketing firm with clients in the electronics industry

Education B.S. in Marketing, Minor in Computer Science, University of Denver, May 2010

Specialized Courses
- Analytical Methods for Marketing
- Strategic Marketing Management
- Stochastics
- Programming Computer Games
- Architecture of Small Computers
- Managing Data Sets

Project for Real Client In Advanced Marketing, we used a telephone survey and focus groups to estimate the potential market for a new hand-held learning game for pre-school children.

Work Experience

Interim Manager, Stan's Electronics, Redvale, Colorado, Summer 2009
- Collaborated with buyer to project product sales and advertising
- Sales increased 15% over the previous year

Sales Intern, Ali Ice Cream Company, Wilke, Colorado, Summer 2008
- Built backroom inventories, stocked cases, and secured endcap displays
- Learned to see consumer marketing from the perspectives of manufacturers and retailers

Tels Marquart, Inc., Redvale, Colorado, Summer 2004
- Receptionist
- Learned to work in a fast-paced office environment

Activities Synchronized Swim Team 2006–2010
- Vice President (Senior Year)
- Planned and directed an hour-long public program for this 50-member club

References Harlan Betrus-Holloway, Professor, Marketing Department, University of Denver, 199 S. University Blvd., Denver, Colorado 80208, 303-871-3418
Sheila Cortez, Professor, Computer Science Department, 199 S. University Blvd., Denver, Colorado 80208, 303-871-1547
Raphael Tedescue, President, Stan's Electronics, 1176 Sunnyside Avenue, Redvale, Colorado 79638, 303-461-9872.

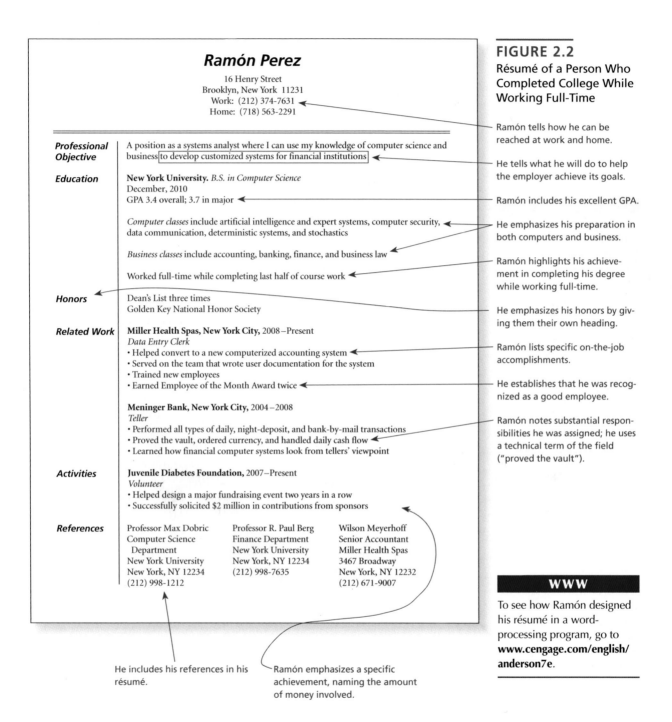

Ramón Perez

16 Henry Street
Brooklyn, New York 11231
Work: (212) 374-7631
Home: (718) 563-2291

Professional Objective	A position as a systems analyst where I can use my knowledge of computer science and business to develop customized systems for financial institutions
Education	**New York University.** *B.S. in Computer Science* December, 2010 GPA 3.4 overall; 3.7 in major *Computer classes* include artificial intelligence and expert systems, computer security, data communication, deterministic systems, and stochastics *Business classes* include accounting, banking, finance, and business law Worked full-time while completing last half of course work
Honors	Dean's List three times Golden Key National Honor Society
Related Work	**Miller Health Spas, New York City,** 2008 – Present *Data Entry Clerk* • Helped convert to a new computerized accounting system • Served on the team that wrote user documentation for the system • Trained new employees • Earned Employee of the Month Award twice **Meninger Bank, New York City,** 2004 – 2008 *Teller* • Performed all types of daily, night-deposit, and bank-by-mail transactions • Proved the vault, ordered currency, and handled daily cash flow • Learned how financial computer systems look from tellers' viewpoint
Activities	**Juvenile Diabetes Foundation,** 2007 – Present *Volunteer* • Helped design a major fundraising event two years in a row • Successfully solicited $2 million in contributions from sponsors
References	Professor Max Dobric Computer Science Department New York University New York, NY 12234 (212) 998-1212 Professor R. Paul Berg Finance Department New York University New York, NY 12234 (212) 998-7635 Wilson Meyerhoff Senior Accountant Miller Health Spas 3467 Broadway New York, NY 12232 (212) 671-9007

He includes his references in his résumé.

Ramón emphasizes a specific achievement, naming the amount of money involved.

FIGURE 2.2
Résumé of a Person Who Completed College While Working Full-Time

Ramón tells how he can be reached at work and home.

He tells what he will do to help the employer achieve its goals.

Ramón includes his excellent GPA.

He emphasizes his preparation in both computers and business.

Ramón highlights his achievement in completing his degree while working full-time.

He emphasizes his honors by giving them their own heading.

Ramón lists specific on-the-job accomplishments.

He establishes that he was recognized as a good employee.

Ramón notes substantial responsibilities he was assigned; he uses a technical term of the field ("proved the vault").

WWW

To see how Ramón designed his résumé in a word-processing program, go to **www.cengage.com/english/ anderson7e**.

When organizing and describing your work experience, follow these guidelines for achieving high impact.

ORGANIZING AND DESCRIBING YOUR WORK EXPERIENCE

- **List your most impressive job first.** If your most recent job is your most impressive, list your jobs in reverse chronological order. To highlight an older job, create a special heading for it, such as "Related Experience." Then describe less relevant jobs in a later section entitled "Other Experience."

- **Put your actions in verbs, not nouns.** Verbs portray you in action. Don't say you were responsible for the "analysis of test data" but that you "analyzed test data." Avoid such weak phrases as "responsibilities included" or "duties were."

- **Use strong verbs.** When choosing your verbs, choose specific, lively verbs, not vague, lifeless ones. Avoid saying simply that you "made conceptual engineering models." Say that you "designed" or "created" the models. Don't say that you "interacted with clients" but that you "responded to client concerns."

- **Use parallel constructions.** When making parallel statements, use a grammatically correct parallel construction. Nonparallel constructions slow reading and indicate a lack of writing skill.

NOT PARALLEL	PARALLEL
• Trained new employees	• Trained new employees
• Correspondence with customers	• Corresponded with customers
• Prepared loan forms	• Prepared loan forms

Learn More

For more information about putting actions in verbs, see Chapter 9, page 271.

Changing *correspondence* to *corresponded* makes it parallel with *trained* and *prepared*.

Activities

At the very least, participation in group activities indicates that you are a pleasant person who gets along with others. Beyond that, it may show that you have acquired certain abilities that are important in the job you want. Notice, for instance, how Jeannie Ryan describes one of her extracurricular activities in a way that emphasizes the essentially managerial responsibilities she held (Figure 2.1).

Emphasis on management responsibilities

Synchronized Swim Team
Vice President (Senior Year)
- Planned and directed an hour-long public program for this 50-member club.

Special Abilities

Let employers know about exceptional achievements and abilities of any sort, using such headings as "Foreign Languages" or "Certifications."

Interests

If you have interests such as golf or skiing that could help you build relationships with coworkers and clients, you may wish to mention them, although a separate section for interests is unnecessary if the information is provided in your activities section.

Personal Data

Federal law prohibits employers from discriminating on the basis of sex, religion, color, age, or national origin. It also prohibits employers from inquiring about matters unrelated to the job for which a person has applied. For instance, employers cannot ask if you are married or plan to marry. Many job applicants welcome these restrictions because they consider such questions to be personal or irrelevant. On the other hand, federal law does not prohibit you from giving employers information of this sort if you think it will help to persuade them to hire you. If you include such information, place it at the end of your résumé, just before your references. It is almost certainly less impressive than what you say in the other sections.

<aside>

Try This

At a website that offers advice to job seekers, find a "model" résumé. Evaluate it in light of the reader-centered advice given in this chapter; what are its strengths and weaknesses?

</aside>

References

When looking for a job, you are going to need to provide references at one time or another. Employers always want them from their top applicants. Some persons with lots of employment experience prefer to omit references from their résumés. They plan to let their current employer know that they are looking for a job only if another organization expresses significant interest in them. For college students, however, it's almost always an advantage. Research shows that most employers want applicants to include references with their résumés (Bowman, 2002). By including your references, you increase your résumé's usability by enabling employers to contact them immediately and directly without having to write to you and then wait for your reply. You also increase your résumé's persuasiveness by letting readers see the names and titles of the impressive people who will speak favorably about you. If you have a short résumé, putting your references on the page fills up the space, making your qualifications look more substantial. On the other hand, if your references won't fit on your page, create a separate page entitled "References" and enclose it with your résumé.

Employers expect three to five references, so include this many. Select a mix of people who, taken together, can describe the range of your qualifications. As appropriate, choose instructors in your major and other key courses, former employers, and advisers of campus organizations. Avoid listing your parents and their friends, who (an employer might feel) are going to say nice things about you no matter what. Provide titles, business addresses, phone numbers, and e-mail addresses. If one of your references has changed jobs so that his or her business address doesn't indicate how the person knows you, provide the needed information: "My supervisor while at Sondid Company." Obtain permission from the people you want to list as references so they aren't taken by surprise by a phone call. Give your résumé to your references so they can quickly review your qualifications when they receive an inquiry from an employer.

Drafting the Text for a Skills Résumé

A skills résumé has the same aims as an experiential one. The chief difference is that in a skills résumé you consolidate the presentation of your accomplishments and experience in a special section located near the beginning rather than weave this information into your sections on education, work experience, and activities. Figure 2.3 shows an example.

For this special section, use a title that emphasizes its contents, such as "Skills" or "Skills and Achievements." Within the section, use subheadings that identify the major areas of ability and experience you would bring to an employer. Typical headings include "Technical," "Management," "Financial," and "Communication." However, the specific headings that will work best for you depend on what employers seek when recruiting people for the kind of job you want. For example, the headings in George Shriver's résumé (Figure 2.3) focus specifically on skills required of managers of technical communication departments.

Because you aggregate your skills and accomplishments in a special "Skills" section, the other sections of your skills résumé should be brief in order to avoid redundancy.

Design Your Résumé's Appearance

At work, good visual design is crucial to achieving communication goals. Nowhere is that more true than with a résumé, where design must support rapid reading, emphasize your most impressive qualifications, and look attractive. The résumés shown in this chapter achieve these objectives through a variety of methods you can use.

DESIGNING YOUR RÉSUMÉ'S VISUAL APPEARANCE	
SHORT, INFORMATIVE HEADINGS	**DIFFERENT TYPEFACES FOR HEADINGS THAN FOR TEXT**
■ Lists	■ White space to separate sections
■ Bullets	■ Ample margins (3/4" to 1")
■ Italics	■ Visual balance
■ Variety of type sizes	■ Bold type for headings and key information

FIGURE 2.3
Skills Résumé

GEORGE SHRIVER

Objective

Senior management position where I can lead a technical communication department that assists a computer manufacturer in achieving high quality and productivity

Skills and Accomplishments

Management Supervise a team of six specialists who create print and on-line user documentation and also develop and deliver training programs for in-house use

Innovation Proposed and oversaw the development of an interactive videodisc training program for process engineers in a small factory that manufactures computer components

Technical Expertise Familiar with latest developments in both hardware and software. Programming knowledge of Visual C++, Java, and various proprietary computer languages

Budgetary Responsibility Manage an annual budget of nearly one-half million dollars

Employment History

Training Director, Saffron Computer Technology, Inc., Anaheim, CA, 2005–Present
Training Specialist, Calpon Software Systems, Deer Park, NJ, 2001–2005

Education

B.A. in Technical and Scientific Communication, Miami University (OH), 2001
Numerous professional development courses

Professional Societies

Society for Technical Communication (Chapter President, 2008)
American Society for Training and Development

Special Qualifications

Fluent in German   Trained in conflict resolution   Certified to teach CPR

References available upon request

1734 Everet Avenue   Pasadena, CA 91101   (314) 417-7787
GShriver@nettlink.com

George names the goal that he will help the employer achieve.

In this skills résumé, George highlights his special qualifications in a separate section.

George uses the present tense in his "Management" and "Budgetary" entries because these are continuing duties; he uses the past tense in his entry about "Innovation" because it describes a completed project.

He describes a major accomplishment.

He provides information about the budget's size.

Because he presented information about substantial on-the-job responsibilities and achievements above, he does not elaborate on his jobs here.

Because he has substantial professional experience, George deemphasizes his college experiences by giving only basic facts.

George shows commitment to continued professional development.

George lists additional qualifications that may interest an employer.

George creates a distinctive design for his résumé by putting his address and phone number at the bottom.

He names a leadership position in his professional society.

WWW

To see how George designed his résumé in a word-processing program, go to **www.cengage.com/english/anderson7e.**

Writer's Tutorial

Using Tables to Design a Résumé

This tutorial tells how to design a multicolumn résumé with Word for Windows 2007. For instructions on using Word for Macintosh, go to www.cengage.com/english/anderson7e and click on Chapter 2. If you use a different word processor or get stuck, click on your program's **Help** menu for assistance. After learning the basic strategies described below, use your creativity to create a résumé that employers will find usable and persuasive.

MAKE A TABLE TO SERVE AS THE VISUAL FRAMEWORK FOR YOUR RÉSUMÉ

1. Create a new **Blank Document**.
2. Under the **Page Layout** tab, click on **Margins**.
3. From the dropdown menu, choose **Normal**.
4. Under the **Insert** tab, click on **Table**.
5. In the dropdown window, highlight 3 squares across and 7 down.
   - You can add or delete rows as necessary later.
6. Left click.

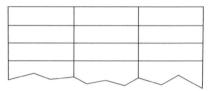

ADJUST THE WIDTHS OF THE COLUMNS

1. Under the **View** menu, click on **Ruler**.
2. Move the cursor over the vertical line that separates the left and center columns so that the cursor changes to this: ◄‖►
3. Hold down the left mouse button.
4. Slide the vertical line to the left until it is at 1.5 inches on the ruler.
   - This narrow column is for your topic headings. ————
5. Release the mouse button.
6. Using the same procedure, move the other vertical as far left as you can.
   - This thin column is a gutter between the heading and text columns.

HIDE THE TABLE'S BORDERS SO THEY WON'T SHOW WHEN YOU PRINT YOUR RÉSUMÉ

1. Highlight the entire table (but not anything else).
2. Under the **Table Tools** tab, click **Design**.
3. Click on the arrow next to **Borders**.
4. In the dropdown menu, click **No Border**.
5. If the borders disappear rather than changing to dashed lines, do the following:
   - Under the **Table Tools** tab, click **Layout**.
   - Click on **View Gridlines** at the left end of the ribbon.
   - The borders show as dashed lines on screen, but will not print.

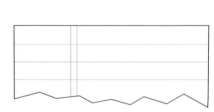

Learn More at the Website

To see tables used to make a variety of résumés, go to www.cengage.com/english/anderson7e.

CREATE A HEADING FOR YOUR RÉSUMÉ

1. Highlight all three columns in the table's top row.
2. Under **Table Tools** tab, click **Layout**.
3. Click **Merge Cells**.
4. Still in the **Layout** section, find the set of nine squares in the **Alignment** area.
5. Click on the center square ("Align Center").
6. In this wide cell, enter your name and contact information.
7. Change your name to a larger type and boldface.

BE CREATIVE

This tutorial's purpose is to teach you to how to use a word-processing program to create an effective résumé, not to suggest that your résumé should look one particular way. Use your creativity.

For example, here are two alternative headings you can make if you don't merge the three cells in the first row. In the bottom example, grid lines for the gutter are moved to the right in the first row only.

ENTER YOUR TOPIC HEADINGS

1. In each cell in the left column, enter a topic heading.
   - If the heading has two words, put a **Return** between them.
   - If the column isn't wide enough for the longest word, enlarge it slightly.
2. Change all the topic headings to bold type.
3. Align the headings on the right side of the column.
   - Highlight all the headings.
   - Under **Table Tools**, click **Layout**.
   - Among the **Alignment** squares, click on the top square in the right-hand column ("Align Top Right").

(continued)

ENTER TEXT

1. In the right-hand column of each row, enter the appropriate text.
2. At the end of each entry, type one **Return.**
   - The **Return** will create a blank line between this entry and the next.
3. If you want to enter multicolumn text in the right-hand column, do the following:
   - Put the cursor in the cell that you want to have columns.
   - Follow the procedure for making a table that is given on the first page of this tutorial.
4. Use bold type, italics, bullets, indentation, and other design elements to create an attractive, easy-to-read design that emphasizes your qualifications.

EXPERIMENT WITH YOUR DESIGN

- Try using a different typeface for your name and the headings.
- Try aligning the headings on the left side of their column.
- Try adding a vertical rule (line).
  1. Highlight the text cells in the right-hand column.
  2. Under the **Table Tools** tab, select **Design**.
  3. Click on the arrow next to **Borders**.
  4. In the dropdown menu, click **Left Border**.
- Try adding one or more horizontal rules (lines).
  1. Highlight the cell that has your heading.
  2. Under the **Table Tools** tab, select **Design**.
  3. Click on the arrow next to **Borders**.
  4. In the dropdown menu, click **Bottom Border**.

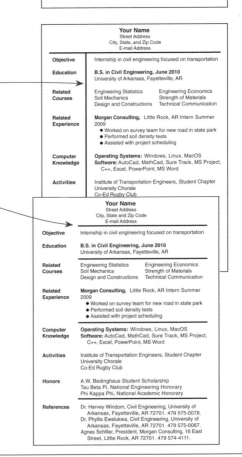

You will probably need to experiment with the design of your résumé to create a balanced, attractive page—one that visually emphasizes your qualifications and makes your résumé seem neither too packed nor too thin. To check visual balance, fold your résumé vertically. Both sides should have a substantial amount of type. Neither should be primarily blank.

Do not rely on a computer program's templates or wizards to design your résumé for you. They are generic packages that don't let you create the most favorable presentation of your qualifications. Each of the various designs shown in this chapter's examples are easy to create using the tables in Microsoft Word. The tutorial on pages 40–42 will get you started. Use knowledge of your readers and your creativity to shape your résumé into a highly persuasive communication.

Revise

After drafting your résumé, but before sending it to the employer, take time to revise. Read it over carefully, trying to see it as the employer would. Then show your draft to other people. Tell them about the job you want so they can read from your readers' perspective, imagining what the employer will look for and how the employer will respond to each feature. Good reviewers for your résumé include classmates, your writing instructor, and instructors in your major department.

After making your revisions, run your spellchecker again, then proofread carefully for errors that spellcheckers don't catch. Employers say repeatedly that even a single error in spelling or grammar can eliminate a résumé from further consideration. The programs some employers use for initial screening of résumés kick out those with errors.

Check, too, for consistency in the use of italics, boldface, periods, bullets, and abbreviations as well as in the format for dates and other parallel items. Finally, check for consistency in the vertical alignment of information that isn't flush against the margin.

Adapting Your Résumé for Different Employers

The reader-centered process emphasizes the importance of tailoring every feature of a communication to the needs and goals of your readers. Does this mean you need to write a new résumé for every job you apply for? Not necessarily. If you are seeking the same type of position with both employers and if both employers are seeking the same qualifications, the same résumé will probably work equally well with both. However, the farther apart the two positions are, the more adjustments you should make. In addition to rewriting your objective, you may need to reshape the descriptions of your education, experience, and activities, as well as reorder your résumé's contents. Even when sending the same résumé to different organizations, tailor each application letter to one specific employer; see page 56.

Figure 2.4 shows a Writer's Guide you can use as you prepare your résumé.

WWW

To see how one student adapted her résumés when applying for two different kinds of jobs, go to **www.cengage.com/english/anderson7e** and click on Chapter 2.

FIGURE 2.4   **Writer's Guide for Résumés**
🌐 To download a copy of this Writer's Guide, go to www.cengage.com/english/anderson7e.

Writer's Guide
RÉSUMÉS

This Writer's Guide describes the basic elements of a résumé. Some of the elements would be organized differently in a chronological résumé than in a skills résumé; see page 38.

Preliminary Research

_____   Determine as exactly as possible what the employer wants.

_____   Learn enough about the job and employer to tailor your résumé to them.

_____   Create a keyword list.

Name and Contact Information

_____   Enable employers to reach you by mail, phone, and e-mail.

Objective

_____   Tailor to the specific job you want.

_____   Emphasize what you will give rather than what your would like to get.

Education

_____   Tell your school, major, and date of graduation.

_____   Provide additional information that shows you are well-qualified for the job you want: academic honors and scholarships, specialized courses and projects, etc.

_____   Use headings such as "Honors" and "Related Courses" to highlight your qualifications.

Work Experience

_____   Identify each employer's name and city, plus your employment dates.

_____   Provide specific details about your previous jobs that highlight your qualifications: accomplishments, knowledge gained, equipment and programs used, responsibilities, etc.

Activities

_____   Describe your extracurricular and community activities in a way that shows you are qualified, responsible, and pleasant.

Interests

_____   Mention personal interests that will help the reader see you as a well-rounded and interesting person.

References

_____   List people who will be impressive to your readers.

_____   Include a mix of references who can speak about your performance in different contexts.

_____   Include title, business address, phone, and e-mail address for each reference.

FIGURE 2.4   *(continued)*

Writer's Guide
RÉSUMÉS
(continued)

_____ Include only people who've given permission to be listed.

_____ Omit personal references (family, friends, etc.).

Prose

_____ Present the most impressive information first.

_____ Express the action in verbs, not nouns.

_____ Use strong verbs.

_____ Use parallel constructions.

_____ Omit irrelevant information.

_____ Use correct spelling, grammar, and punctuation.

Visual Design

_____ Look neat and attractive.

_____ Highlight the facts that will be most impressive to employers.

_____ Use headings, layout, and other design features to help readers to find specific facts quickly.

Ethics

_____ List only experiences, accomplishments, degrees, and job titles you've actually had.

_____ Avoid taking sole credit for things you did with a team.

_____ Avoid statements intended to mislead.

ELECTRONIC RÉSUMÉS: SPECIAL CONSIDERATIONS

Increasingly, job applicants and employers are using electronic résumés rather than paper ones. The following sections provide advice about four types: scannable résumés, résumés submitted via e-mail or websites, and web page résumés.

Scannable Résumés

Many employers ask computers, not people, to be the first readers of your résumé. They feed your résumé to a scanner, which enters it into a computer's database. To find applicants who might be invited for a job interview, the employers ask the computer to search its database for résumés that have words—*keywords*—that the employers believe would appear in the résumés of good candidates for the opening they want to fill. The computer displays a list of the résumés with the most matches, called *hits*. These are the only résumés a person would read.

If you suspect that your résumé might be scanned, call the employer to check or else enclose both a regular résumé and a scannable one. To increase the number of hits your résumé will receive, make a list of keywords that employers are likely to ask their computers to look for. Here are some suggestions.

KEYWORDS FOR RÉSUMÉS

- Words in the employer's ad or job description
- All degrees, certifications, and licenses you've earned: B.A., B.S., R.N. (registered nurse), P.E. (professional engineer license), C.P.A. (certified public accountant)
- Advanced topics you've studied
- Computer programs and operating systems you've mastered: Excel, AutoCAD II, C++, Linux
- Specialized equipment and techniques you've used in school or at work: X-ray machine, PK/PD analysis, ladderlogic
- Job titles and the specialized tasks you performed
- Buzzwords in your field: client server, LAN, low-impact aerobics, TQM (total quality management)
- Names of professional societies to which you belong (including student chapters)
- Other qualifications an employer would desire: leadership, writing ability, interpersonal skills, etc.

To increase the chances a computer will select your résumé to be read by a human, use the strategies listed at the top of the next page. Ramón Perez employed them in the scannable résumé shown in Figure 2.5 (page 48). This résumé is longer than Ramón's print resume (Figure 2.2), but computers don't care how much they read.

- Put your keywords in nouns, even if your scannable résumé becomes wordy as a result.
- Be redundant. For instance, write "Used Excel spreadsheet program" rather than "Used spreadsheet program" or "Used Excel." You don't know which term an employer will use, and if both terms are used, you will have two hits.
- Create a "Keywords" section for keywords you can't easily work into any of your résumé's other sections.
- Proofread carefully. Computer programs don't match misspellings.
- Use 12-point type.
- Use a standard typeface (e.g., Times, Arial, or Helvetica).
- Avoid italics, underlining, and decorative elements such as vertical lines, borders, and shading. Boldface, all caps, and bullets are okay.
- Use blank lines and boldface headings to separate sections.
- Use a single column of text.
- Put your name at the top of every page, on a line of its own.
- Use laser printing or high-quality photocopying on white paper.
- Mail your résumé flat and without staples.

Résumés Submitted via E-mail

More than one-third of human resource managers prefer e-mailed résumés, according to a survey by the Society for Human Resource Managers. The following advice is based on suggestion by certified career coach Kirsten Dixson (2001).

- Follow the employer's directions about whether to include your résumé in an attachment or in the e-mail itself. In either case, follow the advice for scannable résumés because yours may be read by a computer.
- If you are sending your résumé as an attachment, save it as a pdf file so that it will retain the appearance you designed regardless of the computer and printer used by the employer.
- If you are sending your résumé in the e-mail itself, design it the way you would design a scannable résumé. The formatting you create using your own e-mail program may create an eyesore in the employer's program.
- Write a simple subject line. For example: Résumé—Kelly Borowitz-Rioja: Civil Engineering Major.
- Use a straightforward e-mail address for yourself. Avoid cute names. If necessary, set up a free e-mail account at Hotmail, Yahoo, or a similar service.

FIGURE 2.5
Scannable Résumé

Ramón includes a list of keywords that don't appear elsewhere in his résumé.

To assure that a scanner could read his résumé, Ramón uses a single-column format rather than the two-column format he created for his print résumé.

He also eliminates the italics that appeared in his print résumé (see page 35).

Throughout this résumé, Ramón uses 12-point Times, a typeface scanners can read without difficulty.

To make his résumé easy for humans (as well as scanners) to read, Ramón relies on:
- capital letters
- bold type
- bullets
- blank lines

To be read by a scanner, Ramón's résumé requires two pages, but that is fine because scanners don't care how many pages they read.

RAMÓN PEREZ
16 Henry Street
Brooklyn, New York 11231
Work: (212) 374-7631
Home: (718) 563-2291

KEYWORDS
Responsible, financial management, banking experience, accounting, auditing, high motivation, communication ability

PROFESSIONAL OBJECTIVE
A position as a systems analyst where I can use my knowledge of computer science and business to develop customized systems for financial institutions

EDUCATION
New York University
B.S. in Computer Science
December 2010
GPA 3.4 overall; 3.7 in major

Computer classes include artificial intelligence and expert systems, computer security, data communication, deterministic systems, and stochastics

Business classes include accounting, banking, finance, and business law

Worked full-time while completing last half of course work

HONORS
Dean's List three times
Golden Key National Honor Society

RELATED WORK
Miller Health Spas, New York City, 2008–Present
Data Entry Clerk
- Helped convert to a new computerized accounting system
- Served on the team that wrote user documentation for the system
- Trained new employees
- Earned Employee of the Month Award twice

Meninger Bank, New York City, 2005–2008
Teller
- Performed all types of daily, night-deposit, and bank-by-mail transactions
- Proved the vault, ordered currency, and handled daily cash flow
- Learned how financial computer systems look from tellers' viewpoint

ACTIVITIES
Juvenile Diabetes Foundation, 2007–Present
Volunteer
- Helped design a major fund-raising event two years in a row
- Successfully solicited two million dollars in contributions from sponsors

(Ramón uses a second page for his references)

Résumés Submitted via Websites

Some employers ask applicants to submit their résumés by completing online forms at the employers' websites. These résumés will almost surely be read into a computer, so follow the advice given above for scannable résumés. To complete an online application form, draft your text in a word processing program, then review and proofread it carefully before pasting it into the employer's form.

Web Page Résumés

Although most employers do not read web page résumés, a few do. If you create a résumé accessible via the World Wide Web, consider the following advice:

- **Link to samples of your work and fuller descriptions of your experiences and capabilities.**
- **Keep your design simple, uncluttered.** Don't let fancy embellishments distract employers' attention from your qualifications.
- **Include a "mail-to" link to your e-mail address.**
- **Post your online résumé only at secure services.** Some services have no security, making it possible for other people to alter your résumé.

WWW

For additional advice about creating web page résumés, go to **www.cengage.com/ english/anderson7e** and click on Chapter 2.

Learn More

Chapter 20 provides additional advice about creating effective web pages.

A READER-CENTERED APPROACH TO WRITING YOUR JOB APPLICATION LETTER

Your job application letter can be a powerfully persuasive part of the package you submit to employers. In response to a survey of 150 executives from the nation's 1,000 largest employers, 60 percent indicated that a job application letter is just as or more important than the résumé (Patterson, 1996).

Reflecting the approach of many employers, corporate recruiter Richard Berman reports that if a letter doesn't grab him, he merely gives the enclosed résumé a quick scan to see if it might turn his opinion around. In contrast, when he discovers a particularly effective letter, he goes immediately to colleagues to announce that he's "found a good one" (Patterson, 1996).

Define Your Letter's Objectives

Your job application letter has somewhat different work to do than does your résumé. Although the readers are the same, the usability and persuasive objectives differ.

The usability objectives are different because readers look for different information in a letter than in a résumé. Their major letter-reading questions are as follows.

WHAT EMPLOYERS ASK WHEN READING APPLICATION LETTERS

- **Why do you want to work for me instead of someone else?** Employers are more responsive to applicants who know about their organization and can express specific reasons for wanting to work for it.
- **How will you contribute to my organization's success?** In the Wells, Spinks, and Hargrave survey, 88 percent of the personnel directors "agreed" or "strongly agreed" that job application letters should tell how the jobseeker's qualifications match the organization's needs.
- **Will you work well with my other employees and the persons with whom we do business?** Because every job involves extensive interactions with coworkers, employers want to hire people who excel in these relationships. Of course, interactions with clients, vendors, and others outside the employer's organization are also important.

The persuasive objectives of your letter also differ from those of your résumé:

Your job application letter's persuasive objectives

- To respond to the employer's questions listed above in ways that make the employer want to hire you
- To convey a favorable sense of your enthusiasm, creativity, commitment, and other attributes that employers value but can't be communicated easily in your résumé

Plan Your Letter

WWW

For advice about finding employers you might want to work for, go to **www.cengage.com/english/anderson7e** and click on Chapter 2.

Begin planning your job application letter by conducting the research necessary to answer the reader's first question, "Why do you want to work for me instead of someone else?" Seek things you can praise. You'll find it helpful to categorize the facts you discover as either writer-centered or reader-centered. Examples of *writer-centered facts* are the benefits the organization gives employees or the appealing features of its co-op program. You won't gain anything by mentioning writer-centered facts in your letter because whenever you talk about them you are saying, in essence, "I want to work for you because of what you'll give me." When employers are hiring, they are not looking for people to give things to.

In contrast, *reader-centered facts* concern things the organization is proud of: a specific innovation it has created, a novel process it uses, a goal it has achieved. These are facts you can build on to create a reader-centered letter.

In addition, research the goals and activities of people who hold the job you want. The more you know about that job, the more persuasively you will be able to answer the reader's second question, "How will you contribute to my organization's success?"

Here are several ways to obtain specific, reader-centered information about an employer.

Learn More

For more detailed advice about how to conduct research, see Chapter 6 and the Writer's Reference Guide: Using Five Reader-Centered Research Methods (page 165).

LEARNING ABOUT EMPLOYERS

- **Draw on your own knowledge.** If you've already worked for the company, you may already know all you need to know. Remember, however, that you need information related to the specific area of activity in which you'd like to work.
- **Ask an employee, professor, or other knowledgeable person.** People often have information that isn't available in print or online.
- **Contact the company.** It might send publications about itself.
- **Search the web.** Start with the employer's own website. Also use resources such as Business Week Online (www.businessweek.com), LexisNexis (www.lexisnexis.com), and the sites for other major newspapers and magazines.
- **Consult your campus's placement office.**
- **Visit the library.** Business newspapers and magazines, as well as trade and professional journals, are excellent sources of information.

WWW

For links to other online resources for researching employers, go to **www .cengage.com/english/ anderson7e** and click on Chapter 2.

Draft Your Letter

Business letters have three parts: introduction, body, and conclusion. Within this general framework, an effective job application letter does the following:

1. Identifies the job you want
2. Demonstrates your knowledge of the employer's company or organization
3. Explains why you have chosen to apply to this employer
4. Explains how your qualifications enable you to contribute to the employer's success
5. Indicates the next step you will take

There is no single best way to combine these five elements within the three-part framework of a letter. Use your creativity to make your letter distinctive. The following advice will help you write an original letter that observes the general conventions of business correspondence. Figures 2.6 and 2.7 show how Jeannie Ryan and Ramón Perez applied this advice in their letters.

Jeannie addresses a specific person.

She shows specific knowledge of the company and offers praise related to it.

Jeannie states that she wishes to pursue a goal that the employer pursues.

Jeannie explains the relevance of her courses to the job she is seeking.

She describes one way her job experience will benefit the employer.

Jeannie tells how abilities she developed in a college activity will benefit the employer.

She highlights a specific accomplishment.

Jeannie politely requests an interview.

---

325 Foxfire Drive, Apt. 214
Denver, Colorado 70962
February 8, 2010

Ms. Nancy Zwotny, Manager
Employment and Employee Relations
Burdick Marketing
650 Broadway
Denver, Colorado 70981

Dear Ms. Zwotny:

I was very impressed to read in *Marketing News* that last year Burdick added seven national accounts to its client list. I was especially interested to learn that this success is built upon the power and rapid response of computerized tools for marketing analysis developed by your personnel. Moreover, two of these new accounts are in the electronics industry, a special interest of mine. As a senior marketing major at the University of Denver, I would very much like to work for Burdick as a research assistant, helping clients maximize sales. In particular, I would like to contribute to the success of clients in the electronics industry.

To develop a solid background in computerized decision making, I have taken several classes in data analysis, including two classes in quantitative methods, one in data management, and one in survey sampling. In addition, through my minor in computer science, I have gained technical knowledge to work effectively with Burdick's clients selling computer games and similar consumer products.

To learn the practical application of concepts I've studied in school, I have worked in selling to consumers and to retailers. Both jobs taught me how unreliable our intuitive predictions about consumer behavior can be and how important it is to research the market before introducing a new product. In addition, my service as vice president of the university's synchronized swim team has helped me develop my leadership and communication skills that will help me work effectively with other Burdick clients and employees. Last year, I helped this diverse group of fifty students prepare programs that they presented proudly to the public.

I'm sure you realize that a letter and résumé (which I've enclosed) can convey only a limited sense of a person's motivation and qualifications. Next week, I will call your assistant to see if you are able to grant me an interview. I hope that I may look forward to meeting you to discuss my credentials more fully.

Sincerely yours,

Jeannie Ryan

Jeannie Ryan

Enclosure: Résumé

---

16 Henry Street
Brooklyn, New York 11231
March 24, 2010

Estelle Ritter ◄
Financial Systems Division
Medallion Software, Inc.
1655 Avenue of the Americas
New York, New York 11301

Dear Ms. Ritter:

About a year ago, I met David Yang, a systems analyst in your department, who told me about Medallion's highly successful efforts to create integrated computer systems for banks and other financial institutions. I was particularly intrigued when he explained that many of the programs you design for international corporations must conform to government banking regulations—which differ from country to country. What a challenge! As a systems analysis major with a considerable interest in meeting the needs of clients in the financial services industry, I would like to be considered for a position in your division.

Since talking with David, I have sought out courses that would prepare me for exactly the kind of work your division performs. This semester, for example, I am taking a course in computer security and another in international finance. In addition, I have gained considerable insight into the structure and uses of two sophisticated financial systems while working as a data entry clerk for the Miller Health Spas chain and as a teller for the Meninger Bank. I have also completed several classes in written and oral communication, which would help me present Medallion effectively to clients. You will find additional details about my qualifications in the enclosed résumé.

I feel that I am a well-disciplined, highly motivated person with a strong desire to excel. I have taken the last half of my college courses while working full-time to support my wife and young son. After I become settled into a permanent job, I plan to begin graduate study in computer science so that I continue to develop my skills in systems analysis and design.

Would it be possible for me to meet with you to discuss what I have to contribute to Medallion? If so, please call me at (212) 374-7631 during the day.

Sincerely,

Ramón Perez

Ramón Perez

Enclosure

FIGURE 2.7
Letter of Application to Accompany the Résumé Shown in Figure 2.2

Ramón addresses a specific person.

He demonstrates a specific knowledge of the company and conveys his enthusiasm for it.

Ramón states that he wishes to achieve results that the employer desires to achieve.

Ramón conveys his determination to prepare for the job for which he is applying.

Ramón explains how his job experience would benefit the employer.

He also tells how two classes he took would benefit the employer.

He provides specific evidence of personal characteristics he thinks the employer will view favorably.

Ramón asks for an interview.

Introduction

By identifying the job you want in the introduction of your letter, you accommodate the employer's desire to know why you have written. When naming the position, convey your enthusiasm for it.

> I was delighted to see your ad for a chemical engineer in Sunday's newspaper.

To make your enthusiasm credible, explain the *reason* for it. In the following examples, Harlan shaped his explanations to the positions he applied for. To an environmental consulting company, he wrote:

> As an environmental engineering major who will graduate in May, I am eager to help companies assess and remediate hazards.

To a company that makes industrial lubricants, he wrote:

> In my capstone course as a chemical engineering student, I am studying the synthesis of molecules for the same industrial uses that your company's products serve.

Another effective way to begin your letter is to praise an accomplishment, project, or activity you learned about during your research. Praise is almost always welcomed by a reader, provided that it seems sincere. In the following example, Shawana combines the praise with an explanation of her reason for applying to this employer and with her statement of the job she is applying for.

Shawana gives specific praise related to the job she wants.

She introduces herself and expresses her desire to contribute.
She identifies the job she wants.

> While reading the August issue of *Automotive Week,* I learned that you needed to shut down your assembly line for only 45 minutes when switching from making last year's car model to this year's model. This 500 percent reduction in shutdown time over last year is a remarkable accomplishment. As a senior in manufacturing engineering at Western University, I would welcome a chance to contribute to further improvements in the production processes at your plant. Please consider me for the opening in the Production Design Group that you advertised through the University's Career Services Center.

Note that Shawana isn't praising the features of the car or the huge profit the company made. That would be superficial praise anyone could give without having any real understanding of the organization. Instead, Shawana focuses on a specific accomplishment, discovered through her research, that is directly related to her own specialty.

If you have a connection with a current employee or representative of the employer, you might mention it, especially if you can combine this reference with evidence that you are knowledgeable about the organization.

> At last month's Career Fair at the University of Missouri, I spoke with AgraGrow's representative Raphael Ortega, who described your company's novel approaches to no-till agriculture. As an agronomy major, I would welcome the chance to work in your research department.

Body (Qualifications Section)

Sometimes called the *qualifications section,* the body of your letter is the place to explain how your knowledge and experience prepare you to contribute significantly to the employer's organization. Some job applicants divide this section into two parts: one on their education and the other on their work experience or personal qualities. Other applicants devote a paragraph to their knowledge and skills in one area, such as engineering, and a second paragraph to their knowledge and skills in another area, such as communication. Many other methods of organizing are also possible.

No matter how you organize this section, indicate how the specific facts you convey about yourself relate to the demands of the position you want. Don't merely repeat information from your résumé. Instead of merely listing courses you've taken, show also how the knowledge you gained will help you to do a good job for the employer. Rather than listing previous job titles and areas of responsibility, indicate how the skills you gained will enable you to succeed in the job you are seeking.

> In my advanced physical chemistry course, I learned to conduct fluoroscopic and gas chromatographic analyses similar to those your laboratory uses to detect contaminants in the materials provided by your vendors.

What the student learned

> In one course, we designed a computer simulation of the transportation between three manufacturing facilities, two warehouses, and seventeen retail outlets. Through this class, I gained substantial experience in designing the kinds of systems used by your company.

How this will enable the student to contribute

Conclusion

In the conclusion of your letter, look ahead to the next step. If you are planning to follow up your letter with a phone call, indicate that. In some situations, such a call can be helpful in focusing an employer's attention on your résumé. If you are planning to wait to be contacted by the employer, indicate where you can be reached and when.

Use a Conventional Format

Finally, when you draft a letter of application, be sure to use a conventional format. Standard formats are described in Chapter 22 and used in this chapter's sample letters.

Revise Your Letter

A job application letter is an especially challenging communication to write. Look it over very carefully yourself, and ask others to help you determine whether your letter is coming across in the way you intend. The following paragraphs discuss some of the most important things to look for.

Try This

At a website that offers advice to job seekers, find a "model" job application letter for someone who has the same amount of work experience that you have. Compare that letter with the ones in Figures 2.6 and 2.7. In what ways could the model letter be improved?

Review the Personality You Project

Employers will examine everything you say for clues to your personality. When you say what appeals to you about the organization and explain why you are well suited for the job, you are revealing important things about yourself. Notice, for instance, how the first sentence of the second paragraph of Jeannie Ryan's letter (Figure 2.6) shows her to be a goal-oriented person who plans her work purposefully. Also, notice how the first paragraph of Ramón's letter (Figure 2.7) shows him to be an enthusiastic person with a firm sense of direction. When reviewing your own draft, pay special attention to the personality you project. An attribute employers especially value is enthusiasm for their organization and for the work you would do there.

WWW

To see other application letters that follow the advice given in this chapter, go to **www.cengage.com/english/anderson7e** and click on Chapter 2.

Review Your Tone

Some people have difficulty indicating an appropriate level of self-confidence. You want the tone of your letter to suggest to employers that you are self-assured but not brash or overconfident. Avoid statements like this:

Overconfident tone

> I am sure you will agree that my excellent education qualifies me for a position in your Design Department.

The phrase "I am sure you will agree" will offend some readers. And the sentence as a whole seems presumptuous. It asserts that the writer knows as much as the reader about how well the writer's qualifications meet the reader's needs. The following sentence is more likely to generate a favorable response:

More effective tone

> I hope you will find that my education qualifies me for a position in your Design Department.

Achieving just the right tone in the conclusion of a letter is also rather tricky. You should avoid ending your letter like this:

Ineffective, demanding tone

> I would like to meet with you at your earliest convenience. Please let me know when this is possible.

To sound less demanding, the writer might revise the second sentence to read, "Please let me know *whether* this is possible" and add "I *look forward* to hearing from you."

Finally, remember that you must assure that each sentence states your meaning clearly and precisely. And you must eradicate *all* spelling and grammatical errors.

WWW

To see the different letters one student wrote when applying for two different jobs, go to **www.cengage.com/english/anderson7e** and click on Chapter 2.

Adapting Your Application Letter for Different Employers

If you are applying to different employers for the same type of position, you may be able to adapt your application letter by altering only places where you demonstrate your knowledge of each specific company and tell why you are applying to it. However, if you are applying for different kinds of jobs, revise also your explanations of the match between your qualifications and the job you want.

Figure 2.8 shows a Writer's Guide you can use as you prepare your job application letters.

FIGURE 2.8 **Writer's Guide for Job Application Letters**

🌐 To download a copy of this Writer's Guide, go to www.cengage.com/english/anderson7e.

Writer's Guide
JOB APPLICATION LETTERS

Preliminary Research

_____ Determine as exactly as possible what the employer wants.

_____ Learn enough about the job and employer to tailor your letter to them.

Address

_____ Address a specific individual, if possible.

Introduction

_____ Tell clearly what you want.

_____ Persuade that you know specific, relevant things about the reader's organization.

_____ Convey that you like the company.

Qualifications

_____ Explain how the knowledge, abilities, and experiences described in your résumé are relevant to the specific job for which you are applying.

Closing

_____ Sound cordial, yet clearly set out a plan of action.

Prose

_____ Use clear sentences with varied structures.

_____ Use an easy-to-follow organization.

_____ Use a confident but modest tone.

_____ Express the action in verbs, not nouns.

_____ Use strong verbs.

_____ Use correct spelling, grammar, and punctuation.

Appearance

_____ Look neat and attractive.

_____ Include all the elements of a business letter.

Ethics

_____ Describe your qualifications honestly.

_____ Avoid statements intended to mislead.

Overall

_____ Show that you are aware of your reader's goals and concerns when hiring.

_____ Demonstrate that you are a skilled communicator.

ETHICAL ISSUES IN THE JOB SEARCH

One of the most interesting—and perplexing—features of workplace writing is that the ethical standards differ from one situation to another. The résumé and job application letter illustrate this point because the expectations that apply to them are significantly different from those that apply to other common types of on-the-job writing.

For instance, when writing a résumé you are permitted to present facts about yourself in a very selective way that would be considered unethical in most other kinds of workplace writing. For example, imagine that you have suggested that your company reorganize its system for keeping track of inventory. Several managers have asked you to investigate this possibility further, then write a report about it. In this report, your readers will expect you to include unfavorable information about your system as well as favorable information. If you omit the unfavorable information, your employer will judge that you have behaved unethically in order to win approval of your idea. If you were to submit your résumé to these same readers, however, they would not expect you to include unfavorable information about yourself. In fact, they would be surprised if you did.

Furthermore, these readers would expect you to present the favorable information about yourself in as impressive a language as possible, even though they might feel you were ethically bound to use cooler, more objective writing in your proposal about the inventory system. Here are the ethical guidelines that *do* apply to résumés and job application letters.

GUIDELINES FOR AN ETHICAL RÉSUMÉ AND JOB APPLICATION LETTER

- Don't list degrees you haven't earned, offices you haven't served in, or jobs you haven't held.
- Don't list awards or other recognition you haven't actually received.
- Don't take sole credit for things you have done as a team member.
- Don't give yourself a job title you haven't had.
- Don't phrase your statements in a way that is intended to mislead your readers.
- Don't list references who haven't agreed to serve as references for you.

If you are unsure whether you are writing ethically in some part of your résumé, check with your instructor or someone else who is familiar with workplace expectations about this type of communication. (You might also ask for advice from more experienced people whenever you are unsure about the ethical expectations that apply to *any* communication you write at work.)

WRITING FOR EMPLOYMENT IN OTHER COUNTRIES

In other countries, résumés and job application letters may look very different from those used in the United States. The traditional Japanese job application, the *rirekisho,* includes a significant amount of personal information to indicate that the applicant comes from an environment that makes it likely the person will enjoy long-term success in the organization (Lofving & Kennedy-Takahashi, 2000). Although Japanese employers expect less personal information in applications from persons in other cultures, they still want to know the applicant's nationality, age, and marital status. In France, many employers want job application letters written by hand so they can subject the letters to handwriting analysis (Pensot, 2000). French employers believe that this analysis enables them to learn about the applicant's personal traits, which they weigh heavily when making employment decisions. Employers in many countries, including South Africa, want applications to include grades, copies of diplomas, and reference letters, called *testimonials* (Woodburn Mann, 2000).

Because expectations differ so significantly from country to country, conduct reader-centered research about the country to which you will send applications. Among others, Mary Ann Thompson's *Global Résumé and CV Guide* (2000) is an excellent source.

INTERVIEWING EFFECTIVELY AND DISPLAYING YOUR WORK

Your reward for preparing an effective résumé and application letter is, of course, the opportunity to do some additional reader-centered communicating, this time in a job interview. For advice about ways to interview effectively, go to the website for this book and click on Chapter 2.

One way to make yourself stand out in an interview is to bring samples of your work. The website for this book also provides detailed advice on creating print and digital portfolios.

WWW

For information on job interviews and portfolios, click on Chapter 2 at **www.cengage.com/english/anderson7e**.

CONCLUSION

This chapter has demonstrated how the reader-centered approach used by successful workplace communicators can help you create a highly effective résumé and letter of application. The strategies you saw in action here are ones you can use for all your work-related writing:

WWW

For links to websites with information about preparing résumés for jobs in specific countries, go to **www.cengage.com/english/anderson7e** and click on Chapter 2.

- Think continuously about your readers.
- Use your knowledge of your readers to guide all your writing decisions.

The rest of this book is devoted to developing your communication expertise by providing detailed reader-centered advice you can use whenever you write at work.

USE WHAT YOU'VE LEARNED

For additional exercises, go to **www.cengage.com/english/ anderson7e.** *Instructors:* The book's website includes suggestions for teaching the exercises.

EXERCISE YOUR EXPERTISE

1. Find a sample résumé at your college's Career Services Center or in a book about résumé writing. Evaluate it from the point of view of its intended reader. How could it be improved?

2. Using the web, newspapers, or journal articles, locate four or more openings that appeal to you. Create a unified list of the qualifications they specify. Then identify your skills and experiences that match each item on the employers' list.

3. Complete the assignment in Appendix B on writing a résumé and a letter of application.

EXPLORE ONLINE

Using the web, find an employer or online job board that asks you to fill out an online résumé form. Evaluate the extent to which the form helps or hinders you from presenting your qualifications in the most persuasive manner.

COLLABORATE WITH YOUR CLASSMATES

Collaborating with another student in your class, work together on developing a keyword list for each of you that you could use when creating a scannable résumé.

APPLY YOUR ETHICS

Using the library or web, read an article that discusses the attitudes of employers toward unethical résumés. Take notes you can share with your class.

CASE	ADVISING PATRICIA

For additional cases, visit www.cengage.com/english/anderson7e. *Instructors:* The book's website includes suggestions for teaching the cases.

This morning you stopped outside the library to talk with Patricia Norman, a senior who is majoring in marketing. She told you with a mixture of excitement and anxiety that she has finally decided to join the many other seniors who are busily looking for a job. She's even drafted a résumé and begun writing letters to employers listed in a publication she picked up at the Career Services Center.

"Look," she exclaimed. "One of the department store chains I'm writing to is mentioned in this article that Professor Schraff asked us to read." She held out an article from *Retail Management.* "They've begun opening free-standing specialty shops in their stores. The managers order their own merchandise and run their own advertising

campaigns. It's been a huge success. This sounds like such a great place to work—a big chain that welcomes innovators." Then her excitement turned to anxiety. "I'm worried they won't like my résumé and application letter, though. I've gone over both again and again, and my roommate has too. But I'm still worried."

As you tried to reassure her, she pulled out her drafts and put them into your hand. "Take a look at them and tell me what you really think. I need all the help I can get." You had to leave for an appointment, but you agreed to look over her drafts and meet her again this evening.

Now it's afternoon, and you've started to read Patricia's résumé and letter. As you do, you think back over some of the things you know about her. She's an active and energetic person, talkative, and fun to be with. Throughout her years in college, she has spent lots of time with a group called Angel Flight, a volunteer organization that sponsors service activities on campus and off. In fact, this past year you've seen less of her because she has spent so much time serving as the organization's president. "As president, I'm responsible for everything," she once told you. "Everything from running meetings, to getting volunteers, to seeing that the volunteers have done what they said they would." While a junior she held some other office, you recall—also a time-consuming one. But she's like that. In the Marketing Club, she edited the newsletter and handled lots of odd jobs, like putting up posters announcing speakers and meetings. She was also treasurer of the Fencing Club, another of her interests. Once when you marveled at how many things she was able to do, she responded, "It's not so much, if you're organized."

Despite all the time she spends on such activities, Patricia earns good grades, a 3.6 average, she told you once. Although she's had to take lots of business courses, she's also squeezed in a few electives in one of her favorite subjects: art history. One Saturday last year, she even got you to travel 200 miles with her to see an art exhibit—a "major" exhibit, she had assured you.

But the trip you most enjoyed with her was to a shopping center, where she spent more time commenting on how the merchandise was displayed than looking for things to buy. She talked a lot about the way they did things at a Dallas department store where she's worked the past three summers. She must have some interesting opportunities there, you note; after all, one of the people she lists as a reference is the store manager. Her other references are professors who've taught classes that you and Patricia have taken together. They were fun. Everything's fun with Patricia.

YOUR ASSIGNMENT

Decide what you will say to Patricia about her letter (Figure 2.9) and résumé (Figure 2.10). What strengths will you praise? What changes will you suggest? What questions will you ask to determine whether she might include additional information? Assume that her résumé will be read first by a person, not by a scanner.

FIGURE 2.9
Letter of Application for
Use with Case

Box 88
Wells Hall
University of Washington
Seattle, Washington 98195
February 12, 2010

Kevin Mathews, Director
Corporate Recruiting
A. L. Lambert Department Stores, Inc.
Fifth and Noble Streets
San Diego, California 92103

Dear Mr. Mathews:

I saw A. L. Lambert's advertisement in the *College Placement Annual*. I was very
impressed with your company. I hope that you will consider me for an opening in
your Executive Develepment Program.

In June, I will graduate from the University of Washington's retailing program, where I
have focused my study on marketing management. I have learned a great deal about
consumer behavior, advertising, and innovative sales techniques. Furthermore, I have
gained a through overview of the retailing industry, and I have studied successful and
unsuccessful retailing campaigns through the case-study method.

In addition to my educational qualifications, I have experience both in retail sales and
in managing volunteer organizations. While working in a Dallas department store for
the past four summers, I had many opprtunities to apply the knowledge and skills that
I have learned in college. Likewise, in my extracurricular activities, I have gained
experience working and communicating with people. For instance, I have been the
president of Angel Flight, a volunteer service organization at the University of
of Washington. Like a manager, I supervised many of the organization's activities.
Similarly, while holding offices in two campus organizations, I have developed my
senses of organization and responsibility.

I would like to talk with you in person about my qualifications. Please tell me how
that can be arranged.

Sincerely yours,

Patricia Norman

FIGURE 2.10
Résumé for Use with Case

PATRICIA NORMAN
Box 80, Wells Hall
University of Washington
Seattle, Washington 98195
(206) 529-5097

PERSONAL
Born: March 17, 1989
Health: Excellent
Willing to relocate

PROFESSIONAL OBJECTIVE
To work for an innovative and growing retailer.

EDUCATION
University of Washington, Seattle, Washington, B.S. in Retailing,
May 2010.

Earned 23 credit hours in marketing management, obtaining a working knowledge of the factors motivating today's consumer. Also learned how a product is marketed and distributed to the consumer. Took eight credit hours of study focused specifically on principles and problems of retail management.

WORK EXPERIENCE
Danzig's Department Store, Dallas, Texas,
Summers 2006–2009.

Worked as a sales clerk. Helped customers choose their purchases and listened politely to their complaints. Cash register operation. Stocked shelves and racks. Provided assistance to several department managers.

ACTIVITIES
Fencing Club, served as treasurer.
Angel Flight, President.
Marketing Club, member.

REFERENCES
Derek Yoder, *Store Manager*
Danzig's Department Store
11134 Longhorn Drive
Dallas, Texas 75220

Gregory Yule
Pinehurst Hall
University of Washington 98195
(206) 579-9481

Lydia Zelasko
Putnam Hall
University of Washington
Seattle, Washington 98195

Chapter 3   Defining Your Communication's Objectives: Purpose, Reader, Context

3 | Defining Your Communication's Objectives: Purpose, Reader, Context

GUIDELINES

| DEFINING OBJECTIVES |
| PLANNING |
| RESEARCHING |
| DRAFTING |
| REVISING |

This is the first of thirteen chapters that will help you develop expertise in the many activities people perform as they write at work. This chapter begins where writing begins: with your description of what you want your communication to accomplish.

"But wait," you may be thinking. "Won't my objectives for writing be so obvious that I don't need a textbook to know how to describe them?"

Your objectives may be obvious, but your communication's objectives may not. Defining *your communication's* objectives can be one of the most important steps you take when writing at work.

As Chapter 1 explained, whenever you write at work, you will be attempting to change something from the way it is to the way you want it to be. That's *your* objective. However, you can only achieve your objective if your reader responds to your communication in a certain way. A résumé and job application letter, discussed in Chapter 2, provide perfect examples. Your objective when writing them is to obtain an interview. Their objective is to lead employers looking quickly through a stack of applications to conclude that you possess precisely the qualifications they want in the person they will hire for certain, specific positions. Unless your résumé and letter achieve this objective, they don't succeed. Similarly, the success of all the communications you write at work will depend on their ability to create a *particular* response from your *particular* readers. The more specifically you can define what your communication must do to create this response, the more likely you are to succeed in creating a communication that achieves its objectives. This chapter leads you through the process of defining a communication's objectives in a truly helpful way.

WWW

For additional information and examples related to this chapter's guidelines, go to **www.cengage.com/english/ anderson7e** and click on Chapter 3.

A communication's objective is to create a certain response from its readers.

Three factors that influence a reader's response: the reader's purpose, the reader's characteristics, and the context in which he or she will read.

Guideline 1 | Create a mental movie of your reader in the act of reading

Let's begin with the way you can use a specific, detailed definition of a communication's objectives. While writing, you must make a multitude of decisions—what to include, what to leave out, how to organize the contents, what to present in text and what in graphics, how to design the pages—indeed, how to handle every detail of the message you will give your readers. As Chapter 1 suggested, an excellent way to make these decisions is to imagine your reader's response to each one. Make a mental movie of your reader in the act of reading that will enable you to predict how he or she will react to the alternatives you are considering (Elbow, 1998). If the reader's reaction to a particular organization, style, word choice, or other feature leads to the overall response you want your communication to create, then go with it. If not, try something else.

Of course, you can never be absolutely sure how someone will react to anything you write or say. That's because the reader's reaction depends on a complex set of factors. To make the best predictions possible, the script for the mental movie you create of your reader must take these factors into account to the extent you are able. In general, these factors fall into three groups: those related to your reader's purpose in reading, your reader's characteristics, and the context in which your reader will read.

Purpose, reader, context. You may be familiar with these elements of the writing situation from a course in some other kind of writing. They are just as important in workplace writing. This chapter presents two guidelines that will help you define the *purpose* of your communications in a reader-centered way that is perfect for the workplace. The next three guidelines will help you create a profile of your *reader* that includes the characteristics that are likely to shape his or her response to what you've written. Then one more guideline will help you identify elements of the *context* that are likely to influence your reader's reactions.

To help you see the practical benefits of defining your communication's objectives in a reader-centered way, the rest of this chapter shows how this approach helped Stephanie, a college student, identify effective strategies for writing a memo to her boss, Ms. Land, at the nonprofit organization that gave Stephanie a summer job. The organization provides Braille translations for books, magazines, and other reading material requested by people who are blind. All translations are prepared by volunteers who work at their homes. Stephanie observed that some urgently needed translations, such as those of textbooks that students require for their courses, aren't completed on time because all translations are assigned to the volunteer translators on a rotating basis rather than their ability to complete a translation rapidly. Stephanie decided to write to Ms. Land, recommending that urgently needed translations be assigned to the fastest volunteers rather than the next ones on the volunteer list.

To define her memo's objectives, Stephanie used the Writer's Guide for Defining Your Communication's Objectives, shown in Figure 3.1. This guide can also help you. To download a copy, go to www.cengage.com/english/anderson7e. A copy of the guide that Stephanie completed is shown at the end of this chapter (Figure 3.4, page 91). As you'll see, the guide challenges your creativity, imagination, and problem-solving ability in ways that help you identify effective strategies for writing on-the-job communications.

The guide helps you look at your communication from your reader's perspective rather than your own. After all, the success of your communications depends on the reader's response, not your desires. You begin, of course, by thinking about what you want to achieve. Therefore, the guide asks why you are writing, what you want the outcome to be, and who your reader is. However, the guide then shifts your focus to your reader by asking, "What outcomes does your reader desire?" Stephanie answered that Ms. Land wanted "the best possible service to the persons who request translations." Although Stephanie had the same goal, her previous thinking about her memo had focused on the ways her proposed system would work. As important as the process was, she realized the major theme of her memo should be improving service to the organization's clients. The guide's remaining questions, which mirror Guidelines 2 through 9, helped Stephanie identify many strategies for crafting an effective memo to Ms. Land.

Stephanie wants to write a memo that will persuade her boss to adopt a new system for assigning Braille translations.

Try This

How do people who are blind use computers? Try searching the web for information about assistive devices and computer programs designed for people with visual impairments.

WWW

To download the Writer's Guide for Defining Your Communication's Objectives, go to **www.cengage.com/english/anderson7e** and click on Chapter 3.

How Stephanie answered the Writer's Guide's question about the outcome her reader desired—and how the answer helped her plan her memo

FIGURE 3.1 Writer's Guide for Defining Your Communication's Objectives

To download a copy of this Writer's Guide, go to Chapter 3 at www.cengage.com/english/anderson7e.

Writer's Guide
DEFINING YOUR COMMUNICATION'S OBJECTIVES

YOUR PURPOSE
1. What are you writing?
2. What outcome do you desire?
3. Who is your reader?

PURPOSE OF YOUR COMMUNICATION   ⟵—— See Guidelines 2 and 3, pages 71 and 75
1. What task will your communication help your reader perform? (Usability)
    What is your reader's purpose for reading?
    What information does your reader want? (What questions will your reader ask?)
    How will your reader search for the information? (May use more than one strategy)
       ____ Sequential reading from beginning to end
       ____ Reading for key points
       ____ Reference reading
       ____ Other (describe)
    How will your reader use the information?
       ____ Compare alternatives (What will be the points of comparison?)
       ____ Determine how the information will affect him or her (or the organization)
       ____ Perform a procedure (following instructions step by step)
       ____ Other (describe)
2. How do you want your communication to change your reader's attitudes? (Persuasiveness)
    What is your reader's attitude toward your subject? What do you want it to be?
    What is your reader's attitude toward you? What do you want it to be?
    What is your reader's attitude toward your organization? What do you want it to be?

READER'S PROFILE   ⟵—— See Guidelines 4, 5, and 6, pages 76, 79, and 84
1. Role in the organization
2. Role as a reader
3. Familiarity with your subject
4. Familiarity with your specialty
5. Relationship to you
6. Other personal characteristics you should take into account
7. Cultural characteristics you should take into account
8. Who else might read your communication?

CONTEXT   ⟵—— See Guidelines 7 and 8, pages 87 and 89
1. What features of the context may affect the way your reader reads your communication?
2. What expectations, regulations, or other factors constrain the way you can write?

ETHICAL Treatment of Stakeholders   ⟵—— See Guideline 9, page 90
1. Who, besides your reader, are stakeholders in your communication?
2. How will they be affected by it?

GUIDELINES FOR DESCRIBING YOUR COMMUNICATION'S PURPOSE

Guidelines 2 and 3 focus on reader-centered ways to define a communication's purpose. They correspond to the two qualities that, as Chapter 1 explained, are essential for effective workplace writing: usability and persuasiveness.

Guideline 2 | Describe the task your communication will help your reader perform

As you learned in Chapter 1, people at work read in order to complete tasks that are part of their job responsibilities. To be effective, communications must help them perform those tasks. That is, they must be *usable*—from the *reader's* perspective. The distinction between less usable and more usable communications can be illustrated by the evolution of computer manuals. Early manuals were organized around a computer's features. For instance, they would tell what each menu choice or each F-key would do. This organization frustrated users. They wanted to know how to do something, not what the F8 key would do. Computer companies made their manuals much more usable when they took the time to learn about the tasks users wanted to perform and then began organizing their manuals around those tasks.

> Usability is the ease with which readers can use your communication to perform their tasks.

Similarly, to make your communications usable, you must begin by learning about the tasks—whether mental, physical, or a combination of the two—that your reader will try to perform while reading what you've written. The following four-step procedure will enable you to identify these tasks.

PROCEDURE FOR IDENTIFYING YOUR READER'S TASKS

1. Identify your reader's purpose for reading.
2. Identify the information your reader wants from your communication.
3. Determine how your reader will look for this information.
4. Determine how he or she will use the information while reading.

Identify Your Reader's Purpose for Reading

First, identify your reader's purpose when picking up your communication to start to read it. Almost certainly, the reader will be trying to gain information he or she can use in some practical way. Managers read reports to gather information needed to make a decision. Electronic engineers consult a product brochure to find the best battery to use in a newly designed cell phone.

In many situations, your reader's purpose will be close to your own. If you are designing the product brochure for cell phone batteries, you and your reader will

---

Consumers assembling a product, executives making decisions, and supervisors managing factory production need many different things from the communications they read. To write effectively, you must identify and meet the needs of the particular persons to whom you are writing.

have the same purpose in mind: to help the reader find the best match with the cell phone he or she is designing. In other situations, your reader's purposes will be quite different from yours. When you and your reader have different purposes, you must help your reader achieve his or hers in order for you to be able to achieve yours. Stephanie's case is an example.

Stephanie's purpose in writing is to persuade Ms. Land to change the system for assigning Braille translations. Ms. Land's purpose in reading will be to compare the current system with the one Stephanie recommends. The more difficulty Ms. Land has comparing the two systems, the less likely she will be to notice the superiority of Stephanie's recommendation. Thus, to achieve her goal of persuading Ms. Land, Stephanie must create a communication that helps Ms. Land easily compare the two systems.

Identify the Information Your Reader Wants

By identifying your reader's purpose, you advance a long way toward identifying the information the reader will want to find in your communication. You can pinpoint this information more specifically by imagining the questions your reader will have in mind when picking up your communication. Focusing on your reader's questions in this way has the great, reader-centered benefit of concentrating your attention on what your reader wants to read, not what you want to say.

When Stephanie began thinking about what to include in her memo, she immediately thought of the arguments she wanted to make on behalf of her proposed system: It would improve service to persons who request urgent translations. It would be easy to implement. It would not take any additional time for the office staff to manage.

However, when she began to fill in the section of the Writer's Guide that asked her to list her reader's questions, Stephanie realized that her memo had to include several kinds of information she would have omitted if she thought of the memo only from her own perspective, not Ms. Land's. Here are questions she predicted Ms. Land would ask.

Stephanie gained insights into topics she should address in her memo by filling out the Writer's Guide section about her reader's questions.

- What evidence do you have that there's a problem with the current system?
- Who would determine which translations deserve highest priority?
- How would we decide which translators are placed in our top group?
- Will your system really work? Are other agencies using it successfully?

Determine How Your Reader Will Look for Information

By identifying the ways your reader will look for information, you can design your communication to match their search strategy. Here are three search strategies readers often use on the job, together with some writing strategies you can use to assist readers who use them.

THREE WAYS READERS SEARCH FOR INFORMATION

- **Thorough, sequential reading.** When writing for readers who will read each sentence and paragraph in turn, you can systematically build ideas from one sentence, paragraph, and section to the next.
- **Reading for key points only.** When writing to readers who will scan for key points, you can use lists, tables, boldface, and page design to make the key points stand out.
- **Reference reading.** When you write to readers who will seek only specific pieces of information, you can use headings, tables of contents, and indexes to guide your readers rapidly to the information they seek.

Learn More

For more information on graphics and page design, see Chapters 13 and 14.

When Stephanie worked on the Writer's Guide section about the ways her reader would look for information in her memo, she guessed that Ms. Land would read the memo sequentially. But she also thought that Ms. Land might want to refer back to the memo to remind herself of some fact it included. Consequently, she decided to use headings that identified the information included in each of the memo's sections. The headings would help Ms. Land see the organization of the memo and also help her locate specific pieces of information she might want to review in the future.

How Stephanie described the way her reader would search for information in the memo—and the plans Stephanie was able to make as a result

Determine How Your Readers Will Use the Information While Reading

Once readers find the information they want, they start using it. Many of their uses occur after reading, for instance in discussions with other people. When you are writing, you can make your communications more usable by thinking instead about the ways your reader will use your information *while he or she is reading it.* Here are three common ways people use information while reading at work, together with example writing strategies appropriate to each:

THREE WAYS READERS USE INFORMATION AT WORK

- **To compare alternatives.** Usually, you can help readers compare alternatives by organizing your information about the alternatives around the readers' criteria.

- **To determine how the information will affect them and their organization.** You can assist these readers by organizing around the kinds of concerns they have, whether budget and efficiency or prestige and parking spaces.

- **To perform a procedure.** In addition to putting the steps in sequential order, you can help readers by putting the steps in a numbered list rather than paragraphs.

How Stephanie said Ms. Land would use the information in her memo and how this description helped Stephanie plan her memo

By answering the question on the Writer's Guide that asked how Ms. Land would use the information in her memo, Stephanie gained two important insights. First, she realized that Ms. Land would want to compare, point by point, the current and proposed procedures for assigning Braille translations. This insight told her that she should organize her memo around points of comparison so that Ms. Land could find the comparable information about both systems in one place. Previously, Stephanie had been planning to describe the current system, with its weaknesses, in one section and then her proposed system, with all its good points, in another. That organization would have made point-by-point comparison difficult for Ms. Land.

Stephanie's second insight came when she read the Writer's Guide's prompt that asked her to list her reader's criteria. Stephanie discovered that she hadn't yet identified Ms. Land's criteria. After some thought, she determined that they would be cost, efficiency, and impact on the morale of the volunteers, especially those who were not chosen as the most efficient. Further, she realized that these criteria should be the points of comparison around which she should organize her information about the two systems. She also realized that if she wanted Ms. Land to take her recommendation seriously, she would have to include information that addressed these criteria for both procedures.

Thus, the Writer's Guide's questions about how her readers would read her communication helped Stephanie identify the best way to organize her memo and also helped her identify important information that she would have failed to include if she had continued thinking about her memo primarily from her own perspective, not Ms. Land's.

Guideline 3 | Describe the way you want your communication to alter your reader's attitudes

As you learned in Chapter 1, persuasiveness is the second of the two essential qualities of all on-the-job writing. Because you always know the end point, the attitude you want your reader to have after reading your communication, it may seem unnecessary to spend time describing the persuasive purpose of your communications. However, the writing strategies that are most likely to achieve the outcome a writer desires depend also on the starting point, that is, on the reader's attitude right now, before reading. If the reader's initial attitude toward your topic is positive, your communication's persuasive purpose will be to reinforce that attitude, making it even stronger than before. Because you can build on the reader's existing attitude and expect little resistance from the reader, you can probably succeed by presenting only positive points without trying to rebut arguments against your position. If you were to discuss negative points, it would be for the purpose of persuading the reader that you have thoroughly investigated the topic. In contrast, if your goal is to reverse the reader's initial attitude, you can expect resistance. To persuade, you must make positive points *plus* address the objections the reader will raise.

To define the way you want a communication to alter your reader's attitudes, describe the reader's present attitude and the attitude you want the reader to have after reading your communication.

When Stephanie responded to the Writer's Guide's questions about the way she wanted her communication to change Ms. Land's attitudes, she realized that she wanted it to reverse them. To write effectively, Stephanie would have to anticipate and address Ms. Land's counterarguments.

How Stephanie said she wanted her memo to change Ms. Land's attitudes

The Writer's Guide also asks a writer to describe the reader's attitude toward the writer. This question reminded Stephanie that Ms. Land has indicated on several occasions that she believes college students like Stephanie who have summer jobs with the organization lack practical knowledge about its complexities. Thus, in the Writer's Guide, Stephanie wrote the following.

> Ms. Land thinks that I am a good summer employee. But she thinks all summer employees are ignorant concerning policy issues. I want her to think that I am knowledgeable enough to have a recommendation worth serious consideration.

How Stephanie described the way she wanted to change Ms. Land's attitudes toward her

To bring about this change of attitude, Stephanie decided to discuss thoroughly all aspects of the current and her proposed systems for assigning Braille translations. Without those details, she thought, Ms. Land might think she was too uninformed to be taken seriously.

The Writer's Guide also asked Stephanie to describe her reader's attitude toward Stephanie's organization. Although this question can be helpful when writer and reader work for different employers or in different departments of the same organization, Stephanie skipped it because she and Ms. Land work for the same organization.

When completing the Writer's Guide sections on her reader's attitudes (Guideline 3) and reading task (Guideline 2), Stephanie relied on her personal knowledge of Ms. Land. In many cases at work, you will be writing to people you don't know. In those situations, it's helpful to ask others, such as your coworkers,

Ask others to help you understand the persons to whom you are writing about your readers. Sometimes you can introduce yourself to the readers themselves. They are likely to tell you what kinds of information they want to find in your communication, as well as how they will look for and use it. As you talk with them, listen for indications about their attitudes that can guide the way you write to them.

GUIDELINES FOR CREATING A PROFILE OF YOUR READER

Different people respond differently to the same writing strategies. What one person finds to be highly usable and persuasive, another may not. You can greatly increase your ability to predict your particular reader's responses by creating a profile of your reader that includes the characteristics most likely to affect his or her reactions. Guidelines 4, 5, and 6 lead you through the process of creating such a profile.

Guideline 4 | Describe your reader's professional characteristics

To a large extent, people's ways of approaching any question, problem, or communication at work depend on the small set of characteristics asked about in the Writer's Guide for Defining Your Communication's Objectives. Each is discussed in the following sections.

Professional Specialty

People with different professional specialties ask different questions and use the answers differently. When reading a report on the industrial emissions from a factory, an environmental engineer working for the factory might ask, "How are these emissions produced, and what can be done to reduce them?" whereas the corporate attorney might ask, "Do the emissions exceed Environmental Protection Agency limits; if so, how can we limit our fines for these violations?"

The most obvious clues to your readers' specialties are their job titles: systems analyst, laboratory technician, bank cashier, director of public relations. However, don't settle for obtaining only a general sense of your readers' specialties. Determine as precisely as possible why each person will be reading your communication and exactly what information he or she will be looking for.

Organizational Role

Regardless of the particularities of their professional specialties, people at work usually play one of three organizational roles: decision maker, adviser, or implementer. Each role leads to a different set of questions.

Decision makers   The decision makers' role is to say how the organization will act when it is confronted with a particular choice. Decision makers determine what the company should do in the future—next week, next month, next year. They usually ask questions shaped by their need to choose between alternative courses of action. In her memo, Stephanie will address Ms. Land as a decision maker.

THREE WAYS READERS USE INFORMATION AT WORK

- **What are your conclusions?** Decision makers want your conclusions, not the raw data you gathered or the details about your procedures. Conclusions can serve as the basis for decisions; details cannot.
- **What do you recommend?** Decision makers usually ask you about a topic because you have special knowledge of it. This knowledge makes your recommendation especially valuable to decision-making readers.
- **What will happen?** Decision makers want to know what will occur if they follow your recommendations—and what will happen if they don't. How much money will be saved? How much will production increase? How will customers react?

Advisers   Advisers provide information and advice for decision makers to consider when deciding what the organization should do. Unlike decision makers, advisers are very interested in details. They need to analyze and evaluate the evidence supporting your general conclusions, recommendations, and projections.

Consequently, advisers ask questions that touch on the thoroughness, reliability, and impact of your work.

TYPICAL QUESTIONS ASKED BY ADVISERS

- Did you use a reasonable method to obtain your results?
- Do your data really support your conclusions?
- Have you overlooked anything important?
- If your recommendation is followed, what will be the effect on other departments?
- What kinds of problems are likely to arise?

Implementers   Decisions, once made, must be carried out by someone. Implementers are these individuals. Their most important questions are these.

- **What do you want me to do?** Whether you are writing step-by-step instructions, requests for information, or policies that others must follow, implementers want you to provide clear, exact, easy-to-follow directions.

- **Why do you want me to do it?** To produce satisfactory results, implementers often must know the reason for the policy or directive they are reading. Imagine, for instance, the situation of the managers of a factory who have been directed to cut by 15 percent the amount of energy used in production. They need to know whether they are to make long-term energy savings or compensate for a short-term shortage. If the latter, they might take temporary actions, such as altering work hours and curtailing certain operations. However, if the reduction is to be long-term, they might purchase new equipment and modify the factory building.

- **How much freedom do I have in deciding how to do this?** People often devise shortcuts or alternative ways of doing things. They need to know whether they have this freedom or whether they must do things exactly as stated.

- **What's the deadline?** To be able to adjust their schedules to include a new task along with their other responsibilities, implementers need to know when the new task must be completed.

Familiarity with Your Topic

Your readers' familiarity with your topic—company inventory levels, employee morale on the second shift, problems with new software—will determine how much background information you must provide to make your communication understandable and useful to them.

Knowledge of Your Specialty

To use the information you provide, readers need to understand the terms and concepts you employ. A person unfamiliar with your specialty would want you to explain what *zeroing* and the *Z-axis* are. On the other hand, if you provided those explanations for readers who are familiar with this specialty, they would ask, "Why is this writer making me read about things I already know?" Your goal is to learn enough about your reader to be able to strike the right balance between too little and too much explanation.

Relationship with You

When you are having a conversation, you adjust your speech to your relationship with the other person. You talk with a friend more informally than you do with a college instructor you don't know well—and both your friend and your

---

Try This

Think of a recent letter, e-mail, or instant message you wrote to a friend or family member. How did you adjust your content and style to your purpose, reader, and context? How would you have changed your message if you wrote on the same topic to a different reader? What if your purpose or the context had been different?

---

instructor might be startled if you didn't make such adjustments. Similarly, at work you should write in a way that reflects the relationships you have with your readers.

For Stephanie, answering the Writer's Guide question about her relationship with Ms. Land highlighted Ms. Land's preference for a formal relationship with summer employees. This insight highlighted for Stephanie the importance of using a formal, respectful style throughout her memo. Stephanie also knew that Ms. Land had set up the current system years ago, which might cause her to react defensively toward a suggestion that a better system might exist. Focusing on this possibility, Stephanie began searching for a way to argue for her system without criticizing the current one. She decided to say that she was recommending a slight adjustment to Ms. Land's system that would make it "even more effective at doing what it now does so well."

What Stephanie learned about how to write her memo by answering the Writer's Guide's question about her relationship with her reader.

Personal Preferences

A variety of personal characteristics can also influence readers' responses to your writing. You may be writing to an individual who detests the use of certain words or insists on particular ways of phrasing certain statements. Or you might be writing to someone who is keenly interested in information you would not have to supply to most people. Some readers favor brief messages, and some insist on more detail. It only makes sense to accommodate such personal preferences where feasible.

Silatul Rahim Dahman, who is 100 percent blind, writes and reads at his computer at work. He is Information Communication and Technology Manager at the Malaysian Association for the Blind.

Special Considerations

This is a catchall category. It is a reminder that each reader is unique. You should always be on the lookout for reader characteristics you would not normally need to consider. For example, you may be addressing individuals with an especially high or low reading level, weak eyesight, or color blindness—and it will be essential for you to take these characteristics into consideration when writing.

Guideline 5 | Global Guideline: Describe your reader's cultural characteristics

One of the great pleasures for many college graduates is the opportunity to work with persons who live in or have been raised in cultures different from their own. One reason for this is the spread of international companies and organizations that have employees, suppliers, and customers in many other nations. In many countries, even small companies conduct business with people on other continents. In many parts of the world, employees in the same building come from many countries and cultural backgrounds. No matter what country you live in, this diversity will enable you to learn about other ways of life and other ways of knowing and being in the world.

Cultural Differences That Affect Communication

Among the many possible differences among people from different cultures are differences in what they expect in writing communications and how they respond to what they read. When writing to someone from a cultural background different from yours, you may need to use different writing strategies than would succeed with persons from your own culture. Consequently, you may need to add cultural information to the profile you use to make your mental movie of your reader in the act of reading your communication.

There's no magical set of cultural characteristics that applies in every case. To achieve the necessary understanding of your reader, you may need to conduct some research; a later section in this chapter suggests ways to do that. When conducting research, it can be helpful to know some of the kinds of differences you may find. The following sections discuss six.

- Amount of detail expected
- Distance between the top and bottom of organizational hierarchies
- Individual versus group orientation
- Preference for direct or indirect statements
- Basis of business decisions
- Interpretation of images, gestures, and words

Some cultural differences that can affect the ways readers respond to communications in the workplace.

As you read about these differences, remember that each contains a range of possibilities. Between the poles are many gradations.

Learn More

For information on communicating between high-context and low-context cultures, see Chapter 4, page 109.

Amount of detail expected   One of the pioneers in studying intercultural communication, Edward Hall (1976) distinguished among cultures on the basis of the amount of detail they provide in a communication. In countries such as Japan, communications provide a small amount of detail because writers and readers both assume that readers can fill them in by drawing on their existing knowledge. Hall called these high-context cultures because successful communication depends on the large amount of contextual information the readers bring to the message.

In contrast, in low-context cultures, such as the United States and much of Northern Europe, writers and readers both assume that readers are responsible for bringing very little contextual knowledge to a communication. Consequently, writers typically provide extensive detail, trying to cover thoroughly every aspect of their topic.

Knowledge of these differences and the ability to accommodate communications appropriately is important to you if you are writing from a high- or low-context culture to readers in the other kind of culture. Readers in a high-context culture can be offended if a writer includes more detail than they expect and need. The extra detail would seem to imply that they do not know the things they should know. Similarly, a low-context reader could feel that the writer who has provided a high-context amount of detail was not considering their needs because much of the expected information (even if not needed) wasn't provided.

Distance between the top and bottom of organizational hierarchies   Through research that included the study of managers at IBM facilities in forty countries, Geert Hofstede (2001) distinguished cultures according to the distance they maintain between people at the bottom and the top of an organization's hierarchy. In the United States and in some European cultures, the distance is very small. In other cultures, such as the Japanese, the distance is much greater. This information is helpful to writers because Hofstede also found that, in general, where the distance is greatest, communication styles are most formal whether the communications are written by people in lower ranks to people above them or vice versa.

Individual versus group orientation   Hofstede also distinguished cultures that focus on the individual from those that focus on the group. Individualistic cultures honor personal achievement and expect individuals to take care of themselves. The dominant cultures in the United States and northern Europe provide examples. In group-oriented cultures, success belongs to the group and people pursue group goals rather than individual ones. Many Asian cultures are group oriented.

Writers often can increase their effectiveness by adjusting to the individualistic or group orientation of their readers' culture. For example, in persuasive communications it can be helpful to highlight benefits to the individual, whereas an emphasis on benefits to the readers' organization can be more persuasive to readers in group-oriented cultures.

Preference for direct or indirect statements   Cultures also vary in the directness with which people typically make requests, decline requests, and express their opinions, particularly negative ones. For example, in U.S. and northern European cultures writers typically decline a request directly. They may apologize and offer an explanation for their decision. But they will state the denial explicitly. In contrast, Japanese and Korean cultures prefer an indirect style (Gudykunst & Ting-Toomey, 1998). Instead of declining a request explicitly, they might say that fulfilling it would be difficult or that they need time to think about how to reply. In this way, they save the requester the humiliation of being denied, while readers in those cultures know that the request will not be fulfilled.

Using one culture's style when writing to people in another culture can create troublesome misunderstandings. To many U.S. readers, an indirect refusal might be misinterpreted. Because they didn't hear the direct denial they expected, these U.S. readers could believe that the request could be fulfilled so they may persist in asking. On the other hand, readers in the Japanese and Korean cultures may interpret the direct U.S. style as rude and inconsiderate.

Basis of business decisions   There are also differences among cultures in the ways business decisions are made. In the United States and many European cultures, organizations typically choose among alternatives on the basis of impersonal evaluations and data analyses. In Arab and other cultures, these same decisions are often made on the basis of relationships. For instance, whereas a U.S. company might choose a company to build a new plant or supply parts for its products by carefully studying detailed proposals from the competitors, an Arab company might select

Learn More

For more information on communicating between cultures that have large and small distances from top to bottom of the organizational hierarchy, see Chapter 9, page 268.

Learn More

For more information on communicating between individualistic and group-oriented cultures, see Chapter 5, page 140.

Learn More

For more information on communicating between cultures that have different preferences about direct and indirect statements, see Chapter 8, page 215.

Learn More

For more information on communicating between cultures that have different ways of making business decisions, see Chapter 5, page 140.

the company represented by a person with whom it would like to do business. These are differences that writers in either kind of culture would need to keep in mind when trying to win business, maintain business relationships, and even respond to complaints from organizations in the other kind of culture.

Learn More

Chapter 13 (page 352) has more information on the interpretation of images, and Chapter 9 (page 281) has more on the interpretation of words.

Interpretation of images, gestures, and words   An image, gesture, or word can elicit markedly different responses in different cultures. Even the same words have different meanings in different countries and cultures. Photographs, drawings, and other pictures sometimes depict relationships between people that seem ordinary in one culture but violate the cultural customs of another. Gestures likewise have different meanings in different cultures. When people in the United States signal "Okay" by joining a thumb and forefinger to form a circle, they are making a gesture that is offensive in Germany and obscene in Brazil (Axtell, 1998). To avoid the risk of unintentionally offending others, some experts advise technical communicators to avoid showing hands in graphics that will be read by people in other cultures.

Applying Cultural Knowledge When You Write

Learn More

For information about ways that cultural differences can affect writing teams, see Chapter 18, page 470. For information related to oral presentations to people from other cultures, see Chapter 19, page 494.

As the preceding discussion indicates, gaining knowledge of your readers' culture can greatly increase your success in creating a communication that your readers will find useful and persuasive. The discusion is not, however, intended to suggest that gaining general knowledge about a culture can provide you with a recipe for adapting your communication strategies to the needs and expectations of your specific readers. As intercultural communication expert Ron Scollon (1999) says, "Cultures don't talk to each other. People do." When writing to people in other cultures, your reader-centered goal is the same as when addressing people in your own culture: to understand as fully as possible the relevant facts about the specific readers you are addressing.

In addition to the fact that every reader is an individual, a primary reason why you can't rely solely on general cultural information is that you are likely to be addressing your readers in their roles as employees of a company, government agency, or other organization. And organizations have cultures, too (Hofstede, 2002; Scollon & Scollon, 2001). In general, organizational cultures reflect the culture of the nation or region in which they are located. However, some organizational cultures have significant features that modify or run counter to the general culture in the region. Subsidiaries and branch facilities in any region of the world are likely to develop some—but not all—characteristics of the corporate headquarters, whether the headquarters are in Asia, the Middle East, or North America.

Researchers have also found that readers are influenced by the culture of their professions. Webb and Keene (1999) found that the worldwide culture of aeronautical engineering and science possesses characteristics that transcend geographical boundaries. Similarly, scientists, doctors, and teachers share many cultural traits with their counterparts around the globe.

Finally, when you are writing to individuals or organizations in another culture, there's a good chance that they have also made an effort to learn about your culture and to accommodate their communications to your cultural characteristics. Some researchers maintain that when people and organizations from two cultures communicate regularly, they create a third context as they develop their ways of understanding one another (Bolten, 1999; Steier, 1999).

How To Gain Knowledge about Your Intercultural Readers

A variety of resources can help you learn about your readers' cultural characteristics. The most helpful are people—including your coworkers—who are familiar with your readers' regional and organizational culture.

You can also consult many helpful print and online sources that present broad descriptions of cultures around the world. Some of the information they provide concerns such topics as holidays, political systems, and wedding customs, so you may want to focus your attention on topics relevant to writing work-related communications. On the other hand, learning about these other topics can increase your general understanding of other cultures. Research has demonstrated a positive association between intercultural communication competence and awareness of the other culture (Wiseman, Hammer, & Nishida, 1989). To gain additional insight into the ways other cultures differ from yours, look at the information these sources provide people in other cultures about your own. Self-awareness is another characteristic that correlates intercultural competence (Gudykunst, Yang, & Nishida, 1987). Here are some sources you can consult:

- Cyborlink website: www.cyborlink.com
- Global EDGE website: www.globaledge.msu.edu
- U.S. Department of State Background Notes: www.state.gov/r/pa/ei/bgn/

You can also learn from the response you receive each time you write to readers in other cultures (Brownell, 1999). If the opportunity arises, ask your readers about their reactions to your communications. When preparing instructions or other communications that will be read by many people in another culture, conduct a user test (described in Chapter 16) with members of your target audience, if at all possible.

WWW

For references and links to additional sources of information about cultures, go to **www.cengage.com/ english/anderson7e** and click on Chapter 3.

Importance of Your Attitudes

Having positive attitudes toward your readers is essential to success in creating reader-centered communications whether the readers are in your own culture or another one. However, your attitudes toward your readers and their cultures are especially important when you are writing to readers in other cultures. Research has shown that competence in intercultural communication has positive associations with open-mindedness (Adler, 1975), a nonjudgmental attitude (Ruben, 1976), empathy toward others (Chen & Tan, 1995; Ruben, 1976), and a positive attitude toward the other culture (Randolph, Landis, & Tzeng, 1977). On the other side, competence in intercultural communication has a negative association with ethnocentrism (Neuliep & McCroskey, 1997; Nishida, Hammer, & Wiseman, 1998). The more you focus on learning about your readers' cultures and the less you engage in judging them, the more successfully you will be able to write.

Guideline 6 | Learn who *all* your readers will be

So far, this chapter has assumed that you will know from the start just who your readers will be. That may not always be the case. Communications you prepare on the job may find their way to many people in many parts of your organization. Numerous memos and reports prepared at work are routed to one or two dozen people—and sometimes many more. Even a brief communication you write to one person may be copied or shown to others. To write effectively, you must learn who *all* your readers will be so you can keep them all in mind when you write. The following discussion will help you identify readers you might otherwise overlook.

Phantom Readers

The most important readers of a communication may be hidden from you. That's because at work, written communications addressed to one person are often used by others. Those real but unnamed readers are called *phantom readers*.

Phantom readers are likely to be present behind the scenes when you write communications that require some sort of decision. One clue to their presence is that the person you are addressing is not high enough in the organizational hierarchy to make the decision your communication requires. Perhaps the decision will affect more parts of the organization than are managed by the person addressed, or perhaps it involves more money than the person addressed is likely to control.

Much of what you write to your own boss may actually be used by phantom readers. Many managers accomplish their work by assigning it to assistants. Thus, your boss may sometimes check over your communications, then pass them along to his or her superiors.

After working at a job for a while, employees usually learn which communications will be passed up the organizational hierarchy. However, a new employee may be chagrined to discover that a hastily written memo has been read by executives at very high levels. To avoid such embarrassment, identify your phantom readers, then write in a way that meets their needs as well as the needs of the less influential person you are addressing.

Future Readers

Your communications may be put to use weeks, months, or even years after you imagined their useful life was over. Lawyers say that the memos, reports, and other documents that employees write today are evidence for court cases tomorrow. Most company documents can be subpoenaed for lawsuits concerning product liability, patent violation, breach of contract, and other issues. If you are writing a communication that could have such use, remember that lawyers and judges may be your future readers.

Your future readers also may be employees of your company who may retrieve your old communications for information or ideas. By thinking of their needs, you may be able to save them considerable labor. Even if you are asked to write something "just for the record," remember that the only reason to have a record is to provide some future readers with information they will need to use in some practical way that you should understand and support.

Complex Audiences

Writers sometimes overlook important members of their audience because they assume that all their readers have identical needs and concerns. Actually, audiences often consist of diverse groups with widely varying backgrounds and responsibilities.

That's partly because decisions and actions at work often impact many people and departments throughout the organization. For instance, a proposal to change a company's computer system will affect persons throughout the organization, and people in different areas will have different concerns: some with recordkeeping, some with data communication, some with security, and so on. People in each area will examine the proposal.

Even when only a few people are affected by a decision, many employers expect widespread consultation and advice on it. Each person consulted will have his or her own professional role and area of expertise, and each will play that role and apply that expertise when studying your communication.

When you address a group of people who will be reading from many perspectives, you are addressing a *complex audience.* To do that effectively, you need to write in a way that will meet each person's needs without reducing the effectiveness of your communication for the others. Sometimes you may have to make a tradeoff by focusing on the needs and concerns of the most influential members of your audience. In any case, the first step in writing effectively to a complex audience is to identify each of its members or groups.

Identifying Readers: An Example

When Stephanie considered whether there might be phantom readers for her memo, she decided that Ms. Land probably would share it with one or two of the other permanent staff. However, she also concluded that these readers would ask the same questions and apply the same criteria as would Ms. Land. As Stephanie's situation illustrates, having multiple readers does not necessarily mean that you have a complex audience. An audience is complex if you need to develop different strategies to meet the needs of different readers.

To see one way that a writer might identify the members of a complex audience and adjust his or her communication accordingly, consider Thomas McKay's situation. McKay was writing on behalf of his employer, Midlands Research Incorporated, to request compensation from another company, Aerotest Corporation, which had sold Midlands faulty equipment for testing smokestack emissions. McKay addressed his letter to Robert Fulton, Aerotest's Vice President for Sales, but realized that Fulton would distribute copies to many others at Aerotest. To identify these other readers, McKay asked himself who at his own employer's company, Midlands, would be asked to read such a letter if it received one. In this way, McKay identified the following phantom readers:

- Engineers in the department that designed and manufactured the faulty equipment, who would be asked to determine whether Aerotest's difficulties really resulted from flaws in the design.
- Aerotest's lawyers, who would be asked to determine the company's legal liability.
- Personnel in Aerotest's repair shop, who would be asked to examine the costs Midlands said it had incurred in repairing the equipment.

Learn More

For more information on modular designs, see Guideline 4 in Chapter 4 (page 106).

This chapter's main point:
When defining your communication's objectives, focus on your readers, not yourself.

To meet the needs of the diverse readers in his complex audience, McKay created a letter with a modular design, a commonly used workplace strategy in which different, readily distinguishable parts of a communication each address a distinct group of readers. Modular designs are very common at websites, where home pages often have links for different kinds of users. For example, your college's home page may have separate links for current students, prospective students, faculty, and graduates. Figure 3.2 shows the modular design of the home page example from a company that makes surgical equipment. To create a modular design for his letter to Aerotest, McKay wrote a one-page letter that provided background information for all his readers (Figure 3.3). He also attached enclosures addressed to specific groups in his complex audience. Two enclosures contained detailed accounts of the problems Midlands encountered with the emissions testers. With these enclosures, McKay provided evidence that the problems encountered by Midlands were, in fact, caused by poor work on Aerotest's part. McKay's third enclosure was a detailed statement of the repair expenses that would enable Aerotest's repair technicians to see that the reimbursement Midlands requested was fully justified. Of course, neither McKay nor the website designers could have developed a modular design if they had been unaware of the complexity of their audiences. When defining your communication's objectives, take similar care to identify all of your readers.

GUIDELINE FOR UNDERSTANDING YOUR READER'S CONTEXT

The following guideline will help you consider the many contextual factors that can affect your reader's response to your communication.

This website for Ethicon, a manufacturer of medical products, has a modular design with separate areas for four distinct groups of visitors. This modular design enables each group of readers to find quickly the information they want.

FIGURE 3.2
Modular Design of a Website

- **Job seekers.** Their link leads to information about the kinds of company's openings, benefits, and locations.

- **Patients.** Their link provides information about their condition and the treatment they would receive if their doctor were to use the company's products.

- **Persons wanting information about the company itself.** These persons include investors, news reporters, and others interested in information about the company itself.

- **Physicians and other health care professionals.** The link for them provides information about the company's products as well as online training to their use.

Guideline 7 | **Describe the context in which your reader will read**

At work, people interpret what they read as a chapter in an ongoing story. Consequently, they respond to each message in light of prior events as well as their understanding of the people and groups involved. Fill out your mental portrait of your readers by imagining how the following circumstances might influence their response to your communication:

- **Recent events related to your topic.** Maybe you are going to announce the reorganization of a department that has just adjusted to another major organizational change. You'll need to make a special effort to present the newest change in a positive light. Or maybe you are requesting money to attend an important professional meeting. If your department has just been reprimanded for excessive travel expenses, you will have to make an especially strong case.

- **Interpersonal, interdepartmental, and intraorganizational relationships.** If you are requesting cooperation from a department that has long competed with yours for company resources, you will need to employ special diplomacy.

FIGURE 3.3
Letter to a Complex Audience (Enclosures Not Shown)

In his letter requesting compensation for expenses caused by faulty equipment, McKay addresses one person (Robert Fulton) but designed his message for the complex audience he knows will read his letter.

In the first two paragraphs, McKay provides background information of interest to all members of his complex audience at Aerotest.

In the third paragraph, McKay discusses his three major points: the equipment failure resulted from faulty design and construction, repairs were costly for his company, and his company wants compensation.

McKay created a modular design by enclosing several items, each useful primarily to one part of his complex audience at Aerotest.

- For Aerotest's engineers and lawyers, two accounts of the problems that are intended to persuade that the difficulties really did arise from Aerotest's faulty equipment.

MIDLANDS RESEARCH INCORPORATED
2796 Buchanan Boulevard   Cincinnati, Ohio 45202

Mr. Robert Fulton   October 17, 2010
Vice President for Sales
Aerotest Corporation
485 Connie Avenue
Sea View, California 94024

Dear Mr. Fulton:

In August, Midlands Research Incorporated purchased a Model Bass 0070 sampling system from Aerotest. Our Environmental Monitoring Group has been using—or trying to use—that sampler to fulfill the conditions of a contract that MRI has with the Environmental Protection Agency to test for toxic substances in the effluent gases from thirteen industrial smokestacks in the Cincinnati area. However, the manager of our Environmental Monitoring Group reports that her employees have had considerable trouble with the sampler.

These difficulties have prevented MRI from fulfilling some of its contractual obligations on time. Thus, besides frustrating our Environmental Monitoring Group, particularly the field technicians, these problems have also troubled Mr. Bernard Gordon, who is our EPA contracting officer, and the EPA enforcement officials who have been awaiting data from us.

I am enclosing two detailed accounts of the problems we have had with the sampler. As you can see, these problems arise from serious design and construction flaws in the sampler itself. Because of the strict schedule contained in our contract with the EPA, we have not had time to return our sampler to you for repair. Therefore, we have had to correct the flaws ourselves, using our Equipment Support Shop, at a cost of approximately $15,000. Because we are incurring the additional expense only because of problems with the engineering and construction of your sampler, we hope that you will be willing to reimburse us, at least in part, by supplying without charge the replacement parts listed on the enclosed page. We will be able to use those parts in future work.

Thank you for your consideration in this matter.

Sincerely,

Thomas McKay
Vice President
Environmental Division

Enclosures: 2 Accounts of Problems
             1 Statement of Repair Expenses
             1 List of Replacement Parts

- For Aerotest's repair shop, a statement of the repair expenses intended to persuade that the costs he said his company incurred were reasonable.

- For Aerotest's shipping department, a list of replacement parts to use when sending the parts to McKay's company.

Political conflicts between individuals and groups can also create delicate writing situations in which certain ways of expressing your message can appear to support one faction and weaken another even if you have no intention of doing so.

When she reached the question about factors in her reader's context that might influence Ms. Land's reaction to her communication, Stephanie recalled a recent conversation. While she walked to the parking lot at the end of the day with two permanent employees, one said she'd heard that at least one member of the organization's Board of Directors had told Ms. Land it was time for her to retire. Even if this were inaccurate gossip, Stephanie thought that she should be especially careful not to seem to be criticizing Ms. Land or suggesting that a younger person might have new and better ideas.

Stephanie thought of a possible contextual factor that could guide the way she would write to Ms. Land.

GUIDELINES FOR OTHER IMPORTANT CONSIDERATIONS

Although you should focus primarily on your reader when defining a communication's objectives, it is important for you to think about any constraints that restrict the way you can write and to identify people other than your readers who might be affected by your communication.

Guideline 8 | **Identify any constraints on the way you write**

So far, this chapter has focused on developing a full understanding of your readers as you define your communication's objectives. As you gather the information that will form the basis for the way you craft your communication, you should also learn about any expectations, regulations, or other factors that may constrain what you can say and how you can say it. In the working world, expectations and regulations can affect any aspect of a communication—even tone, use of abbreviations, layout of tables, size of margins, and length (usually specifying a maximum length, not a minimum).

Some constraints come directly from your employer, reflecting such motives as the company's desire to cultivate a particular corporate image, to protect its legal interests (because any written document can be subpoenaed in a lawsuit), and to preserve its competitive edge (for example, by preventing employees from accidentally tipping off competitors about technological breakthroughs). In addition, most organizations develop writing customs—"the way we write things here." Writing constraints also can originate from outside the company—for instance, from government regulations that specify how patent applications, environmental impact reports, and many other types of documents are to be prepared. Similarly, scientific, technical, or other professional journals have strict rules about many aspects of the articles they publish.

Some companies publish style guides that set forth their regulations about writing. Ask if your employer has one. You can also learn about these constraints by asking coworkers and reading communications similar to yours that your coworkers have written in the past.

WWW

To view sample style guides, go to **www.cengage.com/ english/anderson7e** and click on Chapter 3.

Guideline 9 | Ethics Guideline: Identify your communication's stakeholders

There are many strategies for assuring that on-the-job writing is ethical. Some writers use an *alarm bell strategy*. They trust that if an ethical problem arises in their writing, an alarm bell will go off in their heads. Unless they hear that bell, however, they don't think about ethics. Other writers use a *checkpoint strategy*. At a single, predetermined point in the writing process, they review their work from an ethical perspective.

In contrast, this book teaches a more active and thorough *process strategy* for ethical writing. In it, you integrate an ethical perspective into *every* stage of your work on a communication. It's important to follow a process strategy because at every step of writing you make decisions that shape the way your communication will affect other people. Accordingly, at every step you should consider your decisions from the viewpoint of your personal ethical beliefs about the ways you should treat others.

Stakeholders

In a process approach to ethical writing, no step is more important than the first: defining objectives. When you are defining your communication's objectives, you are (in part) identifying the people you will keep in mind throughout the rest of your writing effort. When you follow the reader-centered approach to writing that is explained in this book, you begin by identifying your readers.

To write ethically, you must also identify another group of people: the individuals who will gain or lose because of your message. Collectively, these people are called *stakeholders* because they have a stake in what you are writing. Only by learning who these stakeholders are can you assure that you are treating them in accordance with your own ethical values.

How to Identify Stakeholders

Because communications written at work often have far-reaching effects, it's easy to overlook some stakeholders. If that happens, a writer risks causing accidental harm that could have been avoided if only the writer had thought through all the implications of his or her communications.

To identify the stakeholders in your communications, begin by listing the people who will be directly affected by what you say and how you say it. In addition to your readers, these individuals may include many other people. For instance, when Craig was preparing a report for his managers on the development of a new fertilizer, he realized that his stakeholders included not only the managers but also the farmers who would purchase the fertilizer and the factory workers who would handle the chemicals used to manufacture it.

Next, list people who will be affected indirectly. For example, because fertilizers run off the land into lakes and rivers, Craig realized that the stakeholders of his report included people who use these lakes and rivers for drinking water or recreation. Indeed, as is the case with many other communications, the list of indirect stakeholders could be extended to include other species (in this case, the aquatic life in the rivers and lakes) and the environment itself.

Finally, think of the people who may be remotely affected. These people may include individuals not yet born. For example, if Craig's fertilizer does not break down into harmless elements, the residue in the soil and water may affect future generations.

When Stephanie came to the Writer's Guide's question about her stakeholders, she realized that so far she had thought about her plan only from the perspective of persons requesting rapid Braille translations. She had neglected to consider her plan's impact on the volunteer translators. Therefore, she decided that she would have to modify her plan to ensure that none of them suffered a loss of self-esteem because they were not included in the group who would be assigned urgent translations. As Stephanie's example illustrates, thinking about stakeholders sometimes leads to improvements not only in the communications that are being written but also in the actions that are discussed in them.

By identifying the stakeholders for her memo, Stephanie thought of an addition to her original plan that would help one group be treated more considerately.

Stephanie's completed Writer's Guide is shown in Figure 3.4.

FIGURE 3.4 **Stephanie's Completed Writer's Guide for Defining Your Communication's Objectives.**

Writer's Guide
DEFINING YOUR COMMUNICATION'S OBJECTIVES

YOUR PURPOSE
1. What are you writing?
   A proposal for completing more of the urgent Braille translations on time

2. What outcome do you desire?
   Adoption of a system that assigns urgent translations to the quickest volunteers

3. Who is your reader?
   Ms. Land

PURPOSE OF YOUR COMMUNICATION ⟵—— See Guidelines 2 and 3, pages 71 and 75
1. What task will your communication help your reader perform? (Usability)
   What is your reader's purpose for reading?
   She wants the best possible service to the persons who request translations.

   What information does your reader want? (What questions will your reader ask?)
   What evidence do you have that there's a problem with the current system?
   Who would determine which translations deserve highest priority?
   How would we decide which translators are placed in our top group?
   Will your system really work? Are other agencies using it successfully?

How will your reader search for the information? (May use more than one strategy)

__X__ Sequential reading from beginning to end

_____ Reading for key points

__X__ Reference reading

_____ Other (describe)

How will your reader use the information?

__X__ Compare alternatives (what will be the points of comparison?)

Cost, efficiency, and impact on the morale of the volunteers, especially those who were not chosen as the most efficient

_____ Determine how the information will affect him or her (or the organization)

_____ Perform a procedure (following instructions step by step)

_____ Other (describe)

2. How do you want your communication to change your reader's attitudes? (Persuasiveness)

What is your reader's attitude toward your subject? What do you want it to be?

She believes that the current system for assigning Braille translations is the best one possible. I want her to see that mine is better.

What is your reader's attitude toward you? What do you want it to be?

Ms. Land thinks that I am a good summer employee. But she thinks all summer employees are ignorant concerning policy issues. I want her to think that I am knowledgeable enough to have a recommendation worth serious consideration.

What is your reader's attitude toward your organization? What do you want it to be?

Not relevant to this memo

READER'S PROFILE   ⟵———— See Guidelines 4, 5, and 6, pages 76, 79, and 84

1. Role in the organization

She is director of the translation department.

2. Role as a reader

Decision maker

3. Familiarity with your subject

Very familiar

4. Familiarity with your specialty
   Very familiar

5. Relationship to you
   She is my boss and likes to maintain a formal superior–subordinate relationship.

6. Other personal characteristics you should take into account
   She designed the current system and may feel defensive if I suggest that it could be improved.

7. Cultural characteristics you should take into account
   None

8. Who else might read your communication?
   A few experienced employees. They will have the same perspective that Ms. Land has.

CONTEXT   ←——— See Guidelines 7 and 8, pages 87 and 89
1. What features of the context may affect the way your reader reads your communication?
   Ms. Land may have been told it is time for her to retire.

2. What expectations, regulations, or other factors constrain the way you can write?
   None

ETHICAL TREATMENT OF STAKEHOLDERS   ←——— See Guideline 9, page 90
1. Who, besides your reader, are stakeholders in your communication?
   Persons requesting urgent translations; volunteers not selected for the top group

2. How will they be affected by it?
   Persons making requests may benefit. The volunteers may feel they are not valued.

CONCLUSION

The major lesson to take away from this chapter is simple: You can greatly increase your ability to write successfully if you define your communications' objectives by focusing on your reader, not yourself. Imagine the tasks your readers will want to perform while reading. Helping your reader perform these tasks is the first of your communication's purposes. Think too about the ways you want your communication to alter your reader's attitudes. Bringing about these changes is your communication's second purpose. Next, learn enough about your reader and your reader's context to make a mental movie of him or her in the act of reading. You can use the movie to predict the way your reader is likely to respond to ways you might write your communication. Using your imagination and creativity in the same way, you can ensure that you are writing ethically: Identify your communication's stakeholders and the ways your communication might affect them.

Stephanie's example illustrates the many *immediate* insights you can gain by taking this reader-centered approach. But the benefits will continue. The rest of this book's chapters tell how you can use your reader-centered understanding of a communication's objectives throughout all your work on it.

USE WHAT YOU'VE LEARNED

For additional exercises, go to www.cengage.com/english/ anderson7e. *Instructors*: The book's website includes suggestions for teaching the exercises.

EXERCISE YOUR EXPERTISE

1. Find an example of a communication you might write in your career. Following the guidelines in this chapter, define its objective. Be sure to identify each of the following items:

    The readers and their characteristics
    The stakeholders and the ways they might be affected by the communication
    The final result the writer desires
    The communication's usability goals
    The communication's persuasive goals

    Then explain how the communication's features have been tailored to fit its objectives. If you can think of ways the communication might be improved, make recommendations.

2. Using the Writer's Guide shown in Figure 3.1, define the objectives of an assignment you are preparing for your technical communication class. To download a copy of the worksheet, go to Chapter 3 at www.cengage.com/english/anderson7e.

EXPLORE ONLINE

Find a web page that could be used as a resource for a person in the profession for which you are preparing.

Describe the target readers, usability objectives, and persuasive objectives that the web page's creators might have had for it. Evaluate their success in achieving these objectives.

COLLABORATE WITH YOUR CLASSMATES

Working with another student, pick a technical or scientific topic that interests you both. Next, one of you should locate an article on the topic in a popular magazine such as *Time* or *Discover*, and the other should locate an article on the topic in a professional or specialized journal. Working individually, you should each study the ways your article has been written so that its target audience will find the article to be usable and persuasive. Consider such things as the way your article opens, the language used, the types of details provided, and the kind of visuals included. Next, meet together to compare the writing strategies used to meet the needs and interests of the two audiences. Present your results in the way your instructor requests.

APPLY YOUR ETHICS

A variety of websites present case studies that describe ethical issues that arise in business, engineering, science, and other fields. Locate, read, and respond to one such case. For links to ethics cases, go to Chapter 3 at www.cengage.com/english/anderson7e.

For additional cases, go to www.cengage.com/english/anderson7e. *Instructors*: The book's website includes suggestions for teaching the cases.

As you sit in your office of the large, one-story building owned by your employer, you look up from the draft of an e-mail message that company president C. K. Mitchell sent you a few minutes ago. He asked you for your opinion of it.

The e-mail announces a new no-smoking policy for the company, which employs about 250 people. Smoking has been permitted in a smoking lounge located in your building between the manufacturing area, where most of the employees work, and the area that houses the offices for company managers and executives. The lounge is used by employees from both areas. There is also a smoking lounge in a nearby building that C. K. purchased a year ago to expand manufacturing capacity. C. K. recently announced that the continued growth of orders for the company's products necessitates closing the lounge in your building next month. The space is needed for additional manufacturing equipment.

Because this smoking lounge will be closed, the company must establish a new policy about smoking. C. K. has appointed a committee of top managers, including you, to make a recommendation to him. A vocal group of nonsmokers has long advocated a complete smoking ban inside your building. They complain that smoke escapes from the lounges and pollutes their work areas. Therefore, the committee has considered allowing people who work in your building to smoke only outside the door that leads from the manufacturing area to the parking lot. However, some nonsmokers complained that the constant opening of the door would enable smoke to invade their work area. Additionally, managers who smoked complained that this outdoor area is too far from their offices. In response to the managers' complaints, the committee almost recommended that C. K. allow people to smoke in their private offices, provided they shut the door and use an air purifier. However, Maryellen Rosenberg, Director of Personnel, observed that smokers who don't have private offices would complain that the policy discriminated against them. Smokers in both the manufacturing area and the office area also have objected to a total ban on smoking in your building, pointing out the unfairness of prohibiting them from smoking while employees in the newly acquired building could continue to use its smoking lounge.

In a meeting with C. K. yesterday, the committee recommended that C. K. ban smoking anywhere on company property. C. K. has accepted the recommendation. "What could be more appropriate," he asked, "for a company that tries to make people healthier?" Your company designs and markets exercise equipment for homes and fitness centers. After prompting from you and several others, C. K. decided to hire a consulting firm that offers a course to help employees stop smoking. The course will be offered to employees free of charge.

"C. K.'s e-mail announcing this policy had better be good," you think as you prepare to review his draft. Although the employees know that C. K. has been meeting with you and others to work out a policy, the committee's discussions have been kept secret. Nonetheless, feelings about the new policy have been running high. Some employees have talked of quitting if they can't smoke at work, and others have talked of quitting if smoking isn't prohibited. The controversies are harming morale, which is already low because C. K. recently reduced the company's contribution to the profit-sharing plan despite a steady rise in revenues. The money saved is being used to buy the building next door, a move that will save the company money in the long run. However, C. K.'s action is widely viewed as yet another example of his heavy-handed, insensitive style.

Your Assignment

First, to fashion a reasonable set of objectives for C. K.'s e-mail message, fill out a Writer's Guide for Defining Your Communication's Objectives (Figure 3.1; a downloadable copy is available at www.cengage.com /english /anderson7e). Remember that the message must somehow satisfy all the employees. Second, evaluate C. K.'s draft (Figure 3.5) in light of the objectives you have established, and then revise the draft to make it more effective. Be prepared to explain your revisions to C. K.

FIGURE 3.5
Draft E-mail for Use with
Case

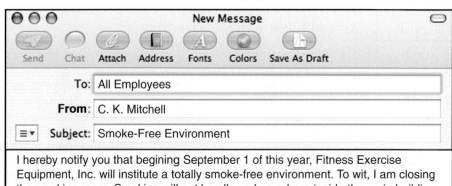

To: All Employees

From: C. K. Mitchell

Subject: Smoke-Free Environment

I hereby notify you that begining September 1 of this year, Fitness Exercise Equipment, Inc. will institute a totally smoke-free environment. To wit, I am closing the smoking room. Smoking will not be allowed anywhere inside the main building or the satellite building next door.

The delay between this announcement and the begining date for this new policy will allow any employees who smoke the chance to enroll in courses which, I hope, will help them break or curtail their habit. In accordance with our concern for the wellness of all our employees, we will enhance our working environment by prohibiting all smoking.

All employess are thanked for their cooperation, understanding, and dedication to better health.

PART III

PLANNING

4 | Planning for Usability

CHAPTER 4

DEFINING OBJECTIVES
PLANNING
RESEARCHING
DRAFTING
REVISING

WWW

For additional information and links to relevant websites, go to **www.cengage.com/english/anderson7e** and click on Chapter 4.

This chapter and the next one address a question that you and all writers face when taking a reader-centered approach to writing at work. Consider Toni, for example. A recent college graduate, she has been asked by her employer, a large construction company, to prepare a report that evaluates several high-end computer programs it is considering for distribution to all of its civil engineers. Toni understands that she must make her report both usable and persuasive *in her readers'* eyes, just as Chapter 1 explains. That is, she knows that her goal is to create a report that her readers can use easily when comparing the computer programs, and she knows that the report must persuade her readers that she has provided all the information they need to be able to select the best program. In addition, by following Chapter 3's advice, Toni has learned enough about her readers to understand what will make her report usable and persuasive to them. "Now," she asks, "what's the best way for me to move from knowing what my report must do to actually creating the report I need to write?"

This chapter begins to answer by describing efficient, effective strategies she and you can use to plan for usability. The next chapter completes the answer by focusing on planning for persuasiveness.

PROCESS FOR PLANNING FOR USABILITY

The border between defining objectives and planning a communication is fuzzy, and for good reason. When developing your understanding of a communication's purpose, readers, and context, you will often think immediately about their implications for the way you should write. Similarly, to plan reader-centered communication, you must refer back continuously to its reader-centered objectives.

It should be no surprise, then, that the three-step process for planning highly usable communication corresponds with Chapter 3's three-step process for defining a communication's usability objectives:

Advice in Chapter 3 provides the foundation for this chapter's Guidelines 1 through 3.

- Identifying the information your readers want (page 101) is the first step in creating a complete communication that includes all the information they need (Guideline 1).
- Determining how your readers will use the information you provide (page 101) enables you to organize around their tasks (Guideline 2).
- Describing how your reader will look for the information (page 105) helps you plan for ways to help them find it (Guideline 3).

The discussions in this chapter's first three guidelines offer additional advice to the implications for planning that were described in Chapter 3. Guidelines 4 through 8 describe five techniques you can use during the planning process. Guideline 9 discusses an ethical consideration linked to planning for usability.

Guideline 1 | Identify the information your readers need

Readers can't use what isn't there. Consequently, your first job when planning for usability is to ensure that your communication will be *complete*, that it will include all the information your readers need in order to perform their tasks.

Begin by listing the questions your readers will bring to your communication or ask while reading it. To make your communication usable to them, you will need to include the answers, as suggested by Chapter 3's Guideline 2.

To identify *all* the information your readers will need, however, you must go beyond the questions you can predict that they will ask and the kinds of answers you believe will be most helpful to them. You may know important things that your readers don't realize they should ask about.

For example, while evaluating the high-end computer programs her employer might distribute to its engineers, Toni learned that the company that produces one program might soon be purchased by a larger rival. When such purchases are made, the smaller company's products often languish. They are still sold, but no longer upgraded, a distinct problem for companies that rely on them. Although her readers are not likely to ask about takeover possibilities, this information is relevant to the decision Toni's readers will make. She should include it in her report.

On the other hand, Toni should avoid including information just because she finds it interesting or wants to demonstrate how much she knows. Such information only makes it more difficult for readers to locate and use the information they require. In a reader-centered communication, it's just as important to omit information your readers don't need as it is to provide all that they do.

Learn More

For detailed advice about identifying the information your readers need, see Chapter 3's Guidelines 2 and 4 (pages 71 and 76).

Guideline 2 | Organize around your readers' tasks

After you've identified the information your reader needs, you should organize it so that your readers will find it easy to use. As explained in Chapter 1, reading involves a moment-by-moment interaction between reader and text. In this interaction, readers perform a sequence of mental tasks. In some cases, these mental tasks are mixed with physical tasks, as when people read instructions. Your communications will be most usable if you organize them to mirror and support the readers' tasks.

Of course, these tasks vary from situation to situation. Consequently, you should think about your specific reader's tasks each time you write. At the same time, keep in mind that some mental tasks are common enough that certain organizational strategies almost always increase usability. Here are three.

- Organize hierarchically.
- Group together the items your readers will use together.
- Give the bottom line first.

These strategies are described in the following paragraphs.

Learn More

For advice about imagining your reader's tasks, see Chapter 3's Guideline 2 (page 71).

Learn More

For more information on creating a mental portrait of your readers, see Chapter 1 (page 19) and Chapter 3's Guideline 1 (page 68).

Organize Hierarchically

When we read, we encounter small bits of information one at a time: First, what we find in this sentence, then what we find in the next sentence, and so on. One of our major tasks is to build these small bits of information into larger structures of meaning that we can store and work with in our own minds. Represented on paper, these mental structures are hierarchical. They look like outlines (Figure 4.1) or tree diagrams (Figure 4.2, page 104), with the overall topic divided into subtopics and some or all of the subtopics broken down into still smaller units. For readers, building these hierarchies can be hard work, as we've all experienced when we've had to reread a passage because we can't figure out how its sentences fit together.

The easiest and surest way to help your readers build mental hierarchies is simply to present your information organized in that way. This doesn't mean that you need to start every writing project by making an outline (see Guideline 6). Often, you may be able to achieve a hierarchical organization without outlining. But you should always organize hierarchically.

Group Together the Items Your Readers Will Use Together

Creating a hierarchy involves *grouping* facts. You assemble individual facts into small groups and then gather these groups into larger ones. When grouping your facts, be sure to group together the information your readers will use together. An example in Chapter 3 shows how doing so increases usability. Stephanie wanted to write a memo that would persuade her boss, Ms. Land, to adopt a new system for assigning Braille translations to volunteers. Stephanie realized that Ms. Land would want to compare the proposed

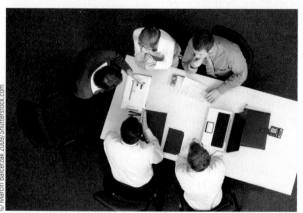

People use communications written at work to perform a wide variety of tasks.

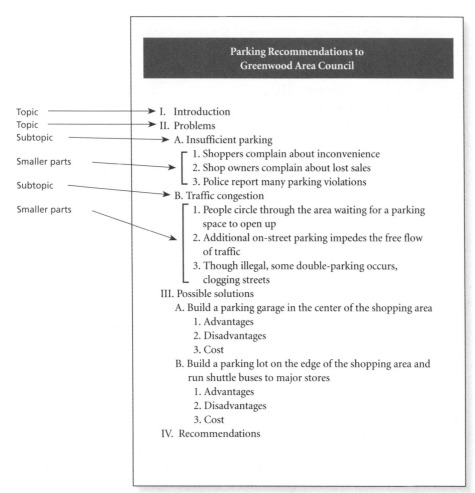

FIGURE 4.1
Outline Showing the
Hierarchical Organization
of a Report

**Parking Recommendations to
Greenwood Area Council**

Topic ⟶ I. Introduction
Topic ⟶ II. Problems
Subtopic ⟶ A. Insufficient parking
Smaller parts ⟶ ⎡ 1. Shoppers complain about inconvenience
2. Shop owners complain about lost sales
⎣ 3. Police report many parking violations
Subtopic ⟶ B. Traffic congestion
Smaller parts ⟶ ⎡ 1. People circle through the area waiting for a parking
space to open up
2. Additional on-street parking impedes the free flow
of traffic
3. Though illegal, some double-parking occurs,
⎣ clogging streets
III. Possible solutions
A. Build a parking garage in the center of the shopping area
1. Advantages
2. Disadvantages
3. Cost
B. Build a parking lot on the edge of the shopping area and
run shuttle buses to major stores
1. Advantages
2. Disadvantages
3. Cost
IV. Recommendations

and current systems point by point. Consequently, she decided to organize around Ms. Land's criteria (cost, efficiency, and so on), putting all the relevant information about both systems in the same sections. By grouping together the information Ms. Land would use together, Stephanie made her memo more usable than it would otherwise had been. If, instead, Stephanie had grouped all the facts about the current system in one place and all the facts about the other system in another place, Ms. Land would have had to flip back and forth among the pages for a point-by-point comparison.

A second example illustrates how the strategy of grouping around the readers' tasks can increase usability in a much different situation. Daniel is preparing a report that summarizes hundreds of research studies on the effects of sulfur dioxide (SO_2) emissions from automobiles and factories. Daniel could group his information in several ways. For instance, he could group together all the research published in a given year or span of years. Or he could discuss all the studies

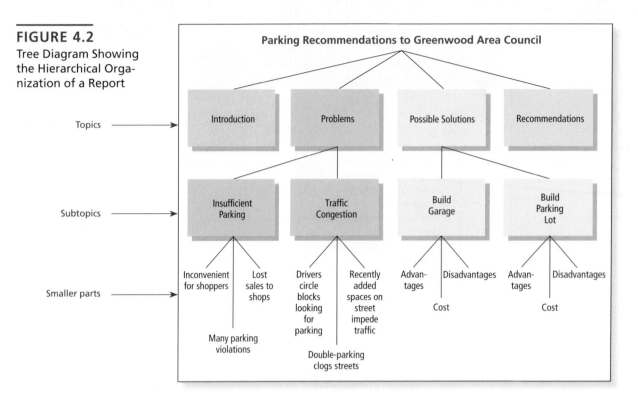

Topics

Subtopics

Smaller parts

conducted in Europe in one place, all conducted in North America in another place, and so on. However, Daniel is employed by a federal task force. His report will be read by members of the U.S. Congress as they decide how to structure legislation on SO_2 emissions. Daniel knows that these readers will want to read about the impacts of SO_2 emissions on human health separately from the effects on the environment. Therefore, he organizes around kinds of impacts, not around the dates or locations of the studies.

Like Daniel, you can organize your communications in a usable, task-oriented manner by grouping together the information your readers will use together.

Give the Bottom Line First

Readers at work often say that their most urgent reading task is to find the writer's main point. The most obvious way to make it easy for them to find your main point is to put it first. Indeed, this strategy for task-oriented organization is the subject of one of the most common pieces of advice given in the workplace: "Put the bottom line first." The bottom line, of course, is the last line of a financial statement. Literally, the writers are being told, "Before you swamp me with details on expenditures and sources of income, tell me whether we made a profit or took a loss." However, this advice is applied figuratively to many kinds of communication prepared at work.

Of course, you don't always need to place your main point in the very first sentence, although this is sometimes the most helpful place for it. When appropriate, you can always provide relevant background information before stating your main point. But don't keep your readers in suspense. As soon as possible, get it out: Is the project on schedule, or must we take special action to meet the deadline? Will the proposed design for our product work, or must it be modified?

To some writers, it seems illogical to put the most important information first. They reason that the most important information is generally some conclusion they reached fairly late in their thinking about their subject. Consequently, they think, it is logical to describe the process by which they arrived at the conclusion before presenting the conclusion itself. However, such a view is writer-centered. It assumes that information should be presented in the order in which the writer acquired it. In most workplace situations such an organization runs counter to the sequence that readers would find most helpful.

Focus on use, not logic, when you organize.

There are, nevertheless, some situations in which it really is best to withhold the bottom line until later in a communication. You will find a discussion of such situations in the fifth guideline of Chapter 5. As a general rule, however, you can increase your readers' satisfaction and reading efficiency by creating a task-oriented organization in which you give the bottom line first.

Guideline 3 | Identify ways to help readers quickly find what they want

At work, readers often want to find a particular piece of information without reading the entire document that contains it. Perhaps they are reading a communication with many kinds of information, only some of which are relevant to their responsibilities. Perhaps they previously read the entire document and now want to refresh their memories concerning certain facts. In these and similar situations, one of the reading tasks they perform is to locate the desired information.

To devise your accessibility strategies, turn once again to your mental portrait of your readers. Picture the various circumstances under which they would want to locate some particular subpart of your overall message. What will they be looking for? How will they search for it? With the answers to these two questions, you can plan the pathways that will guide your readers through your communication to the information they want.

For memos, letters, reports, proposals, instructions, and other printed documents, you can construct these pathways with headings, topic sentences, tables of contents, and other devices. If your communication has a complex audience, one that includes readers whose responsibilities and interests differ, you can also increase accessibility by planning a modular design. In a modular communication, different parts are addressed to different readers or groups of readers.

Chapter 3 provided an example. To request reimbursement for repairs that his company made to faulty systems for sampling toxic gases emitted from industrial smokestacks, Thomas McKay created a communication that had four parts: a letter that contained general information for all readers, two attachments for engineers and lawyers, and one attachment for readers in a repair shop.

Try This

How usable are the syllabuses for your courses? Why not evaluate one?. How easily can you find each day's assignment? Due dates for papers? Instructions for assignments? Other information you want? What features increase the syllabus's usability for students? What revisions would you recommend?

Learn More

To learn more about McKay's reason for creating a modular design and to see the letter he wrote, go to page 88.

Learn More

For more on modular designs, turn to Chapter 3's Guideline 8 (page 89), including Figures 3.2 and 3.3.

Learn More

For advice about creating site maps, see Chapter 20.

WWW

To view other websites that illustrate good use of the guidelines in this chapter, go to **www.cengage.com/english/anderson7e** and click on Chapter 4.

The standard organization for long reports and proposals is also a modular design. For example, many reports and proposals are written for both decision makers and advisers. Usually such reports have two parts: (1) a very brief summary—called an *executive summary* or *abstract*—at the beginning of the report, designed for decision makers who want only the key information; and (2) the body of the report, designed for advisers, who need the details. Typically, the executive summary is only a page or a few pages long, whereas the body may exceed a hundred pages. The body of the report might be divided into still other modules, one addressed to technical experts, one to accountants, and so on. Reports and proposals often include appendixes that present technical details.

You can use similar strategies for helping users quickly locate the information they want when you are designing a website. Menus, links within the text, and clickable images and icons provide the same kind of assistance that headings provide in print communications. A site map serves the same purpose as a table of contents. If your site will be visited by persons wanting different kinds of information, you can also use a modular design to help readers find the information they want. For example, the Epilepsy Project knew that its website would be visited by several distinct groups of people: adults with epilepsy, their family members, kids and teens with epilepsy, and professionals who treat persons with epilepsy, among others. For each of these groups, the Project created a special area with appropriate information. All of these areas can be accessed from links on the home page (see Figure 4.3).

All of these strategies for helping readers and website users quickly locate information are discussed in detail elsewhere in this book. The important point now is that when planning your communication, even before you begin drafting, you should begin devising ways to make your communication accessible to your readers.

TECHNIQUES FOR PLANNING FOR USABILITY

The next five guidelines suggest techniques that people in the workplace often find helpful as they plan the content and organization of their communications along with their strategies for making the information in them accessible to their readers.

Guideline 4 | **Look for a technical writing superstructure you can adapt**

At work, you will often write in a situation that closely resembles circumstances encountered by other people many times before. Like other people, you may need to report on a business trip, tell someone how to operate a piece of equipment, or request funds for a project you would like to conduct. For many of these recurring situations, writers employ conventional patterns for constructing their communications. These patterns are sometimes called *genres*. Here, they are named *superstructures* (van Dijk, 1980).

At work, you will encounter many superstructures: the business proposal, budget report, computer manual, feasibility report, project proposal, and environmental

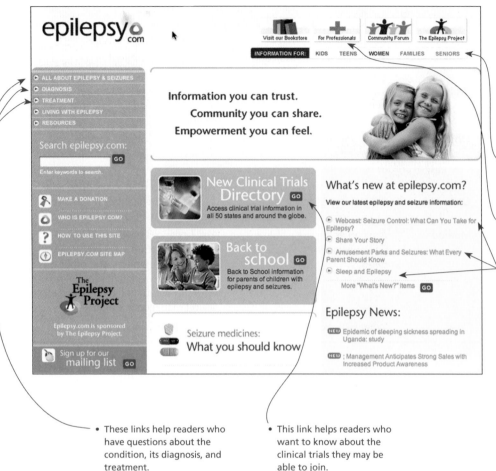

FIGURE 4.3
Website Showing
Modular Design

To help each visitor find quickly the information most relevant to him or her, the Epilepsy Project used a modular design for its website.

The row of colored links provide "Information For" five different groups of readers, ranging from "kids" to "seniors."

There is also a separate link for professionals who work with persons who have epilepsy.

This very usable, reader-centered site also provides readers with access to answers to questions they may have.

• These links help readers who have questions about the condition, its diagnosis, and treatment.

• This link helps readers who want to know about the clinical trials they may be able to join.

impact statement, to name a few. Carolyn R. Miller (1984), Herrington and Moran (2005), and other researchers suggest that each superstructure exists because writers and readers agree that it provides an effective pattern for meeting some particular communication need that occurs repeatedly. Figure 4.4 shows how the superstructure for proposals accommodates readers (by answering their questions) and also writers (by providing them with a framework for presenting their persuasive claims).

Because superstructures serve the needs of both readers and writers, they are ideal for writing reader-centered communications. They can be especially helpful at the planning stage because they suggest the kinds of information your readers probably want and the manner in which the information should be organized and presented. However, superstructures are not surefire recipes for success. Each represents a general framework for constructing messages in a typical situation. But no two situations are exactly alike. Moreover, for many situations no superstructure exists. To use superstructures effectively, look for an appropriate one; if you find one, adapt it to your particular purpose and readers.

FIGURE 4.4
Superstructure for
Proposal

SUPERSTRUCTURE FOR PROPOSALS		
TOPIC	**READERS' QUESTION**	**WRITER'S PERSUASIVE POINT**
Introduction	What is this communication about?	Briefly, I propose to do the following.
Problem	Why is the proposed project needed?	The proposed project addresses a problem, need, or goal that is important to you.
Objectives	What features will a solution to this problem need in order to be successful?	A successful solution can be achieved if it has these features.
Product or Outcome	How do you propose to do those things?	Here's what I plan to produce and how it has the features necessary for success.
Method	Are you going to be able to deliver what you describe here?	Yes, because I have a good plan of action (method), the necessary facilities, equipment, and other resources, a workable schedule, appropriate qualifications, and a sound management plan.
Costs	What will it cost?	The cost is reasonable.

Each topic in the superstructure for proposals answers a particular question from the reader in a way that readers will find persuasive.

Chapters 22 through 27 describe the general superstructures for letters, memos, reports, proposals, and instructions. These chapters also describe ways of adapting the superstructures to specific readers and purposes. At work, you may encounter more specialized superstructures developed within your profession or industry— or even within your own organization.

Guideline 5 | **Plan your graphics**

On-the-job communications often use graphics such as charts, drawings, and photographs rather than text to convey key points and information. Graphics convey certain kinds of information more clearly, succinctly, and forcefully than words. When planning a communication, look for places where graphics provide the best way for you to show how something looks (in drawings or photographs), explain a process (flowcharts), make detailed information readily accessible (tables), or clarify the relationship among groups of data (graphs).

Chapter 13 provides detailed advice about where to use graphics and how to construct them effectively. However, don't wait until you read that chapter to begin planning ways to increase your communication's usability with graphics.

Guideline 6 | **Outline, if this will be helpful**

When they talk about planning a communication, many people mention outlining and ask, "Is outlining worth the work it requires?" No single answer to this question is valid for all writers and all situations. The following general observations about outlining practices at work may help you determine when (and if) outlining might be worthwhile for you.

On the job, outlining is rarely used for short or routine messages. However, many writers outline longer, more complex communications, especially if they expect to have difficulties in organizing. They use outlining as a way of experimenting with alternative ways of structuring their message before they begin to invest time in drafting. Similarly, if they encounter problems when drafting, some writers will try to outline the troublesome passage.

Also, writers sometimes wish to—or are required to—share their organizational plans with a superior or coworker. Outlining provides them with a convenient way of explaining their plans to such individuals.

Finally, outlining can help writing teams negotiate the structure of a communication they must create together (see Chapter 18).

When you create an outline, remember that the kind of tidy structure shown in Figure 4.1 represents the product of the outlining process. Many writers begin organizing by making lists or diagrams, drawing arrows and pictures, and using many other techniques to think through possible ways of putting their information together for their readers.

Using Computers to Outline

Most desktop publishing programs include special tools for outlining. With these tools, you can, for instance, make an outline and then convert it immediately into the headings for your document. These tools will also allow you to convert from the normal view of your document to an outline view, so you can review the organizational structure that is evolving as you write. When you move material in the outline, the program automatically moves the corresponding parts of your full text, which can sometimes make for an extremely efficient way to revise a draft. Figure 4.5 shows some of the features of one widely used desktop publishing program's outlining tool.

WWW

For more on outlining, go to **www.cengage.com/english/anderson7e** and click on Chapter 4.

Guideline 7 | Global Guideline: Determine your readers' cultural expectations about what makes a communication usable

Among cultures, there are many different assumptions about what makes a communication usable. As an example, consider differences in the amount of detail that readers in different cultures expect writers to provide. In some cultures, writers provide a large amount of detail to spell out fully the writers' meaning. Today, researchers follow Edward T. Hall (1976) in calling these *low-context* cultures. German, Scandinavian, and U.S. cultures are examples. Typical readers in low-context cultures expect a communication to provide plenty of detail. If they don't find that detail in a communication, these readers are likely to feel that the writer has ignored their need for a thorough discussion.

In other cultures, writers omit many details. They and their readers both expect the readers to supply the necessary details by drawing on their knowledge of the

Learn More

For more information on writing to readers in other cultures, go to Chapter 3, page 79.

FIGURE 4.5
Outlining Functions of
a Desktop Publishing
Program

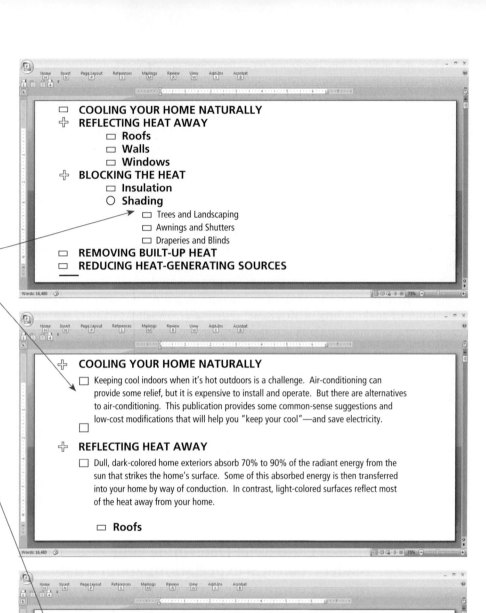

The outlining function of
Microsoft Word enables you to:

• Create your outline.

• Add text under your headings.

• Shift instantly to a finished
communication.

You can easily switch between
viewing the outline and the full
text.

You can reorganize your communication simply by moving a
heading; the subheadings and
text automatically move also.

situation, the relevant facts and cultural conventions, and the history and nature of their relationship with the writer. These are *high-context* cultures. Arab and Asian cultures are examples. If a communication is filled with details that readers could supply on their own, the readers may feel insulted because the writer appears to assume that they lack the appropriate knowledge.

To illustrate the difference between technical communications written in high- and low-context cultures, researcher Daniel Ding (2003) compared Chinese and U.S. instructions for installing a water heater. The U.S. instructions meet the expectations of a low-context culture by telling the user how to perform each step and substep in the installation process. For instance, to explain how to connect the hot and cold water pipes to the heater, the instructions say: "Put two or three turns of teflon tape or pipe joint compound around the threaded ends of the ¾" × 3" nipples and, using a pipe wrench, tighten the nipples into the 'HOT' and 'COLD' fittings of the water heater."

Reports, instructions, and web pages are among the many kinds of technical communications that use graphics extensively.

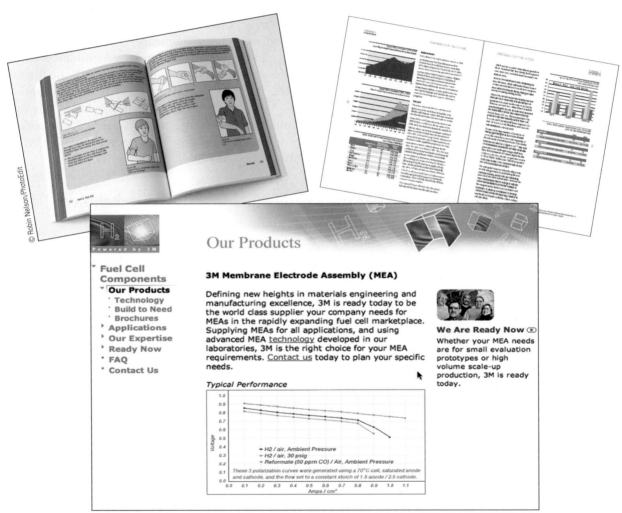

In contrast, the Chinese instructions meet the expectations of readers in a high-context culture by explaining the outcomes of each major step but not the actions needed to achieve the outcome. For instance, the Chinese instructions say that heater should be connected to the water pipes, but they don't describe actions necessary to connect them. There's no mention of turning the teflon tape three times, using pipe joint compound, or tightening with a pipe wrench. These details are considered to be part of the contextual knowledge the reader possesses.

When planning ways to make your communications usable to readers in other cultures, you should learn what these readers consider to be usable.

Guideline 8 | Check your plans with your readers

At the core of the guidelines you have just read is one common strategy: Focus on your readers as you plan. For that reason, the guidelines urged you to refer continuously to various sources of insight into what will make your communication usable in your readers' eyes. These include the understanding of your readers that you gained when defining your objectives and the superstructures that others have found successful in similar circumstances. Nevertheless, such planning involves guesswork about what will really work with your readers. Therefore, whenever you have the opportunity, bring your plans to your readers. Ask for their responses and requests. If it isn't feasible to check your plans with your readers, share them with someone who understands your readers well enough to help you find possible improvements. The better your plans, the better your final communication.

Guideline 9 | Ethics Guideline: Investigate stakeholder impacts

As Chapter 3 explained, the first step in writing ethically is to identify your communication's stakeholders—the people who will be affected by what you say and how you say it. Next, learn how your communication will impact these people and how they feel about these potential effects.

Asking Stakeholders Directly

The best way to learn how your communication will affect its stakeholders is to talk to these individuals directly. In many organizations, such discussions are a regular step in the decision-making process that accompanies the writing of reports and proposals. When an action is considered, representatives of the various groups or divisions that might be affected meet together to discuss the action's potential impacts on each of them.

Similarly, government agencies often solicit the views of stakeholders. For example, the federal agencies that write environmental impact statements are required to share drafts of these documents with the public so that concerned citizens can express their reactions. The final draft must respond to the public's comments.

Even in situations that are traditionally viewed as one-way communications, many managers seek stakeholder inputs. For instance, when they conduct annual employee evaluations, some managers draft an evaluation, then discuss it with the employee before preparing the final version.

Action You Can Take

If you are writing a communication for which there is no established process for soliciting stakeholders' views, you can initiate such a process on your own. To hear from stakeholders in your own organization, you can probably just visit or call. To contact stakeholders outside your organization, you may need to use more creativity and also consult with your coworkers.

If you already know the stakeholders well, you may be able to find out their views very quickly. At other times, an almost impossibly large amount of time would be needed to thoroughly investigate stakeholders' views. When that happens, the decision about how much time to spend can itself become an ethical decision. The crucial thing is to make as serious an attempt as circumstances will allow. There's all the difference in the world between saying that you can spend only a limited amount of time investigating stakeholders' views and saying that you just don't have *any* time to find out what the stakeholders are thinking.

Speaking for Others

To save time, you might try to imagine what the stakeholders would say if you spoke to them directly. Clearly this course of action is superior to ignoring stakeholders altogether. But you can never know exactly what others are thinking. The greater the difference between you and the stakeholders—in job title, education, and background—the less likely you are to guess correctly. So it's always best to let stakeholders speak for themselves. When you do this, avoid the mistake of assuming that all the stakeholders will hold the same view. The persons affected by a communication may belong to many different groups, and opinions can vary even within a single group. When letting stakeholders speak for themselves, seek out a variety of persons.

At work, you might need authorization from your managers to seek stakeholder views. If you encounter reluctance to grant you this permission, you have an opportunity to open a conversation with your managers about the ethical dimensions of the communication you are writing. Even if they aren't swayed this first time, you will have introduced the issue into the conversation at your workplace.

Learn More

For an additional discussion of ways to address an organization's reluctance to consider ethical issues, see the ethics guideline on page 300.

Seeking Stakeholder Views

In sum, on many issues, different people hold different views about what is the most ethical course of action. Asking stakeholders for their input does not guarantee that they all will be happy with what you ultimately write. But it does guarantee that you have heard their opinions. Without this step, you can scarcely attempt to take their needs and concerns into account.

CONCLUSION

This chapter has described the reader-centered strategies that successful workplace communicators use in order to plan communications that are highly usable. As you have seen, to develop expertise in using these strategies you must possess the detailed

FIGURE 4.6   **Writer's Guide for Planning Your Usability Strategies**

To download a copy of this Writer's Guide, go to www.cengage.com/english/ anderson7e and click on Chapter 4.

Writer's Guide
PLANNING YOUR USABILITY STRATEGIES

Plan Content

1. What questions will your readers want your communication to answer?
2. What additional information do your readers need?
3. What information do you need to gather through research?

Organize

1. What will be your readers' reading tasks?
2. If there is a technical writing superstructure you can adapt, which one is it?
3. What organization will best support your readers as they perform these tasks?

   ☐ Group together the information your readers will use together.
   ☐ Create a hierarchical organization.
   ☐ Place the bottom line first.
   ☐ Use a modular design, if you have a variety of readers with different needs.
   ☐ Outline your organizational plan, if this will be helpful.

Provide Quick Access to Information

1. What headings would help your readers find the information they want?
2. Would a table of contents help the readers?
3. If you are creating a website, sketch a reader-centered site map and identify the navigation aids you would provide.
4. What other strategies will help your readers find information?

Plan Graphics

1. What tables, graphs, drawings, or other graphics increase the usability and persuasiveness of your communication?

Write Ethically

1. Who, besides your readers, are the stakeholders in your report?
2. How might they be affected by it?
3. How can you assure that your readers and other stakeholders are treated ethically?

---

knowledge of your readers that you gained by following Chapter 3's reader-centered advice for defining your communication's objectives. The Writer's Guide shown in Figure 4.6 summarizes this chapter's advice for using your knowledge of your readers to plan communications that are usable (complete, task-oriented, and accessible) as well as ethical.

USE WHAT YOU'VE LEARNED

For additional exercises, go to www.cengage.com/english/anderson7e. *Instructors:* The book's website includes suggestions for teaching the exercises.

EXERCISE YOUR EXPERTISE

Imagine that you are employed full-time and have decided to take a course at a local college. First, name a course you might like to take. Next, by following Guideline 1, list the information you would include in a memo in which you ask your employer to pay your tuition and permit you to leave work early two days a week to attend class.

EXPLORE ONLINE

Develop a list of questions that high school students might ask about studying in your major at your college. Then try to answer these questions by going to your college's website. How accessible are the answers? How complete are they from the viewpoint of high school students? How could the website be made more usable for these students?

COLLABORATE WITH YOUR CLASSMATES

Imagine that you have been hired to create a brochure or website that presents your college department in a favorable light to entering first-year students and students who are thinking of changing their majors. Working with another student and following the guidelines in this chapter, generate a list of things that you and your partner would want to say. How would the two of you group and order this information?

APPLY YOUR ETHICS

Think of a policy of your college or employer that you would like to change. Imagine that you are going to write a report recommending this change. After identifying all the stakeholders, tell how you could gain a complete understanding of the relevant concerns and values of each stakeholder or stakeholder group. Which stakeholders' concerns and values would be most difficult for you to learn about? Why? Would you have more difficulty listening sympathetically to some stakeholders than others? If so, who and why? How would you ensure that you treat these stakeholders ethically?

CASE	FILLING THE DISTANCE LEARNING CLASSROOM

For additional cases, visit www.cengage.com/english/anderson7e. *Instructors:* The book's website includes suggestions for teaching the cases.

You knock on the office door softly, unsure why the Chair of the English Department has asked to meet with you.

"Welcome," Professor Rivera says, immediately putting you at ease. "I'm glad you could come. Professor Baldwin suggested you might be able to help me with a sort of marketing project."

What a relief. At least you haven't done anything wrong.

"Have you heard about our new distance learning classroom?" Professor Rivera asks. You tell her you know that a large room in the English building is being redesigned as a computerized classroom.

"Well, in addition to computers, the room will have video cameras, projection equipment, special software, and a very large screen. By means of this equipment, it can be paired with a similar classroom elsewhere in the world so that students in both locations can become part of the same class.

"In fact, we're getting ready to conduct a few experimental sections of First-Year English in which students here will be paired with students in a similar course at a university in Brazil. A team of faculty from both schools has planned a common syllabus and paired sections here and in Brazil to meet at the same hours. When the Brazilian instructor or students speak, the students and instructor here will be able to see and hear them, and vice versa. Also, students in the two locations will exchange drafts via the Internet, and they will even be able to project a draft of a student's paper in both locations so that all the students can talk jointly about it."

"But isn't Portuguese the native language of Brazil?" you ask.

"Yes," Professor Rivera replies. "All of the Brazilian students will be studying English as a second language, although they will also have studied several years of it before attending their university. Of course, English is a second language for many students here at our own university as well."

"What an interesting project," you observe.

"Yes," Professor Rivera responds. "The faculty members involved think it will have several benefits for students. It's increasingly important for college graduates to be able to work in an international economic and intellectual environment. This course will help prepare them for such work. In addition, many businesses are now using exactly the kind of technology we have placed in the distance learning classroom to hold meetings that involve employees or clients located at various points around the globe. Students in the experimental classes will gain experience with such technology that could give them a tremendous boost in the job market."

"Sounds great," you say. "But what does it have to do with me? I've already completed that course."

"What we'd like you to do is to write a pamphlet we can send this summer to incoming students to entice some of them to sign up for the experimental distance learning sections of First-Year English."

"Sounds easy," you say.

"Well, there are some complications," Professor Rivera explains. "This is an experiment of sorts. Many tests will be run to determine the effects of the distance learning environment. The experiment is important because some faculty members are skeptical about the program, worried that it might focus on technology in ways that will detract from the quality of the writing instruction.

"Because this is an experiment, the faculty members who designed the program want the students in the distance learning sections to represent a cross section of students in First-Year English. It can't be populated with just students who have a high interest in technology or international studies. Therefore, they have asked the Admissions Office to generate a list of 240 randomly selected incoming students to whom the pamphlet will be sent. The aim of the pamphlet is to persuade at least half of the recipients to enroll in six special sections."

"In that case, writing the pamphlet will be a challenge," you agree.

"Will you accept the challenge?" Professor Rivera asks.

"Sure. Why not?"

Your Assignment

Following the guidelines in Chapter 3, define the objectives for your pamphlet. Then, by following the guidelines in this chapter, list the things you would say in the pamphlet to the incoming students. Identify the things you would need to investigate further before writing.

5 | Planning Your Persuasive Strategies

GUIDELINES

DEFINING
OBJECTIVES

PLANNING

RESEARCHING

DRAFTING

REVISING

As explained in Chapter 1, all on-the-job writing aims to persuade its readers in some way. At work, you will engage in two related but distinguishable kinds of persuasion, each with a distinctive goal.

PERSUASION TO INFLUENCE ATTITUDES AND ACTION

WWW

For additional information about persuasion, go to **www.cengage.com/english/ anderson7e** and click on Chapter 5.

The goal of one type of persuasion is to influence other people's attitudes and actions.

Your goal when using the first kind of persuasion is to *influence other people's attitudes and actions.* This is the goal that comes most readily to mind when we think of persuasion. At work, you'll use this type of persuasion when you want your readers to purchase your company's product, fund the project you are proposing, or adopt a policy you advocate. This kind of persuasion will also be important to you when you are preparing many kinds of communication we don't ordinarily think of as persuasive. A report on an engineering test may seem to convey facts only. However, if the report doesn't persuade other engineers that the writer used an appropriate test method, they are likely to conclude that the test results can't be trusted. Even instructions have a persuasive element. If they don't persuade readers to follow their directions, their entire purpose is undermined.

PERSUASION TO HELP A TEAM EXPLORE IDEAS COLLABORATIVELY

The goal of another type of persuasion is to enable people to collaborate while exploring ideas and possible courses of action.

Your goal when using the second kind persuasion isn't to influence others but to collaborate with them in *exploring ideas or possible courses of action.* This kind of persuasion is very common in the workplace. For example, a team of engineers at Osage Corporation was preparing a proposal to build a billion-dollar high-speed transit system between two cities. In planning meetings, the team explored ideas about every aspect of the system, from the location of the passenger stations to the design of the transit cars. This ability of all team members to write and speak persuasively was crucial to the team's success. If a team member had a good idea but couldn't present it persuasively, the team could miss the best way to design some part of the system. If another team member spotted a problem with an idea but couldn't persuade others that the problem was serious, the team might pursue a flawed plan.

At work, the success of group planning and problem-solving sessions can depend on the participants' skill at persuasion.

This second type of persuasion is central to many types of workplace activities. It is required when scientists gather to discuss new data, when technicians team up to solve a technical problem, and when medical personnel confer about the best way to address the outbreak of an illness. It is also the type of persuasion in which you engage when you work on a team project in your technical communication course or other classes.

HOW PERSUASION WORKS

Over twenty years ago, researchers Petty and Cacioppo (1986) explained the process by which we influence one another in both competitive and cooperative situations. The key, they demonstrated, is to concentrate on shaping our readers' attitudes.

> To persuade, concentrate on shaping your readers' attitudes.

At work, you will be concerned with your readers' attitudes toward a wide variety of subjects, such as products, policies, actions, and other people. As Chapter 3 explains, you may use your persuasive powers to change your readers' attitudes in any of the following ways:

- **Reverse** an attitude you want your readers to abandon.
- **Reinforce** an attitude you want them to hold even more firmly.
- **Shape** their attitude on a subject about which they currently have no opinion.

> Ways to change readers' attitudes

What determines a person's attitude toward an idea, object, or action? Petty and Cacioppo (1981, 1986) have found that it isn't a single thought or argument, but the *sum* of the various thoughts a person associates with the idea, object, or action under consideration. Consider the following example.

Edward is thinking about purchasing a particular copy machine for his department. As he deliberates, a variety of thoughts cross his mind: the amount of money in his department's budget, the machine's special features (such as the ability to make copies directly from computer files), its appearance and repair record, his experience with another product made by the same company, and his impression of the manufacturer's marketing representative. Each of these thoughts will make Edward's attitude more positive or more negative. He may think his budget has enough money (positive) or too little (negative); he may like the salesperson (positive) or detest the salesperson (negative); and so forth.

Some of these thoughts will probably influence Edward's attitude more than others. For instance, Edward may like the machine's special features and color, but those two factors may be outweighed by reports that the machine's repair record is very poor. If the sum of Edward's thoughts associated with the machine is positive, he will have a favorable attitude toward it and may buy it. If the sum is negative, he will have an unfavorable attitude and probably not buy it.

> The sum of a reader's positive and negative thoughts about an object, action, or person determines his or her attitude.

THE SOURCES OF THIS CHAPTER'S ADVICE

Although we have learned a great deal about the ways attitudes are shaped in the past several decades, much of the best advice about how to write persuasively has been developed over a much longer period. After carefully observing the persuasive

WWW

For a fuller explanation of Aristotle's thoughts on persuasion, go to **www.cengage.com/english/anderson7e** and click on Chapter 5.

strategies used by citizens of ancient Greece more than 2400 years ago, Aristotle classified them into three main types: *logos* (appeals to logic), *pathos* (appeals to emotion), and *ethos* (for our purposes, roughly equivalent to credibility). Because of their effectiveness—and despite their age—Aristotle's concepts can serve you well in your workplace writing. Other highly effective persuasive strategies have developed since Aristotle's time, many of them through scientific studies conducted in the past 75 years. This chapter's nine reader-centered guidelines draw together the most helpful ideas gained over the long period from ancient Greece to our day.

Guideline 1 | **Listen—and respond flexibly to what you hear**

To communicate persuasively, you must listen well. What you learn about your readers by listening to them and to people who tell you about them provides the basis for all your writing decisions.

Successful persuasion depends on how well you listen as well as how well you write or speak.

When you want to influence others' attitudes and actions, you need to listen very carefully to their goals, needs, aims, and concerns. In collaborative persuasive situations, you will need to listen carefully to what your team members are saying in order to contribute productively to the group's discussions.

Equally important to the way you listen will be the way you respond to what you hear. Be flexible. The things you hear may not only help you determine how to strengthen your communication but also suggest ways to improve the action or idea you are advocating.

The next two guidelines will help you determine what to listen for and how to make the most productive use of what you hear.

Guideline 2 | **Focus on your readers' goals and values**

A powerful way to prompt your readers to experience favorable thoughts about the ideas, actions, process, or product you advocate is to tell them how it will help them accomplish some goal they want to achieve. For instance, if Edward's goal is to be sure that his department produces polished copies, a writer could influence him to experience favorable thoughts toward a particular copier by pointing out that it prints high-quality images. If his goal is to keep costs low, a writer could influence him to view the copier favorably by emphasizing that it is inexpensive to purchase and run.

Similarly, at work you can increase the persuasiveness of communications on any topic by focusing on the goals that guide your readers' decisions and actions. Here is a three-step process for doing so:

Process for developing persuasive strategies based on your readers' goals

1. Identify your readers' goals.
2. Determine how the ideas or actions you are recommending can help your readers achieve their goals.
3. In your communication, focus on the ways your recommended actions and ideas can help your readers achieve their goals.

In cooperative situations, your readers' goals may be identical or similar to your own, though that is not always the case. In both cooperative and competitive situations, the best way to identify your readers' goals is to ask them—and listen carefully. It's also helpful to talk with coworkers who know your readers. You will also find it helpful to consider the ways that your readers' personal goals may be shaped by the following sources:

- Objectives of the organizations that employ your readers
- Values your readers derive from their organizations or from their personal lives
- Your readers' desires for achievement and growth

Sources of readers' goals

Organizational Goals

On the job, you will often write to coworkers and to employees in other organizations. Because they are hired to advance their employer's interests, employees usually adopt their employer's goals as their own. Consequently, one way to identify your readers' goals is to examine the goals of the organization for which they work.

In any organization, some goals are general and some specific. At a general level, many organizations share the same goals: to increase revenue, operate efficiently, keep employee morale high, and so on. However, each company also has its own unique goals. A company might be seeking to control 50 percent of its market, to have the best safety record in its industry, or to expand into ten states in the next five years. Moreover, each department within a company has specific objectives. Thus, the research department aims to develop new and improved products, the marketing department seeks to identify new markets, and the accounting department strives to manage financial resources prudently. When identifying goals to serve as the basis for persuasion, choose specific goals over general ones.

Organizations have goals at many levels.

Also, as you uncover an organization's various goals, focus on the ones that are most closely related to the specific idea, action, or process you are advocating. Then craft your communication to emphasize the ways that doing what you recommend will help your readers achieve the specific outcomes they desire.

Figure 5.1 shows a marketing brochure that uses organizational objectives to persuade its intended readers (purchasing managers for supermarkets and supermarket chains) to carry a certain product. Notice the boldface statements that proclaim how the product will build sales volume, stimulate impulse buying, and deliver profits.

WWW

To view additional communications that employ persuasive strategies described in this chapter, go to **www.cengage.com/english/anderson7e** and click on Chapter 5.

Values-Based Goals

Most companies have goals that involve social, ethical, and aesthetic values not directly related to profit and productivity. Some companies spell out these values in corporate credos; Figure 5.2 on page 123 is an example. Even in companies that do not have an official credo, broad human and social values may provide an effective foundation for persuasion, especially when you are advocating a course of action that seems contrary to narrow business interests. For example, by pointing to the benefits to be enjoyed by a nearby community, you might be able to

WWW

To see other corporate credos, go to **www.cengage.com/english/anderson7e** and click on Chapter 5.

FIGURE 5.1
Brochure That Stresses
Organizational Objectives

The production processes for many ordinary products, including foods and beverages, are created by chemists, manufacturing engineers, and other technical and scientific specialists.

In this brochure, Welch's tells food retailers how selling one of its products will help them achieve three organizational goals: greater sales volume, more impulse purchases, and increased profits.

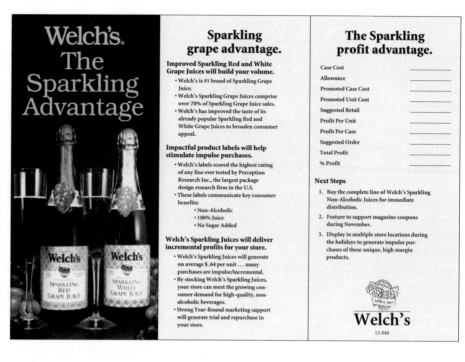

Deficiency needs are needs that disappear when satisfied, as the need for higher pay is satisfied with a pay raise.

Growth needs are needs that are always present, like the need for personal development and enjoyment in one's work.

persuade your employer to strengthen its water pollution controls beyond what is required by law.

In addition, individual employees bring to work personal values that can influence their decisions and actions. Especially when writing to only one or a few readers, learn whether they have personal values that are related to what you are recommending.

Achievement and Growth Goals

In studies of employee motivation that are still highly respected today, Abraham Maslow (1970) and Frederick Herzberg (1968) highlight another type of benefit you can use to persuade. Both researchers confirmed that, as everyone knew, employees are motivated by such considerations as pay and safe working conditions. However, the studies also found that after most people feel they have an adequate income and safe working conditions, they become less easily motivated by these factors. Consequently, such factors are called *deficiency needs:* they motivate principally when they are absent.

Once their deficiency needs are met—and even before—most people are motivated by so-called *growth needs,* including the desire for recognition, good relationships at work, a sense of achievement, personal development, and the enjoyment of work itself.

Many other studies have confirmed these findings. For instance, a study of supervisors showed that achievement, recognition, and personal relationships are

FIGURE 5.2
Corporate Credo that
Includes Many Values-
Based Goals

Our Credo

We believe our first responsibility is to the doctors, nurses, and patients, to mothers and fathers and all others who use our products and services. In meeting their needs, everything we do must be of high quality. We must constantly strive to reduce our costs in order to maintain reasonable prices. Customers' orders must be serviced promptly and accurately. Our suppliers and distributors must have an opportunity to make a fair profit.

We are responsible to our employees, the men and women who work with us throughout the world. All employees must be considered as individuals. We must respect their dignity and recognize their merit. They must have a sense of security in their jobs. Compensation must be fair and adequate, and working conditions clean, orderly, and safe. We must be mindful of ways to help our employees fulfill their family responsibilities. Employees must feel free to make suggestions and complaints. There must be equal opportunity for employment, development and advancement for those qualified. We must provide competent management, and their actions must be just and ethical.

We are responsible to the communities in which we live and work and to the world community as well. We must be good citizens—support good works and charities and bear our fair share of taxes. We must encourage civic improvements and better health and education. We must maintain in good order the property we are privileged to use, protecting the environment and natural resources.

Our final responsibility is to our stockholders. Business must make a sound profit. We must experiment with new ideas. Research must be carried on, innovative programs developed, and mistakes paid for. New equipment must be purchased, new facilities provided, and new products launched. Reserves must be created to provide for adverse times. When we operate according to these principles, the stockholders should realize a fair return.

The company describes its values-based goals for the way it treats the people who use its products.

The company continues by stating its values-based goals for treating its employees.

It includes supporting employees' development and advancement—two growth needs—among its values- based goals. See page 122.

The company states its values-based goals for respecting the communities in which it has facilities.

The statement concludes with its values-based goals for its treatment of stockholders.

much more powerful motivators than money (Munter, 1987). Another found that computer professionals are less motivated by salary or status symbols than by the opportunity to gain new skills and knowledge in their field (Warren, Roth, & Devanna, 1984).

These studies demonstrate that one powerful persuasive strategy is to show how the decisions or actions you are advocating will help your readers satisfy their desires for growth and achievement. When you request information or cooperation from a coworker, mention how much you value his or her assistance. When you evaluate a subordinate's performance, discuss accomplishments as well as shortcomings. When you ask someone to take on additional duties, emphasize the challenge and opportunities for achievement that lie ahead.

Try This

How do criminals persuade people to fall for their scams? At www.scambusters.org, try to find a scam that appeals to people's deficiency needs and another that appeals to their growth needs. Can you find one that appeals to their values-based goals? Their organizational goals?

Figure 5.3 shows a web page that General Electric uses to attract new employees. Notice the many ways it appeals to college graduates by emphasizing the opportunities they would have with the company to assume responsibility, take on challenges, and grow professionally.

On this reader-centered web page, GE appeals to the desire for achievement and growth of potential employees. The page also appeals to potential employees' desire to work for a company that shares their ethical values.

Throughout this page and the additional pages to which readers can link from it, the company addresses potential employees as "you," emphasizing the opportunities "you" would receive at GE.

FIGURE 5.3
Page from a Recruiting Website that Focuses on Growth Needs

Try This

Which of the eight sections under "About Us" describe a benefit of working for GE that appeals to you? Which don't? Why or why not? What other benefits might GE offer that would attract your interest? Which of the benefits GE lists are related to deficiency needs and which to growth needs?

Why GE

‣ Careers at GE

▾ **Why GE**

  Breakthrough Technologies

  Continuous Learning

‣ Exciting Businesses

  Integrity & Citizenship

  Our Culture

‣ Our People

  Our Plan for Growth

  Rewards & Benefits

GE is a place where you can live your dreams. Our global presence, innovation, and financial strength help to make GE a dynamic place to work, giving you the advantage of a large company, with the agility of a small company, where your voice is heard.

Our 11 different businesses give you flexibility for change and the opportunity to learn about new industries and markets, as well as provide unparalleled career options.

About Us

Breakthrough Technologies
Be on the cutting edge of technology and the future. Learn about our history of innovation and some of the things we're working on now to help shape your world.

Rewards & Benefits
We provide a competitive benefits and incentive packages for our employees.

Continuous Learning
With more than $1 billion spent annually in training and development of employees, GE is committed to learning.

Integrity & Citizenship
At GE, we believe that it's important to give back...to our community, to our neighborhoods, to our environment.

Exciting Businesses
With 11 businesses, GE offers you continued career growth and an advantage few other companies can match.

Our Culture
GE has a performance-based culture that is grounded in a set of core values.

Our People
GE people are some of the best in the world ...with a shared desire to learn and stretch beyond their limits.

Our Plan for Growth
Our growth strategy is to build positions in new markets where we can achieve superior growth and returns.

Guideline 3 | Address—and learn from—your readers' concerns and counterarguments

Although it is tempting to focus your attention on the reasons that your readers should agree with you, you can gain a great benefit from learning the reasons they may be reluctant to do so.

To begin, learning your readers' concerns and counterarguments can help you improve the project you are proposing or refine your ideas. That's because some of these concerns and counterarguments may be valid. By following Guideline 1's advice to remain flexible you can build on the insights of your readers.

When you are satisfied that your proposed idea or action is as sound as possible, use your insights into your readers' concerns and counterarguments to help you increase your communication's persuasiveness. Gaining additional insights into the nature of persuasion will help you do so.

Research shows that when people read, they not only pay attention to the writer's statements but also generate their own thoughts. Research further shows that whether readers' self-generated thoughts are favorable or unfavorable can have a greater influence on readers' attitudes than any point you make in the communication itself (Petty & Cacioppo, 1986). Consequently, when you are trying to persuade, avoid saying anything that might prompt negative thoughts.

One way to avoid arousing negative thoughts is to answer all the important questions your readers are likely to ask while reading your communication. Most of these questions will simply reflect the readers' efforts to understand your position or proposal thoroughly. "What are the costs of doing as you suggest?" "What do other people who have looked into this matter think?" "Do we have employees with the education, experience, or talent to do what you recommend?" If you answer such questions satisfactorily, they create no problem whatever. But if you ignore them, they may prompt the reader to think that you have overlooked some important consideration or ignored evidence that might weaken your position.

In addition to asking questions while reading, your readers may generate arguments against your position. For example, if you say there is a serious risk of injury to workers in a building used by your company, your readers may think to themselves that the building has been used for three years without an accident. If you say a new procedure will increase productivity by at least 10 percent, your readers may think to themselves that some of the data on which you are basing your estimate are inaccurate. Readers are especially likely to generate counterarguments when you attempt to reverse their attitudes. Research shows that people resist efforts to persuade them that their attitudes are incorrect (Petty & Cacioppo, 1986).

To deal effectively with counterarguments, you must offer a reason for relying on your position rather than on the opposing position. For example, imagine that you are proposing the purchase of a certain piece of equipment and that you predict your readers will object because it is more expensive than a competitor's product they believe to be its equal. You might explain that the competitor's product, though less expensive to purchase, is more expensive to operate and maintain.

> Your readers' counterarguments can help you improve the project or position you are advocating.

Try This

Think of a policy or regulation at your school or workplace that you want to see changed. List the good reasons someone supporting the policy or regulation could have for retaining it.

One way to learn about the questions and counterarguments your readers might raise while reading is to ask them about their reasons for holding their present attitudes. Why do they do things the way they do? What arguments do they feel most strongly support their present position? You can also follow Chapter 3's advice for identifying your readers' questions, but focus specifically on questions that skeptical readers might pose.

Figure 5.4 shows a letter in which the marketing director of a radio station attempts to persuade an advertising agency to switch some of its ads to her station. Notice how she anticipates and addresses possible counterarguments.

FIGURE 5.4
Letter that Addresses
Counterarguments

WGWG

12741 Vienna Boulevard
Philadelphia, PA 19116

August 17, 2010

Mr. Roger L. Nordstrom
Llat Marketing, Inc.
1200 Langstroth Avenue
Philadelphia, Pennsylvania 19131

Dear Mr. Nordstrom:

What a lucky coincidence that we should have met at Shelby's last week. As I told you then, I believe you can improve the advertising service you provide many of your clients by switching to our station some of the radio advertising that you presently place for your clients with WSER and WFAC. Since we met, I have put together some figures.

The primary advantage of advertising with us is that you will extend the size of your audience considerably. As an example, consider the reach of the campaign you are now running for Fuller's Furniture Emporium. You currently buy 45 spots over four weeks on the other two stations. According to Arbitron market analysis, you thereby reach 63% of your target market—women in the ages 25–54 demographic group—an average of 16 times per week. However, there is a lot of overlap in the audiences of those two stations: many listeners switch back and forth between them.

In contrast, we serve a very distinct audience. By using only 30 spots on those two stations and giving 15 to us, you would increase your reach to 73% of the ages 25–54 group. It's true, of course, that this change would reduce from 16 to 13 the number of times the average listener hears your message. However, this is an increase of 8% in the portion of your target audience that you would reach.

But there's more. At the same time that you increase your audience, you would also decrease your costs. For example, if you switched the Fuller's Furniture Emporium spots in the way described, your costs would drop from $6,079 to $4,850. Our audience is smaller than the audiences on those stations, but our rates are only about half of theirs.

Writer states a counterargument that might occur to the reader.

Writer specifies a benefit to the reader that will outweigh the objection.

FIGURE 5.4
(continued)

Roger L. Nordstrom   – 2 –   August 17, 2010

As you consider these figures and others that might be given to you by other radio stations, evaluate them carefully. As we are all aware, anyone can manipulate statistics to their advantage. We recently discovered that one of our competitors was circulating data regarding the portion of the audiences of Philadelphia stations that are "unemployed." In that analysis, our station had a significantly higher portion of "unemployed" listeners than did either WSER or WFAC.

> Writer states a counterargument the reader is likely to hear from competitors.

However, after some investigation, we discovered that in those statistics, "unemployed" meant "unemployed outside the home." That meant that people "unemployed" are not necessarily breadwinners who are unable to provide an income for their families. They could be housewives with considerable power over family decisions about spending. In fact, surveys indicate that housewives are often the *primary* decision makers about such things as household furnishings—a point that is quite important with respect to your account with Fuller's Furniture Emporium.

> Writer exposes faulty reasoning underlying the counterargument.

I suggest that we get together soon so that we can sketch out the details of some radio advertising schedules that would include WGWG. I'll call you next week.

Cordially,

Ruth Anne Peterson

Ruth Anne Peterson
Marketing Director

Guideline 4 | Reason soundly

In today's workplace, as in ancient Greece, one of the most favorable thoughts your readers and listeners could possibly have is, "Yeah, that makes sense." One of the most unfavorable is, "Hey, there's a flaw in your reasoning."

Sound reasoning, the equivalent of Aristotle's *logos,* is especially important when you are trying to influence your readers' decisions and actions. In addition to identifying potential benefits that will appeal to your readers, you also must persuade them that the decision or action you advocate will actually bring about these benefits—that the proposed new equipment really will reduce costs enough to pay for itself in just eighteen months or that the product modification you are recommending really will boost sales 10 percent in the first year.

Sound reasoning is also essential when you are describing conclusions you have reached after studying a group of facts, such as the results of a laboratory experiment or consumer survey. In such cases, you must persuade your readers that your conclusions are firmly based on the facts.

Notice that in each of the situations just mentioned (as well as in any other you might encounter on the job), you must not only use sound reasoning, but also convince your readers that your reasoning is sound. The ability to do so is one of the most valuable writing skills you can develop.

Stephen Toulmin has developed a way of thinking about reasoning that will help you reason soundly and also show your readers that you do (Toulmin, Rieke, & Janik, 1984). You will also find Toulmin's concepts useful when you are analyzing and evaluating the reasoning in other people's communications.

How Reasoning Works

Sound reasoning involves your *claim,* your *evidence,* and your *line of reasoning.* The diagram illustrates the relationship among them.

Relationship among claim, evidence, and line of reasoning

EVIDENCE	CLAIM
The facts, observations, and other evidence that support your claim	The position you want your readers to accept

LINE OF REASONING
The connection linking your claim and evidence; the reason your readers should agree that your evidence supports your claim

Imagine that you work for a company that manufactures cloth. You have found out that one of your employer's competitors recently increased its productivity by using newly developed computer programs to manage some of its manufacturing processes. If you were to recommend that your employer develop similar programs, your argument could be diagrammed as follows:

■ CHAPTER 5 **Planning Your Persuasive Strategies**

EVIDENCE Our competitor has increased productivity by using new computer programs.		CLAIM By using new computer programs, we will increase productivity.

The line of reasoning explains why the evidence provides support for the claim.

LINE OF REASONING
Experience has shown that actions that increase profits for one company in an industry will usually increase profit of other companies in the same industry.

To accept your claim, readers must be willing to place their faith in both your evidence and your line of reasoning. The next two sections offer advice about how to persuade them to do so.

Present Sufficient and Reliable Evidence

First, you must convince your readers that your evidence is both sufficient and reliable.

To provide *sufficient* evidence, you must furnish all the details your readers will want. For instance, in the example of the textile mills, your readers would probably regard your evidence as skimpy if you produced only a vague report that the other company had somehow used computers and saved some money. They would want to know how the company had used computers, how much money had been saved, whether the savings had justified the cost of the equipment, and so forth.

Your evidence is sufficient if it includes all the details your readers want.

To provide *reliable* evidence, you must produce the type of evidence your readers are likely to accept. The type of evidence varies greatly from field to field. For instance, in science and engineering, certain experimental procedures are widely accepted as reliable, whereas common wisdom and unsystematic observation usually are not. In contrast, in many business situations, personal observations and anecdotes provided by knowledgeable people often are accepted as reliable evidence. However, the following three types of evidence are widely accepted:

Your evidence is reliable if your readers will accept it as a valid type of evidence.

The types of evidence readers accept as valid differ from context to context.

- **Data.** Readers typically respond very favorably to claims that are supported by numerical data.
- **Expert testimony.** People with advanced education, firsthand knowledge, or extensive experience related to a topic are often credited with special understanding and insight.
- **Examples.** Specific instances can effectively support general claims.

Explicitly Justify Your Line of Reasoning Where Necessary

To argue persuasively, you must not only present sufficient and reliable evidence, but also convince your readers that you are using a valid line of reasoning to link your evidence to your claim. Writers often omit any justification of their line of reasoning in the belief that the justification will be obvious to their readers. In fact, that is sometimes the case. In the construction industry, for example, people generally agree that if an engineer uses the appropriate formulas to analyze the size and shape of a bridge, the formulas will accurately predict whether or not the bridge will be strong enough to support the loads it must carry. The engineer doesn't need to justify the formulas themselves.

Readers sometimes look aggressively for weaknesses in the writer's line of reasoning.

In many cases, however, readers search aggressively for a weak line of reasoning. In particular, they are wary of arguments based on false assumptions. For example, they may agree that if another textile mill like yours saved money by computerizing, then your mill would probably enjoy the same result; your readers, however, may question the assumption that the other mill is truly like yours. Maybe it makes a different kind of product or employs a different manufacturing process. If you think your readers will suspect that you are making a false assumption, offer whatever evidence or explanation you can to dispel their doubts.

Readers also look for places where writers have overgeneralized by drawing broad conclusions from too few specific instances. If you think your readers will raise such an objection to your argument, mention additional cases. Or, narrow your conclusion to better match the evidence you have gathered. For example, instead of asserting that your claim applies to all textile companies in all situations, argue that it applies to specific companies or specific situations.

Guideline 5 | **Organize to create a favorable response**

The way you organize a communication may have almost as much effect on its power to persuade as what you say in it. That point was demonstrated by researchers Sternthal, Dholakia, and Leavitt (1978), who presented two groups of people with different versions of a talk urging that a federal consumer protection agency be established. One version *began* by saying that the speaker was a highly credible source (a lawyer who graduated from Harvard and had extensive experience with consumer issues); the other version *ended* with that information.

The order in which you present information can affect the reader's response to your overall message.

Among people initially opposed to the speaker's recommendation, those who learned about his credentials at the beginning responded more favorably to his arguments than did those who learned about his credentials at the end. Why? This outcome can be explained in terms of two principles you learned in Chapter 1. First, people react to persuasive messages moment by moment. Second (and here's the key point), their reactions in one moment will affect their reactions in subsequent moments. In this experiment, those who learned of the speaker's credentials before hearing his arguments were relatively open to what he had to say. But those who learned of the speaker's credentials only at the end worked more vigorously at creating counterarguments as they heard each of his points. After those counterarguments had been recorded in memory, they could not be erased simply by adding information about the speaker's credibility.

Thus, it's not only the array of information that is critical in persuasion, but also the way readers process the information. The following sections suggest two strategies for organizing to elicit a favorable response: Choose carefully between direct and indirect organizational patterns, and create a tight fit among the parts of your communication.

Choose Carefully between Direct and Indirect Organizational Patterns

As you learned in Chapter 4, the most common organizational pattern at work begins by stating the bottom line—the writer's main point. Communications organized this way are said to use a *direct pattern* of organization because they go directly to the main point and only afterward present the evidence and other information related to it. For example, in a memo recommending the purchase of a new computer program, you might begin with the recommendation and then explain why you think the new program is desirable.

In the direct pattern, you state the main point up front, before presenting the supporting evidence.

The alternative is to postpone presenting your main point until you have presented your evidence or other related information. This is called the *indirect pattern* of organization. For example, in a memo recommending the purchase of a new computer program, you might first explain the problems created by the current program, withholding until later your recommendation that a new one be purchased.

In the indirect order, you postpone the main point.

To choose between the direct and indirect organizational patterns, focus your attention on your readers' all-important initial response to your message. The direct pattern will start your readers off on the right foot when you have good news to convey: "You're hired," "I've figured out a solution to your problem," or something similar. By starting with the good news, you put your readers in a favorable frame of mind as they read the rest of your message.

The direct pattern also works well when you are offering an analysis or recommending a course of action that you expect your readers to view favorably—or at least objectively—from the start. Leah is about to write such a memo, in which she will recommend a new system for managing the warehouses for her employer, a company that manufactures hundreds of parts used to drill oil and gas wells. Leah has chosen to use the direct pattern shown in the left-hand column of Figure 5.5. This is an appropriate choice because her readers (upper management) have expressed dissatisfaction with the present warehousing system. Consequently, she can expect a favorable reaction to her initial announcement that she has designed a better system.

Use the direct pattern of organization when you expect a favorable response from your readers.

The direct pattern is less effective when you are conveying information your readers might view as bad, alarming, or threatening. Imagine, for example, that Leah's readers are the people who set up the present system and that they believe it is working well. If Leah begins her memo by recommending a new system, she might put her readers immediately on the defensive because they might feel she is criticizing their competence. Then she would have little hope of receiving an open and objective reading of her supporting information. By using an indirect pattern,

FIGURE 5.5 Comparison of Direct and Indirect Organizational Patterns for Organizing Leah's Memo

Direct Pattern	Indirect Pattern
I. Leah presents her recommended strategy.	I. Leah discusses the goals of the present system from the *reader's point of view*.
II. Leah explains why her way of warehousing is superior to the present way.	II. Leah discusses the ways in which the present system does and does not achieve the reader's goals.
III. Leah explains in detail how to implement her system.	III. Leah presents her recommended strategy for achieving those goals more effectively, focusing on the ways her recommendation can overcome the shortcomings of the present system.
	IV. Leah explains in detail how to implement her system.

The direct pattern presents the recommendation first.

The indirect pattern delays recommendation; it's for use where the reader may react unfavorably.

Consider using the indirect pattern when you believe your readers will respond negatively to your main point unless you prepare them for it beforehand.

however, Leah can *prepare* her readers for her recommendation by first getting them to agree that it might be possible to improve on the present system. The right-hand column of Figure 5.5 shows an indirect pattern she might use.

You may wonder why you shouldn't simply use the indirect pattern all the time. It presents the same information as the direct organization (plus some more), and it avoids the risk of inciting a negative reaction at the outset. The trouble is that this pattern frustrates the readers' desire to learn the main point first.

In sum, the choice between direct and indirect patterns of organization can greatly affect the persuasiveness of your communications. To choose, you need to follow the basic strategy suggested throughout this book: Think about your readers' moment-by-moment reactions to your message.

Create a Tight Fit Among the Parts of Your Communication

When you organize a communication, you can also strengthen its persuasiveness by ensuring that the parts fit together tightly. This advice applies particularly to longer communications, where the overall argument often consists of two or more subordinate arguments. The way to do this is to review side by side the claims made in the various parts of a communication.

To create a tight fit, be sure all parts of your communication work together to support your overall position.

For example, imagine that you are writing a proposal. In an early section, you describe a problem your proposed project will solve. Here, your persuasive points are that a problem exists and that the readers should view it as serious. In a later section, you describe the project you propose. Check to see whether this description tells how the project will address each aspect of the problem that you described. If it doesn't, either the discussion of the problem or the discussion of the project needs to be revised so the two match up. Similarly, your budget should include expenses that are clearly related to the project you describe, and your schedule should show when you will carry out each activity necessary to complete the project successfully.

Of course, the need for a tight fit applies not only to proposals but also to any communication whose various parts work together to affect your readers' attitudes. Whenever you write, think about ways to make the parts work harmoniously together in mutual support of your overall position.

Guideline 6 | **Build an effective relationship with your readers**

One of the most important factors influencing the success of a persuasive communication is how your readers feel about you. You will remember from Chapter 1 that you should think of the communications you prepare at work as interpersonal interactions. If your readers feel well disposed toward you, they are likely to consider your points openly and without bias. If they feel irritated, angry, or otherwise unfriendly toward you, they may immediately raise counterarguments to every point you present, making it extremely unlikely that you will elicit a favorable reaction, even if all your points are clear, valid, and substantiated. Good points rarely win the day in the face of bad feelings.

The following sections describe two ways you can present yourself to obtain a fair—or even a favorable—hearing from your readers at work: Present yourself as a credible person, and present yourself as a friend, not a foe.

Present Yourself as a Credible Person

Your credibility is your readers' beliefs about whether you are a good source for information and ideas. If people believe you are credible, they will be relatively open to what you say. They may even accept your judgments and recommendations without inquiring very deeply into your reasons for making them. If people do not find you credible, they may refuse to give you a fair hearing no matter how soundly you state your case.

The more credible you seem to be, the more open your readers will be to your message.

Because people's response to a message can depend so greatly on their views of the writer or speaker, Aristotle identified *ethos,* which corresponds roughly to credibility, as one of the three major elements of persuasion.

In a classic experiment, researchers H. C. Kelman and C. I. Hovland (1953) demonstrated how greatly your readers' perception of your credibility affect the way they respond to your message. They asked two groups to listen to a taped speech advocating lenient treatment of juvenile delinquents. Before playing their tape for one group, they identified the speaker as an ex-delinquent out on bail. They told another group that the speaker was a judge. Only 27 percent of those who heard the "delinquent's" talk responded favorably to it, while 73 percent of those who heard the "judge's" talk responded favorably.

Researchers have identified the factors that affect peoples' impression of someone else's credibility. In a summary of this research, Robert Bostrom (1981) identified five such factors: expertise, trustworthiness, group membership, dramatic appeal, and power. By using the following writing strategies, each associated with one of Bostrom's factors, you can build your credibility with your readers.

In Figure 5.6, the president of a biometric security company uses these strategies when requesting an opportunity to demonstrate a new security system.

Several factors that influence credibility

1. **Expertise.** Expertise is the knowledge and experience that readers believe you possess that is relevant to the topic. *Strategies you can use:*
   - Mention your credentials.
   - Demonstrate a command of the facts.
   - Avoid oversimplifying.
   - Mention or quote experts so their expertise supports your position.

2. **Trustworthiness.** Trustworthiness depends largely on your readers' perceptions of your motives. If you seem to be acting from self-interest, your credibility is low; if you seem to be acting objectively or for goals shared by your readers, your credibility is high. *Strategies you can use:*
   - Stress values and objectives that are important to your readers.
   - Avoid drawing attention to personal advantages to you.
   - Demonstrate knowledge of the concerns and perspectives of others.

3. **Group membership.** You will gain credibility if you are a member of the readers' own group or a group admired by your readers. *Strategies you can use:*
   - If you are associated with a group admired by your readers, allude to that relationship.
   - If you are addressing members of your own organization, affirm that relationship by showing that you share the group's objectives, methods, and values.
   - Use terms that are commonly employed in your organization.

4. **Dynamic appeal.** An energetic, enthusiastic person has much higher credibility than a passive, guarded one. *Strategies you can use:*
   - State your message confidently and directly.
   - Show enthusiasm for your ideas and subject.

5. **Power.** For example, simply by virtue of her position, a boss acquires some credibility with subordinates. *Strategies you can use:*
   - If you are in a position of authority, identify your position if your readers don't know it.
   - If you are not in a position of authority, associate yourself with a powerful person by quoting the person or saying that you consulted with him or her or were assigned the job by that individual.

Figure 5.7 (page 136), shows how the International Fund for Animal Welfare uses the strategies on its website to persuade people to participate in and support its efforts.

Present Yourself as a Friend, Not a Foe

Psychologist Carl Rogers (1952) identified another important strategy in fostering open, unbiased communication: Reduce the sense of threat that people often feel when others are presenting ideas to them. According to Rogers, people are likely to feel threatened even when you make *helpful* suggestions. As a result, your readers may see you as an adversary even though you don't intend at all to be one.

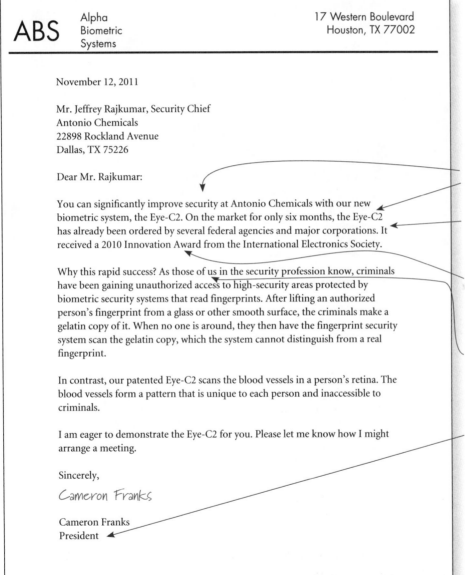

FIGURE 5.6
Letter that Uses Several Strategies for Building Credibility

In this letter, Cameron Franks uses several strategies for building credibility.

Trustworthiness
He aligns his goal with his readers' goal: to improve security.

He reports that his product has been ordered by organizations whose security knowledge is assumed to be great.

Expertise
He indicates that his product is so technically innovative that it has won an award and a patent (third paragraph).

Group membership
He uses the word "us" to identify himself as a member of the same group as his reader: security professionals.

Power
He indicates that he holds a powerful position in his own company.

The letter reads:

ABS — Alpha Biometric Systems

17 Western Boulevard
Houston, TX 77002

November 12, 2011

Mr. Jeffrey Rajkumar, Security Chief
Antonio Chemicals
22898 Rockland Avenue
Dallas, TX 75226

Dear Mr. Rajkumar:

You can significantly improve security at Antonio Chemicals with our new biometric system, the Eye-C2. On the market for only six months, the Eye-C2 has already been ordered by several federal agencies and major corporations. It received a 2010 Innovation Award from the International Electronics Society.

Why this rapid success? As those of us in the security profession know, criminals have been gaining unauthorized access to high-security areas protected by biometric security systems that read fingerprints. After lifting an authorized person's fingerprint from a glass or other smooth surface, the criminals make a gelatin copy of it. When no one is around, they then have the fingerprint security system scan the gelatin copy, which the system cannot distinguish from a real fingerprint.

In contrast, our patented Eye-C2 scans the blood vessels in a person's retina. The blood vessels form a pattern that is unique to each person and inaccessible to criminals.

I am eager to demonstrate the Eye-C2 for you. Please let me know how I might arrange a meeting.

Sincerely,

Cameron Franks

Cameron Franks
President

FIGURE 5.7
Website that Uses Several Strategies for Building Credibility

This website by the International Fund for Animal Welfare uses a variety of strategies for building credibility with readers it hopes to persuade to make donations and support the IFAW in other ways.

Trustworthiness. The row of flags not only enables readers from different countries to read the site in their own language, but also conveys the sense that the IFAW truly is "international."

Dynamic Appeal and Power. The flags also suggest that IFAW is so active and powerful that it is influential on several continents.

Trustworthiness. By showing two images that include both an animal and a human hand, the IFAW suggests that the donations they make will, in fact, be used for animals' direct care and protection.

Dynamic Appeal. By highlighting a rescue mission in its "Features" area, the IFAW indicates that it is involved in practical actions.

Expertise. The "Breaking News" section suggests that the IFAW is a knowledgeable and up-to-date organization.

Credibility and Dynamic Appeal. By announcing that its site has won one of the well-known Webby Awards, the IFAW provides an outside endorsement. By indicating that the award was for an "Activism" website, the IFAW builds its dynamic appeal.

Group Membership. By showing images of many animals, the IFAW indicates that it has the same commitment to animals and the environment that is shared by the readers most likely to provide support. (Note that this site does not attempt to persuade people that they should be commited to animals and the environment, but rather is designed to elicit support from people who already have this commitment.)

Here are four methods, based on the work of Rogers, that you can use to present yourself as nonthreatening to your readers.

To see how to apply this advice, consider the following situation. Marjorie Lakwurtz is a regional manager for a company that hires nearly 1,000 students in various cities during the summer months to paint houses. The company requires each student to pay a $250 deposit to cover the costs of lost equipment. Noting that the equipment isn't worth even $100, the students believe the company demands the deposits so it can invest the money and keep the interest for itself.

Marjorie agrees with the students and decides to write a memo suggesting that the company change its policy. She could argue for the change in a negative way by telling the company that it should be ashamed of its greedy plot to profit by investing the students' money. Or she could take a positive approach, presenting her suggestion as a way the company can better achieve its own goals. In the memo shown in Figure 5.8, she takes the latter course. Notice how skillfully she presents herself as her reader's partner.

Guideline 7 | Decide whether to appeal to your readers' emotions

Of Aristotle's three types of persuasive strategies, the appeal to emotions (*pathos*) is used least in workplace communications. In many communications, it makes no appearance at all. In fact, readers would usually consider appeals to emotion in

FIGURE 5.8
Memo in Which the
Writer Establishes a
Partnership with the
Reader

Marjorie writes her memo's
subject to align with her reader's
goal profit.

She opens by praising the
company.

Marjorie links the student
employees' attitudes with
Speed's goal of making a profit.

She uses "we" and "our"
throughout the memo to
reinforce the sense of
partnership with Speed.

Marjorie presents herself as
Speed's partner in the mutual
goal of keeping student
commitment (and
company profits) strong.

She shows that she understands
Speed's reason for the high deposit.

She states the problem in terms
of Speed's concern with profits.

She states a counterargument to
her suggestion.

She addresses the counterargument
by offering an alternative.

Marjorie closes on a positive note
that focuses on Speed's goal of
making a profit.

PREMIUM STUDENT PAINTERS, INC.

TO   Martin Speed

FROM   Marjorie Lakwurtz ML

DATE   October 7, 2011

RE   **Protecting Company Profits**

During my first year with Premium Student Painters, I have learned an enormous amount about what contributes to our company's incredible success. One key factor is our ability to hire 1000 college students who are willing to work so energetically and diligently all summer long on the house-painting jobs we arrange. In fact, the students' positive attitude is indispensable to our success. If they slack off, our profits drop. If they become careless, splattering and spilling paint, for instance, we lose profits on cleaning up and compensating customers.

Because the students' commitment is so crucial to our success, I have a suggestion for keeping it strong. Students in several cities have complained to me about the $250 equipment deposit we require. They believe the company requires the deposit so it can invest their money during the summer, keeping the profits for itself. Why else, they ask, would the deposit be so much larger than the value of the equipment we give them to use?

I realize that we have good reason for setting the deposit at $250, but I worry that the students' belief could reduce their commitment to us, thereby threatening our profits and customer relations. To avoid this risk, I suggest that we include the interest earned when we return the deposits at summer's end. It would be difficult, I know, to calculate the precise amount of interest earned by each student. All deposits are placed in one account and different students work different numbers of weeks. However, we could pay a flat amount to each student: for example, the interest one deposit would earn in three months.

Returning the interest would create some extra work for our office staff, but it could pay off handsomely in the students' productivity, which is the source of our profits.

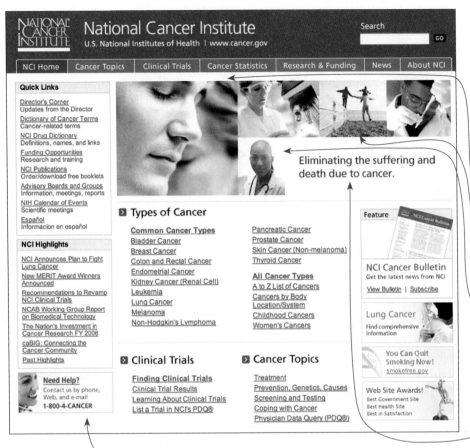

FIGURE 5.9
Websites that Appeal to the Reader's Emotions

On its home page, the National Cancer Institute (NCI) appeals to the emotions of cancer patients, their families, and friends.

In two ways, the site suggests the NCI understands these emotions and offers hope and relief.

1. The large photo portrays an individual overcome by emotions that many people feel when they visit the site. Another photo farther to the right shows three people on a beach, all apparently healthy and strong. Between these images are the NCI employees who will create this transformation: a researcher and a caregiver who looks directly at the reader.

2. The website identifies in words what many of its readers fear: "suffering and death due to cancer." It also promises to "eliminate" them.

Try This

This site uses several strategies for building credibility that are described on page 134. Can you find three?

In addition, the site also appeals to the readers' emotions by promising assistance to those who "need help." The NCI employees who will provide this help are represented by a smiling person who seems happy to hear from callers.

scientific research reports, test reports, and many other types of documents to be highly inappropriate.

In others, appeals to emotion are important. For example, documents on either side of public policy often include them. Organizations that advocate opening new sections of federal land for oil and mineral exploration sometimes include photographs of animals in natural settings to emphasize the measures to be taken to protect them. Organizations opposed to opening the lands sometimes include photographs of damage accompanying such exploration in other areas.

Emotional appeals are also used widely by government agencies and private health providers as they advocate healthy lifestyles. Nonprofit foundations that support research for conditions such as arthritis, breast cancer, and AIDS use emotional appeals liberally. See Figure 5.9.

To recruit college graduates in engineering, science, and other specialized fields, employers often include photographs of happy employees to emphasize the pleasures of employment with them.

Base your decisions about whether—and how strongly—to appeal to your readers' emotions on your knowledge of your readers and their likely response. Situations in which you are using emotional appeals are ones in which it can be especially helpful to ask others to review your drafts.

Learn More

For more on having other people review your drafts, turn to page 415.

Guideline 8 | Global Guideline: Adapt your persuasive strategies to your readers' cultural background

Whenever you are writing to readers in a culture other than your own, you will need to learn about the persuasive practices that succeed there. Like all other aspects of communication, every aspect of persuasion can differ from one culture to another.

Consider, for example, Guideline 2, which advises you to focus on your readers' goals and values. In different cultures, people's goals and values differ in ways that mean the benefits you highlight for readers in one culture are not necessarily those that would be appealing to readers in another culture. Marketing specialists are well aware of these differences. To appeal to car buyers, Volvo focuses on safety and durability in the United States, status and leisure in France, and performance in Germany (Ricks, 2000).

What people in your culture consider to be persuasive may not be persuasive in another culture.

Cultural differences in goals and values influence persuasion in the workplace as well as in consumer marketing. For example, in Japan, where group membership is highly valued, the most effective way to influence attitudes of individual employees is to talk about the success the proposed action will bring to the group. In the United States, where individual achievement is highly valued, appeals to personal goals can be effective in ways they would not be in Japan.

Similarly, different cultures have different views of what counts as evidence that a particular action should be taken. Whereas many Western cultures base their arguments on facts, international communication experts Farid Elashmawi and Philip R. Harris (1998) report that in some Arab cultures people support their positions through emotions and personal relations.

You must adapt to your readers' culture even when following Guideline 1: Listen— and respond flexibly to what you hear. Imagine that you have sent an e-mail to members of your target audience to find out how they feel about an idea you are planning to propose. If your readers are in the United States, they are likely to tell you directly. However, in many Asian cultures, they are not likely respond in that way because they would not want you to lose face (Lustig & Koester, 1993). In this case, you may need to listen to their silence, which is an indirect way of indicating disagreement in many Asian cultures.

Guideline 9 | Ethics Guideline: Employ ethical persuasive techniques

The ethical dimensions of on-the-job writing are never more evident than when you are trying to persuade other people to take a certain action or adopt a certain attitude. Here are four guidelines for ethical persuasion that you might want to keep in mind: Don't mislead, don't manipulate, open yourself to your readers' viewpoint, and argue from human values.

Don't Mislead

When you are writing persuasively, respect your readers' right to evaluate your arguments in an informed and independent way. If you mislead your readers by misstating facts, using intentionally ambiguous expressions, or arguing from false premises, you deprive your readers of their rights.

Don't Manipulate

The philosopher Immanuel Kant originated the enduring ethical principle that we should never use other people merely to get what we want. Whenever we try to influence our readers, the action we advocate should advance their goals as well as our own.

Under Kant's principle, for instance, it would be unethical to persuade readers to do something that would benefit us but harm them. High-pressure sales techniques are unethical because their purpose is to persuade consumers to purchase something they may not need or even want. Persuasion is ethical only if it will lead our readers to get something they truly desire.

Open Yourself to Your Readers' Viewpoint

To keep your readers' goals and interests in mind, you must be open to understanding their viewpoint. Instead of regarding their counterarguments as objections you must overcome, try to understand what lies behind their concerns. Consider ways of modifying your original ideas to take your readers' perspective into account.

In this way, rather than treat your readers as adversaries, accept them as your partners in a search for a course of action acceptable to you all (Lauer, 1994). Management experts call this the search for a "win–win" situation—a situation in which all parties benefit (Covey, 1989).

Argue from Human Values

Whenever you feel that human values are relevant, don't hesitate to introduce them when you are writing to persuade. Many organizations realize the need to consider these values when making decisions. In other organizations, human values are sometimes overlooked—or even considered to be inappropriate topics—because the employees focus too sharply on business objectives. However, even if your arguments based on human values do not prevail, you will have succeeded in introducing a consideration of these values into your working environment. Your action may even encourage others to follow your lead.

> **Try This**
>
> Have you ever felt manipulated by a salesperson or someone urging you to take a certain action? What was it about that individual's arguments that made you feel that you were being manipulated to do something you didn't want to do?

CONCLUSION

This chapter has focused on writing persuasively in both competitive and cooperative situations. As you can see, nearly every aspect of your communication affects your ability to influence your readers' attitudes and actions. Although the guidelines in this chapter will help you write persuasively, the most important persuasive strategy of all is to keep in mind your readers' needs, concerns, values, and preferences whenever you write.

Figure 5.10 provides a guide you can use when planning your persuasive strategies.

FIGURE 5.10   Writer's Guide for Planning Your Persuasive Strategies

To download a copy of this Writer's Guide, go to www.cengage.com/english/anderson7e and click on Chapter 5.

Writer's Guide
PLANNING YOUR PERSUASIVE STRATEGIES

Understand Your Readers' Goals   ⟵   Guideline 2

1. What organizational goals affect your readers and are related to your topic?

2. What are your readers' values-based goals that are related to your topic?

3. What are your readers' achievement and growth goals?

Respond to Your Readers' Concerns and Counterarguments   ⟵   Guideline 3

1. What concerns or counterarguments might your readers raise?

2. How can you respond persuasively to these concerns and counterarguments?

3. Where do you need to be careful about inspiring negative thoughts?

Reason Soundly   ⟵   Guideline 4

1. What kinds of evidence will your readers consider to be reliable and persuasive support for your claims?

2. Where will you need to justify your line of reasoning?

3. Where do you need to be cautious about avoiding false assumptions and overgeneralizations?

Organize to Create a Favorable Response   ⟵   Guideline 5

1. Would a direct or indirect organizational pattern be most effective?

2. What specific strategies can you use to create a tight fit among the parts?

Build a Relationship with Your Readers   ⟵   Guideline 6

1. What strategies can you use to present yourself as a credible person?

2. What strategies can you use to present yourself as a friend, not a foe?

Appeal to Your Readers' Emotions   ⟵   Guideline 7

1. Will your readers think that an appeal to their emotions is appropriate?

2. If so, what kind of emotional appeal would be effective in this situation with your readers?

Adapt to Your Readers' Cultural Background   ⟵   Guideline 8

1. Do you need to check your draft with people who share your readers' cultural background?

Persuade Ethically   ⟵   Guideline 9

1. Avoid misleading or manipulative statements.

2. Consider your readers' viewpoint openly.

3. Assure that the relevant human values are considered.

USE WHAT YOU'VE LEARNED

For additional exercises, go to **www.cengage.com/english/anderson7e.** *Instructors:* The book's website includes suggestions for teaching the exercises.

EXERCISE YOUR EXPERTISE

1. Find a persuasive communication that contains at least 25 words of prose. It may be an advertisement, marketing letter, memo from school or work, or a web page. What persuasive strategies are used in the prose and any images that are included? Are the strategies effective for the intended audience? Are all of them ethical? What other strategies might have been used? Present your responses in the way requested by your instructor.

2. Do the "Unsolicited Recommendation" assignment given in Appendix B.

EXPLORE ONLINE

Study the website of a company or organization related to your major or career. What attitudes toward the organization and its products or services does the site promote? What persuasive strategies are used in the text and images in order to persuade readers to adopt the desired attitudes? How effective are these strategies? Are all of them ethical?

COLLABORATE WITH YOUR CLASSMATES

Working with one or two other students, analyze the letter shown in Figure 5.11 on page 144, identifying its strengths and weaknesses. Relate your points to the guidelines in this chapter. The following paragraphs describe the situation in which the writer, Scott Houck, is writing.

Before going to college, Scott worked for a few years at Thompson Textiles. In his letter, he addresses Thompson's Executive Vice President, Georgiana Stroh. He is writing because in college he learned many things that made him think Thompson would benefit if its managers were better educated in modern management techniques. Thompson Textiles could enjoy these benefits, Scott believes, if it would offer management courses to its employees and if it would fill job openings at the managerial level with college graduates. However, if Thompson were to follow Scott's recommendations, it would have to change its practices considerably. Thompson has never offered courses for its employees and has long sought to keep payroll expenses low by employing people without a college education, even in management positions. (In a rare exception to this practice, the company has guaranteed Scott a position after he graduates.)

To attempt to change the company's policies, Scott decided to write a letter to one of the most influential people on its staff, Ms. Stroh. Unfortunately for Scott, throughout the three decades that Stroh has served as an executive officer at Thompson, she has consistently opposed company-sponsored education and the hiring of college graduates. Consequently, she has an especially strong motive for rejecting Scott's advice: She is likely to feel that, if she agreed that Thompson's educational and hiring policies should be changed, she would be admitting that she had been wrong all along.

APPLY YOUR ETHICS

Find an online or print communication you feel is unethical. Identify the specific elements in the text images that you feel are untrue, misleading, or manipulative. Present your analysis in the way your instructor requests.

FIGURE 5.11
Letter for Collaborate
with Your Classmates
Exercise

616 S. College #84
Oxford, Ohio 45056
April 16, 2010

Georgiana Stroh
Executive Vice President
Thompson Textiles Incorporated
1010 Note Ave.
Cincinnati, Ohio 45014

Dear Ms. Stroh:

As my junior year draws to a close, I am more and more eager to return to our company, where I can apply my new knowledge and skills. Since our recent talk about the increasingly stiff competition in the textile industry, I have thought quite a bit about what I can do to help Thompson continue to prosper. I have been going over some notes I have made on the subject, and I am struck by how many of the ideas stemmed directly from the courses I have taken here at Miami University.

Almost all of the notes featured suggestions or thoughts I simply didn't have the knowledge to consider before I went to college! Before I enrolled, I, like many people, presumed that operating a business required only a certain measure of commonsense ability—that almost anyone could learn to guide a business down the right path with a little experience. However, I have come to realize that this belief is far from the truth. It is true that many decisions are common sense, but decisions often only appear to be simple because the entire scope of the problem or the full ramifications of a particular alternative are not well understood. A path is always chosen, but often is it the BEST path for the company as a whole?

In retrospect, I appreciate the year I spent supervising the Eaton Avenue Plant because the experience has been an impetus to actually learn from my classes instead of just receiving grades. But I look back in embarassment upon some of the decisions I made and the methods I used then. I now see that my previous work in our factories and my military experience did not prepare me as well for that position as I thought they did. My mistakes were not so often a poor selection among known alternatives, but were more often sins of omission. For example, you may remember that we were constantly running low on packing cartons and that we sometimes ran completely out, causing the entire line to shut down. Now I know that instead of haphazardly placing orders for a different amount every time, we should have used a forecasting model to determine and establish a reorder point and a reorder quanity. But I was simply unaware of many of the sophisticated techniques available to me as a manager.

I respectfully submit that many of our supervisory personnel are in a similar situation. This is not to downplay the many contributions they have made to the company. Thompson can directly attribute its prominent position in the industry to the devotion and hard work of these people. But very few of them have more than a high school education or have read even a single text on management skills. We have always counted on our supervisors to pick up their

FIGURE 5.11
(continued)

Georgiana Stroh   Page 2

management skills on the job without any additional training. Although I recognize that I owe my own opportunities to this approach, this comes too close to the commonsense theory I mentioned earlier.

The success of Thompson depends on the abilities of our managers relative to the abilities of our competition. In the past, EVERY company used this commonsense approach, and Thompson prospered because of the natural talent of people like you. But in the last decade, many new managerial techniques have been developed that are too complex for the average employee to just "figure out" on his or her own. For example, people had been doing business for several thousand years before developing the Linear Programming Model for transportation and resource allocation problem-solving. It is not reasonable to expect a high school graduate to recognize that his or her particular distribution problem could be solved by a mathematical model and then to develop the LP from scratch. But as our world grows more complex, competition will stiffen as others take advantage of these innovations. I fear that what has worked in the past will not necessarily work in the future: We may find out what our managers DON'T know CAN hurt us. Our managers must be made aware of advances in computer technology, management theory, and operations innovations, and they must be able to use them to transform our business as changing market conditions demand.

I would like to suggest that you consider the value of investing in an in-house training program dealing with relevant topics to augment the practical experience our employees are gaining. In addition, when management or other fast-track administrative positions must be filled, it may be worth the investment to hire college graduates whose coursework has prepared them to use state-of-the-art techniques to help us remain competitive. Of course, these programs will initially show up on the bottom line as increased expenses, but it is reasonable to expect that, in the not-so-long run, profits will be boosted by newfound efficiencies. Most important, we must recognize the danger of adopting a wait-and-see attitude. Our competitors are now making this same decision; hesitation on our part may leave us playing catch-up.

In conclusion, I believe I will be a valuable asset to the company, in large part because of the education I am now receiving. I hope you agree that a higher education level in our employees is a cause worthy of our most sincere efforts. I will contact your office next week to find out if you are interested in meeting to discuss questions you may have or to review possible implementation strategies.

Sincerely,

Scott Houck

"Have you read your e-mail yet this afternoon?" Hal asks excitedly on the telephone. The two of you started working at Life Systems, Inc. in the same month and have become good friends even though you work in different departments. By now you've both risen to management positions.

"Not yet. What's up?"

"Wait 'til you hear this," Hal replies. "Tonti's sent us all a message announcing that she's thinking about starting a company drug-testing program." The person Hal calls "Tonti" is Maria Tonti, the company president, who has increased the company's profitability dramatically over the past five years by spurring employees to higher productivity while still giving them a feeling they are being treated fairly.

"A what?"

"Drug testing. Mandatory urine samples taken at unannounced times. She says that if she decides to go ahead with the program, any one of our employees could be tested, from the newest stockroom clerk to Tonti herself."

"You're kidding!" you exclaim as you swivel around and look out your window and across the shady lawn between your building and the highway.

"Nope. And she wants our opinion about whether she should start the program or not. In the e-mail she says she's asking all of her 'key managers' to submit their views in writing. She wants each of us to pick one side or the other and argue for it, so that she and her executive committee will know all the angles when they discuss the possibilities next week."

Hal pauses, waiting for your reaction. After a moment, you say, "Given the kinds of products we make, I suppose it was inevitable that Tonti would consider drug testing

sooner or later." Life Systems, Inc. manufactures highly specialized, very expensive medical machines, including some that keep patients alive during heart and lung transplant operations. The company has done extremely well in the past few years because of its advanced technology and—more important—its unsurpassed reputation for reliable machines.

You add, "Does she say what started her thinking about drug testing? Has there been some problem?"

"I guess not. In fact, she says she hasn't had a single report or any indication that any employee has used drugs. I wonder what kinds of responses she'll get."

You speculate, "I bet she gets the full range of opinions—all the way from total support for the idea to complete rejection of it."

Just then someone knocks at your door. Looking over your shoulder, you see it's Scotty with some papers you asked him to get.

"I've got to go, Hal. Talk to you later."

"Before you hang up," Hal says, "tell me what you're going to say in your reply to Tonti."

"I don't know. I guess I'll have to think it over."

YOUR ASSIGNMENT

Write a reply to Maria Tonti's e-mail message. Begin by deciding whether you want to argue for or against the drug-testing program. Then think of all the arguments you can to support your position. Next, think of the strongest counterarguments that might be raised against your position. In your memo, argue persuasively on behalf of your position, following the guidelines in Chapters 4 and 5. Remember to deal with the arguments on the other side.

CONDUCTING RESEARCH

6 | Gathering Reader-Centered Information

| DEFINING OBJECTIVES |
| PLANNING |
| CONDUCTING RESEARCH |
| DRAFTING |
| REVISING |

At work, good writing often depends on good research. Because of the knowledge and skills you are acquiring in college, your employer, client, or another person will ask you a difficult but important question. The answer you provide will influence his or her decisions and actions—but only if you present it in a usable and persuasive manner. In Chapters 4 and 5, you learned how to determine what will make a particular communication usable and persuasive to its specific readers. This chapter and the next tell how to conduct the research needed to create reader-centered communications that answer research questions.

Conducting this research will be one of the most enjoyable parts of your career. It will give you an opportunity to use the knowledge and skills you've developed in your major. Like a detective, you will use these abilities to search for clues that will help you answer your readers' question. This chapter will guide you through the information gathering that is often part of this detective work. Once you've gathered many facts and ideas, you become a puzzle solver who must figure out how the various bits of information you've assembled fit together, what they mean, and how that meaning can help your reader. The next chapter guides you through this process of analyzing the information you've assembled. Between the two chapters is a Writer's Reference Guide (see page 165) that will help you employ five widely used methods for gathering information.

SPECIAL CHARACTERISTICS OF ON-THE-JOB RESEARCH

The initial step in developing this expertise is understanding that research at work differs significantly from research in school. First, the *purposes* are very different. As explained in Chapter 1, your college instructors assign writing projects in order to advance your personal, intellectual, and professional development. Typically, your college research goal is to gain either a general overview or a comprehensive understanding of a topic that will be useful *to you* sometime *in the future.* In contrast, at work you will write for practical purposes, such as helping others—managers, coworkers, or clients—perform practical tasks and make good decisions immediately. Consequently, your research goal will be to develop ideas, information, and arguments that *your readers* will find to be valuable *right now.*

Second, in the workplace it is much more important to be able to *conduct research efficiently,* without taking extra time to travel down avoidable dead ends or study material that is irrelevant to your readers' current situation. You will need to produce your results quickly because your readers will have an immediate need for them and because you will also have many other responsibilities and tasks to complete.

Third, in the workplace some (but not all) of the *ethical principles* concerning research differ from those that apply in school.

WWW

To read additional information and access links related to this chapter's guidelines, go to **www.cengage.com/ english/anderson7e** and click on Chapter 6.

WHAT READERS WANT

In a reader-centered approach to research, you focus on providing your readers with what they need and want: information they can use confidently as the basis for their decisions and actions. In general, readers will need and want the information you gathered to have four qualities.

QUALITIES OF READER-CENTERED RESEARCH RESULTS

- **Complete.** Readers will want you to gather all the information they must have to make a good decision or take productive action.
- **Unbiased.** Readers will want you to evaluate the information you've gathered to assure that it is from unbiased sources or to allow you to take the bias into account when considering it.
- **Credible.** Readers will want your information to be from sources they consider to be credible. However, what counts as a credible source varies from one profession to another.
- **Divisible.** In many cases, readers will want your information to permit you and them to look at it in parts. Election polling provides an example. In addition to being able to say that a certain percentage of eligible voters favor one political candidate over another, the pollsters can also divide their overall data into subgroups that enable them to say what percentage of women favor that candidate, what percentage of people in a certain region, and so on.

GUIDING YOU THROUGH THE READER-CENTERED INFORMATION-GATHERING PROCESS

This chapter's first four guidelines will lead you through the three major activities of the research process, focusing on ways you can gather information that is complete, unbiased, credible, and divisible.

- Defining your research objectives (Guideline 1)
- Identifying potential sources for information you need (Guideline 2)
- Gathering information from those sources (Guidelines 3 and 4)

Research process

Three additional guidelines will help you conduct research efficiently and address important ethical and legal issues in workplace research.

To illustrate the reader-centered research process, this chapter follows the research activities of three engineering students—Anna, Chen, and Terry—who

Anna, Chen, and Terry
conducted research to
evaluate their college's
advising system.

were asked by their dean to evaluate and, if appropriate, recommend changes to the advising system the dean had instituted three years ago. In the old system, the dean's staff provided career and academic advising for students in all engineering departments. In the new system, each engineering faculty member advises a small group of students throughout their four years of study. The dean believed that department faculty would give better advice because they would know more than the dean's staff about their department's courses and about career opportunities for the department's graduates. Now, the dean has asked Anna, Chen, and Terry to help determine whether the new system was as effective as had been predicted.

Guideline 1 | Define your research objectives

You can streamline your research by defining in advance what you want to find. After all, you are not trying to dig up everything that is known about your subject. You are seeking only information and ideas needed to achieve your communication's objectives.

Base your research objectives
on your communication's
usability and persuasive
goals.

In fact, the most productive way to begin your research is by reviewing your communication's usability and persuasive goals (see Chapter 3). Your research objectives should be built squarely on them.

First, identify the information you need in order to write a communication your readers will find to be highly usable. Review the questions your readers will ask about your topic. The questions you can't answer immediately are the ones you need to research. Also review the kinds of answers your readers will want you to provide. For example, remind yourself whether they will want general information or specific details, introductory overviews or technical explanations. In addition, consider the point of view from which your readers will read your communication. Are they going to want it to meet their needs as engineers or accountants, consumers or producers? The answers to these questions will help you determine not only the kinds of information you must obtain but also where you should look for it.

Learn More

For detailed advice about identifying the information your readers need, see Chapter 3's Guidelines 2 and 4 (pages 71 and 76).

Second, identify the additional questions that will assist your readers even though your readers don't know to ask them. The list of these questions is likely to grow during your research as you learn more about your subject.

Third, identify the kinds of information and arguments that will make your communication most persuasive in your readers' eyes. Given your readers' goals, values, and preferences, what kinds of information and arguments are most likely to influence their attitudes and actions? For example, are they primarily interested in efficiency or profitability, safety or consumer acceptance? Consider also the types of evidence and kinds of sources your readers will find most compelling. For instance, are they more likely to be swayed by quantitative data or by testimonials from leaders in their field?

Although you should define your research objectives at the outset, be prepared to revise them as you proceed. Research is all about learning. One thing you may learn along the way is that you need to investigate something you hadn't originally thought important—or even thought about at all.

As they began their research on the engineering school's advising system, Anna, Chen, and Terry mapped out the general array of questions they knew their readers, the dean and the dean's advisers, would want them to answer: Are students getting all the advising they should have? Is the advising of high quality? What are the strengths of the system? What are its weaknesses? How could it be improved? Is it working well enough to be kept or should it be replaced?

Initial questions identified by Anna, Chen, and Terry

Guideline 2 | Identify the full range of sources that may have helpful information

After you've defined your research objectives, pause to identify the full range of sources you should investigate. This step in research is too often skipped by researchers. Many concentrate so exclusively on the first information source they think of that they neglect other sources. To provide your readers with a complete and reliable answer to their questions, you must take into account the variety of perspectives that people and organizations have on the topic. You are very unlikely to find that variety in the first—or first few—sources you consult.

How can you locate all of these perspectives? The answer varies from one situation to another. A method that is often helpful is to consider gathering information from four types of sources.

- Persons affected
- Persons involved
- Other organizations or groups engaged in similar efforts
- Professional publications

Sources that often have helpful information

When Anna, Chen, and Terry first thought about their research for the dean, they planned to gather information only from engineering students (persons affected). However, after the dean suggested that they also think about the other three types of sources, the student team realized that the faculty (persons involved) might have helpful insights to share and that similar engineering programs (other organizations) might provide information about the effectiveness of similar advising systems. They also wondered whether schools that had developed especially effective advising efforts might have published articles about them in the many journals on engineering education (professional publications).

Anna, Chen, and Terry used the list of the four types of sources to identify valuable sources they would otherwise have missed.

Identifying sources in all four categories can help you conduct thorough research on many topics, including ones that are much more technical than the effectiveness of a student advising system. Imagine, for instance, that your employer has asked you to investigate ways to make some of its processes "greener"—more environmentally friendly. You could gain ideas from employees who perform the processes, other organizations that have similar environmental goals, and publications for specialists in your field.

After you've identified the types of sources you will consult, gather information broadly from each of them. Look widely enough to uncover disagreements and

Look widely enough to uncover disagreements and controversies or to feel certain none exist.

controversies or to feel certain none exist. Different people and organizations can have very conflicting views on many topics. If you learn the views of only a few, you risk presenting your readers with incomplete information. For example, in research on greener processes your employer might use, you would probably find contrasting views on some of them. Perhaps a manufacturing process that removes harmful chemicals from solid waste products will discharge warm water into a river, damaging the ecological system for miles downstream. Only by looking broadly can you present your readers with a full answer to the questions they've asked you.

Try This

Can you name two other groups of students Anna, Chen, and Terry should have included in their survey to assure that they gather the full range of student views on the advising system?

Anna, Chen, and Terry followed this advice when planning a survey they would give to engineering students. To discover the full range of views on the advising system, they decided to distribute their survey to students in each of the engineering majors, to students in each of the four years of the program (first-year, sophomore, and so on), and to both men and women. Similarly, they decided to gather information from faculty in all the engineering departments, not just the one in which they were studying.

Gathering a breadth of information has another, critical benefit: It helps you present your readers with information they can trust. No matter how alert you may be to the biases of the people, organizations, and publications you consult, the biases of some sources can be nearly impossible to detect unless you compare their information with information from other sources. Even when biases don't come into play, gathering information broadly can help you determine which sources provide the information that is most accurate, relevant, up to date, and consistent with the evidence. This is the information you want to give your readers when answering their questions.

The biases of some sources are almost impossible to detect unless you compare their information with information from other sources.

Guideline 3 | Gather information that can be analyzed in subgroups

In your career, you will often conduct research whose goal is to help your employer, client, or customer address a complex decision that involves many components or subparts. The new advising system that the engineering dean has asked Anna, Chen, and Terry to evaluate is an example. This system has two separate *goals* (advising on courses and on careers), uses a *process* that involves several separate steps (preparing faculty to give advice, getting students to know who their advisers are, getting them to meet with their advisers, and so forth), and serves several separate subgroups of *people* (students majoring in each department; men and women; first-year students, second-year students, and so on). Because the dean wanted all students to receive excellent advising about both courses and careers, he wanted Anna, Chen, and Terry's report to do much more than tell him that "Yes, the new advising system works" or "No, it doesn't." Like the employers, clients, and customers to whom you will write in your career, he desired a much more fine-grained analysis, one that would pinpoint ways to improve the system.

As mentioned previously, Chapter 7 will teach you how to analyze information, including subgroup information. However, you can't analyze subgroup information unless you have first gathered it, and gathering it often requires special planning. For example, Anna, Chen, and Terry decided to include questions about each student's major, year in college, and sex. This information would enable them to determine whether the advising system was working better for some subgroups of students than for others—for instance, for some majors more than others and for one sex more than the other. If they discovered such differences, Anna, Chen, and Terry would be able to recommend targeted actions focused on improving advising for subgroups that needed it most. In fact, as you will learn in Chapter 7, some differences did turn up. Anna, Chen, and Terry were able to detect them only because they recognized the value of gathering information that can be analyzed in subgroups.

Learn More

To see what Anna, Chen, and Terry learned by analyzing their subgroup data, go to Guideline 4 in Chapter 7.

A focus on gathering information that can be analyzed in subgroups can be equally important with other kinds of research. When reading articles about advising in professional journals, Anna, Chen, and Terry used their list of subgroups of information to help them spot and take notes on relevant pieces of information they might otherwise miss. The list also helped them ask helpful questions when they interviewed people at other engineering schools about their advising systems.

For each subgroup you identify, collect enough information to be reasonably sure that you can make generalizations about it and that you have tapped the major views regarding it. For example, if Anna, Chen, and Terry had obtained 100 completed surveys, they would probably have had enough to make generalizations about the advising system overall. However, if that 100 included only 4 from chemical engineering majors, they would probably have needed to gather more surveys from chemical engineering students to be able to conduct a meaningful analysis of responses from that subgroup.

For each subgroup, collect enough information to be able to make generalizations about it.

The three categories of subgroups that Anna, Chen, and Terry used (goals, processes, and people) work in many situations. In others, you will be able to identify subgroups of information that will be much more helpful to your readers. The important point is that, as you plan your research, you determine the kinds of analyses you would like to perform so that you can collect the information you will need later.

Guideline 4 | Create an efficient and productive research plan

Many people conduct research haphazardly. They dash off to the library or log onto the Internet in the hopes of quickly finding just the right book or website. If the first source fails, they scoot off to another one. Such an approach can waste time and cause you to miss very helpful sources, including (perhaps) the one that includes exactly the information you need.

You will conduct your research most efficiently and productively if you begin by making a plan. Guidelines 1 through 3 explained how to define what you are looking for. The following strategies can increase the efficiency and productivity of your search.

- **Consult general sources first.** By gaining a general view of your subject, you increase the ease with which you can locate, comprehend, and interpret the more detailed facts you are seeking. Useful general sources include encyclopedias, review articles that summarize research on a particular subject, and articles in popular magazines.

- **Focus on sources that will be most persuasive to your readers.** For instance, scientists are more likely to be persuaded by information you've gathered from an article in a research journal than by information from Internet sites. Scientists reason that research journals assure that each article is scientifically sound before it is published. No such evaluation exists for most items on the Internet.

- **Conduct preliminary research when appropriate.** For example, before interviewing a technical specialist or upper-level manager, conduct the background research that will enable you to focus the interview exclusively on facts this person alone can supply. Similarly, before conducting a survey, determine what other surveys have learned and study the techniques they used.

- **Make a schedule.** Establish a deadline for completing all your research that leaves adequate time for you to draft and revise your communication. Then set dates for finishing the subparts of your research, remembering to complete general and preliminary research before proceeding with other sources and methods.

- **Study the research methods you are going to use.** In your schedule, include time to study research methods you haven't used before. Also, provide time to learn advanced techniques of methods, such as searching the Internet, whose basics you already know. The Writer's Reference Guide that follows this chapter provides detailed advice for skillfully using five research methods that are very helpful on the job.

Expertise in using research sources and methods is as important as expertise in any other writing activity.

Guideline 5 | Carefully evaluate what you find

Your readers will depend on you to provide information they can trust as they make decisions and take action. To meet their needs, you must evaluate carefully every one of your sources. Six questions will help you do that.

- Is it **relevant** to my readers' needs?
- Will it be **credible** in my readers' eyes?
- Is it **accurate**?
- Is it **complete**?
- Is it **current** and up to date?
- Is it **unbiased**?

To answer the questions about relevance and credibility, refer to the imaginary portrait you created of your readers when defining your communication's objectives. Ask yourself, "Can I imagine my readers using this information as they perform the job my research is supposed to help them perform?" If you can't, the source is irrelevant. Your portrait of your readers will also help you learn whether the source will be credible to them. As you learned in Chapter 4, what people in one profession consider credible may lack credibility in another.

To review the process for creating an imaginary portrait of your reader, see Chapter 3's Guideline 1 (page 68).

The questions about accuracy, completeness, and currency can be difficult when you are researching a subject that is new to you. Follow Guideline 2's advice: Gather information from lots of sources and compare them with one another. Contradictions and inconsistencies may indicate that you should conduct additional research to learn which source to trust. Although information from old publications or other sources can still be up to date, it's always good to check some recent ones as well.

As Guideline 2 points out, comparing sources can also help you detect bias. When your sources disagree, ask whether one source, such as a person or company, will benefit if people believe the information it provides. The source's information may still be valid, but try to discern whether the source has ignored unfavorable information. Also, review the reasonableness of its interpretations of evidence. Even more suspicious would be the absence of evidence to support the source's claims. The language used can also be a tip-off. Strong positive or negative language may indicate that the source is relying on emotions rather than facts. Note, however, that bias does not necessarily mean that a source should be dismissed. Sometimes your readers will want to know conflicting views. Chapter 7 discusses ways to analyze information in such situations.

Detecting bias

When evaluating sources, be as cautious about your own biases as you are about any biases your sources may possess. Don't dismiss a source simply because it contradicts your views or presents data that fail to support your conclusions. Your readers depend on your thoroughness and integrity.

Guideline 6 | Take careful notes

A simple but critical technique for conducting productive, efficient research is to take careful notes every step of the way. When recording the facts and opinions you discover, be sure to distinguish quotations from paraphrases so you can properly identify quoted statements in your communication. Also, clearly differentiate between ideas you obtain from your sources and your own ideas in response to what you find there.

In addition, make careful bibliographic notes about your sources. Include all the details you will need when documenting your sources (see Guideline 7). For books and articles, record the following details.

BOOKS	ARTICLES
Author's or editor's full name	Author's or editor's full name
Exact title	Exact title
City of publication	Journal title
Year of publication	Volume (and issue unless pages are numbered consecutively throughout the volume)
Edition	Year of publication
Page numbers	Page numbers

For interviews, record the person's full name (verify the spelling!), title, and employer, if different from your own. Special considerations apply when your sources are on the Internet; they are described in Appendix A.

It is equally important for you to record the information you will need if you later find that you need to consult this source again. For instance, when you are working in a library, jot down the call number of each book; when interviewing someone, get the person's phone number or e-mail address; and when using a site on the World Wide Web, copy the universal resource locator (URL).

As you proceed, be sure to keep a list of sources that you checked but found useless. Otherwise, you may find a later reference to the same sources but be unable to remember that you have already examined them.

Guideline 7 | Ethics Guideline: Observe intellectual property law and document your sources

Once you've completed your research and decided which resources you would like to use in your communication, you have two questions to answer:

- Do I need permission to use this material?
- Do I need to document this source in my communication?

The answers to these questions overlap, but they are not identical. To understand their relationship, you need to consider the laws concerning intellectual property as well as the ethical guidelines for acknowledging sources.

Intellectual Property Law

Broadly speaking, intellectual property law includes the following areas:

Three areas of intellectual property law

- **Patent law.** Governs such things as inventions and novel manufacturing processes.
- **Trademark law.** Pertains to such things as company and product names (Microsoft, Pentium), slogans ("We bring good things to life"), and symbols (the Nike "swoosh").

- **Copyright law.** Deals with such things as written works, images, performances, and computer software.

When you are writing at work, copyright law will probably be the most important to you. Copyright law was created to encourage creativity while also providing the public with an abundant source of information and ideas. To achieve these goals, copyright law enables the creators of a work to profit from it while also allowing others to use the work in limited ways without cost.

Any communication, such as a report, letter, e-mail, photograph, or diagram, is copyrighted as soon as it is created. If the creator generated the work on his or her own, that individual owns the copyright to it. If the creator made the work while employed by someone else, the copyright probably belongs to the employer. Whether the copyright owner is an individual or an organization, the owner has the legal right to prohibit others from copying the work, distributing it, displaying it in a public forum, or creating a derivative work based on it. When copyright owners grant others permission to do any of these things, they may charge a fee or make other contractual demands. The copyright owner has these rights even if the work does not include the copyright notation or the copyright symbol: ©.

The copyright law does, however, place limits on the copyright owner's rights. First, copyright expires after a certain number of years, which varies depending on the date of publication. Second, the law provides that other people, including you, may legally quote or reproduce parts of someone else's work without their permission if your use is consistent with the legal doctrine of *fair use*. Whether your use is "fair" depends primarily on the following four factors (Stanford University, 2005):

- **Purpose of the use.** The law provides more liberal use for educational purposes than for commercial ones. Thus, copyright restrictions are generally stricter at work than in school.
- **Proportion of the work used.** Your quotation of a few hundred words from a long book is likely to be considered fair use, but quotation of the same number of words from a short pamphlet may not.
- **Publication status.** The law gives greater protection to works that a copyright owner has not published or distributed than to ones the owner has.
- **Economic impact.** If your use will diminish the creator's profits, it is unlikely that the law would consider it to be fair use.

WWW

For more information about fair use, go to **www.cengage.com/english/anderson7e.**

It is also legal for you to use other people's work without their permission if the work is in the *public domain.* Such works include those created by or for the U.S. government and similar entities and works whose copyright has expired. Also, private individuals and organizations sometimes put their work in the public domain. The owners of websites that offer free use of clip art are an example.

Finally, you can generally use work that other people working for your employer created as part of their job responsibilities. In fact, in the workplace it is very common for employees to use substantial parts of communications created by other

WWW

To learn more about the public domain, go to **www.cengage.com/english/anderson7e** and click on Chapter 6.

employees. For instance, when you are creating the final report on a project, you may incorporate portions of the proposal written to obtain the original authorization for the project as well as parts of progress reports written during the project.

Copyright law permits people and organizations to share their work by relinquishing some of their rights without giving up ownership, an increasingly popular practice. For example, many professional journals allow faculty and students to print copies of their articles as long as this is done for educational, not commercial, purposes. A nonprofit organization called Creative Commons supports these efforts in many ways.

The following guidelines on copyright coverage will help you observe intellectual property law.

Learn More

For details about ways copyright owners can share their work, go to **www.cengage.com/english/anderson7e** and click on Chapter 6.

OBSERVING COPYRIGHT

TEXT

Ask for permission except in the following circumstances:

- You created the source yourself.
- Someone else at your employer's created it.
- The source is in the public domain.
- The copyright owner explicitly includes a statement with the source that it may be used without permission.
- You are using the text for a course project that will not be published on paper or on the web.

  Honor fair use restrictions. Don't use larger amounts of someone else's work than fair use allows.

  When in doubt, ask someone. Intellectual property laws are complicated. If you are uncertain about what to do, consult your instructor, your employer's legal department, or the resources available at the website for this book, **www.cengage.com/english/anderson7e**.

GRAPHICS

Note: Each graphic is separately copyrighted. Consequently, the principle of "fair use" does not apply. You always need permission to use a graphic, even if it is only a small part of a larger work.

Obtain permission for all graphics unless

- You created the graphic.
- Your employer is the copyright holder.
- The graphic is in the public domain.
- You are using the graphic for a course project that will not be published on paper or on the web.

Note: Many people mistakenly believe that anything on the web is in the public domain. Actually, all web content is copyrighted by its creator.

Get permission to use web material unless

- The site belongs to your employer.
- The material you are going to use from the site is in the public domain.
- The site declares that its contents are available for free use.
- You are using the material for a course project that will not be published on paper or on the web.

Ethical Guidelines for Documenting Sources

On the job, as in college, you have an ethical obligation to credit the sources of your ideas and information by citing them in a reference list, footnotes, or bibliography. Failure to do so is considered plagiarism. However, standards for deciding exactly which sources need to be listed at work differ considerably from the standards that apply at school. By asking the questions listed below, you can usually determine whether you need to credit sources when writing on the job. Note, however, that ethical standards for citing sources differ from culture to culture. The questions below apply in the United States, Canada, and Europe. If you are working in another part of the world, ask your coworkers to help you understand the standards that apply where you are.

DETERMINING WHETHER YOU NEED TO DOCUMENT A SOURCE AT WORK

- **Did you obtain permission from the copyright owner?** If you obtained the copyright owner's permission, you must document the source.
- **Is the information I obtained from this source common knowledge?** Both in college and at work, you must indicate the source of ideas and information that (1) you have derived from someone else, and (2) are not common knowledge.

  However, what's considered common knowledge at work is different from what's considered common knowledge at school. At school, it's knowledge every person possesses without doing any special reading. Thus, you must document any material you find in print.

  At work, however, common knowledge is knowledge that is possessed by or readily available to people in your field. Thus, you do not need to acknowledge material you obtained through your college courses, your textbooks, standard reference works in your field, or similar sources.

Learn More

You will find information about how to write bibliographic citations in Appendix A.

- **Does my employer own it?** As explained above, employers own the writing done at work by their employees. Consequently, it is usually considered perfectly ethical to incorporate information from one proposal or report into another without acknowledging the source.

- **Am I taking credit for someone else's work?** On the other hand, you must be careful to avoid taking credit for ideas that aren't your own. In one case, an engineer was fired for unethical conduct because he pretended that he had devised a solution to a technical problem when he had actually copied the solution from a published article.

- **Am I writing for a research journal?** In articles to be published in scientific or scholarly journals, ethical standards for documentation are far more stringent than they are for on-the-job reports and proposals. In such articles, thorough documentation is required even for ideas based on a single sentence in another source. Thus, you must document any information you find in print or online. In research labs where employees customarily publish their results in scientific or scholarly journals, even information drawn from internal communications may need to be thoroughly documented.

- **Whom can I ask for advice?** Because expectations about documentation can vary from company to company and from situation to situation, the surest way to identify your ethical obligations is to determine what your readers and employer expect. Consult your boss and coworkers and examine communications similar to the one you are preparing. For clarification about what sources you need to document for your class, consult your instructor.

CONCLUSION

This chapter's seven reader-centered guidelines apply to all your research efforts regardless of the research method you employ. They are summarized in Figure 6.1. Following this chapter is a Writer's Reference Guide that provides additional advice for using five research techniques that are frequently employed in the workplace.

FIGURE 6.1 **Writer's Guide for Conducting Reader-Centered Research**

To download a copy of this Writer's Guide, go to at www.cengage.com/english/anderson7e and click on Chapter 6.

Writer's Guide
CONDUCTING READER-CENTERED RESEARCH

Review your communication's objectives.

1. Its usability objectives

2. Its persuasive objectives

Create an efficient plan for research.

1. Identify the full range of sources that might be helpful.

2. Gather broad, credible information from each source.

3. Gather information that can be analyzed in subgroups.

4. Consult general sources first.

5. Conduct preliminary research when appropriate.

6. Master each of the research methods you are going to use.

7. Check each source for leads to other sources.

8. Evaluate each source for its relevance to your readers' needs, credibility in your readers' eyes, accuracy, completeness, currency, and freedom from bias.

Take careful notes.

1. Facts and ideas your readers will find useful or persuasive

2. Details about your sources

Obtain permission from copyright holders, if necessary.

Document all sources you need to document in order to write ethically.

USE WHAT YOU'VE LEARNED

For additional exercises, go to www.cengage.com /english / anderson7e. *Instructors*: The book's website includes suggestions for teaching the exercises.

EXERCISE YOUR EXPERTISE

1. Choose a concept, process, or procedure that is important in your field. Imagine that one of your instructors has asked you to explain it to first-year students in your major. (See the Writer's Reference Guide that follows this chapter for guidelines for using each of the following research methods.)

    a. Use brainstorming or freewriting to generate a list of things you might say in your talk.
    b. Use a flowchart, matrix, cluster sketch, or table to generate a list of things you might say.
    c. Compare your two lists. What inferences can you draw about the strengths and limitations of each technique?

2. Imagine that a friend wants to purchase some item about which you are knowledgeable (for example, a motor-cycle, MP3 player, or sewing machine). The friend has asked your advice about which brand to buy. Design a matrix in which you list two or three brands and also at least six criteria you recommend your friend use to compare them. Fill in the matrix as completely as you can. Each box you can't fill indicates an area you must research. Describe the methods you would use to gather the additional information. (See the Writer's Reference Guide that follows this chapter for advice about using a matrix as a research tool.)

3. Imagine that you have been asked by the chair of your major department to study student satisfaction with its course offerings. Devise a set of six or more closed questions and four open-ended questions you could use in interviews or a survey. (For information about interview and survey questions, see pages 186–187 and 189–191 in the Writer's Reference Guide that follows this chapter.)

4. Create a research plan for a project you are preparing for your technical communication course.

EXPLORE ONLINE

1. Use two search engines and an Internet directory to look for websites on a topic related to your major. How many hits does each produce? Compare the first ten results from each search in terms of the quality of the sites and the amount and kind of information the search engine or Internet directory provides about each one. (For information about using search engines and Internet directories, see pages 172–175 in the Writer's Reference Guide that follows this chapter.)

2. Using a search engine and online library resources, identify three websites, two books, and two articles on a topic related to your field. Which would you find most interesting? Which would be most helpful if you were writing a paper on the topic for a class? Which would be most helpful if you were writing a report on the topic for your employer? (For information about using search engines and online library resources, see pages 156–175 in the Writer's Guide that follows this chapter.)

COLLABORATE WITH YOUR CLASSMATES

Working with another student, choose a topic that interests you both. Find five websites that provide substantial information on your topic. Which sites are most appealing to you initially? Following the advice on pages 172–183, evaluate each site, and then compare the results with your initial impression of it.

APPLY YOUR ETHICS

Create a bibliography of sources concerning an ethical issue related to your major or career. Include four websites, one book, and two journal articles that you believe would help you understand various approaches to this issue.

USING FIVE READER-CENTERED RESEARCH METHODS

CONTENTS

WWW

For additional information related to this reference guide, go to the section for this guide at **www.cengage.com/english/anderson7e.**

This Guide explains how to use five research methods commonly employed on the job. Often, two or more of the methods are used in combination. Each method is discussed separately to create a reference source that gives you quick access to advice about the particular method you need to use.

No matter which method you use, follow Chapter 6's guidelines for conducting reader-centered research. They will help you work efficiently and produce information that establishes your credibility, meets your readers' needs, and persuades them to take the action you advocate.

In addition, keep Chapter 6's discussion of intellectual property, copyright, and plagiarism in mind (page 158). If you have any questions about how these guidelines apply to a project, ask your instructor or boss for advice.

EXPLORING YOUR OWN MEMORY AND CREATIVITY

Almost always, your best research aids will include your own memory and creativity. The following sections discuss four ways you can exploit the power of these mental resources. Each can be useful at the beginning of your research and at many points along the way:

- Brainstorm.
- Freewrite.
- Draw a picture of your topic.
- Create and study a table or graph of your data.

Brainstorm

When you brainstorm, you generate thoughts about your subject as rapidly as you can, through the spontaneous association of ideas, writing down whatever thoughts occur to you.

Brainstorming lets your thoughts run free.

Brainstorming's power arises from the way it unleashes your natural creativity. By freeing you from the confines imposed by outlines or other highly structured ways of organizing your ideas, brainstorming lets you follow your own creative lines of thought.

Brainstorming is especially effective at helping you focus on the core ideas you want to communicate to your readers, whether in an overall message or in one part of a longer message. It also works well in group writing projects: When the members brainstorm aloud, the ideas offered by one person often spark ideas for the others.

The key to brainstorming is to record ideas quickly without evaluating any of the thoughts that come to mind. Record everything. If you shift your task from generating ideas to evaluating them, you will disrupt the free flow of associations

BRAINSTORMING PROCEDURE

1. Review your knowledge of your readers and your communication situation.

2. Ask yourself, "What do I know about my subject that might help me achieve my communication purpose?"

3. As ideas come, write them down as fast as you can, using single words or short phrases. As soon as you list one idea, move on to the next thought that comes to you.

4. When your stream of ideas runs dry, read back through your list to see whether your previous entries suggest any new ideas.

5. When you no longer have any new ideas, gather related items in your list into groups to see whether this activity inspires new thoughts.

© Masterfile Royalty Free

on which brainstorming thrives. Similarly, when brainstorming in a group, each person should concentrate on adding new ideas without pausing to evaluate anyone else's contribution.

Here's the first part of a brainstorming session by Nicole. She wanted to write to her boss about ways to improve the quality control procedures at her company, which makes machines that keep patients alive during organ transplant operations. Sitting at her computer, Nicole began by simply typing out her initial thoughts on her subject as they came to her.

Writing teams at work often brainstorm together to plan their communications.

While she was brainstorming, Nicole sometimes moved from one idea to a related one.

Ideas for Quality Control Recommendations

Problem: Present system is unreliable
Everyone is "supposed" to be responsible for quality ◄
No one has specific responsibilities ◄
People's lives are at stake ◄
Near-fatal failure last year in Tucson ◄
Need procedures to test critical components when they arrive from supplier
Workers follow their own shortcuts, using personal assembly techniques
Don't realize harm they could do ◄
Product's overall record of performance is excellent ◄

At other times, she jumped to a completely unrelated thought.

After running out of ideas, Nicole grouped related items under common headings. Note how organizing her notes around her three main topics spurred her to develop additional thoughts.

Ideas for Quality Control Recommendations

Importance of Quality Control
    Lives depend on it (near-fatal failure in Tucson)

New ideas suggested by outlining ⟶
    Avoid product liability lawsuits
    Keep sales up

Present System
    Overall record is excellent but could easily fail
    Everyone is supposed to be responsible for quality, but no one is
    Workers follow their own shortcuts and personal assembly techniques

New idea ⟶
    Only consistent test is after the machine is built (some parts may cause the machine to fail due to flaws that are not detectable in the whole-machine test)

Strategies for a Better System
    Test critical parts as they are delivered from the supplier

New idea ⟶
    Insist that standard assembly procedures be followed

New idea ⟶
    Assign specific responsibilities—and rewards

Freewrite

Freewriting is much like brainstorming. Here, too, you tap your natural creativity, free from the confines of structured thought. You rapidly record your ideas as they pop into your mind. Only this time, you write prose rather than a list. The goal is to keep your ideas flowing.

FREEWRITING PROCEDURE

1. Review your knowledge of your readers and your communication situation.
2. Ask yourself, "What do I know about my subject that will help me achieve my communication purpose?"
3. As ideas come, write them down as sentences. Follow each line of thought until you come to the end of it, then immediately pick up the next line of thought that suggests itself.

4. Write rapidly without making corrections or refining your prose. If you think of a better way to say something, start the sentence anew.

5. Don't stop for gaps in your knowledge. If you discover that you need some information you don't possess, note that fact, then keep on writing.

6. When you finally do run out of ideas, read back through your material to select the ideas worth telling your readers. The rest you throw away.

Freewriting is especially helpful when you are trying to develop your main points. Use it when writing brief communications or parts of a long communication.

Below is a sample freewriting done by Miguel, an employee of a company that makes precision instruments. He had spent two weeks investigating technologies to be placed aboard airplanes for detecting microbursts and wind shear, two dangerous atmospheric conditions that have caused several crashes. Miguel wrote this freewriting draft when deciding what to say in the opening paragraph of his report.

Wind shear and microbursts blamed for several recent airline crashes (find out which ones—Dallas?). How detect these conditions? Then pilots can fly around them. Technology could be used on the ground or in planes. Onboard would be more helpful to pilots. Equipment would need approval by the Federal Aviation Administration. Would be high demand for onboard devices from airlines. Many companies are working on them. Lots of profit if we develop right instrument first. Need to pick most promising technology and develop it.

In Miguel's freewriting, ideas are jumbled together, often in incomplete sentences, but they are now recorded so he can sort through and organize them.

The following passage shows the paragraph Miguel wrote after freewriting. Note that it further develops some ideas from his freewriting, and it omits others. It is a fresh start—but one built on the ideas Miguel generated while freewriting.

Freewriting produces ideas, not a draft.

We have a substantial opportunity to develop and successfully market instruments that can be placed aboard airplanes to detect dangerous wind conditions called wind shear and microbursts. These conditions have been blamed for several recent air crashes, including one of a Lockheed L-1011 that killed 133 people. Because of the increasing awareness of the danger of these wind conditions, airlines are eager for detection equipment. We could make a large profit by identifying the most promising technology and then developing it rapidly. In this report, I evaluate the four major technologies.

Through freewriting, Miguel realized that profitability was a key point and so placed it first in his subsequent draft.

His ideas are now organized and clearly developed.

Draw a Picture of Your Topic

Another effective strategy for exploiting your memory and creativity is to explore your topic visually. Here are four kinds of diagrams that writers at work have found to be useful.

Flowchart

When you are writing about a process or procedure, try drawing a flowchart of it. Leave lots of space around each box in the flowchart so you can write notes next to it. Here is a flowchart that Nicole used to generate ideas for a report recommending improved quality control procedures in the manufacture and delivery of the medical equipment sold by her employer.

Above the flowchart, Nicole wrote the ideas that occurred to her as she studied the chart.

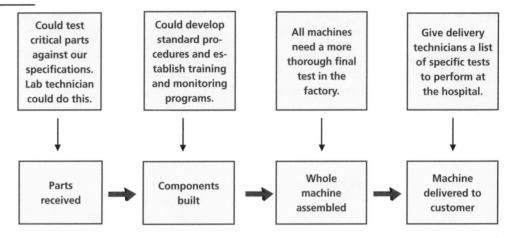

Matrix

A matrix is a table used to generate and organize ideas.

When you are comparing two or more alternatives in terms of a common set of criteria, drawing a matrix can aid you in systematically identifying the key features of each item being compared. Make a table in which you list the alternatives down the left-hand side and write the topics or issues to be covered across the top. Then, fill in each cell in the resulting table by brainstorming. Blank boxes indicate information

FIGURE WG1.1
Matrix Used by Miguel to Develop Ideas

	How It Works	Limitations	Potential Competition
Doppler radar	Detects rapidly rotating air masses, like those found in wind shear	Technology still being researched	General Dynamics Hughes
Infrared detector	Detects slight increases in temperature that often accompany wind shear	Temperature doesn't always rise	None—Federal Aviation Administration suspended testing
Laser sensor	Sudden wind shifts affect reflectivity of air that lasers can detect	Provides only 20-second warning for jets traveling at a typical speed	Walton Electronic Perhaps Sperry

By filling in the cells of this matrix, Miguel identified some of the key information he needed to include in his report.

you need to obtain. Miguel created the matrix shown in Figure WG1.1 on his computer, but he could also have made it with pencil and paper.

Cluster Sketch

Creating a cluster sketch is a simple, powerful technique for exploring a topic visually. Write your overall topic in a circle at the center of a piece of paper, then add circles around the perimeter that identify the major issues or subtopics, joining them with lines to the main topic. Continue adding satellite notes, expanding outward as far as you find productive. Figure WG1.2 shows a cluster sketch created by Carol, an engineer who is leading a team assigned to help a small city locate places where it can drill new wells for its municipal water supply.

A variation of the cluster sketch is the idea tree, shown in Figure WG1.3. At the top of a sheet of paper, write your main topic. Then list the main subtopics or issues horizontally below, joining them to the main topic with lines. Continue this branching as long as it is fruitful.

Create and Study a Table or Graph of Your Data

Often at work you will need to write communications about data, such as the results of a test you have run, costs you have calculated, or production figures you have gathered. In such cases, many people find it helpful to begin their writing process by making the tables or graphs that they will include in their communication. Then they can begin to interpret the data arrayed before them, making notes about the data's meaning and significance to their readers.

Try This

Make a cluster sketch or idea tree to generate a list of your abilities that you would like to highlight in an employment interview. Identify general areas of ability and specifics related to each one.

FIGURE WG1.2
Cluster Sketch

An engineer created this cluster sketch in order to identify the topics she would need to cover in a report on the possibility of developing new wells for her town's water supply.

To identify topics for her report, the engineer might have used an idea tree like this rather than the cluster sketch shown in Figure WG1.2.

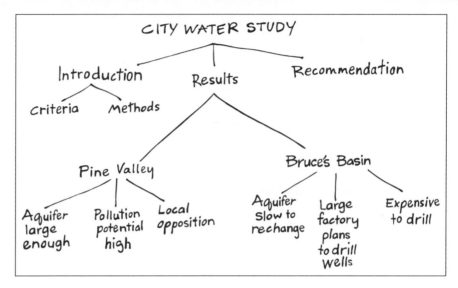

SEARCHING THE INTERNET

The Internet has created a rich and continuously evolving resource for your on-the-job research. Wherever you are, the Internet lets you read technical reports from companies such as IBM, view pictures taken by NASA spacecraft in remote areas of the solar system, or join online discussions on an astonishing array of topics with people around the world. Figure WG1.4 lists just a few of the resources the Internet makes available to you.

As a resource for research, the Web also presents you with significant challenges. Its sheer size can make it difficult to find the exact information your readers need. Search engines are helpful, but not perfect. In addition, no one monitors the Web's content to weed out biased, misleading, or inaccurate information.

To conduct web research successfully on the job, you must search through its vast contents efficiently so you don't waste time you should devote to other tasks, and you must evaluate carefully what you find so you don't report wrong information to your readers.

WWW

For links to a variety of search engines and Internet directories, visit the section for this Writer's Reference Guide at **www.cengage.com/english/anderson7e.**

Use Both Search Engines and Internet Directories

In addition to search engines like Google and Yahoo!, there is another very useful tool for web research: Internet directories. Search engines and Internet directories provide distinctly different kinds of research support, so it is often helpful to use both.

Search engines produce a large number of potentially helpful results but don't evaluate them for bias, completeness, or accuracy.

Search engines produce the more comprehensive results. Using computer technology, they scour a great deal of the Web (though not all of it), looking for

INTERNET RESOURCE	Examples
Corporate reports and information	IBM posts technical documents, Microsoft offers detailed information on its products, and the World Wildlife Fund reports on its environmental projects. Thousands of other profit and nonprofit organizations do the same.
	Examples
	IBM Research Papers on Networking http://www.research.ibm.com
	Microsoft product information and downloading http://www.microsoft.com
	World Wildlife Fund for Nature http://www.panda.org
Technical and scientific journals	Many technical and scientific journals are available online, some for free and some for a fee.
	Examples
	Journal of Cell Biology http://www.jcb.org
	IEEE Transactions on Software Engineering http://www.computer.org/tse
Government agencies	Many government agencies have websites at which they provide reports, regulations, forms, and similar resources.
	Examples
	NASA http://www.nasa.gov National Institutes of Health http://www.nih.gov National Park Service http://www.nps.gov

FIGURE WG1.4

Some Major Internet Resources for Research

resources that include the same words that you enter into the search line. They also determine the order in which they will place their results in the list that appears on your computer screen. For both searching and ranking, they apply a combination of criteria, such as the number of times the words appear on a web page, their location on the page, the number of websites that link to the source, and so on. Because different search engines use different combinations of criteria, they produce different numbers of results with different rankings. On the same day, a Google search for basking shark (the second largest of all sharks) returned 286,000 results. Yahoo! produced almost four times as many (1,250,000) and placed different results at the top of its list.

Internet directories are created by people, not computers, who search the Internet for resources they judge to be particularly valuable. Consequently, they yield much smaller, more sharply focused results organized in a hierarchical framework that

Try This

Conduct a side-by-side test to see some of the differences between search engines. Open two browser windows. Using a different search engine in each one, search for information on a topic related to your major or personal interests. Which search engine produces the most hits? Which has the most helpful links at the top of the results?

Internet directories provide a small number of results that people have judged to be valuable resources, but their criteria for selection may or may not match yours.

WWW

For links to a variety of search engines and Internet directories, visit the section for this Writer's Reference Guide at **www.cengage.com/english/ anderson7e.**

enables you to search systematically for the information you need to obtain for your readers. Using Yahoo!'s directory, you would first choose the category "Science," then "Animals, Insects, and Pets," then "Fish," then "Marine and Anadromous Fish," then "Sharks," then "Basking Shark." You would find eight results, not hundreds of thousands.

Which is better, a search engine or an Internet directory? There's no absolute answer. Search engines can provide overwhelming numbers of results, and they don't distinguish between sites created by experts and those created by second graders. Internet directories simplify your search by reducing the number of results, but they miss many useful results, and their selection criteria may not match yours. The Yahoo! directory's eight results for "basking shark" omitted important websites maintained by wildlife and scientific organizations, and they included a site for elementary school children. These and other differences explain why it's often worthwhile to use more than one search engine and more than one directory in the same search.

Also, the Web's information on many subjects is incomplete even when all its resources are combined. For some topics, you'll need to look elsewhere, for instance in books and reference journals, to obtain a thorough, balanced understanding of the topic you are researching on your readers' behalf.

Use Search Engines Efficiently

A web search is basically a word game. You can significantly increase your skill at it by employing the following strategies.

Ways to increase the efficiency of web searches

- **Use as few words as possible.** With the exception of conjunctions, prepositions, and a few other types of words, each word you enter becomes part of the search. For most searches, fewer words are helpful.

- **Use the words most likely to be used on the site you want.** Instead of "stomach hurts," use "stomachache" because this term is more likely to appear on the sites you want.

- **Use precise words.** Especially when researching technical or scientific topics, use the words specialists would use. While "heart attack" is the common term, medical specialists use "myocardial infarction."

- **Try different words.** Keep your vocabulary flexible. If your search isn't producing the results you want, use synonyms. "Gene splicing" will produce different results than "genetic engineering."

- **Reorder words.** Word order influences the number and ranking of search results. A search for "Mars life" produces results focused on space science. One for "life Mars" yields many kinds of results, including those about an old television comedy.

- **Narrow results by adding words.** To sharpen an initial search, add another word you would expect to appear on the websites you are targeting. Adding "science" or "comedy" significantly focuses the results of a search on "life Mars."

- **Narrow results by using quotation marks.** Quotation marks signal most search engines that you want sites where the enclosed words appear exactly as you have written them. Putting quotation marks around "basking sharks" cuts the number of results in half for both Google and Yahoo!

- **Narrow results by using "advanced search" features.** Advanced searches enable you to focus your search in many ways. Among other techniques, you can specify words that should not appear at any site you want and the type of file you want (.doc, .xls, .jpg). Another useful option is to limit the search to certain domains such as only those for educational institutions (.edu) and governments (.gov), but not businesses (.com and .biz).

For a brief tutorial on conducting efficient web searches, go to page 176.

Evaluating Your Search Results

As you learned from Guideline 5 in Chapter 6, you should carefully evaluate the information you obtain from any source. However, research on the Web requires special scrutiny because people can post anything there, whether it's true or false.

Begin by examining the URL (web address) of each search. The most helpful part of a URL is often the site's domain. Different kinds of organizations are assigned to different domains on the Web. For instance, in the following URL, the three-letters "edu" identify the "education" domain, meaning that the site belongs to an educational institution, in this case the University of Florida ("ufl").

© Courtesy of Rackable Systems

Google uses thousands of PCs to store and search information about websites its computers have visited.

http://www.flmnh.ufl.edu/fish/Gallery/Descript/baskingshark/baskingshark.html

Location of a site's domain in its address

Other domains often used by researchers include:

.com   Commercial (sites for businesses)
.gov   Government (sites for local, state, and federal governments)
.org   Not-for-profit organizations

Depending on what you are looking for, sites in any of these domains may be good sources or bad ones. For example, a .com site may provide useful details about the features of its products but biased information about product quality or lawsuits against the corporation.

When reviewing a site's URL, look also for a tilde (~), which often indicates that the site is a personal site. For instance, an official website created for Miami University (Ohio) is:

www.muohio.edu/aboutmiami

Writer's Tutorial

Three Ways to Search Efficiently on the Internet

By narrowing your search to specific parts of the Internet, you can increase the efficiency of your research by focusing your energy on websites your readers will find credible. The directions below show how to narrow a search with Google and Yahoo!. Other search engines have similar options.

USE AN ADVANCED SEARCH

To obtain sharply focused results, use an advanced search to limit the kinds of web pages a search engine looks for.

1. Choose **Advanced Search.**
2. Enter the words you want the search engine to look for or to avoid.
3. Limit the search to web pages that are:
   - Written in a certain language
   - Created in a certain file format (e.g., doc, xls)
   - Updated in the last 3 months, 6 months, etc.
   - Located in a certain domain (e.g., edu, org, com)

USE A FOCUSED SEARCH

Save time by telling the search engine to look only for a certain type of source.

You can also gain credibility by focusing the search on types of sites your readers respect.

For example, Google's Scholar search looks only at sites refereed by experts in their subjects.

1. Go to Google's home page.
2. Click on **more.**
3. Click on **Scholar.**
4. If you wish, add advanced search options that are customized for a Scholar search.

Note: A Google Scholar search includes books and journals that are not on the web. You can follow up with a library search.

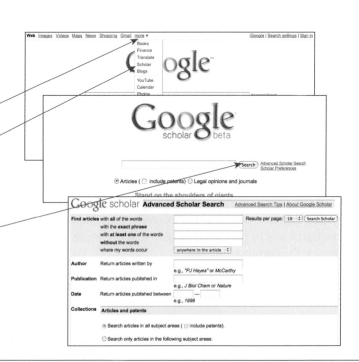

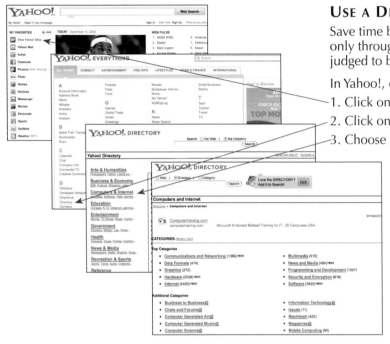

Use a Directory Search

Save time by using an Internet directory to search only through sites that human reviewers have judged to be good sources.

In Yahoo!, do the following:

1. Click on **View Yahoo! Sites** on the home page.
2. Click on **Directory**.
3. Choose the major category in which you will find useful information.
4. As each new screen appears, choose the most appropriate subcategory until you find the list of sites indexed for your search topic.

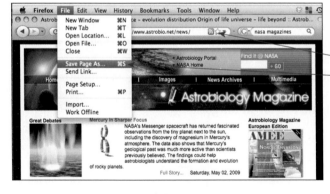

Save a Page for Future Reference

Save time by saving web pages you may want to visit again later in your research. Here are two ways to save a page.

- Create a bookmark in your browser.
- Use your browser's SAVE option. Record the date to include in your documentation.

Note: Copyright law protects information and images at websites. You must ask permission to reproduce an entire site or any image in it unless the site fits an exception in intellectual property law (page 161).

Learn More at the Website

For more advice about Internet research, go to www.cengage.com/english/anderson7e.

In contrast, all files created as part of personal sites by faculty, staff, and students include a tilde:

www.muohio.edu/~filename

Tilda

Finally, if you decide to visit a site, evaluate its contents critically. In addition to applying the evaluative criteria described in Chapter 6 (page 156), determine whether the site identifies the person or organization that created it, whether you can contact the creator, and when the site was last modified (how up to date it is).

Keeping Records

Finally, when conducting Internet research, keep careful records of the sites you find valuable. It's easy to lose your way when searching the Internet, which can make it difficult to relocate a site you need to visit again. Most browsers provide a bookmark feature that lets you add any page you are visiting to a personalized menu of sites you can return to with a single click. Even so, it's best to write down the URL of any site whose information you believe you will provide to your readers.

Bookmark valuable sites and write down their URLs.

Be sure to record the date you visit each site. Sites can change and even disappear suddenly, so this date is a crucial part of your bibliographic citation, as Appendix A explains.

USING THE LIBRARY

When you are preparing to write many communications, the library will be your best source of information and ideas.

Libraries are as much online as on the shelf.

The first step in using the library effectively is to discard the old image of it as a place that primarily houses books and periodicals. Although libraries still feature these publications, most are now as much online as on the shelf. In fact, many libraries are so computerized that you don't even need to enter the buildings to use many of their resources because you can access them through a computer in your classroom, home, or office.

Library resources fall into two broad categories.

Major library resources

Research Aids

Reference librarians	Catalogs, indexes, etc.

Information Sources

Printed books, periodicals, government documents, etc.	Electronic books, periodicals, government documents, etc.	Connections to external sources via the Internet

Generally, your excursions in library research will begin with one of the research aids, which can guide you to the most productive information sources. The following sections will help you use the research aids productively and also introduce some of the information sources with which you may not be familiar.

Reference Librarians

You will rarely find any research aid more helpful than reference librarians. They can tell you about specialized resources that you may not be aware of, and they can explain how to use the time-saving features of these resources.

Reference librarians will be able to give you the best help if you indicate very specifically what you want. In addition to stating your topic, describe what the purpose of your communication is, who your readers are, and how your readers will use your communication.

Tell the reference librarian your communication's objectives.

Library Catalog

The library catalog lists the complete holdings of a library, including books, periodicals, pamphlets, recordings, videotapes, and other materials. In most libraries, the catalog is computerized so you can search for items in several ways. If you are looking for a particular book whose title you know or for work written or edited by a person whose name you know, library catalogs are very simple to use. However, when you begin by looking for information about a particular topic, your success may depend on your ingenuity and knowledge of how to use the computerized catalog that most libraries have.

To search for a specific topic, you have two choices:

© Kzenon 2009/Shutterstock.com

Many employers have their own libraries that include resources directly related to their specialties.

- **Subject search.** To aid researchers, librarians include subject headings in the record for each library item. When you indicate that you want to do a subject search, the computer will prompt you to enter the words that identify the subject you are looking for. The computer will search through all items that have been tagged with the exact words you entered.
- **Word search.** When you indicate that you wish to conduct a word search, you will also be prompted to enter the words that identify your subject. This time, however, the computer will search the entire contents of all its records, including the title, author, and subject lines as well as tables of contents and other information that particular records might have.

Conducting Subject and Word Searches

Subject and word searches in computerized library catalogs are very similar to keyword searches on the Internet. Therefore, all the advice given on pages 174–175 applies. However, there is also one very important difference: When identifying the words used to describe the subject of a book, librarians use a formal and restricted set of terms that are defined in a large volume entitled the *Library of Congress Subject Headings List.* In the discussion about keyword searches on the Internet, you learned that some websites might use the term "gene splicing" and others might use "genetic engineering." In library subject headings, only "genetic engineering" is used. Consequently, while "gene splicing" will produce some results in an Internet search, it won't produce any in a subject search in a library. For help in determining the correct terms for subject searches, you have three resources:

- Many computerized library catalogs will tell you the correct term if you use an incorrect one that it recognizes as a synonym.
- The *Library of Congress Subject Headings List* is available at any library.
- Reference librarians are most willing to assist you.

Many computerized library catalogs will let you choose between abbreviated and extended displays of your search results. Extended displays are usually more helpful because they give you more information to consider as you decide whether to look at the entry for a particular item.

Refining and Extending Your Search

If your initial effort produces too few results, an overwhelming number of them, or an inadequate quality or range of them, there are several ways to refine and extend your search:

- **Look in the catalog entries of books you find for leads to other books.** Catalog entries not only name the subject headings under which a book is cataloged, but they also provide links to lists of other works that also have those subject headings.
- **Use Boolean operators and similar aids.** If you receive too many results, you may narrow the search in many of the ways described in the discussion of Internet searches (see pages 174–175).
- **Use other resources.** Don't limit yourself only to resources you locate through the library catalog. Your best source of information may be a corporate publication or other item not listed there. A reference librarian can help you identify other aids to use.

- **When you go to the library shelves, browse.** Sometimes books that will assist you are located right next to books you found through the library catalog. Don't miss the opportunity to discover them. Browse the shelves.

Indexes

Indexes are research aids that focus on specific topics or specific types of publications. Most catalog the contents of periodicals, but some include television programs, films, and similar items. Many are available online.

To use indexes well, you need to select ones that cover the topic you are researching:

- **General periodical indexes.** They index the contents of publications directed to a general audience. The familiar *Reader's Guide to Periodical Literature* is an example.
- **Specialized periodical indexes.** Almost every field has at least one. Examples are *Applied Science and Technology Index, Biological and Agricultural Index, Business Periodicals Index,* and *Engineering Index.*
- **Newspaper indexes.**

Some indexes not only list articles, but also provide an abstract (or summary) of each one. Figure WG1.5 shows an index entry that includes an abstract. By scanning through an abstract, you can usually tell whether reading the entire article would be worthwhile.

Whether you are using a printed or a computerized index, you can often speed your search by looking at the index's thesaurus. Different indexes use different sets of terms. Most publish this list in a thesaurus that can be accessed through a menu selection in the online version or found in the front or back of the printed version.

Computerized indexes work the same way that library catalogs do. You can search by the same variables: author, title, words. Searches can also be limited in similar fashion, although different indexes do this in different ways, so you should look at the "Help" feature for instructions.

Reference Works

When you hear the term *reference works,* you probably think immediately (and quite correctly) of encyclopedias, dictionaries, and similar storehouses of knowledge, thousands of pages long. What you may not realize is that many of these resources, such as the *Encyclopedia Americana,* are now available online or on CD-ROM, so that finding information in them can be very quick and easy.

In addition to such familiar reference works as the *Encyclopedia Britannica,* thousands of specialized reference works exist. Some surely relate to your specialty. For example, there are the *Encyclopedia of the Biological Sciences, McGraw-Hill*

Try This

Using the online resources in your college's library, find an article related to your major or a personal interest in a journal you haven't heard of before. Can you find articles in two?

Some indexes include abstracts.

Find out what indexing terms are used.

Use "Help" to customize a search.

FIGURE WG1.5
Abstract from an Abstracting Index

This abstract illustrates the research help you can receive when using online databases and indexes.

Full citation. In addition to using this information to locate this item, you can copy and paste the citation into your notes for possible inclusion in your references.

Links to other works by the author. These links can help you find other publications that may also relate to your research.

Abstract. This summary helps you decide whether it would be worthwhile to read the entire publication article.

Subject codes. This list identifies the keywords used to index this article. You can use it to link to other publications that were coded with these words.

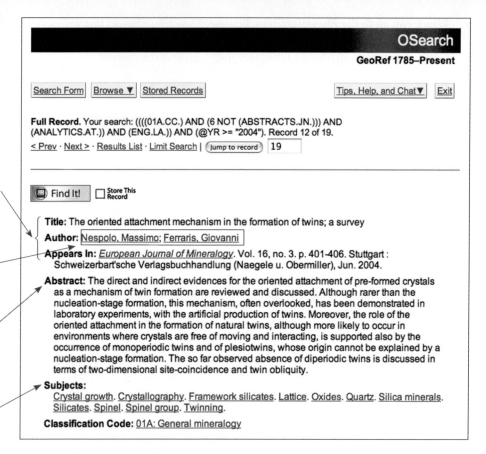

Encyclopedia of Science and Technology (20 volumes), Elsevier's Medical Dictionary, Harper's Dictionary of Music, and the Petroleum Dictionary.

Government Documents

Every year, the U.S. Government Printing Office distributes millions of copies of its publications, ranging from pamphlets and brochures to periodicals, reports, and books. Some are addressed to the general public, while others are addressed to specialists in various fields. Sample titles include Acid Rain, Chinese Herbal Medicine, Poisonous Snakes of the World, and A Report on the U.S. Semiconductor Industry.

Government publications that may be especially useful to you are reports on research projects undertaken by government agencies or supported by government grants and contracts. Annually, the National Technical Information Service acquires more than 150,000 new reports on topics ranging from nuclear physics to the sociology of Peruvian squatter settlements. Chances are great that some relate to your subject.

The following indexes are especially helpful. A reference librarian can help you find many others.

INDEXES TO U.S. GOVERNMENT PUBLICATIONS

- Monthly Catalog of U.S. Government Publications
  Publications handled by the Government Printing Office
  http://catalog.gpo.gov:80/F?RN=94090418
- Lists of Publications by Specific Agencies
  EPA http://www.epa.gov/ncepihom/
  NASA http://www.nasa.gov/news/reports/index.html
  National Institutes of Health http://www.nih.gov

Computerized Full-Text Sources

Many libraries also offer computer access to the full text of various sources. These include standard reference works, such as the *Encyclopedia Britannica,* and specialized publications such as scientific journals. These utilities allow you to search for topics in the same sorts of ways described in the discussion of the library catalog. When you locate an article of interest, you can read the text on your computer screen, download it to your computer's memory, or print a copy of it.

You can download full texts.

INTERVIEWING

At work, your best source of information will often be another person. In fact, people will sometimes be your only source of information because you'll be researching situations unique to your organization or its clients and customers. Or you may be asking an expert for information that is not yet available in print or from an online source.

The following advice focuses on face-to-face interviews, but it applies also to telephone interviews, which are quite common in the workplace.

Preparing for an Interview

Preparing for an interview involves three major activities:

- **Choose the right person to interview.** Approach this selection from your readers' perspective. Pick someone you feel confident can answer the questions your readers are likely to ask in a way that your readers will find useful and credible. If you are seeking someone to interview who is outside your own organization, the directories of professional societies may help you identify an appropriate person.

Take a reader-centered approach to selecting your interviewee.

Writer's Tutorial

Conducting Efficient Library Research

By wisely using various features of online library catalogs and indexes, you can increase your research efficiency.

FINDING BOOKS

Online library catalogs typically show an opening page that invites you to search by one of four topics: keywords, author, title, and subject.

However, you can often research more efficiently by using an advanced search to sharpen your focus.

1. Activate Boolean operators between your keywords (see page 180). Each "and" can be changed to "or" or "not."

2. Limit the search to sources that are:
   - Written in a certain language.
   - A certain type of resource (e.g., book, periodical).
   - Published before or after a certain date.

3. Examine the search results to determine which items are worth clicking on for more details.

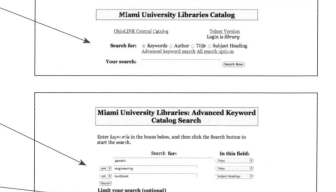

FOLLOWING LEADS TO OTHER SOURCES

Use the detailed description of a source to find other helpful items.

1. Click on the author's name to see other items written by this person.

2. Note whether the item has a bibliography you can review for leads to other sources.

3. Click on the relevant subject headings to see other items indexed the same way (see page 179).

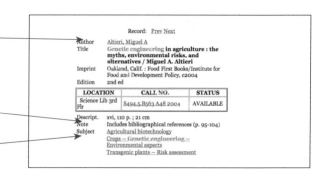

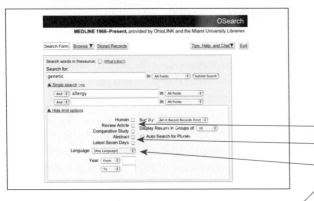

FINDING PERIODICAL SOURCES

Save time using online indexes.

1. Ask a librarian to help you choose the best index or indexes for your topic.

2. Use the advanced search functions to search for items that are:

   • A certain type (e.g., a review article).

   • Published before or after a certain date.

   • Written in a certain language.

3. Click on items in the search results list that look as though they might be useful (see page 176).

4. Read the abstract to decide whether it would be worthwhile to get the full item.

5. Click on the subject terms that are relevant to your research in order to view other articles indexed the same way (see page 179).

Some let you read sources instantly on screen.

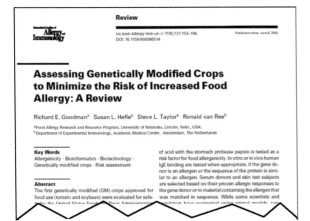

For more advice about library research, go to www.cengage.com/english/anderson7e.

Learn More at the Website

A well-planned interview can be productive and enjoyable for both interviewer and interviewee.

Do only 10 to 20 percent of the talking.

Learn More

For advice about succeeding in employment interviews, go to **www.cengage.com/english/anderson7e** and click on Bonus Chapters.

- **Make arrangements.** Contact the person in advance to make an appointment. Let the person know the purpose of the interview. This will enable him or her to start thinking about how to assist you before you arrive. Be sure to say how long you think the interview will take. This will enable your interviewee to carve out time for you. If you would like to record the interview, ask permission in advance.

- **Plan the agenda.** As the interviewer, you will be the person who must identify the topics that need to be discussed. Often, it's best simply to generate a list of topics to inquire about. But if there are specific facts you need to obtain, identify them as well. To protect against forgetting something during the interview, bring a written list of your topics and specific questions. For advice on phrasing questions, see the section on surveys (pages 189–191).

Conducting the Interview

Unless you are seeking a simple list of facts, your goal in an interview should be to engage the other person in a conversation, not a question-and-answer session. In this conversation, your goal should be to have the other person do 80 to 90 percent of the talking—and to have him or her focus on the information you need. To achieve these goals, you will need to ask your questions well and maintain a productive interpersonal relationship with your interviewee. Here are practical steps that you can take.

CONDUCTING A PRODUCTIVE INTERVIEW

Establish rapport.
- Arrive on time.
- Thank the person for agreeing to meet with you.

Explain your goal.
- Tell what you are writing and who your readers will be.
- Explain the use your readers will make of your communication.
- Describe the outcome you desire.

Ask questions that encourage discussion.
- Use questions that ask the interviewee to explain, describe, and discuss. They can elicit valuable information that you might not have thought to ask for. Avoid closed questions that request a yes/no or either/or response.

Closed question: Does the present policy create any problems?

Open question: What are your views of the present policy?

- Use neutral, unbiased questions.

   Biased question: Don't you think we could improve operations by making this change?

   Neutral question: If we made this change, what effect would it have on operations?

- Begin with general questions, supplemented by more specific follow-up questions that seek additional details important to you.

   General question: Please tell me the history of this policy.

   Follow-up question: What role did the labor union play in formulating the policy?

Show that you are attentive and appreciative.

- Maintain eye contact and lean forward.
- Respond with an occasional "uh-huh" or "I see."
- Comment favorably on the interviewee's statements.

   Examples: "That's helpful." "I hadn't thought of that." "This will be useful to my readers."

Give your interviewee room to help you.

- If the interviewee pauses, be patient. Don't jump in with another question. Assume that he or she is thinking of some additional point. Look at him or her in order to convey that you are waiting to hear whatever he or she will add.
- If the interviewee begins to offer information out of the order you anticipated, adjust your expectations.

Keep the conversation on track.

- If the interviewee strays seriously from the topic, find a moment to interrupt politely in order to ask another question. You might preface the question by saying something like this: "My readers will be very interested to know . . ."

Be sure you understand and remember.

- If anything is unclear, ask for further explanation.
- On complicated points, paraphrase what your interviewee has said and then ask, "Have I understood correctly?"

Try This

Write two biased questions intended to get students to give a favorable evaluation of a cafeteria, bookstore, or other facility at your school. Next, write two parallel questions designed to elicit unfavorable evaluations. Finally, write versions of the same questions that are not biased.

- ■ Take notes. Jot down key points. Don't try to write down everything because that would be distracting and would slow down the conversation.
- ■ Double-check the spelling of names, people's titles, and specific figures.

It's especially important that you assume leadership for guiding the interview. You are the person who knows what information you need to obtain on your readers' behalf. Consequently, you may need to courteously redirect the conversation to your topics.

Concluding the Interview

During the interview, keep your eye on the clock so that you don't take more of your interviewee's time than you requested. As the time limit approaches, do the following:

- Check your list. Make sure that all your key questions have been answered.
- **Invite a final thought.** One of the most productive questions that you can ask near the end of an interview is, "Can you think of anything else I should know?"
- **Open the door for follow-up.** Ask something like this: "If I find that I need to know a little more about something we've discussed, would it be okay if I called you?"
- **Thank your interviewee.** If appropriate, send a brief thank-you note by letter, memo, or e-mail.

CONDUCTING A SURVEY

While an interview enables you to gather information from one person, a survey enables you to gather information from *groups* of people.

At work, surveys support practical decision making.

At work, surveys are used as the basis for practical decision-making. Manufacturers survey consumers when deciding how to market a new product, and employers survey their employees when deciding how to modify personnel policies or benefit packages. While some surveys require the use of specialized statistical techniques that are beyond the scope of this book, you will usually be able to construct surveys that provide a solid basis for on-the-job decision-making simply by following the advice provided in the following sections, which lead you through the survey process.

Decide What To Ask About

The first step in writing survey questions is to decide exactly what you want to learn.

- **Review your research objectives by focusing on the decisions your readers must make.** Roger worked for a small restaurant chain that asked him to study the feasibility of opening a premium pastry and coffee shop next to a college campus. His readers' question, Roger knew, would relate primarily to whether there would be enough business to make the shop profitable.

- **Identify the full range of information your readers will find helpful.** Thinking about the information his readers would want, Roger realized that his survey should ask about the full range of variables that could influence the shop's profitability, such as location, hours of operation, products offered, and pricing.

- **Gather the information needed for analyzing information from subgroups.** Because different groups answer survey questions differently, Roger asked about the respondents' sex, age, income, relationship to the college (student, employee, or not affiliated), and other characteristics. By analyzing responses from various demographic groups, Roger could help his readers understand more precisely the potential market for the shop.

Write the Questions

More than anything else, the success of your survey depends on the skill with which you write your questions. The following suggestions will help you create an effective questionnaire that provides useful information and elicits the cooperation of the people you ask to fill it out.

- **Avoid ambiguity.** The greatest threat to a survey's value is ambiguity in the questions. If different people interpret a question differently, they will, in fact, be answering different questions, making your data worthless. If they all interpret your question differently than you do, then your interpretation of the results will be erroneous. Take the reader-centered approach of asking yourself how the people responding to the survey might misunderstand each question. Pilot test your questions by asking a few people to tell you what they think each is asking. In survey questionnaires, as in all communications, what matters isn't what you mean but what your readers think you mean.

- **Mix closed and open questions.** *Closed questions* allow only a limited number of possible responses. They provide answers that are easy to tabulate. *Open questions* allow the respondent freedom in devising the answer. They provide respondents an opportunity to react to your subject matter in their own terms. See Figure WG1.6.

Learn More

For more detailed advice about the questions to include in a survey, see Guidelines 1 through 4 in Chapter 6 (pages 152–156.)

Ambiguity is the greatest threat to a survey's value.

Try This

If one of your professors asked you to help him or her improve a course by writing a survey for students to take, what three closed questions and what three open-ended questions would you include in it. Why?

	Closed Questions
Forced Choice	▪ Respondents must select one of two choices (yes/no, either/or).
	Example Would you buy pastries at a shop near campus, yes or no?
Multiple Choice	▪ Respondents select from several predefined alternatives. How many times a month would you visit the shop? _____ 1 to 2 _____ 3 to 4 _____ 5 or more
Ranking	▪ Respondents indicate an order of preference.
	Example Please rank the following types of pastry, using a 1 for your favorite, and so on.
Rating	▪ Respondents pick a number on a scale.
	Example Please circle the number on the following scale that best describes the importance of the following features of a pastry shop: Music Unimportant 1 2 3 4 5 Important Tables Unimportant 1 2 3 4 5 Important
	Open Questions
Fill in the Blank	▪ Respondents complete a statement.
	Example When deciding where to eat a late-night snack, I usually base my choice on _____.
Essay	▪ Respondents can frame responses in any way they choose.
	Example Please suggest ways we could make a pastry shop that would be appealing to you.

You may want to follow each of your closed questions with an open one that simply asks respondents to comment. A good way to conclude a survey is to invite additional comments.

- **Ask reliable questions.** A *reliable* question is one that every respondent will understand and interpret in the same way. For instance, if Roger asked, "Do you like high-quality pastries?" different readers might interpret the term "high-quality" in different ways. Roger might instead ask how much the respondents would be willing to pay for pastries or what kinds of snacks they like to eat with their coffee.

- **Ask valid questions.** A *valid* question is one that produces the information you are seeking. For example, to determine how much business the pastry shop might attract, Roger could ask either of these two questions:

  | | • How much do you like pastries? | | Invalid |
  | • How many times a month would you visit a pastry shop located within three blocks of campus? | | Valid |

  The first question is invalid because the fact that students like pastries does not necessarily mean that they would patronize a pastry shop. The second question is valid because it can help Roger estimate how many customers the shop would have.

- **Avoid biased questions.** Don't phrase your questions in ways that seem to guide your respondents to give a particular response.

  | • Wouldn't it be good to have a coffee shop near campus? | | Biased |
  | • How much would you like to have a coffee shop near campus? | | Unbiased |

- **Place your most interesting questions first.** Save questions about the respondent's age or similar characteristics until the end.
- **Limit the number of questions.** If your questionnaire is lengthy, people may not complete it. Decide what you really need to know and ask only about that.
- **Test your questionnaire.** Even small changes in wording may have a substantial effect on the way people respond. Questions that seem perfectly clear to you may appear puzzling or ambiguous to others. Before completing your survey, try out your questions with a few people from your target group.

Select Your Respondents

At work, writers sometimes present their survey questions to every person who belongs to the group whose attitudes or practices they want to learn about. For example, an employee assigned to learn what others in her company feel about a proposed change in health care benefits or a switch to flextime scheduling might send a survey questionnaire to every employee.

However, surveys are often designed to permit the writers to generalize about a large group of people (called a *population*) by surveying only a small portion of individuals in the group (called a *sample*). To ensure that the sample is truly representative of the population, you must select the sample carefully. Here are four types of samples you can use:

Your sample should reflect the composition of the overall group.

- **Simple random sample.** Here, every member of the population has an equal chance of being chosen for the sample. If the population is small, you could put the name of every person into a hat, then draw out the names to be included in your sample. If the population is large—all the students at a major university, for example—the creation of a simple random sample can be difficult.

- **Systematic random sample.** To create a systematic random sample, you start with a list that includes every person in the population—perhaps by using a phone book or student directory. Then you devise some pattern or rule for choosing the people who will make up your sample. For instance, you might choose the fourteenth name on each page of the list.

Convenience samples can give unreliable results.

- **Convenience sample.** To set up a convenience sample, you select people who are handy and who resemble in some way the population you want to survey. For example, if your population is the student body, you might knock on every fifth door in your dormitory, or stop every fifth student who walks into the library. The weakness of such samples is obvious: From the point of view of the attitudes or behaviors you want to learn about, the students who live with their parents or in apartments may be significantly different from those who live in dorms, just as those who don't go to the library may differ in substantial ways from those who do.

- **Stratified sample.** Creating a stratified sample is one way to partially overcome the shortcomings of a convenience sample. For instance, if you know that 15 percent of the students in your population live at home, 25 percent live in apartments, and 60 percent live in dormitories, you would find enough representatives of each group so that they constituted 15 percent, 25 percent, and 60 percent of your sample.

Use enough respondents to persuade your readers.

Even if you can't choose the people in each group randomly, you would have made some progress toward creating a sample that accurately represents your population.

When creating your sample, you must determine how many people to include. On one hand, you want a manageable number; on the other hand, however, you also want enough people to form the basis for valid generalizations. Statisticians use formulas to decide on the appropriate sample size, but in many on-the-job situations, writers rely on their common sense. One good way to decide is to ask what number of people your readers would consider to be sufficient.

Contact Respondents

© Lisa F. Young/Shutterstock.com

In face-to-face interviews, the interviewer carefully avoids facial expressions, comments, or other indications that he or she wants a particular response to a question.

There are three methods for presenting your survey to your respondents:

- **Face-to-face.** In this method, you read your questions aloud to each respondent and record his or her answers on a form. It's an effective method of contacting respondents because people are more willing to cooperate when someone asks for their help in person than they are when asked to fill out a printed questionnaire. The only risk is that your intonation, facial expressions, or body language may signal that you are hoping for a certain answer. Research shows that respondents tend to give answers that will please the questioner.

- **Telephone.** Telephone surveys are convenient for the writer. However, it can sometimes be difficult to use a phone book to identify people who represent the group of people being studied.
- **Mail or handout.** Mailing or handing your survey forms to people you hope will respond is less time-consuming than conducting a survey face to face or by telephone. Generally, however, only a small portion of the people who receive survey forms in these ways actually fill them out and return them. Even professional survey specialists typically receive responses from only about 20 percent of the people they contact.

Interpret Your Results

Of courses, survey results don't speak for themselves. You need to analyze and interpret them in order to make the results meaningful and useful to your readers. Chapter 7 guides you through this process using an extended example that involves the analysis, interpretation, and presentation of survey data.

Learn More

See Chapter 7 for detailed suggestions for interpreting survey results.

7 | Analyzing Information and Thinking Critically

GUIDELINES

Imagine that by following the guidelines in Chapter 6, you have gathered the information needed to answer a question that your employer, a client, or other person has asked. The next step in your research is determining what to tell that person about it. In some cases, the information's significance to your readers will be obvious. No special effort will be required to discern its meaning. Often, however, you will gather a large amount of information that you must carefully analyze before you are ready to report your results.

DEFINING
OBJECTIVES

PLANNING

CONDUCTING
RESEARCH

DRAFTING

REVISING

CHAPTER 7

Think back to the research that Anna, Chen, and Terry conducted for the dean, as described in Chapter 6. The dean asked them to evaluate the new system for providing engineering students with academic and career advising. In the past, the dean's staff provided the advising. In the new system, professors provide this guidance, with each one assigned to advise a few students throughout their four years of study. Anna, Chen, and Terry gathered relevant information by surveying students, talking with professors, and consulting journals on engineering education. From all the opinions, facts, and ideas they've discovered, they must draw conclusions about the strengths and weaknesses of the advising system as well as suggest ways to improve it. Furthermore, the conclusions and recommendation must be ones the dean can use confidently as the basis for his decisions and actions. In other words, they must be:

- Focused on supporting the readers' decisions and actions
- Evidence-based
- Carefully reasoned

Essential qualities of conclusions based on workplace research

These are the three essential qualities of the conclusions resulting from on-the-job research.

GUIDING YOU THROUGH THE READER-CENTERED PROCESS FOR ANALYZING INFORMATION AND DATA

This chapter guides you through the process of analyzing a collection of facts and ideas like the ones Ann, Chen, and Terry gathered. The major steps in the process correspond to this chapter's seven guidelines.

1. Review your research objectives
2. Arrange your information in an analyzable form
3. Look for meaningful relationships in the information
4. Interpret each relationship for your readers
5. Explain why each relationship is important to your readers

© Mike Agliolo/Corbis/Jupiter Images

6. Recommend actions based on your analysis

7. Think critically throughout your analysis

No matter what your major and no matter what kind of information and data is involved in your research, these guidelines will help you draw conclusions that have the three essential qualities of on-the-job research. Guideline 1, "Review your research objectives," sharpens your focus on drawing conclusions that support your readers' decisions and actions. Guidelines 2 through 6 describe a method for drawing conclusions that are based solidly on the evidence. Guideline 7, "Think critically throughout your analysis," helps you draw conclusions that are carefully reasoned.

Your work at analyzing information leads seamlessly to drafting your report. As you follow each of the middle five guidelines (2 through 6), you will be creating content for one of the major elements of the kind of report most often used to describe research. This tight relationship between analyzing and writing is illustrated in Figure 7.1.

FIGURE 7.1
Relationship between
Steps in the Analysis
Process and Report
Element

RELATIONSHIP BETWEEN STEPS IN THE ANALYSIS PROCESS AND REPORT ELEMENT

STEPS IN RESEARCH	SECTIONS OR CHAPTERS OF AN EMPIRICAL RESEARCH REPORT
Gathering information (Chapter 6)	
Define objectives of research	→ Introduction
Plan information gathering Collect information	→ Method
Analyzing information (Chapter 7)	
■ Arrange your information in analyzable form (Guideline 2) ■ Find meaningful relationships in the information (Guideline 3) ■ Examine subgroups of information (Guideline 4)	→ Results
■ Interpret the relationships for your readers (Guideline 5)	→ Discussion
■ Identify the significance of the relationships for your readers (Guideline 6)	→ Conclusion
■ Recommend actions based on your analysis (Guideline 7)	→ Recommendations

Guideline 1 | Review your research objectives

Analysis, like every other aspect of writing, begins with your readers. While gathering research information, it's easy to be distracted from your readers' needs. You can become enthralled by fascinating topics or controversies that are irrelevant to your readers' needs. Your hard work at obtaining information can tempt you to analyze pieces of it that are useless to your readers. By reviewing your research objectives, you can refocus your attention on the way your research is intended to assist your readers.

While gathering information for their study, Anna, Chen, and Terry heard a few professors complain vigorously about the new advising system and said the dean should return to the old one. Because they liked these professors, Anna, Chen, and Terry started to think that they should help them by making their complaints a major focus of the report. Reviewing their research objectives helped them recall that the ultimate goal of their research was to provide the dean with practical advice for improving the system, not for replacing it. It also reminded them that the opinion of these few professors needed to be considered in the context of all the other information they had gathered, not as a major consideration all by itself.

Guideline 2 | Arrange your information in an analyzable form

To analyze research information, you study it, looking for patterns, connections, and contrasts that can help your readers make decisions or take actions that will help them achieve their goals.

Perhaps surprisingly, the most effective way to look for these patterns doesn't involve mental action, but a physical one: Arrange your data in a table, chart, graph, or other visual display that helps you "see" the relationships. For engineers and scientists in many fields, this is standard practice. To analyze their data, they make a "picture" of them that they can then study.

When analyzing your information, make these visual displays quickly, without the refinements you would use if you were preparing the final draft of your report. Some of these displays will yield insights. Others won't. There's no need to get fancy with them at this point. Chart A is one of the many "exploratory" charts Anna, Chen, and Terry used as they began analyzing results of their student survey. To save time while analyzing, they used abbreviations in the chart and left off a label for the y-axis.

In addition to charts and graphs, tables are often a useful way of displaying information for analysis, especially if your information is verbal rather than numeric. Freewriting, idea trees, and flowcharts can also be effective. The key point is that it's often helpful to look, literally, at your information.

Learn More

For information on using charts, graphs, idea trees, and similar aids to research, see the Writer's Reference Guide for Using Five Reader-Centered Research Methods (page 165).

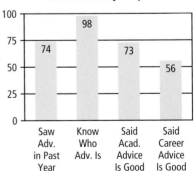

CHART A
Student Survey Responses

Guideline 3 | Look for meaningful relationships in the information

The reason for arranging research information in visual displays is to help you discover meaningful relationships, ones that can guide the actions of your readers.

For example, looking at the two right-hand bars in Chart A, Anna, Chen, and Terry saw a substantial difference between the percentage of students who said they received "good" academic advice from their faculty advisers (73%) and those who said they received "good" career advice (59%). This gap was surely an important one to report to the dean because it could support a decision to focus specifically on building professors' skill at providing career advice.

Some relationships raise questions for further examination. Looking at the left-hand bars in Chart A, Anna, Chen, and Terry saw the large difference between the percentage of students who knew who their adviser was (98%) and those who had visited their adviser in the past year (68%). If students knew who their advisers were, why didn't they consult with them?

Strategies for Finding Meaningful Relationships

Not every chart or other visual display will disclose a meaningful relationship. The activity of analyzing information often involves many cycles of choosing possible relationships to look for, creating a visual display, finding no meaningful relationships, and choosing another possible relationship to investigate. The process inevitably involves some hits and some misses.

However, your knowledge of your subject matter and your readers can suggest many likely pieces of information to explore. Also, the following strategies are often fruitful.

STRATEGIES FOR FINDING RELATIONSHIPS MEANINGFUL TO YOUR READERS

- Compare information on related outcomes (for example, academic advising and career advising).
- Compare information about different steps in a process (for example, getting students to know who their advisers are and getting them to meet with their advisers).
- Compare information on the same topic from different sources (for example, surveys, interviews, and books or articles).
- Consider possible cause-and-effect relationships.
- Consider possible correlations.
- Look for relationships where your instinct leads you.

Examining Subgroups of Information

Examination of subgroups of information is a major characteristic of high-quality research. It provides you with a deeper, more insightful understanding of the information you've gathered. In turn, it enables you to help your readers make more effective, targeted actions than would otherwise be possible.

For example, as Anna, Chen, and Terry looked at the left-hand bar in Chart A, they realized that it would be worth determining whether all subgroups of students had the same low rate of consulting with their advisers. If there were differences, they reasoned, this information could help the dean plan targeted action aimed at increasing the consultation rate for the groups that had the lowest. In fact, they did find the important difference shown in Chart B. The percentage of electrical engineering students who saw their advisers in the past year was much higher than that of students in any other department. Also, the percentage of nuclear engineering students was much lower than that of any other department. This discovery formed the basis for their recommending a targeted action, as described in Guideline 6.

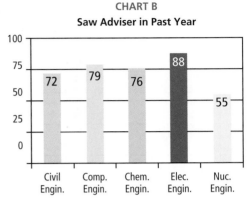

CHART B
Saw Adviser in Past Year

By examining subgroups of information, Anna, Chen, and Terry found many other relationships that shaped their final recommendations to the dean. They were able to discover these relationships only because they had followed Guideline 3 in Chapter 6: "Gather information that can be analyzed in subgroups." If they had omitted questions about the students' major, year in school, and sex from their survey questionnaire, they would have missed a great deal of information that was important to their reader, the dean.

Guideline 4 | Interpret each relationship for your readers

After you've found a meaningful pattern, connection, or other relationship, you need to interpret it for your readers. Interpretations may take many forms, including generalizations, explanations, and comparisons, along with exceptions and counterevidence. Figuring out how to interpret a relationship accurately, precisely, and helpfully can sometimes require careful thought. Two of the challenges you may face are choosing among alternative interpretations and dealing with uncertainty.

Choosing Among Alternative Explanations

There are always alternative ways of understanding a set of data. For example, looking at Chart B, one might say that the students in the nuclear engineering department, who have the worst record of consulting advisers, are less responsible than other engineering majors. They just don't take advantage of the advising that is offered. Alternatively, one might say that the nuclear engineering department is less effective than others at attracting the students to advising. Which interpretation would be most helpful to the dean? Because the dean wants to improve the advising system, it's better to use the second interpretation, which focuses on the department rather than the students. The dean and professors can't control the students' sense of responsibility. They can change the ways they attract students to advising.

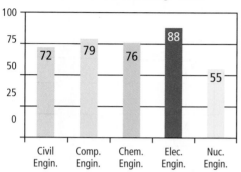

CHART C

Said Academic Advising "Good"

Dealing with Uncertainty

Sometimes, you may need to take a broad and creative look at your information to find a truly helpful interpretation. Trying to understand why students in some departments visit their advisers more often than students in other departments, Anna, Chen, and Terry compared the data in Chart B with those in Chart C, which shows the percentage of students who said they received good academic advice from their faculty advisers. The overall patterns are very similar, with nuclear engineering the lowest and electrical engineering the highest in both.

Having found this relationship, Anna, Chen, and Terry faced a dilemma: How should they interpret it? They could say that the more pleased students are with the academic advice they receive, the more likely they are to consult their advisers. Or they could say that the more often students see their advisers, the more pleased they are with the advice they receive.

Which of these two interpretations is correct? From the data they had, Anna, Chen, and Mike couldn't tell for sure. Their uncertainty raises an important point about interpretation: When you can't be certain, signal your uncertainty to your readers. One way to do that is simply to tell your readers that alternative interpretations are possible and present them both. Another is to say that your information "suggests" or "may indicate" that a relationship exists rather than that the information "demonstrates" or "proves" its existence. In their report to the dean, Anna, Chen, and Terry wrote: "The data suggest that students are more likely to visit their advisers if they feel they receive good academic advice when they do." Note that their statement discusses how good the students "feel" the advice is. It doesn't state whether the advice is actually good (or bad) in any department. Anna, Chen, and Terry didn't have evidence about the quality of the advice itself, only of the students' perception of its quality.

Your Interpretations Are Your Conclusions

The interpretations you make are often called conclusions. They state what you conclude from the research information you gathered. Frequently, they are presented in section of a research report that is titled "Conclusions."

Guideline 5 | Explain why each relationship is important to your readers

To be sure your readers know what each of your conclusions means to them, you should explain its significance explicitly. One way to do that is to indicate the kind of action that your readers might take. For instance, Anna, Chen, and Terry wrote,

"The survey results suggest building professors' skills at career advising would increase student satisfaction with advising and the frequency with which students visit their advisers."

Guideline 6 | Recommend actions based on your analysis

In almost all cases, your readers will want you to recommend a decision or action based on your research. The more specific your recommendations, the more helpful they will be.

A weak recommendation that Anna, Chen, and Terry might have made is that the dean help professors in the departments with low advising satisfaction become better advisers. It's weak because it merely rephrases one of their conclusions (Guideline 5). A much more helpful recommendation would be to suggest specific ways in which this assistance might be provided.

The suggestions would be even more helpful if a specific suggestion were also backed by research indicating that the suggested action is likely to achieve its goal. Anna, Chen, and Terry turned to the information they had gathered from other engineering schools to find specific strategies the dean could use to enhance the advising skills of the professors. They recommended that training sessions be established and that the departments whose students were more satisfied with advising share their strategies with the other departments.

Guideline 7 | Think critically throughout your analysis

Critical thinking is essential in all aspects of your research, but especially when you are analyzing the information you've gathered. At every step, you need to explore alternatives, evaluate possibilities in light of all the relevant evidence, and identify underlying assumptions—including your own. You need to consider the quality of each piece of information you obtain by considering the perspectives and biases of the sources. Four ways to do that are to let go of your anchors, value counterevidence and exceptions, avoid unintentional personal or organizational biases, and think critically about your sources.

Let Go of Your Anchor

One inhibitor of good analysis is a psychological phenomenon called *anchoring* (Tversky & Kahneman, 1974). When people have made an initial commitment of any kind, they tend to view future actions from that vantage point. It's what keeps gamblers at the roulette wheel, thinking that they've got to "win back" the money they've already lost. Thinking rationally, they would know that their odds of winning and losing are the same for each new spin, no matter how much they won or lost on previous spins. For people engaged in research, anchoring can mean that they keep finding evidence to confirm their initial intuitions while ignoring other possibilities. Often, these intuitions are formed even as researchers are gathering information—before they begin can their analysis. The way to free yourself from anchoring is to consciously and actively look for alternative ways of seeing and understanding your research information.

Value Counterarguments, Counterevidence, and Exceptions

In most research, you are likely to find contradictions. Your sources may disagree with one another. A few pieces of information may be inconsistent with everything else you've found. These conflicts can be very valuable. They may signal the need for you to look more deeply for alternative explanations or conclusions—cutting the line to your anchor. They may indicate the need to hedge your conclusions, letting your readers know that there's some uncertainty. They may even be the link to something valuable you can tell your readers. When Anna, Chen, and Terry discovered that the electrical engineering department received exceptionally high ratings for its academic advising, they saw that this department probably had a lot to teach the others, a point they made to the dean.

Avoid Personal or Organizational Biases

Without any malicious or selfish intent, it's easy for all of us to see the merits of interpretations and recommendations that would benefit us, our employer's organization, or our department in it. To serve your readers well, you need to check whether the natural self-interest we all feel is influencing your analysis. The easiest way to do this is to look at your information from your readers' perspective and those of the other stakeholders in the topic your are studying.

Thinking Critically About Your Sources

The advice given so far about thinking critically focuses on your careful examination of your own thoughts. It has focused on you, not your sources, because we all find that the hardest lapses to detect in an analysis are often our own. Of course, you also need to carefully consider the conscious and unconscious biases of the sources from which you have gathered your information. You need to consider the skill with which they have examined their experience and evidence. Most often, you'll find that the biases and lapses you discover mean that you need to adjust for possible distortions, not simply dismiss the source altogether.

CONCLUSION

Analyzing research information requires the creativity to find and explain meaningful relationships and the critical thinking needed to ensure that the relationships you believe you see actually exist and are the most useful ones to tell your readers about.

This chapter used an example involving survey data to illustrate the process of analyzing research information. You would follow the same steps for most other types of research: You examine the evidence that you've gathered, looking for relationships and testing them against exceptions and counterevidence in an effort to interpret your findings in ways that can help your readers. This is the general process for most research done in engineering and science labs, reading nonstatistical articles and books, and talking with people.

USE WHAT YOU'VE LEARNED

For additional exercises, go to www.cengage.com/english/anderson7e. *Instructors*: The book's website includes suggestions for teaching the exercises.

EXERCISE YOUR EXPERTISE

1. While analyzing their survey data, Anna, Chen, and Terry compared responses of men with responses of women to this question: "Have you visited your faculty adviser this year?" They displayed their results in the chart shown in Figure 7.2. Interpret this result in two or more ways, following the advice in Guideline 6. Which of your interpretations would be the most helpful to the dean of the engineering school? Why? Next, write a sentence that explains the significance of your interpretation to the dean (Guideline 6). Finally, make a recommendation to the dean to the dean based on your interpretation (Guideline 7).

2. Work on the "Increasing Organ Donations" Case on page 232.

EXPLORE ONLINE

Using the newspapers available online, find a story that includes a graph. In 200 words, describe the topic of the graph, what the story says it shws, and the additional analysis (such as analysis of subgroups of information) that would provide a fuller understanding of the topic.

COLLABORATE WITH YOUR CLASSMATES

1. In their survey, Anna, Chen, and Terry gathered information from the engineering students that let them analyze subgroup data from students according to their major, year in school, and sex. If you were conducting a survey about academic advising at your school, what additional subgroups of students would you want to identify? Why?

2. Imagine that the chair or head of your department has asked you and one or two of your classmates to conduct a survey of students in your major that is exactly twelve questions long. As a group, name the survey topic and

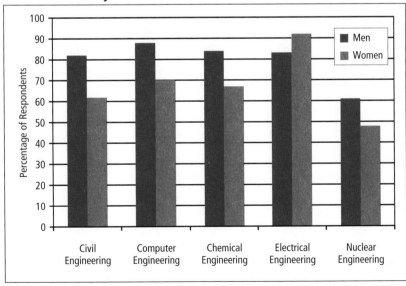

Comparison of Men and Women Who Reported That They had Consulted Their Advisers in the Past Year

FIGURE 7.2
Chart for Expertise Exercise 1

the chair's reason for wanting the information you would obtain through the survey. List the twelve questions you would ask. Explain how the data obtained from each question will help the chair achieve his or her goal.

APPLY YOUR ETHICS

One ethical principle for analysis is to use methods that avoid accidentally reaching incorrect conclusions. Even if they are reached by accident, inaccurate analyses can have serious consequences when they are used to make decisions and take action. By conducting an Internet search, find a case where a person or organization is accused of using "misleading statistics." In 100 words, summarize the topic of the statistics, explain why they are said to be misleading, and identify the harmful consequences that could result or have resulted from the use of these statistics. Tell whether you think the statistics really were misleading and why. If you believe they were misleading, did the problem arise by accident or intention?

PART V
DRAFTING PROSE ELEMENTS

205

8 | Drafting Paragraphs, Sections, and Chapters

GUIDELINES

| DEFINING OBJECTIVES |
| PLANNING |
| CONDUCTING RESEARCH |
| **DRAFTING** |
| REVISING |

This chapter marks a major transition in your study of on-the-job writing. In Chapters 3, 4, and 5, you learned how to define the objectives of a communication in a reader-centered way and then plan reader-centered strategies for achieving these objectives. This chapter is the first of seven that will help you develop your expertise in transforming these plans into action as you draft the communication's prose, graphics, and graphic design.

FROM PLANNING TO DRAFTING

This chapter's advice follows directly from the guidelines in Chapters 4 and 5, which helped you decide what to say in your communication and how to organize that content. This chapter helps you take the next step by showing you how to design the paragraphs and groups of paragraphs that correspond with the topics identified in your plan.

THE SIMILARITIES AMONG PARAGRAPHS, SECTIONS, AND CHAPTERS

WWW

To read additional information, see more examples, and access links related to this chapter's guidelines, go to **www.cengage.com/english/anderson7e** and click on Chapter 8.

This chapter's reader-centered guidelines apply to all parts of a communication that are longer than a few sentences, including paragraphs, groups of paragraphs that make up the sections and chapters of longer communications, and even whole communications. For convenience's sake, this chapter uses the word *segments* to designate these variously sized prose units.

How can the same guidelines apply with equal validity to segments that range in size from a few sentences to an entire communication that may be tens or hundreds of pages long? There are two reasons, one related to each of the two indispensable qualities of an effective workplace communication: usability and persuasiveness.

- **Usability.** You may have heard a paragraph defined as a group of sentences about the same subject. With only slight variation, that definition applies equally well to larger segments: A section or chapter is a group of paragraphs on the same subject, and an entire communication is a group of sections or chapters on the same topic.

  To understand any segment, whether a short paragraph or an entire communication, readers do the same things to determine what the topic is and figure out how its parts fit together. Because all segments demand the same mental work for readers, the same strategies can increase the usability of them all.

- **Persuasiveness.** Regardless of a segment's size, readers mentally process its persuasive claims and evidence in the same way. For example, as Chapter 5 explains, they look for benefits to their organizations and themselves, and they spontaneously raise counterarguments. Because readers read all segments in the same way, the same strategies can increase the persuasiveness of all segments.

GUIDELINES FOR BEGINNING A SEGMENT

Guidelines 1 and 2 describe ways to begin a segment that are usually effective. Suggestions for handling some of the exceptions are included in the discussion of Guideline 2.

Guideline 1 | **Begin by announcing your topic**

Although we often think of topic sentences in association with paragraphs, they can usually increase the usability of segments of any size.

How Topic Statements Increase Usability

How do topic statements increase usability? A key element in usability is the ease with which readers can understand your message. This task requires them to discern how each sentence in a paragraph relates to the others, how each paragraph relates to the other paragraphs in a section, and so on. According to researchers, readers perform two distinct activities in order to identify these relationships.

- **Bottom-up processing.** In bottom-up processing, readers proceed in much the same way as people who are working a jigsaw puzzle *without* being able to see a picture that tells them whether the finished puzzle will show a garden, city street, or three cats. As they read, they try to guess how the small bits of information they gather from each sentence fit together with the information from the other sentences to form the segment's general meaning.
- **Top-down processing.** In top-down processing, readers proceed like people who can see a picture of the finished puzzle. Once they see the communication's overall structure, they know how the information they obtain from each sentence fits into the larger meaning of the segment.

Research shows that although readers engage continuously in both processes, the more top-down processing they can perform, the more easily they can understand and remember the message.

Moreover, topic statements are especially helpful when placed at the *beginning* of a segment. In a classic experiment, researchers John D. Bransford and Marcia K. Johnson (1972) demonstrated how much this placement helps readers understand

Topic statements are especially helpful to readers when placed at the beginning of a segment.

and remember a message. They asked people to listen to the following passage being read aloud.

Passage used in an experiment that demonstrated the importance of top-down processing

> The procedure is actually quite simple. First you arrange things into different groups. Of course, one pile may be sufficient depending on how much there is to do. If you have to go somewhere else due to lack of facilities, that is the next step; otherwise you are pretty well set. It is important not to overdo things. That is, it is better to do too few things at once than too many. In the short run this may not seem important, but complications can easily arise. A mistake can be expensive as well. At first the whole procedure will seem complicated. Soon, however, it will become just another facet of life. . . . After the procedure is completed, one arranges the materials into different groups again. Then they can be put into their appropriate places. Eventually they will be used once more and the whole cycle will then have to be repeated. However, that is part of life.

The researchers told one group the topic of this passage in advance; they told the other group afterward. Then they asked both groups to write down everything they remembered from what they had heard. People who had been told the topic (washing clothes) before hearing the passage remembered many more details than those who were told afterward.

As Bransford and Johnson's study suggests, by stating the topic at the beginning of your segments you increase the usability of your prose for readers who want to understand your communication thoroughly and remember what you said. You also increase your communication's usability for readers who are skimming for particular facts. The first sentence of each segment tells them whether the segment is likely to contain the information they are seeking.

Here are three of the most common and effective ways to provide topic statements at the beginning of your segments.

INDICATING THE TOPIC OF A SEGMENT

- **Use a sentence.** An example is the sentence that introduced this list of strategies ("Here are three of the most common and effective ways to provide topic statements at the beginning of your segments").
- **Use a single word.** When a paragraph or other segment begins with "First," you know that you are starting a discussion with two or more parts. "Second" tells you that you are moving to the next part. Words such as "However," "Finally," and "Therefore" serve the same function.
- **Use a question.** The question ("How do topic statements increase usability?") that begins the second paragraph of the discussion of this guideline told you that you were about to read a segment explaining the ways topic statements contribute to usability.

You can provide your readers with additional assistance in understanding your communications by creating an interlocking set of easy-to-spot topic statements that would form an outline if all the other sentences were removed. Chandra, a zoo manager, did this when writing the report whose first page is shown in Figure 8.1. Here is the outline that the topic statements on her first page would make. Note that in Chandra's report, these topic statements flow right into her prose.

I. Budget Crisis
   A. The crisis first surfaced last August.
   B. What is causing the crisis?
      1. The crisis is not caused by rising costs.
      2. The crisis is caused by declining revenues.

Outline corresponding to the page shown in Figure 8.1

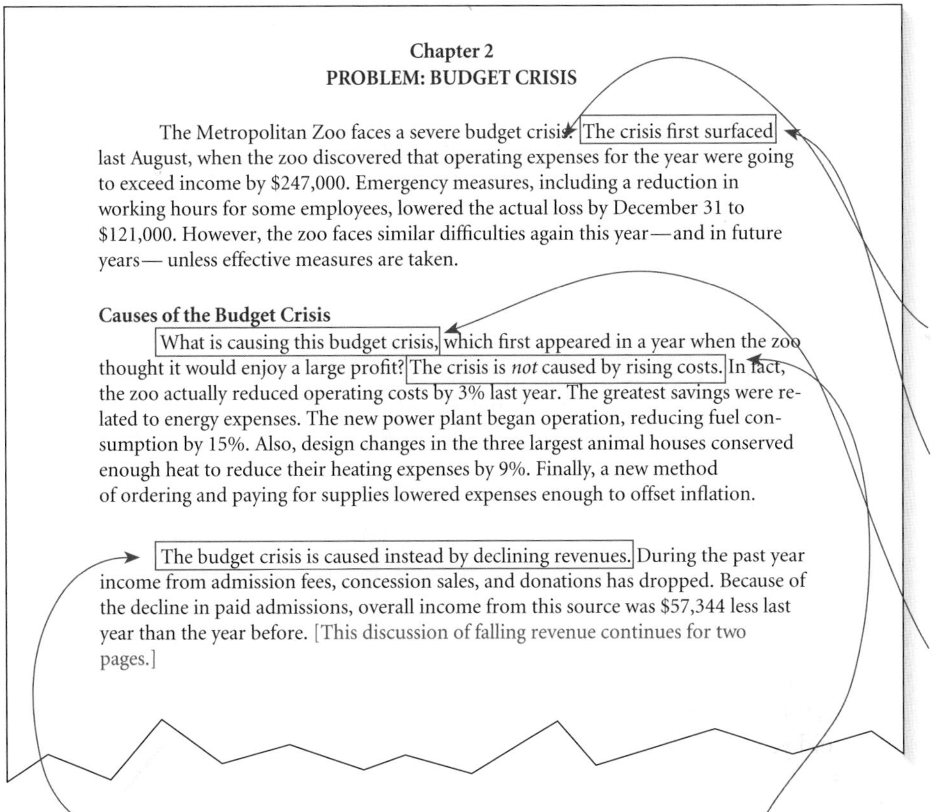

Chapter 2
PROBLEM: BUDGET CRISIS

The Metropolitan Zoo faces a severe budget crisis. The crisis first surfaced last August, when the zoo discovered that operating expenses for the year were going to exceed income by $247,000. Emergency measures, including a reduction in working hours for some employees, lowered the actual loss by December 31 to $121,000. However, the zoo faces similar difficulties again this year—and in future years—unless effective measures are taken.

Causes of the Budget Crisis
What is causing this budget crisis, which first appeared in a year when the zoo thought it would enjoy a large profit? The crisis is *not* caused by rising costs. In fact, the zoo actually reduced operating costs by 3% last year. The greatest savings were related to energy expenses. The new power plant began operation, reducing fuel consumption by 15%. Also, design changes in the three largest animal houses conserved enough heat to reduce their heating expenses by 9%. Finally, a new method of ordering and paying for supplies lowered expenses enough to offset inflation.

The budget crisis is caused instead by declining revenues. During the past year income from admission fees, concession sales, and donations has dropped. Because of the decline in paid admissions, overall income from this source was $57,344 less last year than the year before. [This discussion of falling revenue continues for two pages.]

She provides the topic statement for the second subtopic (falling revenue is the cause).

She provides the topic statement for the first subtopic (rising costs are not the cause).

FIGURE 8.1
Interlocking Topic Statements (see above for the outline of this passage)

The writer has created an interlocking set of topic statements that correspond to the outline shown above.

She announces the topic for the entire chapter in its first sentence (budget crisis).

In the second sentence, she provides the topic statement for the first section for the chapter (history of the crisis). Note that a topic statement does not need to be the first sentence of a paragraph.

The writer provides the topic statement for the second section of the chapter (cause of the budget crisis). Here, she places the topic statement in the first sentence of the entire section.

Guideline 2 | Present your generalizations before your details

In many of the segments you write at work, you will present detailed facts about your topic in order to explain or support a general point. You can usually increase the usability and persuasiveness of these segments by not only stating the *topic* but also stating the *main point* you want to make at the beginning.

How Initial Generalizations Make Writing Easier to Understand and Use

When you present your generalizations first, you save your readers the work of trying to figure out what your general point is. Imagine, for instance, that you are a manager who finds the following sentences in a report:

Details without an initial generalization

> Using the sampling technique just described, we passed a gas sample containing 500 micrograms of VCM through the tube in a test chamber set at 25°C. Afterward, we divided the charcoal in the sampling tube into two equal parts. The front half of the tube contained approximately 2.3 of the charcoal, while the back half contained the rest. Analysis of the back half of the tube revealed no VCM; the front half contained the entire 500 micrograms.

As you read these details, you probably find yourself asking, "What does the writer want me to get from this?" You would have been saved that labor if the writer had begun the segment with the following statement:

> We have conducted a test that demonstrates the ability of our sampling tube to absorb the necessary amount of VCM under the conditions specified.

By placing the generalization at the head of the paragraph, the writer would also help you use the segment more efficiently as you performed the managerial task of determining whether the writer's conclusion is valid. Because you would already know the writer's generalization, you could immediately assess whether each detail does, indeed, provide adequate support for the conclusion that the sample tube absorbs the necessary amount of VCM. In contrast, if the writer waited to reveal his or her conclusion after presenting the details, you would have to recall each detail from memory—or even reread the passage—in order to assess the strength of its support for the writer's conclusion.

How Initial Generalizations Make Writing More Persuasive

Left to themselves, readers are capable of deriving all sorts of generalizations from a passage. Consider the following sentences:

> Richard moved the gas chromatograph to the adjacent lab.

> He also moved the electronic balance.

> And he moved the experimental laser.

One reader might note that everything Richard moved is a piece of laboratory equipment and consequently might generalize that "Richard moved some laboratory equipment from one place to another." Another reader might observe that everything Richard moved was heavy and therefore might generalize that "Richard is strong." A member of a labor union in Richard's organization might generalize that "Richard was doing work that should have been done by a union member, not by a manager" and might file a grievance. Different generalizations lead to different outcomes.

A key point is that readers naturally formulate generalizations even if none are provided. Of course, when you are writing persuasively, you want your readers to draw one particular conclusion—and not other possible ones. You increase your chances of succeeding by stating your desired generalization explicitly and by placing that generalization ahead of your supporting details so that your readers encounter it before forming a different generalization on their own.

Present your generalization before your readers begin to formulate contradictory ones.

Sometimes You Shouldn't Present Your Generalizations First

Although you usually strengthen your segments by stating your generalizations before your details, you will encounter situations where delaying the generalizations will be more effective. As explained in Chapter 5, if you begin a segment with a generalization that is likely to provoke a negative reaction from your readers, you may decrease your communication's persuasiveness. In such cases, you usually increase your communication's persuasiveness by postponing your general points until *after* you've laid the relevant groundwork with your details by using the indirect organizational pattern (described in Chapter 5).

How Guideline 2 Relates to Guideline 1

Taken together, Guidelines 1 and 2 suggest that in most cases you announce your topic and state your main point about it at the beginning of each segment. You can devote a separate sentence to each purpose. In the following example from the beginning of a two-page test report, the first sentence announces the topic and the second states the engineer's main conclusion.

> We conducted tests to determine whether the plastic resins can be used to replace metal in the manufacture of the CV-200 housing. The tests showed that there are three shortcomings in plastic resins that make them unsuitable as a replacement for the metal.

Separate sentences state the topic and the writer's generalization.

Often, however, you can make your writing more concise and forceful by stating both your topic and your main point in a single sentence.

> Our tests showed three shortcomings in plastic resins that make them unsuitable as a replacement for metal in the manufacture of the CV-200 housing.

Topic and generalization are combined in one sentence.

GUIDELINES FOR ORGANIZING THE INFORMATION IN YOUR SEGMENTS

Guidelines 1 and 2 discussed strategies for beginning your segments. The next three guidelines focus on ways to organize the material that follows your opening sentence or sentences.

Guideline 3 | **Move from most important to least important**

In some segments, you will present parallel pieces of information, such as a list of five recommendations or an explanation of three causes of a problem. Whether you devote a single sentence or several paragraphs, to each item you can usually increase your segment's usability by presenting the most important item first and proceeding in descending order of importance. This organization helps readers who scan to locate your key points. It also increases your communication's persuasiveness by presenting the strongest support for your arguments in the most prominent spot.

To identify the most important information, consider your communication from your readers' viewpoint. What information will they be most interested in or find most persuasive? For example, in the segment on the three shortcomings of plastic resins, readers will certainly be more interested in the major shortcoming than the minor ones. Similarly, if the readers must be persuaded that plastic resins are not a good substitute for metal, they are sure to find the major shortcoming more compelling than minor ones.

Occasionally, you may encounter situations where you must ignore this guideline in order to present your overall message clearly and economically. For instance, to explain clearly the multiple causes of a flooding along a river, you may need to describe events chronologically even though the event that occurred first was not the one with the greatest impact. In general, though, presenting the most important information first will be most helpful and persuasive to your readers.

Guideline 4 | **Consult conventional strategies when having difficulties organizing**

Every writer occasionally gets stumped when trying to organize a particular paragraph, section, or chapter. Often the problem is one that many others have faced, such as how to describe a certain process or how to explain the causes of a particular event. For many commonly encountered organizational problems, there are conventional strategies for arranging material in ways that will be understandable and useful to readers. By consulting these strategies, you will often find a quick and effective solution to your own problem.

The Writer's Reference Guide following this chapter suggests ways to use seven of these strategies that are especially useful on the job. As you study the strategies, remember to use them only as guides. To make them work in your particular context, you will need to adapt them to your purpose and the needs of your readers.

Guideline 5 | Global Guideline: Consider your readers' cultural background when organizing

The advice you have read so far in this chapter is based on the customs of readers in the United States and other Western countries where readers expect and value what might be called a *linear* organization. In this organizational pattern, writers express their main ideas explicitly and develop each one separately, carefully leading readers from one to another. As international communication experts Myron W. Lustig and Jolene Koester explain, this pattern can be visualized as "a series of steps or progressions that move in a straight line toward a particular goal or idea" (1993, p. 218).

In other cultures, writers and readers are accustomed to different patterns. For example, the Japanese use a nonlinear pattern that many researchers call a *gyre* (Connor, 1996). The writer approaches a topic by indirection and implication because in Japanese culture it's rude and inappropriate to tell the reader the specific point being conveyed. Communication specialist Kazuo Nishiyama (1983) gives an example: When a Japanese manager says, "I'd like you to reflect on your proposal for a while," the manager can mean, "You are dead wrong, and you'd better come up with a better idea very soon. But I don't say this to you directly because you should be able to understand what I'm saying without my being so rude." Similarly, in Hindi (one of the major languages of India), paragraphs do not stick to one unified idea or thought, as they do in the United States and many other Western nations. Linguist Jamuna Kachru (1988) explains that in the preferred Hindi style, the writer may digress and introduce material related to many different ideas.

Because of these cultural differences, serious misunderstandings can arise when your readers are employed by other companies in another country, work for your own company in another country, or even work in your own building but were raised observing the customs of another culture. Such misunderstandings cannot be avoided simply by translating the words of your communication: The whole message must be structured to suit the customs of your readers' culture.

GUIDELINES FOR HELPING YOUR READERS SEE THE ORGANIZATION OF YOUR SEGMENTS

By following the advice given in Chapters 4 and 5 and the three previous guidelines, you can organize your communications in reader-centered ways that are highly usable and persuasive. However, even in the most skillfully organized communication, the relationships among the parts may not be evident to readers. The next two guidelines suggest ways to help your readers *see* the organization you have created.

Guideline 6 | Add signposts that create a map of your organization

One way to help readers see a communication's organization is to put up signs that explain how each part fits with its neighbors. These organizational signposts are *added* to information about the communication's topic. That's because information about organization can be quite distinct from information about subject matter, as the following diagram indicates.

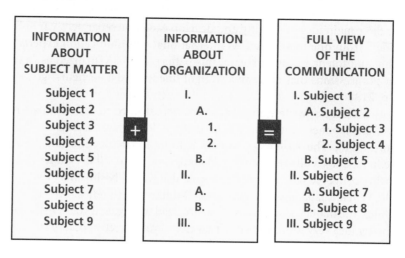

The following sections describe four kinds of signposts you can use to map your communication's organization.

- Forecasting statements
- Transitions
- Headings
- Visual arrangement of your text on the page

Forecasting Statements

Forecasting statements tell the reader the organization of what lies ahead. Often, they appear along with a topic sentence, which they supplement. For instance, here are the first two sentences from a section of a brochure published by a large chain of garden nurseries:

Topic statement followed by a forecasting statement

> Our first topic is the trees found in the American Southwest. <u>Some of the trees are native, some imported</u>.

Forecasting and topic statements are often combined in one sentence:

Topic statement combined with a forecasting statement

> Our first topic is the trees— <u>both native and imported</u>—found in the American Southwest.

Forecasting statements may vary greatly in the amount of detail they provide. The sample sentences above provide both the number and the names of the categories to be discussed. A more general preview is given in the next example, which tells its readers to expect a list of actions but not what these actions are or how many will be discussed:

| To solve this problem, the department must take <u>the following actions.</u> | Forecasting statement

When you are deciding how much detail to include in a forecasting statement, there are three main points to consider.

WRITING FORECASTING STATEMENTS

- **Say something about the segment's arrangement that readers will find helpful.** Usually, the more complex the relationship among the parts, the greater the amount of detail that is needed.
- **Say only as much as readers can easily remember.** A forecasting statement should help readers, not test their memories. When forecasting a segment that will discuss a three-step process, you could name all the steps. If the process has eight steps, state the number without naming them.
- **Forecast only one level at a time.** Don't list all the contents of a communication at its outset. That will only confuse your readers. Tick off only the major divisions of a particular section. If those divisions are themselves divided, provide each of them with its own forecasting statement.

Transitions

When you organize a communication, you understand the connections between adjacent parts—how one topic leads to the second and the second follows from the first. If your readers miss a connection along the way, they will miss part of your meaning.

Transitions include three elements: a reference to the preceding topic, the topic (or topic statement) that is beginning, and the link between the two.

| This large increase in contributions will enable us to expand our free health care program in several ways. | Transition based on cause and effect

This sentence begins with the preceding topic (the increase in contributions) and ends with the next topic (the ways the free health care program might be expanded). The link between the two is that the contributions have created the opportunity for the expansion. Here's another example.

| Having finally reached Dingham Point, the expedition spent the next three days building rafts to cross the river. | Transition based on sequence of events

In this case, the link between the last topic (about the journey to Dingham Point) and the next one (building the raft) is that one followed the other.

Sometimes you can signal transitions without using any words at all. For example, in a report that presents three brief recommendations, you might arrange the recommendations in a numbered list. The numbers themselves would provide the transition from one recommendation to the next. Similarly, in a memo covering

---

several separate topics, the transition from one to the next might be provided by giving each topic a heading (see the discussion of headings that follows).

Headings

Headings are a third kind of signpost for mapping your communication's organization. At work, writers use headings not only in long documents, such as reports and manuals, but also in short ones, such as letters and memos. To see how effectively the insertion of headings can signal organization, look at Figure 8.2, which shows two versions of the same memo, one without headings and one with them. Figure 8.3 highlights the use of headings at a website.

Headings help readers wherever there is a major shift in topic. In much on-the-job writing, such shifts occur every few paragraphs. Avoid giving every paragraph its own heading, which would give your prose a disjointed appearance rather than helping readers see how things fit together. An exception occurs in communications designed to provide readers quick access to specific pieces of information, as in a warranty, troubleshooting guide, reference manual, or fact sheet. In these documents, headings may even label sentence fragments or brief bits of data, turning the communication into something very much like a table of facts.

To be helpful, each heading must unambiguously indicate the kind of information that is included in the passage it labels.

CREATING TEXT FOR HEADINGS

- **Ask the question that the segment will answer for your readers.** Headings that ask questions such as "What happens if I miss a payment on my loan?" or "Can I pay off my loan early?" are especially useful in communications designed to help readers decide what to do.
- **State the main idea of the segment.** This strategy is often used in documents that offer advice or guidance. A brochure on bicycling safety uses headings such as "Ride with the Traffic," "Use Hand Signals," and "Ride Single File."
- **Use a key word or phrase.** This type of heading is especially effective when a full question or statement would be unnecessarily wordy. For instance, in a request for a high-end multimedia production system, the section that discusses prices might have a heading that reads "How Much Will the System Cost?" However, the single word "Cost" would serve the same purpose.

Often, parallel headings make a communication's content easier to access and understand. For instance, when you are describing a series of steps in a process, parallel phrasing cues readers that the segments are logically parallel: "Opening the Computer Program," "Entering Data," and so on. However, in some situations, a mix of heading types will tell readers more directly what each section is about. Figure 8.4 shows the table of contents from a booklet titled *Getting the Bugs Out* that uses all three types of headings. Notice where they are parallel— and where they are not.

FIGURE 8.2
The Same Memo with and without Headings

Garibaldi Corporation
INTEROFFICE MEMORANDUM

MEMO   June 15, 2010

TO   Vice Presidents and Department Managers
FROM   Davis M. Pritchard, President
RE   PURCHASES

Three months ago, I appo
the purchase of computer
I am establishing the follo

The task force was to bala
each department purchas
(2) to ensure compatibilit
create an efficient electror

I am designating one "pre
vendors.

The preferred vendor, YY
unless there is a compelli
purchases from the prefe
purchase price so that ind

Two other vendors, AAA
Garibaldi; both computer
network. Therefore, the s
price of these machines.

We will select one preferr
The task force will choose
when the choice is made

David Pritchard's first version lacked visual cues to the memo's organization.

By adding headings, he helped readers see how his memo is organized. The bold headings tell readers that these are the main parts of the memo.

By indenting, he indicated that these are the two parts of the computer policy.

Garibaldi Corporation
INTEROFFICE MEMORANDUM

MEMO   June 15, 2010

TO   Vice Presidents and Department Managers
FROM   Davis M. Pritchard, President
RE   PURCHASES OF COMPUTER AND FAX EQUIPMENT

Three months ago, I appointed a task force to develop corporate-wide policies for the purchase of computers and fax equipment. Based on the advice of the task force, I am establishing the following policies.

Objectives of Policies
The task force was to balance two possibly conflicting objectives: (1) to ensure that each department purchases the equipment that best serves its special needs and (2) to ensure compatibility among the equipment purchased so the company can create an efficient electronic network for all our computer and fax equipment.

Computer Purchases
I am designating one "preferred" vendor of computers and two "secondary" vendors.

Preferred Vendor: The preferred vendor, YYY, is the vendor from which all purchases should be made unless there is a compelling reason for selecting other equipment. To encourage purchases from the preferred vendor, a special corporate fund will cover 30% of the purchase price so that individual departments need fund only 70%.

Secondary Vendor: Two other vendors, AAA and MMM, offer computers already widely used in Garibaldi; both computers are compatible with our plans to establish a computer network. Therefore, the special corporate fund will support 10% of the purchase price of these machines.

Fax Purchases
We will select one preferred vendor and no secondary vendor for fax equipment. The task force will choose between two candidates: FFF and TTT. I will notify you when the choice is made early next month.

FIGURE 8.3
Headings Used to Indicate
Organization of a Website

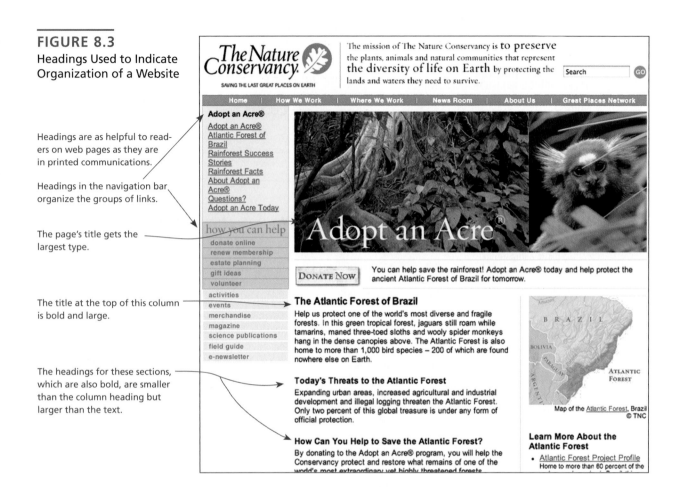

Headings are as helpful to readers on web pages as they are in printed communications.

Headings in the navigation bar organize the groups of links.

The page's title gets the largest type.

The title at the top of this column is bold and large.

The headings for these sections, which are also bold, are smaller than the column heading but larger than the text.

The visual appearance of headings is important to readers. To be useful, headings must stand out visually.

DESIGNING HEADINGS VISUALLY

1. **Make headings stand out from the text, perhaps using these strategies.**
   - Use bold.
   - Use a different color than is used for the text.
   - Place headings at the left-hand margin or center them.
2. **Make major headings more prominent than minor ones. Here are some strategies to consider.**
   - Make major headings larger.
   - Center the major headings and position the others against the left-hand margin.

- Use all capital letters for the major headings and initial capital letters for the others.
- Give the major headings a line of their own and put the others on the same line as the text that follows them.

3. **Give the same visual treatment to headings at the same level in the hierarchy.**

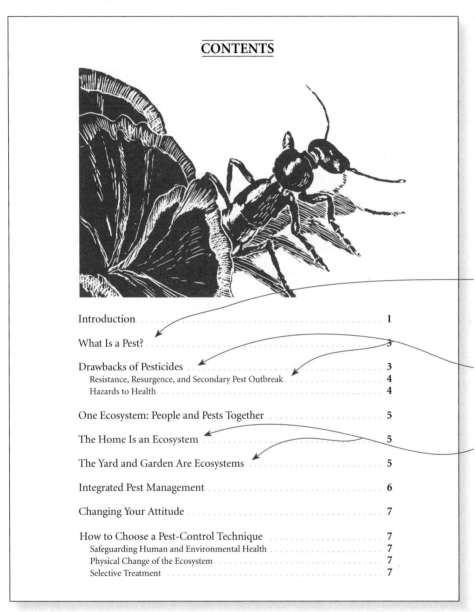

CONTENTS

FIGURE 8.4
Table of Contents from a Booklet that Uses Three Kinds of Chapter and Section Headings

This table of contents shows all the headings used in this short booklet. The writers used several kinds of headings described in this chapter.

The writers used a **question** for this heading to grab the readers' interest. A less interesting title would have been "Definition of a Pest."

The writers used **key words** and phrases for these two topics.

By using **full sentences** as the headings for these sections, the writers ensured that their readers would know the main points of these sections. Because the headings suggest new ways for people to think about their homes and gardens, the writers also raise people's curiosity and thereby increase their interest in reading the sections.

Usually, you can use two or more of these techniques together so they reinforce one another. By doing so, you can readily create two or three easily distinguishable levels of headings that instantly convey the organizational hierarchy of your communication. Section or chapter titles may also function as a type of heading, as in Figure 8.5.

Sometimes, headings include the numbers and letters that are used in an outline. In most circumstances, however, the numbers and letters of an outline diminish the effectiveness of headings by distracting the readers' eyes from the headings' key words.

By convention, headings and the topic statements that follow them reinforce each other, with the topic statement repeating one or more key words from the heading. Topic statements do not contain pronouns that refer to the headings. For example, the heading "Research Method" would not be followed by a sentence that says, "Designing this was a great challenge." Instead, the sentence would read, "Designing the research method was a great challenge."

Visual Arrangement of Text

You can also signal your communication's organization through the visual arrangement of your text on the page.

ARRANGING TEXT VISUALLY TO SIGNAL ORGANIZATION

- **Adjust the location of your blocks of type.** Here are three adjustments you can make:
  - Indent paragraphs lower in the organizational hierarchy. (See Figure 8.6.)
  - Leave extra space between the end of one major section and the beginning of the next.
  - In long communications, begin each new chapter or major section on its own page, regardless of where the preceding chapter or section ended.
- **Use lists.** By placing items in a list, you signal readers that items hold a parallel place in your organizational hierarchy. Usually, numbers are used for sequential steps. Bullets are often used where a specific order isn't required.

Number List	**Bullet List**
The three steps we should take are:	Three features of the product are:
1. _____	■ _____
2. _____	■ _____
3. _____	■ _____

When you are constructing a list, give the entries a parallel grammatical construction: All the items should be nouns, all should be full sentences, or all should

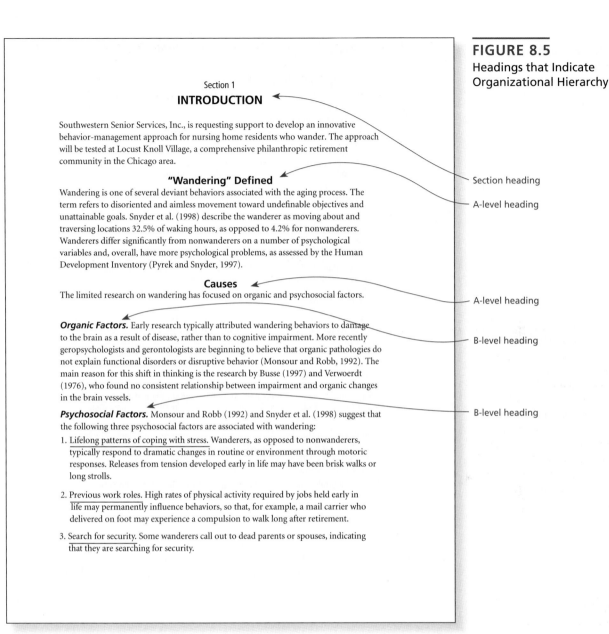

FIGURE 8.5
Headings that Indicate
Organizational Hierarchy

Section 1

INTRODUCTION

Southwestern Senior Services, Inc., is requesting support to develop an innovative behavior-management approach for nursing home residents who wander. The approach will be tested at Locust Knoll Village, a comprehensive philanthropic retirement community in the Chicago area.

"Wandering" Defined

Wandering is one of several deviant behaviors associated with the aging process. The term refers to disoriented and aimless movement toward undefinable objectives and unattainable goals. Snyder et al. (1998) describe the wanderer as moving about and traversing locations 32.5% of waking hours, as opposed to 4.2% for nonwanderers. Wanderers differ significantly from nonwanderers on a number of psychological variables and, overall, have more psychological problems, as assessed by the Human Development Inventory (Pyrek and Snyder, 1997).

Causes

The limited research on wandering has focused on organic and psychosocial factors.

Organic Factors. Early research typically attributed wandering behaviors to damage to the brain as a result of disease, rather than to cognitive impairment. More recently geropsychologists and gerontologists are beginning to believe that organic pathologies do not explain functional disorders or disruptive behavior (Monsour and Robb, 1992). The main reason for this shift in thinking is the research by Busse (1997) and Verwoerdt (1976), who found no consistent relationship between impairment and organic changes in the brain vessels.

Psychosocial Factors. Monsour and Robb (1992) and Snyder et al. (1998) suggest that the following three psychosocial factors are associated with wandering:

1. Lifelong patterns of coping with stress. Wanderers, as opposed to nonwanderers, typically respond to dramatic changes in routine or environment through motoric responses. Releases from tension developed early in life may have been brisk walks or long strolls.

2. Previous work roles. High rates of physical activity required by jobs held early in life may permanently influence behaviors, so that, for example, a mail carrier who delivered on foot may experience a compulsion to walk long after retirement.

3. Search for security. Some wanderers call out to dead parents or spouses, indicating that they are searching for security.

Section heading

A-level heading

A-level heading

B-level heading

B-level heading

FIGURE 8.6
Indentation of Text
Used to Indicate
Organizational Hierarchy

Subordinate material is indented.

To emphasize this caution, the
writer uses an icon and moves
the text margin farther left.

A second level of indentation
signals a second level of
subordination.

Managing Disk Partitions

Use extreme caution when navigating through this next portion of Setup. You can easily create, destroy, and reformat entire disk partitions with a couple of keystrokes. Keep in mind that you're running a cousin of the powerful and potentially destructive DOS FDISK utility.

12. Setup lists hard disks, partitions, and unpartitioned areas on your computer. It then asks you where to install NT 4.0. Use the UP and DOWN ARROW keys to scroll through the list and highlight a partition. Go to step 12a.

    You'll need to find a destination partition with at least 115MB of free space on which to install Windows NT Server.

    In the list, all non-SCSI drives are displayed as "IDE/ESDI Disk." A SCSI drive is displayed as "Disk # at id # on bus # on" followed by the name of the SCSI adapter driver. The ID number is the SCSI ID assigned to the drive.

    Areas of your disks that contain no partition are displayed as "Unpartitioned space." Partitions that have been created but not yet formatted are displayed as "New (Unformatted)" or as "Unformatted or Damaged." Don't worry about the latter. This is just NT's generic way of saying that it doesn't recognize a partition as formatted.

If you're installing Windows NT Server on a computer that contains a previous version of Windows NT and you were using disk stripes, mirrors, or volume sets, these partitions are shown as "Windows NT Fault Tolerance" partitions. Don't delete any of these partitions. See the section entitled "Migrating Fault Tolerance from Windows NT 3.x" in Chapter 19 for details on using existing fault-tolerance partitions.

    If you don't see all of your partitions listed, you may just need to use the UP and DOWN ARROW keys to scroll and display them. Only hard disks are included in this list.

    Don't panic if the drive letter assignments seem out of whack. They probably don't match the drive letters that you see under DOS or even under another version of NT. Once you've got NT up and running, you'll be able to change drive letter assignments very easily.

12a. If you're ready to select a partition on which NT 4.0 will be installed, highlight that partition, press ENTER, and go to step 13. If you're not, go to step 12b to delete an existing partition or step 12c to create a new partition.

    You can select an unformatted partition or an existing formatted FAT or NTFS partition. Unpartitioned space isn't a valid destination for NT installation. You need to partition it first, in step 12c.

    If the partition that you select isn't large enough, Setup will complain and send you back to step 12.

be questions, and so on. Mixing grammatical constructions distracts readers and sometimes indicates a shift in point of view that breaks the tight relationship that should exist among the items.

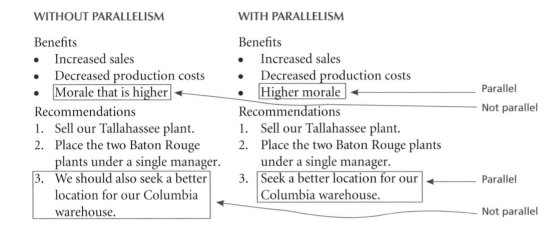

WITHOUT PARALLELISM

Benefits
- Increased sales
- Decreased production costs
- Morale that is higher

Recommendations
1. Sell our Tallahassee plant.
2. Place the two Baton Rouge plants under a single manager.
3. We should also seek a better location for our Columbia warehouse.

WITH PARALLELISM

Benefits
- Increased sales
- Decreased production costs
- Higher morale ←──────── Parallel
                 ──────── Not parallel

Recommendations
1. Sell our Tallahassee plant.
2. Place the two Baton Rouge plants under a single manager.
3. Seek a better location for our ←──── Parallel
   Columbia warehouse.
                 ──────── Not parallel

Guideline 7 | Smooth the flow of thought from sentence to sentence

As people read, they are continuously determining how the sentence they are now reading relates to the sentence they just completed. This connection-making process occurs so rapidly that readers aren't even aware of it until they hit a sentence that doesn't seem to fit. Then, they either stop to figure out the relationship or they push ahead, missing some of the writer's meaning. To avoid these undesirable results, smooth the flow of thought from sentence to sentence, using the following strategies.

- Use transitional words and phrases
- Use echo words
- Keep a steady focus from sentence to sentence

Use Transitional Words and Phrases

Transitional words and phrases can serve as links that tie one sentence to the next. Here are some of the most commonly used transitional words:

Links in time	after, before, during, until, while	Transitional words and phrases
Links in space	above, below, inside	
Links of cause and effect	as a result, because, since	
Links of similarity	as, furthermore, likewise, similarly	
Links of contrast	although, however, nevertheless, on the other hand	

These and many more are undoubtedly familiar to you. The important point is to remember to use them when they will help your readers follow the flow of thought through your communication. When using them, put them where they will help your readers most: at the beginning of sentences. In that position, they immediately signal the relationship between that sentence and the preceding one.

Use Echo Words

Another way to guide your readers from one sentence to the next is to use *echo words*. An echo word is a word or phrase that recalls to the readers' minds some information they've already encountered. For example:

The word echoed is *developed.*
The echo word is *development.*

> A flu vaccine can be developed. Development will take several weeks, however.

In this example, the noun *development* at the beginning of the second sentence echoes the verb in the first.

There are many other kinds of echo words.

- **Pronouns**

In the second sentence, *It* echoes *copier* in the first sentence.

> We had to return the copier. Its frequent breakdowns were disrupting work.

- **Another word from the same "word family" as the word being echoed**

In the second sentence, *Oscilloscope* echoes *lab equipment* in the first sentence.

> I went to my locker to get my lab equipment. My oscilloscope was missing.

- **A word or phrase that recalls some idea or theme expressed but not explicitly stated in the preceding sentence**

In the second sentence, the words *these transactions* echo the purchase and retiring of shares described in the first sentence.

> The company also purchased and retired 17,399 shares of its $2.90 convertible, preferred stock at $5.70 a share. These transactions reduce the number of outstanding convertible shares to 635,200.

Like transitional words and phrases, echo words help readers most when they appear at the beginning of a sentence.

Note that if you use *this* or *that* as an echo word at the beginning of a sentence, you should follow it with a noun. If used alone at the beginning of a sentence, you can leave your readers uncertain about what *this* is.

Original

> Our client rejected the R37 compound because it softened at temperatures about 500°C. This is what our engineers feared.

In this example, the reader would be unsure whether *This* refers to the client's dissatisfaction or the softening of the R37. The addition of a noun after *This* clears up the ambiguity.

Revised

> Our client rejected the R37 compound because it softened at temperatures about 500°C. This softening is what our engineers feared.

Keep a Steady Focus from Sentence to Sentence

The preceding strategies for guiding readers from sentence to sentence involve situations where the two sentences have different topics. In some cases, the topic can remain the same. In these situations, keep a steady focus on that topic by keeping the topic in the subject position of both sentences. You can repeat the same word, use a synonym, or use an echo word (see the discussion of echo words on the previous page).

> The links of the drive chain must fit together firmly. They are too loose if you can easily wiggle two links from side to side more than ten degrees.

These sentences have the same focus because they both have the same subject.

Because "They" in the second sentence is an echo word for "The links" in the first sentence, both sentences have the same subject.

Maintaining a steady focus requires that you consider alternative ways of writing a sentence. For example, you could fashion a sentence in either of the following ways.

> The system has saved thousands of dollars this month alone.

Version A

> Thousands of dollars have been saved by the system this month alone.

Version B

Both versions contain the same information. To decide between them, you could look at the preceding sentence to see which will keep a steady focus. If the preceding sentence had said, "Our company's new inventory system reduces our costs considerably," then Version A would maintain the steady focus because it has the same subject (the system) as the preceding sentence.

> Our company's new inventory system reduces costs considerably. The system saved thousands of dollars this month alone.

Both sentences have the same subject.

Learn More

For more on the active and passive voice, see Chapter 9, page 272.

In order to keep a steady focus, you may sometimes need to use a sentence that has the passive voice. As Chapter 9 explains, it is usually desirable to use the active voice, not the passive. However, the chapter also explains that sometimes the passive is more appropriate, even preferable, to the active. One such time occurs when the passive voice enables you to avoid a needless shift in the topic of two adjacent sentences. Consider the following paragraph from an accident report.

> After lunch on Tuesday, Tom took a shortcut back to his workstation. Fifteen yards above the factory floor, a can of paint slipped off a scaffold and hit him on the left foot. Consequently, he missed seventeen days of work.

The focus shifts from the first sentence to the second.

The subject of the first and third sentences is the same: *Tom* and *he*. However, the second sentence shifts the topic from Tom to the can of paint. Furthermore, because the second sentence shifts, the third does also in bringing the focus back to Tom. The writer could avoid these two shifts by rewriting the second sentence in the passive voice, making Tom its subject, not the can of paint.

> He was hit on the left foot by a can of paint that slipped off a scaffold fifteen yards above the factory floor.

Better second sentence

Guideline 8 | Ethics Guideline: Examine the human consequences of what you're drafting

When drafting their communications, employees sometimes become so engrossed in the technical aspects of their subject that they forget the human consequences of what they're writing. When this happens, they write communications in which their stakeholders are overlooked. Depending on the situation, the consequences can be quite harmful or relatively mild—but they can always lead to the unethical treatment of other people.

Mining Accidents

An example is provided by Beverly A. Sauer (2003), who has studied the reports written by the federal employees who investigate mining accidents in which miners are killed. In their reports, Sauer points out, the investigators typically focus on technical information about the accidents without paying sufficient attention to the human tragedies caused by the accidents. In one report, for example, the investigators describe the path of an underground explosion as it traveled through an intricate web of mineshafts and flamed out of various mine entrances. At one entrance, investigators write, "Debris blown by the explosion's forces damaged a jeep automobile parked near the drift openings." The investigators don't mention in this passage that in addition to damaging the jeep, the explosion killed sixteen miners who were in the mineshafts.

In addition to overlooking the victims of the disasters, the investigators' reports often fail to identify the human beings who created the conditions that caused the accidents. One report says, "The accident and resultant fatality occurred when the victim proceeded into an area of known loose roof before the roof was supported or taken down." This suggests that the miner was crushed to death by a falling mine roof because he was careless. Sauer's research showed, however, that in the same mines ten fatalities had occurred in five years—and seven resulted from falling roofs. Managers of the mines were not following safety regulations, and mine safety inspectors were not enforcing the law.

Writing with Awareness of Human Consequences

Of course, there's nothing the inspectors' reports can do on behalf of the deceased miners or their families. However, the stakeholders in the inspectors' reports include other miners who continue to work in what is the most dangerous profession in the United States. As Sauer points out, the investigators' readers include the federal officials responsible for overseeing the nation's mining industry. If the inspectors wrote in ways that made these readers more aware of the human consequences of mining accidents—and of the human failings that often bring them about—the federal officials might be more willing to pass stricter laws and insist that existing regulations be strictly enforced.

In Europe, where government regulation of mining is much stronger, the death and injury of miners are much rarer events than in the United States. The high accident rate that makes mining so dangerous in the United States, she argues, results in part from the way mining investigators write their reports.

Of course, most people aren't in professions where lives are at stake. In any profession, however, it's possible to become so focused on your technical subject matter that you forget the human consequences of your writing.

To avoid accidentally treating others unethically, you can take the following steps.

Miners in the United States would be safer if the persons who prepare mining accident reports paid more attention to the human consequences of their writing, Dr. Beverly Sauer argues.

AVOIDING ACCIDENTALLY TREATING OTHERS UNETHICALLY

- **When beginning work on a communication, identify its stakeholders** (page 90). Certainly, the stakeholders of the mining disaster reports include miners whose lives are endangered if government officials don't enact and enforce life-saving safety measures.
- **Determine how the stakeholders will be affected by your communication** (page 112).
- **Draft your communication in a way that reflects proper care for these individuals.** Be sure that all your decisions about what to say, what *not* to say, and how to present your message are consistent with your personal beliefs about how you should treat other people.

CONCLUSION

This chapter has suggested that you can increase the usability and persuasiveness of your communications if you begin your segments with topic statements; present generalizations before details; organize from most important to least important; provide readers with a map of your communication's organization with headings, forecasting statements, and similar devices; and consider your readers' cultural background.

Remember that the first seven guidelines in this chapter are suggestions, not rules. The only "rule" for writing segments is to be sure your readers know what you are talking about and how your various points relate to one another. Sometimes you will be able to do this without thinking consciously about your techniques. At other times, you will be able to increase the clarity and persuasiveness of your segments by drawing on this chapter's advice.

To assure that you are always drafting ethically, the eighth guideline is something you *should* treat as a rule: Remember the human consequences of what you are writing.

USE WHAT YOU'VE LEARNED

For additional exercises, go to www.cengage.com/english/anderson7e. *Instructors*: The book's website includes suggestions for teaching the exercises.

EXERCISE YOUR EXPERTISE

1. Circle the various parts and subparts of the passage in Figure 8.7 to show how the smaller segments are contained within larger ones.

FIGURE 8.7
Passage Containing Several Levels of Segments

Importing Insects

By importing insects from other parts of the world, nations can sometimes increase the productivity of their agricultural sector, but they also risk hurting themselves. By importing insects, Australia controlled the infestation of its continent by the prickly pear, a cactus native to North and South America. The problem began when this plant, which has an edible fruit, was brought to Australia by early explorers. Because the explorers did not also bring its natural enemies, the prickly pear grew uncontrolled, eventually rendering large areas useless as grazing land, thereby harming the nation's farm economy. The problem was solved when scientists in Argentina found a small moth, *Cactoblastis cactorum*, whose larvae feed upon the prickly pear. The moth was imported to Australia, where its larvae, by eating the cactus, reopened thousands of acres of land.

In contrast, the importation of another insect, the Africanized bee, could threaten the well-being of the United States. It once appeared that the importation of this insect might bring tremendous benefits to North and South America. The Africanized bee produces about twice as much honey as do the bees native to the Americas.

However, the U.S. Department of Agriculture now speculates that the introduction of the Africanized honey bee into the U.S. would create serious problems arising from the peculiar way the Africanized honey bee swarms. When the honey bees native to the United States swarm, about half the bees leave the hive with the queen, moving a small distance. The rest remain, choosing a new queen. In contrast, when the Africanized honey bee swarms, the entire colony moves, sometimes up to fifty miles. If Africanized bees intermix with the domestic bee population, they might introduce these swarming traits. Beekeepers could be abandoned by their bees, and large areas of cropland could be left without the services of this pollinating insect. Unfortunately, the Africanized honey bee is moving slowly northward to the United States from Sao Paulo, Brazil, where several years ago a researcher accidentally released 27 swarms of the bee from an experiment.

Thus, while the importation of insects can sometimes benefit a nation, imported insects can also alter the nation's ecological system, thereby harming its agricultural business.

2. Identify the topic statements in Figure 8.7 by putting an asterisk before the first word of each sentence that indicates the topic of a segment.

3. Circle all the forecasting statements in Figure 8.5 (page 223). For those segments that lack explicit forecasting statements, explain how readers might figure out the way in which they are organized.

NOTE: Additional exercises are provided at the end of the Writer's Reference Guide that follows this chapter. See page 261.

EXPLORE ONLINE

Examine the ways that this chapter's guidelines are applied by a website that explains technical or scientific topics. For example, study the explanation of a medical topic at the National Cancer Institute (www.nci.nih.gov), the National Institute on Drug Abuse (www.nida.nih.gov), or WebMD (www.WebMD.com). Present your analysis in the way your instructor requests.

COLLABORATE WITH YOUR CLASSMATES

1. Figure 8.8 shows an e-mail whose contents have been scrambled. Each statement has been assigned a number. Working with one or two other students, do the following:

   a. Write the numbers of the statements in the order in which the statements would appear if the e-mail were written in accordance with the guidelines in this chapter. Place an asterisk before each statement that would begin a segment. (When you order the statements,

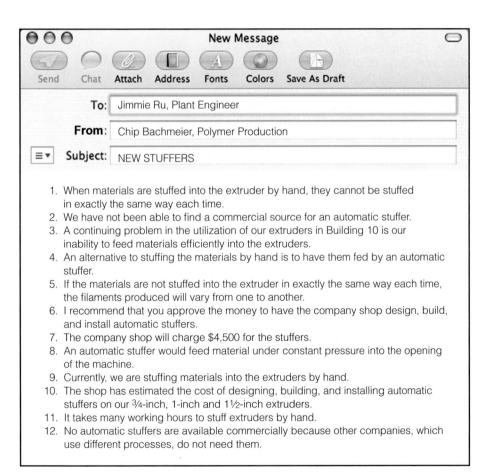

FIGURE 8.8
Memo for Collaboration Exercise

New Message

Send   Chat   Attach   Address   Fonts   Colors   Save As Draft

To: Jimmie Ru, Plant Engineer

From: Chip Bachmeier, Polymer Production

Subject: NEW STUFFERS

1. When materials are stuffed into the extruder by hand, they cannot be stuffed in exactly the same way each time.
2. We have not been able to find a commercial source for an automatic stuffer.
3. A continuing problem in the utilization of our extruders in Building 10 is our inability to feed materials efficiently into the extruders.
4. An alternative to stuffing the materials by hand is to have them fed by an automatic stuffer.
5. If the materials are not stuffed into the extruder in exactly the same way each time, the filaments produced will vary from one to another.
6. I recommend that you approve the money to have the company shop design, build, and install automatic stuffers.
7. The company shop will charge $4,500 for the stuffers.
8. An automatic stuffer would feed material under constant pressure into the opening of the machine.
9. Currently, we are stuffing materials into the extruders by hand.
10. The shop has estimated the cost of designing, building, and installing automatic stuffers on our ¾-inch, 1-inch and 1½-inch extruders.
11. It takes many working hours to stuff extruders by hand.
12. No automatic stuffers are available commercially because other companies, which use different processes, do not need them.

ignore their particular phrasing. Order them according to the information they provide the reader.)

b. Using the list you just made, rewrite the e-mail by rephrasing the sentences so that the finished message conforms with all the guidelines in this chapter.

CASE | INCREASING ORGAN DONATIONS

 For additional cases, visit www.cengage.com/english/anderson7e. Instructors: The book's website includes suggestions for teaching cases.

Case created by Gail S. Bartlett

You've been working for the past six months at Organ Replacement Gives a New Start (ORGANS), a regional clearinghouse for information about organ donation. At any given moment, 30,000 people in the United States are on a waiting list for organ transplants. Seven of them will die today, and every 20 minutes another will join the waiting list. It is believed that from 12,000 to 15,000 potential organ donors die each year in the United States, but only 4,500 of them actually donate. One donor may provide as many as six organs for transplant.

To discover ways of increasing the number of potential organ donors in the area, Eleanor Gaworski, executive director of ORGANS, asked staff member Aaron Nicholson to study and report on a recent Gallup survey of attitudes toward organ donation. However, Aaron suddenly left ORGANS to take another job. Eleanor has asked you to prepare the report using the information he has collected. The information includes the following tables and notes.

YOUR ASSIGNMENT

According to your instructor's directions, do one of the following:

A. For one of the tables, or for a group of the tables identified by your instructor, state the important conclusion or conclusions that you draw, explain the evidence that supports your conclusion, and make a recommendation.

B. Using all the tables or a group identified by your instructor, write a full report to Eleanor. Include an introduction; a brief explanation of the survey method used; the

key conclusions you draw, together with the specific results that support them; and your recommendation.

Notes Aaron Left

Largest survey ever conducted on attitudes about organ donation

National survey—6,127 people interviewed

Procedures used make results representative of the U.S. adult population

Telephone survey

In tables, not all percentages total 100 percent because of rounding.

Organs taken only from people who are brain-dead

Brain-dead means there is no hope of recovery.

Younger people have more organs to donate; as people get older, fewer of their organs are suitable for donation.

People do not need to sign an organ donor card or indicate on their driver's license their desire to donate organs.

Next of kin's permission is always needed even if a person has indicated a desire to donate organs.

1. Do you support or oppose the donation of organs for transplants?

Total Age	Support	Oppose	Don't Know
	85%	6%	9%
18–24	81%	10%	9%
25–34	85	7	8
35–44	91	4	5
45–54	86	4	9
55+	81	7	11

2. How likely are you to want to have your organs donated after your death?

Total Age	Very Likely	Some-what Likely	Not Very/ Not at All Likely	Don't Know
	37%	32%	25%	6%
18–24	33%	41%	22%	5%
25–34	38	40	17	5
35–44	50	28	17	5
45–54	41	36	18	5
55+	27	26	38	9

3. Is there a particular reason you are not likely to want to have your organs donated upon your death? What might that reason be? (Asked of persons who reported they are not likely to want to have their organs donated)

Response Mentioned	% of Times
Medical reasons	13%
Believe I'm too old	10
Don't want body cut up/Want to be buried as a whole person	9
Don't feel right about it	6
Against religion	5
Other	10
No reason/Don't know/ Haven't given much thought	47

4. Have you made a personal decision about whether or not you would want your or your family members' organs donated in the event of your or their deaths?

Percent Who Have Made Decision about:

Total Age	Own Organs	Family Members' Organs
	42%	25%
18–24	40%	23%
25–34	43	26
35–44	49	32
45–54	46	33
55+	36	17

Attitude Toward Organ Donation

Support	45%	27%
Oppose	39	25

5. Have you told some member of your family about your wish to donate your organs after your death? (Asked of respondents who reported themselves likely to wish to become organ donors)

Total Age	Yes
	52%
18–24	39%
25–34	52
35–44	58
45–54	51
55+	52

6. How willing are you to discuss your wishes about organ donation with your family? Would you say very willing, somewhat willing, not very willing, or not at all willing? (Asked of those who have not discussed wishes with family)

Response	Likely to Donate	Not Likely to Donate
Very willing	36%	23%
Somewhat willing	53	35
Not very willing	7	13
Not at all willing	3	23
Don't know	2	4

7. If you had *not discussed* organ donation with a family member, how likely would you be to donate his or her organs upon death?

Total Age	Very/ Somewhat Likely	Not Very/ Not at All Likely	Don't Know
	47%	45%	7%
18–24	44%	52%	4%
25–34	51	44	5
35–44	54	42	4
45–54	51	43	6
55+	38	49	13

Attitude Toward Organ Donation

	Very/ Somewhat Likely	Not Very/ Not at All Likely	Don't Know
Support	52%	41%	7%
Oppose	11	84	5

8. If a family member *had requested* that his or her organs be donated upon death, how likely would you be to donate the organs upon death?

Total Age	Very/ Somewhat Likely	Not Very/ Not at All Likely	Don't Know
	93%	5%	2%
18–24	94%	5%	1%
25–34	94	5	2
35–44	97	2	1
45–54	96	3	1
55+	87	8	3

Attitude Toward Organ Donation

Support	95%	3%	1%
Oppose	69	24	6

9. Most of the people who need an organ transplant receive a transplant.

Total Age	Strongly Agree/Agree	Disagree/ Strongly Disagree	Don't Know
	20%	68%	12%
18–24	32%	54%	14%
25–34	19	72	9
35–44	14	74	12
45–54	12	75	13
55+	24	63	13

Attitude Toward Organ Donation

Support	19%	70%	11%
Oppose	32	55	14

10. It is possible for a brain-dead person to recover from his or her injuries.

Total Age	Strongly Agree/Agree	Disagree/ Strongly Disagree	Don't Know
	21%	63%	16%
18–24	28%	52%	19%
25–34	25	60	15
35–44	20	65	15
45–54	16	71	13
55+	18	64	17

Attitude Toward Organ Donation

Support	20%	65%	15%
Oppose	33	51	15

11. I am going to read you a couple of statements. For each one, please tell me if that statement must be true or not true before an individual can donate his or her organs.

Statement	True	False	Don't Know
The person must carry a signed donor card giving permission.	79%	15%	5%
The person's next of kin must give his or her permission.	58%	34%	8%

12. In the past year, have you read, seen, or heard any information about organ donation?

Total Age	Yes
	58%
18–24	36%
25–34	47
35–44	61
45–54	65
55+	68

Attitude Toward Organ Donation

Support	61%
Oppose	38

Survey results are from the Gallup Organization, Inc., *The American Public's Attitudes toward Organ Donation and Transplantation*, conducted for The Partnership for Organ Donations, Boston, MA, February 1993. Used with permission.

USING SEVEN READER-CENTERED ORGANIZATIONAL PATTERNS

This guide provides detailed, reader-centered advice for using seven patterns for organizing information and arguments. In some brief communications, you may use only one of these patterns, but in most cases, you will weave them together, as described on page 260.

   To use this guide effectively, consult the table below to identify the pattern that will achieve the goal of the paragraph, section, or chapter you are drafting.

CONTENTS

On the job, you will sometimes have to write about what seems to be a miscellaneous set of facts. To organize these facts, you can use a strategy called classification. In *classification*, you arrange your material into groups of related items that satisfy the following criteria:

- **Every item has a place.** In one group or another, every item fits.
- **Every item has only one place.** If there are two logical places for an item, you must describe it in both locations, thereby creating redundancy. Or you will have to describe it in only one location, thereby requiring your readers to guess where to find the information. Neither alternative is desirable from your readers' perspective.

> Even when using formal classification, group items in a way that will be helpful to your readers.

- **The groupings are useful to your readers.** Items that readers will use together should be grouped together.

There are two types of classification: formal classification, which is discussed below, and informal classification, which is described on pages 239–241.

How Formal Classification Works

In formal classification, you group items according to a principle of classification—that is, according to some observable characteristic that every item possesses. Usually, you will have several to choose from. While writing a marketing brochure about sixty adhesives manufactured by her employer, a chemical company, Esther could use any characteristic possessed by all the adhesives, such as price, color, and application (that is, whether it is used to bond wood, metal, or ceramic). To choose among these potential principles of classification, she thought about the way consumers would use her brochure. Realizing that they would look for the adhesive best suited to the particular job they were doing, Esther organized around the type of material each adhesive was designed to bond.

> Esther organized according to the kinds of material the adhesives will bond because she thought her readers would want to find the adhesives that would work with the materials they had.

In classification, large groups can be organized into subgroups. Within her sections on adhesives for wood, metal, and ceramic, Esther subdivided the adhesives according to strength of the bond created by each one.

GUIDELINES FOR FORMAL CLASSIFICATION

1. **Choose a principle of classification that is suited to your readers and your purpose.** Consider the way your readers will use the information you provide.

2. **Use only one principle of classification.** To create a hierarchical organization that has only one place for each item, you must use only one principle of classification at a time. For example, you might classify the cars owned by a large corporation as follows:

> **Learn More**
>
> The formal classification pattern is often combined with other patterns. See page 260.

Cars built in the United States
Cars built in other countries — Valid classification

This grouping uses only one principle of classification—the country in which the car was manufactured. Because each car was built in only one country, each would fit into only one group. Suppose you classified the cars this way:

Cars built in the United States
Cars built in other countries
Cars that are expensive — Faulty classification

This classification is faulty because two principles are being used simultaneously—country of manufacture and cost. An expensive car built in the United States would fit into two categories. You can use different principles at different levels in a hierarchy, as the following classification shows.

Cars built in the United States
    Expensive ones
    Inexpensive ones
Cars built in other countries — Corrected classification
    Expensive ones
    Inexpensive ones

In this case, an expensive car built in the United States would have only one place at each level of the hierarchy.

Classification in Graphics

Tables and graphs often accompany passages organized by formal classification. In the chart below, the principle of classification is the country that owned a successfully launched spacecraft, even if it paid another country to lift the craft into space.

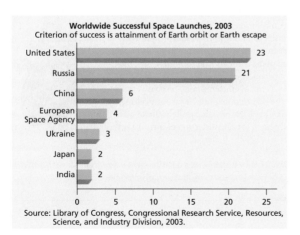

Worldwide Successful Space Launches, 2003
Criterion of success is attainment of Earth orbit or Earth escape

United States	23
Russia	21
China	6
European Space Agency	4
Ukraine	3
Japan	2
India	2

Source: Library of Congress, Congressional Research Service, Resources, Science, and Industry Division, 2003.

Try This

Try classifying the students in your class in two different ways. Include one level of subcategories for each classification. Try to identify someone who could make practical use of information about the students that is organized in each of these ways.

Learn More

For advice about constructing graphics, go to Chapter 13 (page 331) and the Writer's Reference Guide: Creating Eleven Types of Reader-Centered Graphics (page 357).

WWW

To see other communications that use formal classification, go to Writer's Reference Guide: Using Seven Organizational Patterns at **www.cengage.com/english/anderson7e**.

Passage Organized by Formal Classification

In the following passage, the writer uses formal classification to organize his discussion of various methods for detecting coronary heart disease. For his principle of classification, he uses the extent to which each method requires physicians to introduce something into the patient's body. The writer selected this principle of classification because he wants to focus on a new method for which nothing needs to be placed in the patient's body.

The principle of classification is the extent to which the methods involve placing something in the person's body ("invading" it).

The first group of methods involves no invasiveness.

Several examples are listed.

The second group of methods does involve invasiveness.

Two invasive methods are each described in a separate paragraph: the thallium test and cardiac catheterization.

Detecting Coronary Artery Disease

Coronary artery disease (CAD) is the leading cause of death in industrialized countries. In fact, one-third of all deaths in the U.S. are attributable to it. For this reason, the early detection of CAD has long been regarded as among the most vital areas of medical research, and several moderately invasive diagnostic methods have been developed.

Non-invasive methods gather diagnostic information without introducing anything into a patient's body. These methods include traditional physical examinations and history taking, electrocardiography, and echocardiography (ultrasonic imaging).

In contrast, invasive methods involve introducing a substance or object into the person's body. A moderately invasive method is the thallium test, in which a compound of radioactive thallium-201 is injected into the patient and then distributes itself throughout the myocardium (heart muscle) in proportion to the myocardial blood flow. The low-flow regions are detectable as cold spots on the image obtained from a radioisotope camera set over the chest. Although quite sensitive and accurate, the thallium test is also costly and time-comsuming.

As far as invasive techniques are concerned, the most reliable way to diagnose CAD is by means of cardiac catheterization, in which a catheter is inserted into a large artery (usually in the upper arm or thigh) and advanced to the heart. Once the catheter is positioned in the heart, a radio opaque dye is released through it, making it possible to observe the condition of the coronary arteries with X-rays. Although it produces excellent images and definitive diagnoses, cardiac catheterization is expensive, painful, and time-consuming—and it carries an element of risk. For these reasons, an equally accurate, non-invasive method for early detection of coronary artery disease is greatly to be preferred.

[The article continues with a description of a new technology that may provide a highly accurate, non-invasive early detection method.]

© Oguz Aral 2009/Shutterstock.com

Learn More

Exercises for organizing with formal classification are at the book's website and on page 261.

Informal Classification (Grouping Facts)

Informal classification can help you create a reader-centered communication when you need to organize information about a large number of items but find it impossible or undesirable to classify them according to the kind of objective characteristic that is necessary for formal classification.

For example, Calvin needed to organize his analysis, requested by his employer, of advertisements in three trade journals for the heavy equipment industry. Calvin could have created a formal classification by grouping the ads according to an objective characteristic, such as their size or number of words used in them. Instead, however, he classified them according to the type of advertising appeal they made. Obviously, "type of advertising appeal" is not an objective characteristic. Defining an ad's appeal requires subjective interpretation and judgment. Calvin used this informal classification because it best matched his reader's goal, which was to plan the advertising strategies he would use later in the year when he began placing ads in the three journals.

Like formal classification, informal classification enables you to organize communications in a way that achieves the following goals (page 236).

- Every item has a place.
- Every item has only one place.
- The groupings are useful to your readers.

Take a reader-centered approach when using informal classification just as you do when using formal classification.

GUIDELINES FOR INFORMAL CLASSIFICATION

1. **Group your items in a way that is suited to your readers and your purpose.** Calvin organized his analysis around "type of advertising appeal" because he knew his employer was looking for advice about the design of ads.

2. **Create logically parallel groups.** For instance, if you were classifying advertisements, you wouldn't organize into groups like this:

   Focus on price
   Focus on established reputation
   Focus on advantages over a competitor's product
   Focus on one of the product's key features
   Focus on several of the product's key features

   The last two categories are at a lower hierarchical level than the other three. To make the categories parallel, you could combine them in the following way:

   Focus on price
   Focus on established reputation
   Focus on advantages over a competitor's product
   Focus on the product's key features
       Focus on one key feature
       Focus on several key features

Learn More

The informal classification pattern is often combined with other patterns. See page 260.

3. **Avoid overlap among groups.** Even when you cannot use strict logic in classifying items, strive to provide one and only one place for each item. To do this, you must avoid overlap among categories. For example, in the following list the last item overlaps the others because photographs can be used in any of the other types of advertisements listed.

> Focus on price
> Focus on established reputation
> Focus on advantages over a competitor's product
> Focus on the product's key features
> Use of photographs

Table Organized Using Informal Classification

In the following table, the writer used informal classification to organize advice about preventing identity theft.

The types of identity theft described in this table are organized using informal classification.

- There is not a consistent principle of classification. Most of the categories describe what is stolen (e.g., your Social Security number). The last two describe the means used to steal.

- There is overlap among some of the categories. For example, a thief can use a stolen Social Security number to open a credit card account in your name.

Remember that informal classification is not a flaw. Communications organized by informal classification can be very useful.

PROTECT YOURSELF FROM IDENTITY THEFT		
Type of Theft	**What Thieves Do**	**Protection**
Social Security Number	■ Open credit card accounts in your name ■ Obtain loans in your name	■ Give your number only to businesses you know and trust ■ Ask to give an alternative identifier when your SS number is requested
Credit Cards	■ Steal your wallet or purse ■ Intercept information when you are buying online ■ Create fraudulent online businesses ■ Steal preapproved credit card offers mailed to you	■ Check companies with the Better Business Bureau before buying online ■ Use a secure web browser to buy online
Checks	■ Steal your checks or checking account number from your home or office	■ Notify your bank immediately if your checks are stolen
Cellular Telephone Service	■ Open telephone service in your name ■ Use your calling card and PIN to make calls	■ If this occurs, immediately ask your service provider to close your account, and establish another one with a new PIN.
Internet Account Updates	■ Send phony e-mail requesting your credit card information to update or verify a company's records	■ Check with your Internet Service Provider before responding to such a request
Phony Identity Theft Protections Services	■ Request personal information in order to "protect" you	■ Check out the company with the Better Business Bureau before giving personal information

Based on Chicago Better Business Bureau, http://chicago.bbb.org/identitytheft/.

Outline of Report Organized Using Informal Classification

The outline below shows how three writers used informal classification to organize their report on the ways to identify water plants that might be cultivated, harvested, and dried to serve as fuel for power generators. They organized each section around topics that would be most important to the specialists who would read that section.

IDENTIFYING EMERGENT AQUATIC PLANTS THAT MIGHT BE USED AS FUEL
FOR BIOMASS ENERGY SYSTEM

Introduction

Botanical Considerations
    Growth Habitat
    Morphology
    Genetics

Physiological Considerations
    Carbon Utilization
    Water Utilization
    Nutrient Absorption
    Environmental Factors Influencing Growth

Chemical Considerations
    Carbohydrate Composition
    Crude Protein Content
    Crude Lipid Content
    Inorganic Content

Agronomic Considerations
    Current Emergent Aquatic Systems
      Eleocharis dulcis
      Ipomoea aquatica
      Zizania palustris
      Oryza sativa
    Mechanized Harvesting, Collection, Densification, and
      Transportation of Biomass
    Crop Improvement
    Propagule Availability

Ecological Considerations
    Water Quality
    Habitat Disruption and Development
    Coastal Wetlands

Economic Considerations
    Prior Research Efforts
      Phragmites communis
      Arundo donax
    Production Costs for Candidate Species
      Planting and Crop Management
      Harvesting
      Drying and Densification
      Total Costs

Selection of Candidate Species

WWW

To see other communications that use informal classification, go to Writer's Reference Guide: Using Seven Organizational Patterns at **www.cengage.com/english/anderson7e**.

The writers used informal classification to organize the report outlined here. In each section, they included information relevant to one specialty area: botanists, physiologists, chemists, and so on. The writers also used informal classification to organize the subsections.

Learn More

Exercises for organizing with informal classification are at the book's website and on page 261.

Comparison

At work, people write comparisons often, usually for one of the following reasons:

- **To help readers make a decision.** The workplace is a world of choices. People are constantly choosing among courses of action, competing products, alternative strategies. To help them choose, employees often compare the options in writing.

- **To help readers understand research findings.** Much workplace research focuses on differences and similarities between two or more items or groups— of people, animals, climates, chemicals, and so on. To explain the findings of this kind of research, researchers organize their results as comparisons.

Two Patterns for Organizing Comparisons

Learn More

A point of comparison is very much like a principle of classification. See the section on Formal Classification, page 236.

In some ways, comparison is like classification (page 236). You begin with a large set of facts about the things you are comparing, and you group the facts around points of comparison that enable your readers to see how the things are like and unlike one another. In comparisons written to support decision-making, points of comparison are called criteria.

When writing a comparison, you can choose either of two organizational patterns: the divided pattern or the alternating pattern. Lorraine's project illustrates the difference between the two. Her employer wants to replace the aging machines it uses to stamp out metal parts for the bodies of large trucks. Lorraine has been assigned to investigate the two machines the company is considering. She can organize the hundreds of facts she has gathered according to the divided pattern or the alternating pattern.

DIVIDED PATTERN	ALTERNATING PATTERN
Machine A	**Cost**
Cost	Machine A
Efficiency	Machine B
Construction Time	**Efficiency**
Air Pollution	Machine A
Et cetera	Machine B
Machine B	**Construction Time**
Cost	Machine A
Efficiency	Machine B
Construction Time	**Air Pollution**
Air Pollution	Machine A
Et cetera	Machine B
	Et Cetera
	Machine A
	Machine B

When To Use Each Pattern

To make a reader-centered choice between the divided and alternating patterns, consider the way your readers will use your information. Because the alternating pattern is organized around the criteria, it is ideal when readers want to make point-by-point comparisons among alternatives. Lorraine should select this pattern so her readers can find the information about the costs, capabilities, and other characteristics of both stamping machines in one place. The divided pattern would separate the information about each machine into different sections, requiring her readers to flip back and forth to compare the machines in detail.

The divided pattern is well suited to situations where readers want to read all the information about each alternative in one place. Typically, this occurs when both the general nature and the details of each alternative can be described in a short space— say, one page or so. An acoustical engineer used the divided pattern to provide a restaurant manager with information about three sound systems for her business. He described each system in a single page.

Whether you use the alternating or divided pattern, you can usually assist your readers by incorporating two kinds of preliminary information:

Try This

Try creating a list of criteria you would recommend that a friend use when comparing cell phones, MP3 players, videogames, or some other product he or she might be ready to purchase.

- **Description of the criteria.** This information lets your readers know from the start what the relevant points of comparison are.

- **Overview of the alternatives.** This information provides your readers with a general sense of what each alternative entails before they focus on the details you provide.

In both patterns, the statement of criteria would precede the presentation of details.

Taking these additional elements into account, the general structure of the two patterns is as follows:

DIVIDED PATTERN

Statement of Criteria
Overview of Alternatives
Evaluation of Alternatives
    Alternative A
        Criterion 1
        Criterion 2
    Alternative B
        Criterion 1
        Criterion 2
Conclusion

ALTERNATING PATTERN

Statement of Criteria
Overview of Alternatives
Evaluation of Alternatives
    Criterion 1
        Alternative A
        Alternative B
    Criterion 2
        Alternative A
        Alternative B
Conclusion

Comparison (continued)

Learn More

The comparison pattern is often combined with other patterns. See page 260.

Learn More

For advice about constructing graphics, go to Chapter 13 (page 331) and the Writer's Reference Guide: Creating Eleven Types of Reader-Centered Graphics (page 357.).

GUIDELINES FOR USING COMPARISONS

1. **Choose points of comparison suited to your readers and purpose.** When helping readers understand something, focus on major points, not minor ones. When helping readers make a decision, choose criteria that truly make a difference.

2. **Discuss each alternative in terms of all your criteria or points of comparison.** If no information is available concerning some aspect of one alternative, tell your readers. Otherwise, they may assume that you didn't try to obtain it.

3. **Arrange the parts in an order your readers will find helpful.** Often, you can help most by discussing the most significant differences first. Sometimes, it's best to briefly mention criteria on which the alternatives are very similar so that readers can then focus on the criteria that make a difference. In comparisons designed to aid understanding, it's usually best to begin with what's familiar to your readers and then lead them to the less familiar.

4. **Include graphics if they will help your readers understand and use your communication.** Tables, graphs, diagrams, and drawings can be especially helpful.

Graphics in Comparisons

Writers often use tables and graphs to help readers compare alternatives and make decisions. The website below helps vacationers pick a campground in New South Wales, Australia.

In this table, the points of comparison are

• Fees

• Facilities

• Types of camping available

By using icons rather than words, the writers help their readers read this table quickly.

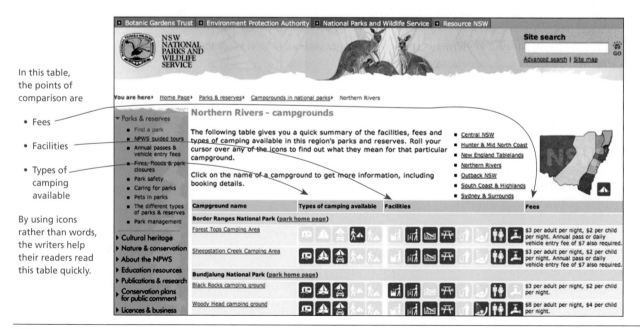

■ Writer's Reference Guide **Using Seven Reader-Centered Organizational Patterns**

Research Findings Organized By Comparison

In the following memo, the writer uses the alternating pattern of comparison to help a decision maker choose among three alternatives.

WWW

To see other communications that explain about comparison relationships, go to Writer's Reference Guide: Using Seven Organizational Patterns at **www.cengage.com/english/anderson7e.**

AdvanceTech
Memorandum

To   Mehash Mehta
From   Kenneth Abney
Date   September 10, 2011

Subject   **Recommendations for Smartphone Purchase**

Last week, Marisol de Silva asked me to provide you with a comparison of the top "smartphones" (cell phones that also serve as handheld computers). He explained that AdvancedTech might purchase smartphones for all 27 sales representatives and all 140 installation and service technicians.

I have studied product capabilities and published reviews for the three smartphones that received the highest ratings by *PCWorld* magazine: Blackberry Pearl 8120, Motozine ZN5, and Omnia.

Kenneth explains his research method, including the method by which he chose the smartphones to compare.

All three provide high-quality phone service and comparable computing capabilities. The key criteria for selection are ease of use and the ability to meet potential needs created by possible future expansion of our business. Here are my recommendations:

Kenneth begins his comparison by explaining what all three alternatives have in common.

- If we continue to do business in North America only, the Omnia is our best choice. Offering both a touchscreen and stylus, along with a large display, it has the longest standby battery life and is the only phone that runs Windows Mobile.

- If we will soon open offices outside North America, the BlackBerry 7230 would be preferable. It supports the Global System for Mobile Communication (GSM) used elsewhere in the world.

Kenneth makes two recommendations, one for each of two possible situations.

The following table compares the phones in detail. If you want more information, please let me know.

	Standby Battery Life	Input	Resolution	Operating System	GSM	Price
BlackBerry Pearl 8120	15 Days	Keyboard Thumb Wheel	240 x 260	Proprietary	Yes	$200
Motozine ZN5	7.8 Days	Keyboard	240 x 320	Proprietary	No	$200
Omnia	20 Days	Touchscreen Stylus	400 x 240	Windows Mobile	No	$200

He provides his reader with additional detail in an easy-to-read table.

Learn More

Exercises for writing comparisons are at the book's website and on page 261.

Description of an Object (Partitioning)

In your career, you will have many occasions to describe a physical object for your readers. If you write about an experiment, you may need to describe your equipment. If you write instructions, you may need to describe the machines your readers will be using. If you propose a new purchase, you may need to describe the object you want to buy.

To organize descriptions, use a strategy called partitioning. Partitioning uses the same basic procedure as does classifying (page 236). Think of the object as a collection of parts. Then identify a principle of classification for organizing the parts into groups of related parts. At work, the principle most useful to your readers will usually be location or function.

Learn More

For more on principles of classification, see the section on Formal Classification, page 236.

Example: Partitioning a Car

Consider, for instance, how you could use location and function to organize a discussion of the parts of a car.

To organize by location, you might talk about the car's interior (passenger compartment, trunk), exterior (front, back, sides, and top), engine compartment, and underside (wheels, transmission, and muffler).

To organize by function, you might focus on parts that provide power and ones that guide the car. The power-producing parts are in several locations: The gas pedal is in the passenger compartment, the engine is under the hood, and the transmission and axle are on the underside. Nevertheless, you would discuss them together because they are related by function.

Of course, other principles of classification are possible.

GUIDELINES FOR DESCRIBING AN OBJECT

Learn More

The partitioning pattern is often combined with other patterns. See page 260.

1. **Choose a principle of classification suited to your readers and purpose.** For instance, when describing a car for new owners who want to learn about the vehicle they've purchased, you might organize according to location. Your readers would learn about the conveniences in the passenger compartment, trunk, and so on. In contrast, when describing the car for auto repair specialists, you could organize according to function or the kinds of problems they might encounter.

2. **Use only one basis for partitioning at a time.** To assure that you have one place and only one place for each part you describe, use only one basis for partitioning at a time, just as you use only one principle of classification at a time (see page 236).

Learn More

Exercises for organizing with partitioning are at the book's website and on page 261.

3. **Arrange the parts of your description in a way your readers will find useful.** When partitioning by location, you might move systematically from left to right, front to back, or outside to inside. When partitioning by function, you might trace the order in which the parts interact during a process that interests your readers.

4. **When describing each part, provide details that your readers will find useful.** For instance, when describing a car tire to a consumer, you might describe the air pressure the tires should have. When describing the same tire to an engineer, you might provide a technical description of the steel belts in the tire's core.

5. **Include graphics if they will help your readers understand and use your information about the object.** Graphics that are especially helpful to readers of an object's description include drawings, diagrams, and photographs.

Learn More

For advice about constructing graphics, go to Chapter 13 (page 331) and the Writer's Reference Guide: Creating Eleven Types of Reader-Centered Graphics (page 357).

Description Organized by Partitioning

This description of new technologies for airplanes is organized by partitioning.

THE HUMAN EYE AND ITS AILMENTS

Several parts of the eye contribute to creation of sharp, clear vision. Disease or damage to any part can distort the light that passes through it or block the light altogether.

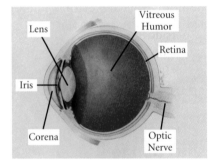

Lens
Vitreous Humor
Retina
Iris
Corena
Optic Nerve

*   **Cornea** The cornea protects the inner eye. Disease, infection, or injury can create scars that block or distort light.

*   **Iris** The iris is the colored part of the external eye. It expands and contracts to regulate the amount of light entering the eye through the pupil. Skin cancer can attack the iris, limiting its ability to function properly.

*   **Lens** A set of muscles attached to the lens control its shape, thereby focusing the light that passes through it. The lens can develop cloudy areas called cataracts.

*   **Vitreous Humor** The vitreous humor is the gel that fills the eyeball. Particles called floaters sometimes drift through it, looking like dust particles or sand in the field of vision.

*   **Retina** Nerve cells in the retina transform light energy into electrical impulses that travel to the brain through the optic nerve. The retina can tear or detach from the lining of the eye, causing flawed vision or blindness.

WWW

To see other communications that use partitioning, go to Writer's Reference Guide: Seven Organizational Patterns at **www.cengage.com/english/anderson7e**.

This description is partitioned according to the steps in the process by which light passes through the eye and is transformed into vision. The organization also corresponds to the location of the parts of the eye.

*   Light enters the eye through the cornea.

*   The amount of light proceeding farther into the eye is controlled by the iris.

*   The light is then focused by the lens.

*   Next, the light passes through the vitreous humor to the retina.

*   The retina transforms the light energy into electrical impulses.

Description of a Process (Segmenting)

A description of a process explains the relationship of events over time. You may have either of two purposes in describing a process:

- **To enable your readers to perform the process.** For example, you may be writing instructions that will enable your readers to analyze the chemicals present in a sample of liver tissue, make a photovoltaic cell, apply for a loan, or run a computer program.

- **To enable your readers to understand the process.** For example, you might want your readers to understand the following:

  - How something is done. For instance, how coal is transformed into synthetic diamonds.

  - How something works. For instance, how the lungs provide oxygen to the bloodstream.

  - How something happened. For instance, how the United States developed the space programs that eventually landed astronauts on the moon.

In either case, you need to help your readers understand the overall structure of the process. To do this, you segment the process. Imagine the entire process as a long line or string of steps. Divide the process (or cut the string) into segments at points where one major group of related steps ends and another begins. If some or all of the major groups are large, you may divide them into subgroups, thereby creating an organizational hierarchy. The outline on the right shows how you might segment the steps in the process for building a cabinet.

Making a Cabinet
Obtaining Materials
Preparing the Pieces
    Cutting the wood
    Routing the wood
Assembling the Cabinet
    Building the base
    Mounting the doors
Finishing the Cabinet
    Sanding
    Applying the stain
    Applying the sealant

Principles of Classification for Segmenting

To determine where to segment the process, you need a principle of classification. Commonly used principles include the time when the steps are performed (first day, second day; spring, summer, fall), the purpose of the steps (to prepare the equipment, to examine the results), and the tools used to perform the steps (for example, table saw, drill press, and so on).

Processes can be segmented by a variety of classification principles. Pick the principle that best supports your readers' goals. For instance, if you were writing a history of the process by which the United States placed a person on the moon, you would segment the process according to the passage of various laws and appropriation bills if your readers were members of the U.S. Congress. If your readers are scientists and engineers, however, you might focus on efforts to overcome various technical problems.

Learn More

For more on principles of classification, see the section on Formal Classification, page 236.

1. **Choose a principle for segmenting suited to your readers and your purpose.** If you are writing instructions, group the steps in ways that support an efficient or comfortable rhythm of work. If you want to help your readers understand a process, organize around concerns that are of interest or use to them.

2. **Make your smallest groupings manageable.** If your smallest groupings include too many steps or too few, your readers will not see the process as a structured hierarchy of activities or events but rather as a long, unstructured list of steps—first one, then the next, and so on.

3. **Describe clearly the relationships among the steps and groups of steps.** To understand and remember the process, readers need to understand how the parts fit together. Where the relationships among the groups of steps will be obvious to your readers, simply provide informative headings. At other times, you will need to explain the relationships in introductory statements, transitions between groups of steps, and, perhaps, in a summary at the end.

4. **Provide enough detail about each step to meet your readers' needs.** Use your understanding of your readers and the ways they will use your communication to determine the appropriate level of detail.

5. **Include graphics if they will help your readers understand and use your information about the process.** Graphics that are especially helpful to readers of a description of a process are diagrams, drawings, photographs, and flowcharts.

Learn More

The segmenting pattern is often combined with other patterns. See page 260.

Try This

Try segmenting an object in the room, apartment, or house where you live.

A Graphic Organized by Segmenting a Process

In the flowchart below, writers at Lawrence Livermore National Laboratory explained the five-step process by which lasers are used to create nuclear fission.

Three hundred million watts of laser energy is projected into a small container that holds a spherical fusion-fuel capsule.	Reflecting from the container's specially coated walls, the laser beams are converted to x-rays that create a 3-million-degree oven.	The heat energy implodes the fusion-fuel cell to 20 times the density of lead.	The fuel core ignites at 100 million kelvins.	A thermonuclear burn spreads through the fuel, creating gain equivalent to the power of miniature star lasting less than a billionth of a second.

The text explains the steps in the process.

The drawings illustrate each step.

Instructions Organized by Segmenting a Process

The pages shown below are from instructions for using a computerized telescope that automatically points to a star or constellation specified by the operator. They are segmented according to the major groups of steps in the process: setting it up, activating the computer, entering the date and location, and so on. See more examples of instructions organized by segmenting on pages 666 and 670.

WWW

To see other communications that use segmenting, go to Writer's Reference Guide: Using Seven Organizational Patterns at **www.cengage.com/english/anderson7e**.

As the headings on these pages indicate, the writers have organized this instruction manual by segmenting the overall process of setting up and using the telescope.

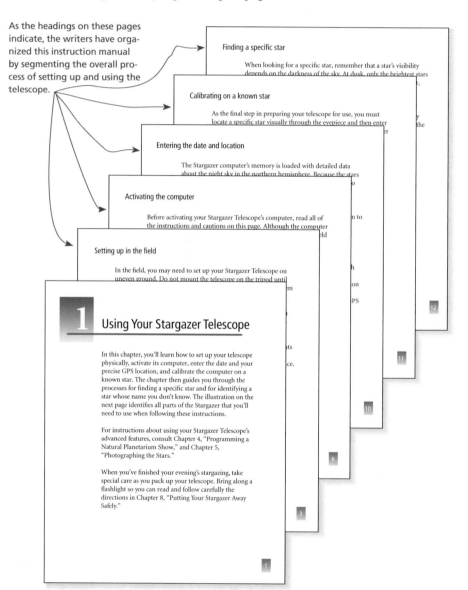

Finding a specific star

When looking for a specific star, remember that a star's visibility depends on the darkness of the sky. At dusk, only the brightest stars

Calibrating on a known star

As the final step in preparing your telescope for use, you must locate a specific star visually through the eyepiece and then enter

Entering the date and location

The Stargazer computer's memory is loaded with detailed data about the night sky in the northern hemisphere. Because the stars

Activating the computer

Before activating your Stargazer Telescope's computer, read all of the instructions and cautions on this page. Although the computer

Setting up in the field

In the field, you may need to set up your Stargazer Telescope on uneven ground. Do not mount the telescope on the tripod until

1 Using Your Stargazer Telescope

In this chapter, you'll learn how to set up your telescope physically, activate its computer, enter the date and your precise GPS location, and calibrate the computer on a known star. The chapter then guides you through the processes for finding a specific star and for identifying a star whose name you don't know. The illustration on the next page identifies all parts of the Stargazer that you'll need to use when following these instructions.

For instructions about using your Stargazer Telescope's advanced features, consult Chapter 4, "Programming a Natural Planetarium Show," and Chapter 5, "Photographing the Stars."

When you've finished your evening's stargazing, take special care as you pack up your telescope. Bring along a flashlight so you can read and follow carefully the directions in Chapter 8, "Putting Your Stargazer Away Safely."

Explanation Organized by Segmenting a Process

In the following passage, the writer describes one theory about the way planets were formed. Notice how the writer uses headings to signal the major phases of the process.

How Planets Are Formed

Theoretical astrophysicists have developed an elaborate model describing the formation of stars like our Sun. In this theory, planet formation is a natural and almost necessary result of the process of star formation.

A Proto-Star Is Born

These astrophysicists believe that the process of star formation begins when a dense region of gas and dust in an interstellar cloud becomes gravitationally unstable. The cloud begins a slow, quasistatic contraction as both its internal turbulent and magnetic support are gradually lost through cooling and the outward diffusion of the magnetic field. The core of this cloud condenses to form a proto-star.

If there is sufficient angular momentum, the collapsing cloud may fragment into two or more smaller pieces, thus leading to the formation of a binary or multiple star.

A Rotating Disc Forms

The remaining gas and dust from the cloud continue to fall inwards. Because this material must conserve the initial angular momentum of the cloud, it forms into a rotating disc around the proto-star in accordance with Kepler's laws, with the inner portion of the disc rotating more rapidly than the outer region.

Planets Are Created

The process of planet formation occurs within this rotating disc. Dust grains collide and stick together, forming larger particles. This collisional growth continues over millions of years and eventually results in the creation of rocky bodies a few kilometers in diameter, known as "planetesimals." At this point the gravitational attraction of the planetesimals begins to dominate their random velocities, increasing their growth rate to the point where planetary cores about a thousand kilometers across can form rapidly.

As the proto-planetary disc continues to evolve, the planetary cores in the inner portion of the disc collide, merge and grow to become terrestrial planets such as Earth and Mars. In the outer disc, the cores grow to larger masses. Once the planetary core reaches about 10 Earth masses, its gravity is sufficient to capture gaseous hydrogen and helium from the disc, forming a gas-giant planet such as Jupiter or Saturn. This rapid-growth phase removes most of the material near the orbit of the growing planet, forming a gap in the disc and effectively shutting off the growth process.

Thus, scientists think that planets are built from their cores outward by accretion processes in the disc of gas and dust that surrounds a star during its creation.

The topic is introduced.

First stage is described.

Alternative outcomes are identified.

Second stage is described.

Headings help to distinguish the stages.

Third stage is described in two parts, with separate paragraphs devoted to planetesimals and to planets.

A concluding sentence summarizes the overall process.

Learn More

Exercises for writing about segmenting are at the book's website and on page 261.

Q uestions about cause and effect are very common in the workplace. Profits slumped or skyrocketed. What caused the change? The number of children diagnosed with autism is increasing. How come? If we clear underbrush from forest floors, how will our action affect animal habitats and the severity of forest fires?

At work, you are likely to write about cause and effect for one of two distinct purposes.

- to help your readers understand a cause-and-effect relationship.
- to persuade your readers that a certain cause-and-effect relationship exists.

The strategies for organizing for these two purposes are somewhat different.

Helping Readers Understand a Cause-and-Effect Relationship

Some cause-and-effect relationships are accepted by people knowledgeable about the topic but need to be explained to those who are not.

For instance, geologists agree about what causes the Old Faithful geyser in the United States' Yellowstone National Park to erupt so regularly. However, many park visitors do not know. Consequently, the park publishes brochures that explain the cause.

Depending on your profession, you may need to explain cause-and-effect relationships to members of the general public; to employees in your own organization, such as coworkers or executives outside your specialty; or to other readers. The following reader-centered guidelines will help you do so effectively in any situation.

Learn More

The cause-and-effect pattern is often combined with other patterns. See page 260.

Learn More

For advice about constructing graphics, go to Chapter 13 (page 331) and the Writer's Reference Guide: Creating Eleven Types of Reader-Centered Graphics (page 357).

GUIDELINES FOR EXPLAINING A CAUSE-AND-EFFECT RELATIONSHIP

1. **Begin by identifying the cause or effect that you are going to explain.** Let your readers know from the beginning exactly what you are going to explain. This advance notice will help them better understand what follows.

2. **Carefully explain the links that join the cause and effect that you are describing.** Remember that you are not simply listing the steps in a process. You want your readers to understand how each step leads to the next one.

3. **If you are dealing with several causes or effects, group them into categories.** Categories help readers to understand a complex set of factors.

4. **Include graphics if they will help your readers understand the relationship you are explaining.** Among the many kinds of graphics that assist readers in visualizing cause-and-effect relationships are those used to describe processes, including flowcharts, diagrams, and drawings.

Sample Explanation of Cause and Effect

The following passage uses a cause-and-effect relationship to explain how CDs work.

How Light Makes Music

When you listen to your favorite songs on a CD, you are enjoying the music that light makes.

Your CD has three layers. On the bottom is a stiff, protective layer of acrylic. Next is a thin layer of reflective aluminum, which is coated with a clear plastic layer that is 1.2 mm thick. Your CD player projects a shaft of light, in the form of a laser beam, at a tiny spot on the reflective aluminum. As your CD spins, the player reads the variations in the intensity of the light that is reflected from the aluminum.

What causes the intensity of the reflected light to vary? Although a CD seems smooth, its ability to make music depends on millions of pits in the clear plastic layer. Only about 0.5 microns deep and several microns long, they are burned into the aluminum side of the clear plastic layer when the CD is manufactured. Despite being "clear," the plastic layer absorbs some light. It absorbs less where the pits are because the plastic is thinner there. As a CD spins, areas with and without pits pass rapidly under the CD player's laser beam. By reading the variations in the intensity of the reflected light, the CD player makes music.

A Laser Puts Music on a CD by Creating Pits in the
Aluminum Side of the Clear Plastic Layer

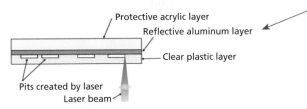

Protective acrylic layer

Reflective aluminum layer

Clear plastic layer

Pits created by laser
Laser beam

This passage explains how a CD player reads music from a CD. The same technology is used for DVDs.

This paragraph provides background that helps readers understand the cause-and-effect explanation given in the next paragraph.

In this explanation, the effect (variations in the intensity of light) is described first. The explanation follows. In other cause-and-effect passages, the order is reversed.

Many kinds of graphics are used with explanations of cause and effect. In this case, the author uses a diagram to provide a visual description of the process by which pits that "make" the music are created on a CD.

Persuading Readers That a Cause-and-Effect Relationship Exists

About many events, people disagree. They may disagree about past events, such as what caused damage to a jet engine's turbine blade or what caused the disappearance of the Inca civilization in Central America. Alternatively, they may disagree about the future. Examples include disagreements about the effects of growing genetically modified crops or the effect on sales that a proposed price reduction would have on company profits.

Whether you are writing about causes of a past event or the effects of a future action, your goal is to persuade your readers to accept your explanation. The guidelines on the next page will help you argue persuasively, as will the guidelines in Chapter 5, "Planning Your Persuasive Strategies."

Cause and Effect (continued)

Learn More

For more advice about persuading readers, see Chapter 5, Planning Your Persuasive Strategies, page 117.

Learn More

For advice about constructing graphics, go to Chapter 13 (page 331) and the Writer's Reference Guide: Creating Eleven Types of Reader-Centered Graphics (page 357).

GUIDELINES FOR PERSUADING READERS TO ACCEPT YOUR VIEW OF CAUSE AND EFFECT

1. **State your claim at the beginning of your passage.** By letting readers know your claim in advance, you will help them see—and evaluate—the way your evidence supports the claim.

2. **Choose evidence your readers will find credible.** As explained in Chapter 5, different kinds of evidence are credible in different situations and industries. You will have to understand your readers in order to determine whether data, expert testimony, or some other form of evidence will be most persuasive for them.

3. **Explain your line of reasoning.** The heart of your explanation is the links you draw between the cause or effect you are discussing. Explain them thoroughly.

4. **Avoid faulty logic.** If your readers find your logic to be faulty, they will reject your explanation. See below for descriptions of two logical fallacies that are common in arguments concerning cause and effect.

5. **Address counterarguments.** To persuade readers to accept your account of a cause-and-effect relationship, you must persuade them to reject alternative explanations. Therefore, you need to identify the other explanations they might consider and explain why yours is better.

6. **Include graphics if they will help your readers understand the relationship you are discussing.** Graphics used to describe processes, such as flowcharts, diagrams, and drawings, can be very helpful.

Logical Fallacies Common in Arguments about Cause and Effect

The following logical fallacies can undermine arguments about cause and effect.

- **Post hoc, ergo propter hoc fallacy.** This fallacy occurs when a writer argues that because an event occurred after another event, it was caused by that event. For example, in an attempt to persuade his employer, a furniture manufacturer, to switch to computerized machinery, Samuel argued that a competitor's profits had risen substantially after making that switch. Samuel's boss pointed out that the competitor's sales increase may have been caused by other changes made at the same time, such as creation of new designs or reconfiguration of sales districts. Samuel's commission of this fallacy didn't mean he was incorrect. However, to persuade his boss that computerization had caused the sales increase, Samuel needed to do more than simply state that it had preceded the increase.

- **Overgeneralization.** Writers overgeneralize when they draw conclusions on insufficient evidence. For instance, writers overgeneralize if they draw a conclusion about the causes of a manufacturing error after examining only 2 percent of the faulty products.

Sample Passage Persuading about a Cause-and-Effect Relationship

In the following passage about the death of the dinosaurs, the writer applies the guidelines for persuading about cause and effect.

WHAT CAUSED THE DEATH OF THE DINOSAURS?

One theory is that a comet, asteroid or other huge extraterrestrial body slammed into the Earth 65 million years ago and ended the 160-million year reign of the dinosaurs. According to this theory, the extraterrestrial body raised a huge dust cloud. Within days the black cloud spread over the Earth, darkening the sun. The air turned cold, and many dinosaurs died. Snow fell. Freezing darkness gripped the Earth for weeks. Plants, cut off from the sunlight that feeds them, couldn't survive. Without plants, the rest of the herbivorous dinosaurs followed, and the carnivores soon afterward. Along with a number of other species, the dinosaurs were gone forever.

Although many leading paleontologists and evolutionary biologists now accept the asteroid-impact theory, and despite popular accounts implying that the question is settled, it is not. A scattering of critics continue to challenge the whole notion.

Still, the theory is compelling. Every few months a new piece of evidence is added to the list, and most, to the critics' consternation, support the idea of an extraterrestrial impact.

Just recently, for example, scientists at the Scripps Institute of Oceanography, in La Jolla, California, found evidence of organic molecules in the layer of sediments laid down at the time the dinosaurs died; the molecules are exceedingly rare on Earth but relatively common in some meteorites and so, presumably, in some asteroids.

To put the discovery in perspective, and to appreciate the arguments on both sides of the impact debate, one must first understand the nature of the original finding.

In 1980, Luis Alvarez, a Nobel laureate in physics, his son Walter, a geologist, both at the University of California, Berkeley, and two associates published the theory that a massive impact took place at the end of the Cretaceous Period.

[continued on the next page]

Effect to be explained is announced.

Possible cause is announced.

Link between effect and cause is explained: An asteroid created a dust cloud that killed the dinosaurs.

Evidence of link is presented: rare molecules in sediment indicate an asteroid may have hit Earth when dinosaurs died.

Additional evidence of link.

WWW

To see other communications that explain or persuade about cause-and-effect relationships, go to Writer's Reference Guide: Using Seven Organizational Patterns at **www.cengage.com/english/anderson7e**.

Sample Passage Persuading about a Cause-and-Effect Relationship (*continued*)

Additional evidence continued. ⟶

Link is restated. ⟶

Additional evidence of link. ⟶

Challenge to link is explained: molecules perhaps from volcano, not asteroid.

Challenge is refuted by new evidence: Other molecules couldn't have come from volcano.

Learn More

Exercises for writing about cause and effect are at the book's website and on page 261.

The team had found a rare substance in the thin layer of sedimentary clay deposited just on top of the highest, and therefore the most recent, stratum of rock contemporary with those bearing dinosaur fossils. It was the element iridium, which is almost nonexistent in the Earth's crust but 10,000 times more abundant in extraterrestrial rocks such as meteorites and asteroids. Deposits above and below the clay, which is the boundary layer separating the Cretaceous layer from the succeeding Tertiary, have very little iridium.

Because the same iridium anomaly appeared in two other parts of the world, in clay of exactly the same age, the Alvarez team proposed that the element had come from an asteroid that hit the Earth with enough force to vaporize, scattering iridium atoms in the atmosphere worldwide. When the iridium settled to the ground, it was incorporated in sediment laid down at the time.

More startling was the team's proposal that the impact blasted so much dust into the atmosphere that it blocked the sunlight and prevented photosynthesis (others suggested that a global freeze would also have resulted). They calculated that the object would have had to be about six miles in diameter.

Since 1980, iridium anomalies have been found in more than 80 places around the world, including deep-sea cores, all in layers of sediment that formed at the same time.

One of the most serious challenges to the extraterrestrial theory came up very quickly. Critics said that the iridium could have come from volcanic eruptions, which are known to bring up iridium from deep within the Earth and feed it into the atmosphere. Traces of iridium have been detected in gases escaping from Hawaii's Kilauea volcano, for example.

The new finding from Scripps appears to rule out that explanation, though, as a source for iridium in the Cretaceous–Tertiary (K–T) boundary layer. Chemists Jeffrey Bada and Nancy Lee have found that the same layer also contains a form of amino acid that is virtually nonexistent on Earth—certainly entirely absent from volcanoes—but abundant, along with many other organic compounds, in a type of meteor called a carbonaceous chondrite.

Problem and Solution

Problems and their solutions will be one of the most frequent topics of your on-the-job writing. The problems you discuss may arise from dissatisfaction with some strategy, product, process, or policy. Alternatively, they may arise from an aspiration to achieve a new goal, such as greater efficiency, or take advantage of a new opportunity, such as the potential to do business in another country. In either case, you will usually be writing proposals for future actions or reports on completed ones.

Proposing Future Action

In proposals to readers outside your organization, you will probably be seeking contracts worth thousands or even millions of dollars. Often, other individuals or organizations will be competing for the same contract. When writing proposals to readers inside your own organization, you might be urging support for a large research or development project or simply suggesting a small alteration in policy or procedure. Always, your goal will be to obtain support or approval for the problem-solving project you propose.

Learn More

Chapter 23 on Writing Reader-Centered Proposals provides additional information on organizing with the problem-and-solution pattern, which provides the overall structure for proposals.

GUIDELINES FOR PERSUADING READERS TO ACCEPT YOUR PROPOSED SOLUTION

1. **Describe the problem in a way that makes it seem significant to your readers.** Remember that your aim is to persuade them to take the action you recommend. They will not be very interested in taking action to solve a problem they regard as insignificant.

2. **Describe your method.** Readers want enough detail to feel confident that it is practical and technically sound. Depending on your reader and the situation, this might require a great deal of information or relatively little.

3. **When describing your method, explain how it will solve the problem.** Provide the evidence and reasoning needed to persuade your readers that your method will, in fact, solve their problem.

4. **Anticipate and respond to objections.** As when reading any persuasive segment, your readers may object to your evidence or your line of reasoning. Devote special attention to determining what those objections are so you can respond to them.

5. **Specify the benefit.** With as much specificity as will interest them, describe the benefits they and their organization will enjoy if they support or fund your proposed action.

6. **Include graphics if they will help your readers understand and approve your proposed solution.** Review opportunities to use graphics that can help your readers understand the problem and your proposed solution and that can highlight the benefits to your readers of supporting the action you propose. Among the graphics often used in problems and their solutions are tables, flowcharts, drawings, photographs, and diagrams.

Learn More

The problem-and-solution pattern is often combined with other patterns. See page 260.

Memo Proposing the Solution to a Problem

In the following memo, the writer uses the problem-and-solution strategy to recommend that her employer investigate the Kohle Reduktion method of steelmaking. Proposals for further study are common in the workplace.

WWW

To see other communications that explain or persuade about problem-and-solution relationships, go to Writer's Reference Guide: Using Seven Organizational Patterns at **www.cengage.com/ english/anderson7e.**

MANUFACTURING PROCESSES INSTITUTE
Interoffice Memorandum

June 21, 2010

To   Cliff Leibowitz

From   Candace Olin

RE   Suggestion to Investigate Kohle Reduktion Process for Steelmaking

As we have often discussed, it may be worthwhile to set up a project investigating steelmaking processes that could help the American industry compete more effectively with the more modern foreign mills. I suggest we begin with an investigation of the Kohle Reduktion method, which I learned about in the April 2005 issue of *High Technology*.

Problem is identified. →

Problem is explained. →

A major problem for American steelmakers is the process they use to make the molten iron ("hot metal") that is processed into steel. Relying on a technique developed on a commercial scale over 100 years ago by Sir Henry Bessemer, they make the hot metal by mixing iron ore, limestone, and coke in blast furnaces. To make the coke, they pyrolize coal in huge ovens in plants that cost over $100 million and create enormous amounts of air pollution.

Solution is announced. →

Solution is explained. →

In the Kohle Reduktion method, developed by Korf Engineering in West Germany, the hot metal is made without coke. Coal, limestone, and oxygen are mixed in a gasification unit at 2500°. The gas rises in a shaft furnace above the gasification unit, chemically reducing the iron ore to "sponge iron." The sponge iron then drops into the gasification unit, where it is melted and the contaminants are removed by reaction of the limestone. Finally, the hot metal drains out of the bottom of the gasifier.

Link between problem and solution is explained. →

The Kohle method, if developed satisfactorily, will have several advantages. It will eliminate the air pollution problem of coke plants, it can be built (according to Korf estimates) for 25% less than conventional furnaces, and it may cut the cost of producing hot metal by 15%.

This technology appears to offer a dramatic solution to the problems with our nation's steel industry: I recommend that we investigate it further. If the method proves feasible and if we develop an expertise in it, we will surely attract many clients for our consulting services

Learn More

Exercises for writing about problems and their solutions are at the book's website and on page 261.

Reporting on a Past Problem-Solving Project

Reports on past problem-solving projects are often used to teach other employees how to approach similar challenges. They are also used to demonstrate a company's capabilities in proposals.

GUIDELINES FOR DESCRIBING PROBLEMS AND THEIR SOLUTIONS

1. **Begin by identifying the problem.** Make the problem seem significant to your readers. Emphasize the aspects of the problem most directly affected by your solution.

2. **Describe your method.** Provide the details needed by your readers.

3. **Describe the results.** Enumerate the benefits produced. Tell what was learned.

4. **Include graphics that will help your readers understand and use your communication.**

Learn More

For advice about constructing graphics, go to Chapter 13 (page 331) and the Writer's Reference Guide: Creating Eleven Types of Reader-Centered Graphics (page 357).

Passage Reporting on a Past Problem-Solving Project

In the passage below, Rashid uses the problem-solution pattern of organization.

Pollution Control System Reduces Cost and Makes Fertilizer

Coal-fired power plants produce 36% of electricity worldwide. They also cause air pollution that creates health hazards and damages the environment. To comply with government regulations in the U.S. and elsewhere, thousands of older plants must greatly reduce their emissions or close permanently.

Because pollution control systems have been specialized, power plants need to install several to meet environmental standards: one for sulfur dioxide, another for mercury, and still others for particulates and nitrous oxides. For older plants, retrofitting with several systems can be prohibitively expensive.

At the 50-year-old Burger Plant in Shadyside, Ohio, Powerspan Corporation has successfully demonstrated its patented Electro-Catalytic Oxidation (ECO) system, which combines the functions of four control technologies. Installed as a single unit, it costs much less than the four systems it would replace.

Courtesy of Powerspan Corporation

ECO Demonstration Facility

The ECO system not only reduced emissions effectively but also produced byproducts sold as feedstock to a fertilizer manufacturer, producing a new (though modest) income stream for the plant.

The ECO system can also reduce fuel costs for older plants in areas like Ohio with abundant local deposits of high-sulfur coal. Decades ago these plants had to replace this inexpensive, local resource with more-expensive low-sulfur coal imported from other regions. The more sulfur in the coal, the better the system works, according to Powerspan.

ECO Pilot Test Restults	
Pollutant	Removal Efficiency
SO2 (sulfur dioxide)	98%
NOx (nitrogen oxides)	90%*
Hg (mercury)	80–90%
*Inlet NOx is 0.4 lb/mmBtu	

Rashid describes the general problem: Coal-fired power plants must reduce emissions.

He then focuses on the main issue: Retrofitting with a collection of specialized systems is very expensive.

He uses a photo to enable readers to see what the facility looks like.

Rashid explains that an integrated system can solve the problem.

He highlights the results by presenting them in a table.

Rashid highlights two other benefits of the system.

Combinations of Patterns

WWW

To see other communications that use combinations of patterns, go to Writer's Reference Guide: Using Seven Organizational Patterns at **www.cengage.com/english/anderson7e.**

This Writer's Reference Guide describes seven organizational patterns in isolation from one another. In practice, you will almost always need to weave two or more together. In a short memo, for instance, you might describe a problem in two sentences, identify its causes in three more, briefly compare alternative solutions, and finally recommend one.

The outline below shows how six patterns were integrated in a 25-page feasibility report.

Like many technical communications, this report interweaves six organizational patterns.

To organize this part, the writers **classified** the companies according to their location.

They organized Chapter II around a **problem** and its possible **solution**.

When explaining why the solar cells collect heat, the writers organized by **cause and effect.**

The writers organized each of these subsections by **partitioning** the component into its major subparts.

They organized their discussion of the construction schedule by gathering the steps into related groups of steps (**segmenting** the process).

The writers used the divided pattern for **comparisons** in their section on cost.

Solar Roofs for New Restaurants
A Feasibility Report for the Brendon's Restaurant Chain

I. Introduction
   A. The Brendon's Restaurant commissioned us to determine the feasibility of covering its restaurant roofs with photovoltaic cells (solar cells) that generate electricity
   B. Background
     1. Brendon's builds 30 new restaurants a year
     2. Brendon's is committed to environmental responsibility
     3. Federal government has goal of one million solar roofs in the U.S. by 2010
     4. Experiences of other companies
       a. Europe
       b. Asia
       c. United States

II. Technical Assessment
   A. Problems
     1. Solar cells do not generate enough electricity to pay for themselves (though this technology is advancing rapidly)
     2. Solar cells collect heat, so additional power would be needed to cool the building
     3. Putting solar cells on top of the regular roof would add 15% to construction time
   B. Potential Solutions (Used by Applebee's Restaurant in Salisbury, North Carolina)
     1. Use the collected heat to preheat the large amount of water used by the restaurant
       a. Collect the heat under the solar cells (instead of venting it)
       b. Use electricity from the solar cells to power fans that blow hot air to the preheating system
     2. Use large solar panels that can serve as the roof itself

III. Components of System
   A. Structure of the Solar Panels
   B. Structure of the Heat Ducts
   C. Structure of the Water Heating System

IV. Construction Schedule

V. Cost Comparison

VI. Conclusion

USE WHAT YOU'VE LEARNED

For additional exercises, go to **www.cengage.com/english/anderson7e**. *Instructors*: The book's website includes suggestions for teaching the exercises.

EXERCISE YOUR EXPERTISE

1. To choose the appropriate principle of classification for organizing a group of items, you need to consider your readers and your purpose. Here are three topics for classification, each with two possible readers. First, identify a purpose that each reader might have for consulting a communication on that topic. Then identify a principle of classification that would be appropriate for each reader and purpose.
   a. *Types of instruments or equipment used in your field*
      Student majoring in your field
      Director of purchasing in your future employer's organization
   b. *Intramural sports*
      Director of intramural sports at your college
      Student
   c. *Flowers*
      Florist
      Owner of a greenhouse that sells garden plants

2. Use a principle of classification to create a hierarchy having at least two levels. Some topics are suggested below. After you have selected a topic, identify a reader and a purpose for your classification. Depending on your instructor's request, show your hierarchy in an outline or use it to write a brief discussion of your topic. In either case, state your principle of classification.

   Have you created a hierarchy that, at each level, has one and only one place for every item?

   Boats   Computers
   Cameras   Physicians
   The skills you will need on the job
   Tools, instruments, or equipment you will use on the job
   Some groups of items used in your field (for example, rocks if you are a geologist, or power sources if you are an electrical engineer)

3. Partition an object in a way that will be helpful to someone who wants to use it. Some objects are suggested below. Whichever one you choose, describe a specific instance of it. For example, describe a particular brand and model of food processor rather than a generic food processor. Be sure that your hierarchy has at least two levels, and state the basis of partitioning you use at each

level. Depending on your instructor's request, show your hierarchy in an outline or use it to write a brief discussion of your topic.

   Aqualung   Microwave oven
   Graphing calculator   Bicycle
   Some instrument or piece of equipment used in your field that has at least a dozen parts

4. Segment a procedure to create a hierarchy you could use in a set of instructions. Give it at least two levels. Some topics are listed below. Show the resulting hierarchy in an outline. Be sure to identify your readers and purpose. If your instructor requests, use the outline to write a set of instructions.

   Changing an automobile tire
   Making homemade yogurt
   Starting an aquarium
   Rigging a sailboat
   Developing a roll of film
   Some procedure used in your field that involves at least a dozen steps
   Some other procedure of interest to you that includes at least a dozen steps

5. Segment a procedure to create a hierarchy you could use in a general description of a process. Give it at least two levels. Some suggested topics are listed below. Show the resulting hierarchy in an outline. Be sure to identify your readers and purpose. If your instructor requests, use the outline to write a general description of the process addressed to someone unfamiliar with it.

   How the human body takes oxygen from the air and delivers it to the parts of the body where it is used
   How television signals from a program originating in New York or Los Angeles reach television sets in other parts of the country
   How aluminum is made
   Some process used in your field that involves at least a dozen steps
   Some other process of interest to you that includes at least a dozen steps

6. One of your friends is thinking about making a major purchase. Some possible items are listed below. Create an outline with at least two levels that compares two or more good alternatives. If your instructor requests, use that outline to write your friend a letter.

   Stereo   Computer
   Binoculars   Bicycle
   CD player

Some other type of product for which you can make a meaningful comparison on at least three important points

7. Think of some way in which things might be done better in a club, business, or some other organization. Imagine that you are going to write a letter to the person who can bring about the change you are recommending. Create an outline with at least two levels in which you compare the way you think things should be done and the way they are being done now.

8. A friend has asked you to explain the causes of a particular event. Some events are suggested below. Write your friend a brief letter explaining the causes.

Static on radios and televisions
Immunization from a disease
Freezer burn in foods
Yellowing of paper

9. Think of a problem you feel should be corrected. The problem might be noise in your college library, shoplifting from a particular store, or the shortage of parking space on campus. Briefly describe the problem and list the actions you would take to solve it. Next, explain how each action will contribute to solving the problem. If your instructor requests, use your outline to write a brief memo explaining the problem and your proposed solution to a person who could take the actions you suggest.

9 | Developing an Effective Style

GUIDELINES

DEFINING
OBJECTIVES

PLANNING

CONDUCTING
RESEARCH

DRAFTING

REVISING

WWW

To read additional information, see more examples, and access links related to this chapter's guidelines, go to Chapter 9 at **www.cengage.com/english/anderson7e**.

This chapter describes ways to develop a writing style that will succeed at work. It covers three kinds of decisions: which words to choose, how to construct your sentences, and what "voice" to use as you address your readers.

The range of factors that can influence your decisions is large. Of course, you will want to choose words and construct sentences that are easy to understand. But what's understandable to one reader may not be for another. You will also want to employ a style that is appropriate in the workplace. But no single style is always appropriate. What is suitable for one circumstance can be completely wrong for another. For these and other reasons, choose your style for each communication in the same way you make all your other writing decisions: Take the reader-centered approach of considering your options in light of their impact on the specific persons who will be your readers. This chapter's guidelines describe strategies that work in the vast majority of cases. Use your creativity, judgment, and knowledge of your readers to decide how—and when—to apply them.

CREATING YOUR VOICE

While reading something you've written, your readers "hear" your voice. Based on what they hear, they draw conclusions about you and your attitudes that can greatly enhance—or detract from—the persuasiveness of your communications. Consequently, the ability to craft and control your voice is an area of expertise essential to your success at writing on the job.

Guideline 1 | Find out what's expected

To a large extent, an effective voice is one that matches your readers' sense of what's appropriate. When successful employees are asked to identify the major weaknesses in the writing of new employees, they often cite the inability to use a tone and style that are appropriate to their readers.

Here are three questions that can help you match your voice to your readers' expectations:

Questions for determining what your readers expect

- **How formal do your readers think your writing should be?** An informal style sounds like conversation. You use contractions (*can't, won't*), short words, and colloquial words and phrasing. A formal style sounds more like a lecture or speech, with longer sentences, formal phrasing, and no contractions.
- **How subjective or objective do your readers believe your writing should be?** In a subjective style, you would introduce yourself into your writing by saying such things as "I believe . . ." and "I observed . . . ." In an objective style, you would mask your presence by stating your beliefs as facts ("It is true that . . .")

and by reporting about your own actions in the third person ("The researcher observed . . .") or the passive voice ("It was observed that . . .").

- **How much "distance" do your readers expect you to establish between them and you?** In a personal style, you appear very close to your readers because you do such things as use personal pronouns (*I, we*) and address your readers directly. In an impersonal style, you distance yourself from your readers—for instance, by avoiding personal pronouns and by talking about yourself and your readers in the third person ("The company agrees to deliver a fully operable model to the customer by October 1").

Here are some major factors that may influence your readers' expectations about style:

- Your professional relationship with your readers (customers? supervisors? subordinates?)
- Your purpose (requesting something? apologizing? advising? ordering?)
- Your subject (routine matter? urgent problem?)
- Type of communication (e-mail? letter? formal report?)
- Your personality
- Your readers' personalities
- Customs in your employer's organization
- Customs in your field, profession, or discipline

To learn what style your readers expect, follow the advice in Chapter 3: Ask people who know (including even your readers) and look for communications similar to the one you are writing.

What If the Expected Style Is Ineffective?

Note that sometimes the expected style may be less effective than another style you could use. For example, in some organizations the customary and expected style is a widely (and justly) condemned style called *bureaucratese*. Bureaucratese is characterized by wordiness that buries significant ideas and information, weak verbs that disguise action, and abstract vocabulary that detaches meaning from the practical world of people, activities, and objects. Often, such writing features an inflated vocabulary and a general pomposity that slows or completely blocks comprehension. Here's an example:

According to optimal quality-control practices in manufacturing any product, it is important that every component part that is constituent of the product be examined and checked individually after being received from its supplier or other source but before the final, finished product is assembled. (45 words)	Bureaucratese

The writer simply means this:

Effective quality control requires that every component be checked individually before the final product is assembled. (16 words)	Plain English

Here is another pair of examples:

Bureaucratese | Over the most recent monthly period, there has been a large increase in the number of complaints that customers have made about service that has been slow. (27 words)

Plain English | Last month, many more customers complained about slow service. (9 words)

Bureaucratese is such a serious barrier to understanding that many states in the United States have passed laws *requiring* "Plain English" in government publications and other documents such as insurance policies. This chapter's guidelines will help you avoid bureaucratese. However, some managers and organizations want employees to use that puffed-up style, thinking it sounds impressive. If you are asked to write in bureaucratese, try to explain why a straightforward style is more effective, perhaps sharing this book. If you fail to persuade, be prudent. Use the style that is required. Even within the confines of a generally bureaucratic style, you can probably make improvements. For instance, if your employer expects a wordy, abstract style, you may still be able to use a less inflated vocabulary.

Try This

Plain English guidelines exist for almost every profession from architecture to statistics. Find some that apply to your major online. If there aren't any for your field, try a closely related one. For example, if you don't find Plain English for "zoology," try "biology."

Guideline 2 | Consider the roles your voice creates for your readers and you

When you choose the voice with which you will address your readers, you define a role for yourself. As manager of a department, for instance, you could adopt the voice of a stern taskmaster or an open-minded leader. The voice you choose also implies a role for your readers. And their response to the role given to them can significantly influence your communication's overall effectiveness. If you choose the voice of a leader who respects your readers, they will probably accept their implied role as valued colleagues. If you choose the voice of a superior, unerring authority, they may resent their implied role as error-prone inferiors—and resist the substance of your message.

Try This

What style do administrators at your college use? Find a letter or notice from one of the offices at your school. Is the style closer to Plain English or bureaucratese? What adjective would you use to describe the writer's voice? Do different offices use different voices? If so, why?

By changing your voice in even a single sentence, you can increase your ability to elicit the attitudes and actions you want to inspire. Consider the following statement drafted by a divisional vice president.

In this draft, the vice president uses a domineering voice. | I have scheduled an hour for you to meet with me to discuss your department's failure to meet its production targets last month.

In this sentence, the vice president has chosen the voice of a powerful person who considers the reader to be someone who can be blamed and bossed around, a role the reader probably does not find agreeable. By revising the sentence, the vice president creates a much different pair of roles for herself and her readers.

In this revision, she creates a supportive voice. | Let's meet tomorrow to see if we can figure out why your department had difficulty meeting last month's production targets.

The vice president transformed her voice into that of a supportive person. The reader became someone interested in working with the writer to solve a problem that stumps them both. As a result of these changes in voice and roles, the meeting is likely to be much more productive.

Guideline 3 | Consider how your attitude toward your subject will affect your readers

In addition to communicating attitudes about yourself and your readers, your voice communicates an attitude toward your subject. Feelings are contagious. If you write about your subject enthusiastically, your readers may catch your enthusiasm. If you seem indifferent, they may adopt the same attitude.

E-mail presents a special temptation to be careless about voice because it encourages spontaneity. As Laura B. Smith (1993) says, "Staring at e-mail can make users feel dangerously bold; they sometimes blast off with emotions that they probably would not use in a face-to-face meeting." Your risk of regretting an e-mail you've written is increased by the ease with which e-mails can be forwarded to readers you didn't intend to see your message. Never include anything in an e-mail that you wouldn't be prepared for a large audience to read. Check carefully for statements that your readers might interpret as having a different tone of voice than the one you intend.

Guideline 4 | Say things in your own words

This guideline urges you to avoid the mistake made by many employees: They puff up their prose with big words and long sentences, believing that this style will make them seem sophisticated, impressive. However, writers who write this way usually come across as pompous and difficult to understand.

Instead, write clear, straightforward sentences containing words you would normally use.

Don't misunderstand this advice. It does *not* mean that you should always use an informal, colloquial style. We all have more than one style. At school, you probably speak differently talking with a professor than when chatting with your friends. Yet, in each case you are able to choose words and express your ideas in ways that feel genuine to you. Similarly, at work choose from among your styles the one that is appropriate to each situation.

> Whether you are writing in an informal or formal style, choose words and express your ideas in ways that feel genuine to you.

To check whether you are using your own voice, try reading your drafts aloud. Where the phrasing seems awkward or the words are difficult for you to speak, you may have adopted someone else's voice—or slipped into bureaucratese, which reflects no one's voice. Reading your drafts aloud can also help you spot other problems with voice—such as sarcasm or condescension.

Despite the advice given in this guideline, it will sometimes be appropriate for you to suppress your own voice. For example, when a report, proposal, or other document is written by several people, the contributors usually strive to achieve a uniform voice so that all the sections will fit together stylistically. Similarly, certain kinds of official documents, such as an organization's policy statements, are usually

> Sometimes it's appropriate to suppress your own voice.

written in the employer's style, not the individual writer's style. Except in such situations, however, let your own voice speak in your writing.

Guideline 5 | Global Guideline: Adapt your voice to your readers' cultural background

Learn More

For more advice about adapting communications to your readers' cultural background, go to Chapter 3, page 79.

From one culture to another, general expectations about voice vary considerably. Understanding the differences between the expectations of your culture and those of your readers in another culture can be especially important because, as Guideline 2 explains, the voice you use tells your readers about the relationship you believe you have with them.

Consider, for instance, the difficulties that may arise if employees in the United States and in Japan write to one another without considering the expectations about voice that are most common in each culture. In the United States and Europe, employees often use an informal voice and address their readers by their first names. In Japan, writers commonly use a formal style and address their readers by their titles and last names. If a U.S. writer used a familiar, informal voice in a letter, memo, or e-mail to Japanese readers, these readers might feel that the writer has not properly respected them. On the other hand, Japanese writers may seem distant and difficult to relate to if they use the formality that is common in their own culture when writing to U.S. readers.

Directness is another aspect of voice. The Japanese write in a more personal voice than do people from the United States, whose direct, blunt style the Japanese find abrupt (Ruch, 1984). Like businesspeople in the United States, the Dutch also use a straightforward voice that causes the French to regard writers from both countries as rude (Mathes & Stevenson, 1991). When writing to people in other cultures, try to learn and to use the voice that is customary there. Library and Internet research provide helpful information about many cultures. You can also learn about the voice used in your readers' culture by studying communications they have written. If possible, ask for advice from people who are from your readers' culture or who are knowledgeable about it.

Guideline 6 | Ethics Guideline: Avoid stereotypes

Let's begin with a story. A man and a boy are riding together in a car. As they approach a railroad crossing, the boy shouts, "Father, watch out!" But it is too late. The car is hit by a train. The man dies, and the boy is rushed to a hospital. When the boy is wheeled into the operating room, the surgeon looks down at the child and says, "I can't operate on him. He's my son."

When asked to explain why the boy would call the deceased driver "Father" and the living surgeon would say "He's my son," people offer many guesses. Perhaps the driver is a priest or the boy's stepfather or someone who kidnapped the boy as a baby. Few guess that the surgeon must be the boy's mother. Why? Our culture's stereotypes about the roles men and women play are so strong that when people think of a surgeon, many automatically imagine a man.

Stereotypes, Voice, and Ethics

What do stereotypes have to do with voice and ethics? Stereotypes are very deeply embedded in a culture. Most of us are prone to use them occasionally, especially when conversing informally. As a result, when we use more colloquial and conversational language to develop our distinctive voice for our workplace writing, we may inadvertently employ stereotypes. Unfortunately, even inadvertent uses of stereotypes have serious consequences for individuals and groups. People who are viewed in terms of stereotypes lose their ability to be treated as individual human beings.

Further, if they belong to a group that is unfavorably stereotyped, they may find it nearly impossible to get others to take their talents, ideas, and feelings seriously. The range of groups disadvantaged by stereotyping is quite extensive. People are stereotyped on the basis of their race, religion, age, gender, sexual orientation, weight, physical handicap, and ethnicity. In some workplaces, manual laborers, union members, clerical workers, and others are the victims of stereotyping by people in white-collar positions.

Learn More

For a discussion of stereotypes and word choice, see page 281.

The following suggestions will help you avoid stereotypes.

AVOIDING STEREOTYPES

- **Avoid describing people in terms of stereotypes.** In your reports, sales presentations, policy statements, and other communications, avoid giving examples that rely upon or reinforce stereotypes. For example, don't make all the decision makers men and all the clerical workers women.

- **Mention a person's gender, race, or other characteristic only when it is relevant.** To determine whether it's relevant to describe someone as a member of a minority group, ask yourself if you would make a parallel statement about a member of the majority group. If you wouldn't say, "This improvement was suggested by Jane, a person without any physical disability," don't say, "This improvement was suggested by Margaret, a person with a handicap." If you wouldn't say, "The Phoenix office is managed by Brent, a hard-working white person," don't say, "The Phoenix office is managed by Terry, a hard-working Mexican-American."

- **Avoid humor that relies on stereotypes.** Humor that relies on a stereotype reinforces the stereotype. Refrain from such humor not only when members of the stereotyped group are present, but at all times.

CONSTRUCTING SENTENCES

Researchers who have studied the ways our minds process information have provided us with many valuable insights about ways to write reader-centered sentences. Based primarily on these research findings, the following six guidelines explain ways to construct highly usable, highly persuasive sentences.

Guideline 1 | Simplify your sentences

The easiest way to increase usability is to simplify your sentences. Reading is *work*. Psychologists say that much of the work is done by short-term memory. It must figure out how the words in each sentence fit together to create a specific meaning. Fewer words mean less work. In addition, research shows that when you express your message concisely, you make it more forceful, memorable, and persuasive (F. Smith, 2004).

SIMPLIFYING SENTENCES

1. **Eliminate unnecessary words.** Look for places where you can convey your meaning more directly. Consider this sentence:

Wordy

> The <u>physical size</u> of the workroom is too small <u>to accommodate</u> this equipment.

With unnecessary words removed in two places, the sentence is just as clear and more emphatic:

Unnecessary words deleted

> The workroom is too small for this equipment.

2. **Avoid wordy phrases.** Unnecessary words can also be found in many common phrases. "Due to the fact that" can be shortened to "Because." Similarly, "They do not pay attention to our complaints" can be abbreviated to "They ignore our complaints." "At this point in time" is "Now."

3. **Place modifiers next to the words they modify.** Short-term memory relies on word order to indicate meaning. If you don't keep related words together, your sentence may say something different from what you mean.

> Mandy found many undeposited checks in the file cabinets, which were worth over $41,000.

Technically, this sentence says that the file cabinets were worth over $41,000. Of course, readers would probably figure out that the writer meant the checks were worth that amount because it is unlikely that the file cabinets were. But readers arrive at the correct meaning only after performing work they would have been saved if the writer had kept related words together by putting *which were worth over $41,000* after *checks* rather than *file cabinets*.

3. **Combine short sentences.** Often, combining two or more short sentences makes reading easier by reducing the total number of words and helping readers see the relationships among the points presented.

Separate

> Water quality in Hawk River declined in March. This decline occurred because of the heavy rainfall that month. All the extra water overloaded Tomlin County's water treatment plant.

Combined

> Water quality in Hawk River declined in March because heavy rainfalls overloaded Tomlin County's water treatment plant.

Guideline 2 | **Put the action in your verbs**

Most sentences are about action. Sales rise, equipment fails, engineers design, managers approve. Clients praise or complain, and technicians advise. Yet, many people bury the action in nouns, adjectives, and other parts of speech. Consider the following sentence:

| Our department accomplished the conversion to the new machinery in two months. | Original

It could be energized by putting the action (*converting*) into the verb:

| Our department converted to the new machinery in two months. | Revised

Not only is the revised version briefer, it is also more emphatic and lively. Furthermore, according to researchers E. B. Coleman and Keith Raynor (1964), when you put action in your verbs, you can make your prose up to 25 percent easier to read.

To create sentences that focus on action, do the following:

FOCUSING SENTENCES ON ACTION

- **Avoid sentences that use the verb *to be* or its variations (*is, was, will be, and so on*).** The verb *to be* often tells what something is, not what it does.

  | The sterilization procedure <u>is a protection</u> against reinfection. | Original
  | The sterilization procedure <u>protects</u> against reinfection. | Revised

- **Avoid sentences that begin with *It is* or *There are.***

  | <u>It is</u> because the cost of raw materials has soared that the price of finished goods is rising. | Original
  | Because the cost of raw materials has soared, the price of finished goods is rising. | Revised
  | <u>There are</u> several factors causing the engineers to question the dam's strength. | Original
  | Several factors cause the engineers to question the dam's strength. | Revised

- **Avoid sentences where the action is frozen in a word that ends with one of the following suffixes: -tion, -ment, -ing, -ion, -ance.** These words petrify the action that should be in verbs by converting them into nouns.

  | Consequently, I would like to make a <u>recommendation</u> that the department hire two additional programmers. | Original
  | Consequently, I <u>recommend</u> that the department hire two additional programmers. | Revised

Although most sentences are about action, some aren't. For example, topic and forecasting statements often introduce lists or describe the organization of the discussion that follows.

| There are three main reasons the company should invest money to improve communication between corporate headquarters and the out-of-state plants. | Topic sentence for which the verb *to be* is appropriate

Guideline 3 | Use the active voice unless you have a good reason to use the passive voice

Another way to focus your sentences on action and actors is to use the *active voice* rather than the *passive voice*. To write in the active voice, place the actor—the person or thing performing the action—in the subject position. Your verb will then describe the actor's action.

Active voice

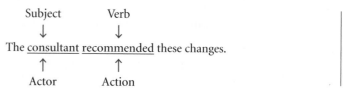

In the passive voice, the subject of the sentence and the actor are different. The subject is *acted upon* by the actor.

Passive voice

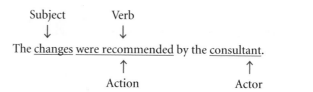

Here are some additional examples:

Passive voice | The Korean ore was purchased by us.

Active voice | We purchased the Korean ore.

Research shows that readers comprehend active sentences more rapidly than passive ones (Layton & Simpson, 1975). Also, the active voice eliminates the vagueness and ambiguity that often characterize the passive voice. In the passive voice, a sentence can describe an action without telling who did it. For example, "The ball was hit" is a grammatically correct sentence but doesn't tell who or what hit the ball. With the active voice, the writer identifies the actor: "Linda hit the ball."

The following sentence illustrates the importance of ensuring that readers understand who the actor is.

Passive voice | The operating temperatures must be checked daily to protect the motor from damage. |

Will the supervisor of the third shift know that he is the person responsible for checking temperatures? In the passive voice, this sentence certainly allows him to imagine that someone else, perhaps a supervisor on another shift, is responsible.

There are some places where the passive voice is appropriate.

Although the passive voice generally reduces readability, it has some good uses. One occurs when you don't want to identify the actor. The following sentence is from a memorandum in which the writer urges all employees to work harder at saving energy but avoids causing embarrassment and resentment by naming the guilty parties.

The lights on the third floor have been left on all night for the past week, despite the efforts of most employees to help us reduce our energy bills.	Passive voice

Also, consider this sentence:

I have been told that you may be using the company telephone for an excessive number of personal calls.	Passive voice

Perhaps the person who told the writer about the breach of corporate telephone policy did so in confidence. If the writer decided that it would be ethically acceptable to communicate this news to the reader without naming the person who made the report, then she has used the passive voice effectively. (Be careful, however, to avoid using the passive voice to hide an actor's identity when it is unethical to do so—for instance, when trying to avoid accepting responsibility for your employer's actions.)

Another good reason for using the passive voice is discussed in Chapter 8, Guideline 7, page 225.

Guideline 4 | Emphasize what's most important

Another way to write clear, forceful sentences is to direct your readers' attention to the most important information you are conveying.

EMPHASIZING WHAT'S MOST IMPORTANT

1. **Place the key information at the end of the sentence.** As linguist Joseph Williams (2005) points out, you can demonstrate to yourself that the end of the sentence is a place of emphasis by listening to yourself speak. Read the following sentences aloud:

   | Her powers of concentration are extraordinary. |
   | Last month, he topped his sales quota even though he was sick for an entire week. |

   As you read these sentences aloud, notice how you naturally stress the final words *extraordinary* and *entire week*.

   To position the key information at the end of a sentence, you may sometimes need to rearrange your first draft.

The department's performance <u>has been superb</u> in all areas.	Original
In all areas, the department's performance <u>has been superb</u>.	Revised

2. **Place the key information in the main clause.** If your sentence has more than one clause, use the main clause for the information you want to emphasize. Compare the following versions of the same statement.

Original	Although our productivity was down, our profits were up.
Revised	Although our profits were up, our productivity was down.

In the first version, the emphasis is on profits because *profits* is the subject of the main clause. The second version emphasizes productivity because *productivity* is the subject of the main clause. (Notice that in each of these sentences, the emphasized information is not only in the main clause but also at the end of the sentence.)

3. **Emphasize key information typographically.** Use boldface and italics. Be careful, however, to use typographical highlighting sparingly. When many things are emphasized, none stand out.

4. **Tell readers explicitly what the key information is.** You can also emphasize key information by announcing its importance to your readers.

> Economists pointed to three important causes of the stock market's decline: uncertainty about the outcome of last month's election, a rise in inventories of durable goods, and— *most important*—signs of rising inflation.

Guideline 5 | Vary your sentence length and structure

If all the sentences in a sentence group have the same structure, two problems arise: Monotony sets in, and (because all the sentences are basically alike) you lose the ability to emphasize major points and deemphasize minor ones.

You can avoid such monotony and loss of emphasis in two ways:

- **Vary your sentence length.** Longer sentences can be used to show the relationships among ideas. Shorter sentences provide emphasis in the context of longer sentences.

In April, many amateur investors believed that another rally was about to begin. Because exports were increasing rapidly, they predicted that the dollar would strengthen in global monetary markets, bringing foreign investors back to Wall Street. Also, unemployment dropped sharply, which they interpreted as an encouraging sign for the economy. They were wrong on both counts. Wall Street interpreted rising exports to mean that goods would cost more at home, and it predicted that falling unemployment would mean a shortage of workers, hence higher prices for labor. Where amateur investors saw growth, Wall Street saw inflation.

This short sentence receives emphasis because it comes after longer ones.

The final sentence is also emphasized because it is much shorter than the preceding one.

- **Vary your sentence structure.** For example, the grammatical subject of the sentence does not have to be the sentence's first word. In fact, if it did, the

English language would lose much of its power to emphasize more important information and to deemphasize less important information.

One alternative to beginning a sentence with its grammatical subject is to begin with a clause that indicates a logical relationship.

<u>After we complete our survey,</u> we will know for sure whether the proposed site for our new factory was once a Native American camping ground.

<u>Because we have thoroughly investigated all the alternatives,</u> we feel confident that a pneumatic drive will work best and provide the most reliable service.

Introductory clause

Introductory clause

Guideline 6 | Global Guideline: Adapt your sentences for readers who are not fluent in your language

The decisions you make about the structure of your sentences can affect the ease with which people who are not fluent in English can understand your message. Companies in several industries, including oil and computers, have developed simplified versions of English for use in communications for readers in other cultures. In addition to limited vocabularies, simplified English has special grammar rules that guide writers to using sentences that will be easy for their readers to understand. Because many readers may not need this degree of simplification, learn as much as possible about your specific readers. Also, remember that simplifying your sentence structure should not involve simplifying your thought.

Learn More

For more advice on adapting communications to your readers' cultural background, go to Chapter 3, page 79 and see the other Global Guidelines throughout this book.

GUIDELINES FOR CREATING SENTENCES FOR READERS WHO ARE NOT FLUENT IN ENGLISH

- **Use simple sentence structures.** The more complex your sentences, the more difficult they will be for readers to understand.
- **Keep sentences short.** A long sentence can be hard to follow, even if its structure is simple. Set twenty words as a limit.
- **Use the active voice.** Readers who are not fluent in English can understand the active voice much more easily than they can understand the passive.

www

For more on simplified versions of English, go to **www.cengage.com/english/ anderson7e**, Chapter 9.

SELECTING WORDS

When selecting words, your first goal should be to increase the usability of your writing by enabling readers to grasp your meaning quickly and accurately. Your word choices also affect your readers' attitudes toward you and your subject matter. Select words that will increase your communication's persuasiveness.

Guideline 1 | Use concrete, specific words

Almost anything can be described either in relatively abstract, general words or in relatively concrete, specific ones. You may say that you are writing on a piece of *electronic equipment* or that you are writing on *a laptop computer connected to a color laser printer.* You may say that your employer produces *consumer goods* or that it makes *cell phones.*

When groups of words are ranked according to degree of abstraction, they form *hierarchies.* Figure 9.1 shows such a hierarchy in which the most specific terms identify concrete items that we can perceive with our senses; Figure 9.2 shows a hierarchy in which all the terms are abstract, but some are more specific than others.

You can increase the clarity, and therefore the usability, of your writing by using concrete, specific words rather than abstract, general ones. Concrete words help your readers understand precisely what you mean. If you say that your company produces television shows for a *younger demographic segment,* they won't know whether you mean *teenagers* or *toddlers.* If you say that you study *natural phenomena,* your readers won't know whether you mean *volcanic eruptions* or the *migration of monarch butterflies.*

Such vagueness can hinder readers from getting the information they need in order to make decisions and take action. Consider the following sentence from a memo addressed to an upper-level manager who wanted to know why production costs were up:

Original | The cost of one material has risen recently.

This sentence doesn't give the manager the information she needs to take remedial action. In contrast, the following sentence, using specific words, tells precisely what

Try This

Pick a group of words used in your major, hobby, or sport. List words in ranked order, moving from more abstract to more concrete.

FIGURE 9.1
Hierarchy of Related Words That Move from Abstract to Concrete

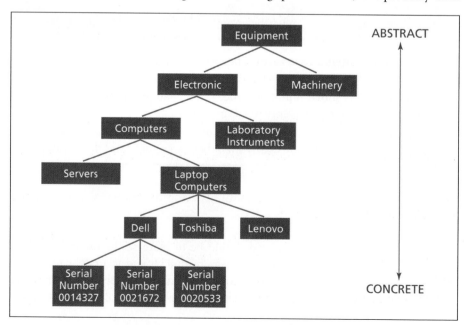

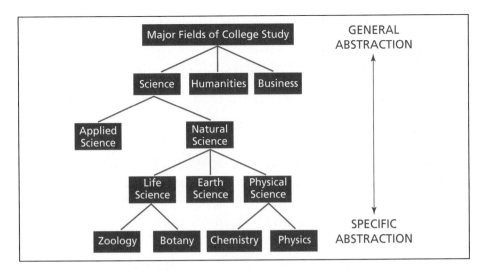

FIGURE 9.2
Hierarchy of Related Words That Move from a General to a Specific Abstraction

the material is, how much the price has risen, and the period in which the increase took place.

| The cost of the bonding agent has tripled in the past six months. | Revised

Of course, abstract and general terms do have important uses. For example, in scientific, technical, and other specialized fields, writers often need to make general points, describe the general features of a situation, or provide general guidance for action. Your objective when choosing words is not to avoid abstract, general words altogether, but rather to avoid using them when your readers will want more specific ones.

Guideline 2 | Use specialized terms when—and only when—your readers will understand them

You can increase the usability and persuasiveness of your writing by using wisely the specialized terms of your own profession.

In some situations, specialized terms help you communicate effectively:

- **They convey precise, technical meanings economically.** Many terms have no exact equivalent in everyday speech.
- **They help you establish credibility.** By using the special terms of your field accurately, you show your fellow specialists that you are adept in it.

However, you should avoid using technical terms your readers won't understand. Consider the following sentence:

> The major benefits of this method are smaller in-gate connections, reduced breakage, and minimum knock-out—all leading to great savings.

Although this sentence would be perfectly clear to any manager who works in a foundry that manufactures parts for automobile engines, it would be unintelligible to most other people because it includes the specialized terms *in-gate connections* and *knock-out.*

How to Explain Unfamiliar Terms If You Must Use Them

Sometimes you may need to use specialized terms even though some people in your audience may not understand them. For instance, you may be writing to a group of readers that includes people in your field and others outside of it, or you may be explaining an entirely new subject to your readers. In such cases, there are several ways to define the terms for readers who are not familiar with them.

DEFINING TERMS YOUR READERS DON'T KNOW

1. **Give a synonym.** Example: On a boat, a rope or cord is called a *line.*
2. **Give a description.** Example: The *exit gate* consists of two arms that hold a jug while it is being painted and then allow it to proceed down the production line.
3. **Make an analogy.** Example: An atom is like a miniature solar system in which the nucleus is the sun and the electrons are the planets that revolve around it.
4. **Give a classical definition.** In a classical definition, you define the term by naming some familiar group of things to which it belongs and then identifying the key distinction between the object being defined and the other members of the group. Examples:

WORD	GROUP	DISTINGUISHING CHARACTERISTIC
A crystal is a	solid	in which the atoms or molecules are arranged in a regularly repeated pattern.
A burrow is a	hole in the ground	dug by an animal for shelter or habitation.

Guideline 3 | **Use words accurately**

Whether you use specialized terms or everyday ones and whether you use abstract, general terms or concrete, specific ones, you must use all your words accurately. This point may seem obvious, but inaccurate word choice is all too common in on-the-job writing. For example, people often confuse *imply* (meaning to *suggest* or *hint,* as in "He implied that the operator had been careless") with *infer* (meaning to draw a *conclusion based upon evidence,* as in "We infer from your report that you do not expect to meet the deadline"). It's critical that you avoid such errors. They distract your readers from your message by drawing their attention to your problems with word choice, and they may lead your readers to believe that you are not skillful or precise in other areas—such as laboratory techniques or analytical skills.

How can you ensure that you use words accurately? There's no easy way. Consult a dictionary whenever you are uncertain. Be especially careful when using words that are not yet part of your usual vocabulary. Pay attention as well to the way words are used by other people.

Guideline 4 | Choose plain words over fancy ones

You can also make your writing easy to understand by avoiding using fancy words where plain ones will do. At work, some writers do just the opposite, perhaps thinking that fancy words sound more official or make them sound more knowledgeable. The following list identifies some commonly used fancy words; it includes only verbs but might have included nouns and adjectives as well.

FANCY VERBS	EQUIVALENT COMMON VERBS
ascertain	find out
commence	begin
compensate	pay
constitute	make up
endeavor	try
expend	spend
fabricate	build
facilitate	make easier
initiate	begin
prioritize	rank
proceed	go
terminate	end
transmit	send
utilize	use

There are two important reasons for preferring plain words over fancy ones:

- **Plain words promote efficient reading.** Research has shown that even if your readers know both the plain word and its fancy synonym, they will still comprehend the plain word more rapidly (F. Smith, 2004).

- **Plain words reduce your risk of creating a bad impression.** If you use words that make for slow, inefficient reading, you may annoy your readers or cause them to conclude that you are behaving pompously, showing off, or trying to hide a lack of ideas and information behind a fog of fancy terms. Consider, for instance, the effect of the following sentence in a job application letter:

> I am transmitting the enclosed résumé to facilitate your efforts to determine the pertinence of my work experience to your opening.

Pompous word choices

Don't misunderstand this guideline. It doesn't suggest that you should use only simple language at work. When addressing people with vocabularies comparable to your own, use all the words at your command, provided that you use them accurately and appropriately. This guideline merely cautions you against using needlessly inflated words that bloat your prose and open you to criticism from your readers.

Guideline 5 | Choose words with appropriate associations

The three previous guidelines for choosing words relate to the literal or dictionary meaning of words. At work, you must also consider the associations your words have for your readers. In particular, be especially sensitive to your words' *connotation* and *register.*

Connotation

Connotation is the extended or suggested meaning that a word has beyond its literal meaning. For example, according to the dictionary, *flatfoot* and *police detective* are synonyms, but they connote very different things: *flatfoot* suggests a plodding, perhaps not very bright cop, while *police detective* suggests a trained professional.

Verbs, too, have connotations. For instance, to *suggest* that someone has overlooked a key fact is not the same as to *insinuate* that she has. To *devote* your time to working on a client's project is not the same as to *spend* your time on it.

Research on the impact of connotation

The connotations of your words can shape your audience's perceptions of your subject matter. Researchers Raymond W. Kulhavy and Neil H. Schwartz (1981) demonstrated those effects in a classic experiment for which they created two descriptions of a company that differed in only seven out of 246 words. In one, the seven words suggested stiffness, such as *required* and *must.* In the other, those seven words were replaced by ones that suggested flexibility, such as *asked* and *should.* None of the substitutions changed the facts of the overall passage. Here's a sentence from the first version:

First version

> Our sales team is constantly trying to locate new markets for our various product lines.

In the second version of this sentence, the researchers replaced the flexible word *trying* with the stiff word *driving.*

Second version

> Our sales team is constantly driving to locate new markets for our various product lines.

The researchers found that people who read the flexible version believed that the company would actively commit itself to the welfare and concerns of its employees, voluntarily participate in affirmative action programs for women and minorities, receive relatively few labor grievances, and pay its employees well. People who read that version also said they would recommend the company to a friend as a place to work. People who read the stiff version reported opposite impressions of the company. That readers' impressions of the company could be affected so dramatically by just seven nonsubstantive words highlights the great importance of paying attention to the connotations of the words you use.

Register

Linguists use the term *register* to identify a second characteristic exhibited by words: their association with certain kinds of communication situations or context. For example, in an ad for a restaurant we might expect to see the claim that it offers *amazingly* delicious food. However, we would not expect to see a research company boast in a proposal for a government contract that it is capable of conducting *amazingly* good studies. The word *amazingly* is in the register of consumer advertising but not in the register of research proposals.

If you inadvertently choose words with the wrong register, your readers may infer that you don't fully grasp how business is conducted in your field, and your credibility can be lost.

Guideline 6 | Global Guideline: Consider your readers' cultural background when choosing words

Take special care in your choice of words when writing to readers in other cultures. Some words whose meaning is obvious in your own culture can be misunderstood or completely mystifying to readers from other cultures. This is true whether your communication will go to your readers in English or whether it will be translated for them. In fact, misunderstanding can even occur when you are writing to readers in other cultures where the native language is English. In the United States, people play football with an oblong object which they try to carry over a goal line or kick through uprights. In England, India, and many other parts of the world, football is played with a round object that people are forbidden to carry and attempt to kick into a net.

The following guidelines will help you choose words your readers will understand in the way you intend. Of course, different readers in other cultures have different levels of facility with English, so follow the guidelines only to the extent that your readers require.

© Yuri Arcurs/Shutterstock.com

At work, even small departments often include a rich diversity of employees.

GUIDELINES FOR CHOOSING WORDS FOR INTERCULTURAL COMMUNICATIONS

- **Use simple words.** The more complex your vocabulary, the more difficult it will be for readers not fluent in English to understand you.

- **Use the same word each time you refer to the same thing.** For instance, in instructions, don't use both "dial" and "control" for the part of a test instrument. In context, those two terms may be synonyms in your language, but they will each be translated into a different word in the other language, where the translated words may not be synonyms.

- **Avoid acronyms your readers won't understand.** Most acronyms that are familiar to you will be based on words in *your* language: AI for Artificial Intelligence; ACL for Anterior Cruciate Ligament.

- **Avoid slang words and idioms.** Most will have no meaning for people in other cultures. Instead of "We want a level playing field," say "We want the decision to be made fairly." Instead of saying "We want to run an idea past you," say "We'd like your opinion of our idea."

Learn More

For more advice on adapting communications to your readers' cultural background, go to Chapter 3, page 79, and see the other Global Guidelines throughout this book.

Guideline 7 | Ethics Guideline: Use inclusive language

When constructing your voice, use language that includes all persons instead of excluding some. For example, avoid sexist language because it supports negative stereotypes. Usually these stereotypes are about women, but they can also adversely affect men in certain professions such as nursing. By supporting negative stereotypes, sexist language can blind readers to the abilities, accomplishments, and potential of very capable people. The same is true of language that insensitively describes people with disabilities, illnesses, or other limitations.

Learn More

For another discussion of stereotypes and ethics, see page 268.

WWW

For additional suggestions about ways to avoid sexist and discriminatory language, go to Chapter 9 at **www.cengage .com/english/anderson7e.**

USING INCLUSIVE LANGUAGE

1. **Use nouns and pronouns that are gender-neutral rather than ones containing the word *man*.**

   Instead of: businessman, workman, mailman, salesman

   Use: businessperson, manager, *or* executive; worker; mail carrier; salesperson

   Instead of: man-made, man hours, man-sized job

   Use: synthetic, working hours, large job

2. **Use plural pronouns or *he or she* instead of sex-linked pronouns when referring to people in general.**

   Instead of: "Our home electronics cater to the affluent shopper. She looks for premium products and appreciates a stylish design."

   Use the plural: "Our home electronics cater to affluent shoppers. They look for premium products and appreciate a stylish design."

   Instead of: "Before the owner of a new business files the first year's tax returns, he might be wise to seek advice from a certified public accountant."

   Use *he or she*: "Before the owner of a new business files the first year's tax returns, he or she might be wise to seek advice from a certified public accountant."

3. **Refer to individual men and women in a parallel manner.**

   Instead of: "Mr. Sundquist and Anna represented us at the trade fair."

   Use: "Mr. Sundquist and Ms. Tokagawa represented us at the trade fair" or "Christopher and Anna represented us at the trade fair."

4. **Revise salutations that imply the reader of a letter is a man.**

   Instead of: Dear Sir, Gentlemen

   Use: The title of the department or company or the job title of the person you are addressing: Dear Personnel Department, Dear Switzer Plastics Corporation, Dear Director of Research

5. **When writing about people with disabilities, refer to the person first, then the disability.**

   <u>Instead of:</u> the disabled

   <u>Use:</u> people with disabilities

CONCLUSION

Your writing style can make a great deal of difference to the success of your writing. The voice you use, the sentence structures you employ, and the words you select affect both your readers' attitudes toward you and your subject matter and also the readability and impact of your writing. This chapter has suggested many things you can do to develop a highly usable, highly persuasive style. Underlying all these suggestions is the advice that you take the reader-centered approach of considering all your stylistic choices from your readers' point of view.

USE WHAT YOU'VE LEARNED

For additional exercises, go to www.cengage.com/english/anderson7e. *Instructors*: The book's website includes suggestions for teaching the exercises.

You can download Expertise Exercises 2 through 5 from Chapter 9 at www.cengage.com/english/anderson7e.

EXERCISE YOUR EXPERTISE

1. Imagine that you are the head of the Public Safety Department at your college. Faculty and staff have been parking illegally, sometimes where there aren't parking spots. Sometimes individuals without handicaps are parking in spots reserved for those with handicaps. Write two memos to all college employees announcing that beginning next week, the Public Safety Department will strictly enforce parking rules—something it hasn't been doing. Write the first memo in a friendly voice and the second in a stern voice. Then compare the specific differences in organization, sentence structure, word choice, and other features of writing to create each voice. (Thanks to Don Cunningham, Auburn University, for the idea for this exercise.)

2. Without altering the meaning of the following sentences, reduce the number of words in them.
   a. After having completed work on the data-entry problem, we turned our thinking toward our next task, which was the processing problem.
   b. Those who plan federal and state programs for the elderly should take into account the changing demographic characteristics in terms of size and average income of the composition of the elderly population.
   c. Would you please figure out what we should do and advise us?
   d. The result of this study will be to make total white-water recycling an economical strategy for meeting federal regulations.

3. Rewrite the following sentences in a way that will keep the related words together.
   a. This stamping machine, if you fail to clean it twice per shift and add oil of the proper weight, will cease to operate efficiently.
   b. The plant manager said that he hopes all employees would seek ways to cut waste at the supervisory meeting yesterday.

c. About 80 percent of our clients, which include over 1,500 companies throughout North and South America and a few from Africa, where we've built alliances with local distributors, find the help provided at our website to be equivalent in most cases to the assistance supplied by telephone calls to our service centers.

d. Once they wilt, most garden sprays are unable to save vegetable plants from complete collapse.

4. Rewrite the following sentences to put the action in the verb.

a. The experience itself will be an inspirational factor leading the participants to a greater dedication to productivity.

b. The system realizes important savings in time for the clerical staff.

c. The implementation of the work plan will be the responsibility of a team of three engineers experienced in these procedures.

d. Both pulp and lumber were in strong demand, even though rising interest rates caused the drying up of funds for housing.

5. Rewrite the following sentences in the active voice.

a. Periodically, the shipping log should be reconciled with the daily billings by the Accounting Department.

b. Fast, accurate data from each operating area in the foundry should be given to us by the new computerized system.

c. Since his own accident, safety regulations have been enforced much more conscientiously by the shop foreman.

d. No one has been designated by the manager to make emergency decisions when she is gone.

6. Create a one-sentence, classical definition for a word used in your field that is not familiar to people in other fields. The word might be one that people in other fields have heard of but cannot define precisely in the way specialists in your field do. Underline the word you are defining. Then circle and label the part of your definition that describes the familiar group of items that the defined word belongs to. Finally, circle and label the part of your definition that identifies the key distinction between the defined word and the other items in the group. (Note

that not every word is best defined by means of a classical definition, so it may take you a few minutes to think of an appropriate word for this exercise.)

7. Create an analogy to explain a word used in your field that is unfamiliar to most readers. (Note that not every word is best defined by means of an analogy, so it may take you a few minutes to think of an appropriate word for this exercise.)

EXPLORE ONLINE

Using your desktop publishing program, examine the readability statistics for two communications. These might be two projects you're preparing for courses, or they might be a course project and a letter or e-mail to a friend or family member. What differences, if any, do you notice in the statistics? What accounts for the differences? Are the statistics helpful to you in understanding and constructing an effective writing style in either case? Read the explanations of the scores that are provided with your desktop publishing program; these may be provided in the program's Help feature. Do the interpretations of your scores agree with your own assessment of your communications? If not, which do you think is more valid? (To learn how to obtain the readability statistics with your program, use its Help feature.)

COLLABORATE WITH YOUR CLASSMATES

Working with another student, examine the memo shown in Figure 9.3. Identify places where the writer has ignored the guidelines given in this chapter. You may find it helpful to use a dictionary. Then write an improved version of the memo by following the guidelines in this chapter.

APPLY YOUR ETHICS

1. Find a communication that fails to use inclusive language and revise several of the passages to make them inclusive.

2. The images in advertising often rely on stereotypes. Find one advertisement that perpetuates one or more stereotypes and one that calls attention to itself by using an image that defies a stereotype. Evaluate the ethical impact of each image. Present your results in the way your instructor requests.

MEMO

July 5, 2010

TO: Gavin MacIntyre, Vice President, Midwest Region

FROM: Nat Willard, Branch Manager, Milwaukee Area Offices

The ensuing memo is in reference to provisions for the cleaning of the six offices and two workrooms in the High Street building in Milwaukee. This morning, I absolved Thomas's Janitor Company of its responsibility for cleansing the subject premises when I discovered that two of Thomas's employees had surreptitiously been making unauthorized long-distance calls on our telephones.

Because of your concern with the costs of running the Milwaukee area offices, I want your imprimatur before proceeding further in making a determination about procuring cleaning services for this building. One possibility is to assign the janitor from the Greenwood Boulevard building to clean the High Street building also. However, this alternative is judged impractical because it cannot be implemented without circumventing the reality of time constraints. While the Greenwood janitor could perform routine cleaning operations at the High Street establishment in one hour, it would take him another ninety minutes to drive to and fro between the two sites. This is more time than he could spare and still be able to fulfill his responsibilities at the High Street building.

Another alternative would be to hire a full-time or part-time employee precisely for the High Street building. However, that building can be cleaned so expeditiously, it would be irrational to do so.

The third alternative is to search for another janitorial service. I have now released two of these enterprises from our employ in Milwaukee. However, our experiences with such services should be viewed as bad luck and not affect our decision, except to make us more aware that making the optimal selection among companies will require great care. Furthermore, there seems to be no reasonable alternative to hiring another janitorial service.

Accordingly, I recommend that we hire another janitorial service. If you agree, I can commence searching for this service as soon as I receive a missive from you. In the meantime I have asked the employees who work in the High Street building to do some tidying up themselves and to be patient.

FIGURE 9.3
Memo for Collaboration
Exercise

10 | Beginning a Communication

GUIDELINES

As you learned in Chapter 1, reading is a dynamic interaction between your readers and your words and graphics. Your readers' response to one sentence or paragraph can influence their reactions to all the sentences and paragraphs that follow. Consequently, the opening sentence or section takes on special importance: It helps to establish the frame of mind readers bring to all the sentences and sections that follow.

In this chapter, you will learn eight reader-centered strategies for beginning your communications in highly usable and highly persuasive ways. Rarely, if ever, will you use all eight at once. To decide which one or combination to use in a particular circumstance, you will need to build on the knowledge of your readers that you gained while defining your communication's objectives (Chapter 3). The chapter's ninth guideline discusses ethical approaches to situations in which people at work sometimes wonder whether they should try communicating at all.

DEFINING
OBJECTIVES

PLANNING

CONDUCTING
RESEARCH

DRAFTING

REVISING

CHAPTER 10

WWW

To read additional information, see more examples, and access links related to this chapter's guidelines, go to Chapter 10 at **www.cengage.com/english/anderson7e.**

INTRODUCTION TO GUIDELINES 1 THROUGH 3

The first three guidelines echo strategies described in Chapter 8 for beginning paragraphs, sections, and chapters: Announce the topic, state the main point, and forecast your communication's organization. These helpful strategies from Chapter 8 are incorporated into this chapter's Guidelines 1 through 3 so you can see how to apply them when beginning your entire communication.

Guideline 1 | Give your readers a reason to pay attention

Sometimes the most important function of a beginning is simply to persuade readers to devote their full attention to the message. At work, people complain that they receive too many e-mails, memos, and reports. As they look at each communication, they ask, "Why should I read this?" Your goal is to convince them not only to pay *some* attention to your message, but also to pay *close* attention. Doing so will be especially important when your communication is primarily persuasive, as when you are writing a proposal or

People are more likely to be persuaded by messages they think deeply about.

recommendation. Research has shown that the more deeply people think about a message while reading or listening to it, the more likely they are to hold the attitudes it advocates, resist attempts to reverse those attitudes, and act upon those attitudes (Petty & Cacioppo, 1986).

To grab your readers' attention, you must do two things at the outset:

- Announce your topic.
- Tell your readers how they will benefit from the information you are providing.

Be sure to do *both* things. Don't assume that your readers will automatically see the value of your information after you have stated your topic. Benefits that appear obvious to you may not be obvious to them. Compare the following sets of statements:

STATEMENTS OF TOPIC ONLY (AVOID THEM)	STATEMENTS OF TOPIC AND BENEFIT (USE THEM)
This memo tells about the new technology for reducing carbon dioxide emissions.	This memo answers each of the five questions you asked me last week about the new technology reducing carbon dioxide emissions.
This report discusses step-up pumps.	Step-up pumps can save us money and increase our productivity.
This manual concerns the Cadmore Industrial Robot 2000.	This manual tells how to prepare the Cadmore Industrial Robot 2000 for difficult welding tasks.

Importance of Subject Lines in E-mails

Learn More

For more on subject lines in e-mail, memos, and letters, go to Chapter 22, pages 547, 550, and 551.

E-mails present a special challenge. They have, in effect, two beginnings: the first sentences and the subject line. No matter how skillfully you craft the opening sentences, they won't be read unless your subject line persuades your intended readers to open your message. Name your topic precisely and indicate that what you have to say about it will benefit your readers. Do the same in the subject lines that memos usually include and letters sometimes do.

This subject line indicates → Subject: Springer Valves
only the writer's topic.

The revised subject line also → Subject: Recommendation for Improving Springer Valves
indicates that the reader
will benefit by finding
recommendations.

Subject: Data

This subject line is vague
about the topic (what data?).

Subject: Analysis of Probe Data

This is more precise: It tells
what kind of data and also
indicates the reader will
benefit by finding an analysis
of the data.

Two Ways to Highlight Reader Benefits

Two strategies are particularly effective at persuading people that they will benefit from reading your communication.

Refer to Your Reader's Request.   At work, you will often write because a coworker, manager, or client has asked you for a recommendation or information. To establish the reader benefit of your reply, simply refer to the request.

As you requested, I am enclosing a list of the steps we have taken in the past six months to tighten security in the Information Technology Department.

References to readers' requests

Thank you for your inquiry about the capabilities of our Model 1770 color laser printer.

Offer to Help Your Readers Solve a Problem.   The second strategy for highlighting reader benefits at the beginning of a communication is to tell your readers that your communication will help them solve a problem they are confronting. Most employees think of themselves as problem solvers. Whether the problem involves technical, organizational, or ethical issues, they welcome communications that help them find a solution.

Communication experts J. C. Mathes and Dwight W. Stevenson (1991) have suggested an especially powerful approach to writing beginnings that builds on readers' concerns with problem solving. First, list problems that are important to the people you are going to address in your communication. From the list, pick one that the information and ideas you will provide can help them solve. When you've done that, you have begun to think of yourself and your readers as partners in a joint problem-solving effort in which your communication plays a critical role. The following diagram illustrates this relationship.

JOINT PROBLEM-SOLVING EFFORT

Once you have determined how to describe a problem-solving partnership between you and your readers, draft the beginning of your communication using the following strategy:

Writer and readers are problem-solving partners.

ESTABLISHING A PROBLEM-SOLVING PARTNERSHIP WITH YOUR READERS

1. **Tell your readers the problem you will help them solve.** Be sure to identify a problem your readers deem important.

2. **Tell your readers what you have done toward solving the problem.** Review the steps you have taken as a specialist in your own field, such as developing a new feature for one of your employer's products or investigating products offered by competitors. Focus on activities that will be significant to your readers rather than listing everything you've done.

3. **Tell your readers how your communication will help them as they perform their part of your joint problem-solving effort.** For example, you might say that it will help them compare competing products, understand a new policy for handling purchase orders, or develop a new marketing plan.

To establish a problem-solving partnership, tell your readers these three things.

The following diagram shows the relationships among the three elements in this type of reader-centered beginning.

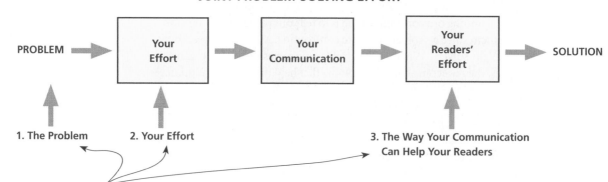

JOINT PROBLEM-SOLVING EFFORT

PROBLEM → Your Effort → Your Communication → Your Readers' Effort → SOLUTION

1. The Problem   2. Your Effort   3. The Way Your Communication Can Help Your Readers

The three elements of a beginning that describes a problem-solving situation

Carla, a computer consultant, used this strategy when she wrote the beginning of a report about her trip to Houston, where she studied the billing system at a hotel her employer recently purchased. First, she identified the problem that her communication would help her readers solve: The Houston hotel was making little money, perhaps because its billing system was faulty. Next, she reviewed what she had done in Houston to help her readers solve this problem: She evaluated the billing system and formulated possible improvements. Finally, she determined how her report would help her readers perform their own problem-solving activities: It would help them choose the best course of action by providing data and recommendations on which they could base their decision.

Here is how Carla wrote the beginning of her report:

Carla names the problem her report will help solve.

Carla describes her work toward solving the problem.

Carla tells how her report will help her readers do their part toward solving the problem.

> Last year, our Houston hotel posted a profit of only 4 percent, even though it is almost always 78 percent filled. A preliminary examination of the hotel's operations suggests that its billing system may generate bills too slowly and it may be needlessly ineffective in collecting overdue payments. After examining the hotel's billing cycle and collection procedures, I identified four ways to improve them. To aid in the evaluation of my recommendations, this report discusses the costs and benefits of each one.

Must You Always Provide All This Information?   Beginnings that use this strategy may be shorter or longer than Carla's. Just make sure your readers understand all three elements of the problem-solving situation. If one or more of the elements will be obvious to them, there's no benefit in discussing them in detail. For example, if the only persons who would read Carla's report were thoroughly familiar with the details of the problem in Houston and the work she did there, Carla could have written a very brief beginning:

> In this report, I evaluate the billing system in our Houston hotel and recommend ways of improving it.

She included more explanation because she knew her report would also be read by people who were hearing about her trip for the first time.

Here are some situations in which a full description of the problem-solving situation usually is desirable:

- **Your communication will be read by people outside your immediate working group.** The more distant some or all of your readers are, the less likely that all of them will be familiar with your message's context.

- **Your communication will have a binding and a cover.** Bound documents are usually intended for large groups of current readers and they are often filed for consultation by future readers. Both groups are likely to include at least some readers who will have no idea of the problem-solving situation you are addressing.

- **Your communication will be used to make a decision involving a significant amount of money.** Such decisions are often made by high-level managers who need to be told about the organizational context of the reports they read.

Circumstances in which you may need to describe the problem-solving situation in detail

Defining the Problem in Unsolicited Communications   In your career, you will have many occasions to make a request or recommendation without being asked to do so. When writing these unsolicited communications, you may need to persuade your readers that a problem even exists. This can require some creative, reader-centered thinking.

Consider the way Roberto accomplished this goal. He works for a company that markets computer programs used to control manufacturing processes. One program contained bugs that Roberto wanted to fix because he sympathized with the customers who called for help in overcoming problems caused by the bugs. However, Roberto knew that his boss did not want to assign computer engineers to this task. Instead, she wanted the engineers to spend all their time on her top priority, which was to develop new products rapidly. Consequently, Roberto wrote his boss a memo that opened by discussing the difficulty the company had been having in releasing new products on time. He then showed how much time the computer engineers needed to spend helping customers overcome problems caused by the bugs rather than working on new products. By tying his request to a problem that his reader found significant (rather than to the problems that actually prompted him to write), Roberto succeeded in being assigned to fix the two most serious bugs.

Try This

Think of a way that one of your professors could improve a course you have taken with him or her. Identify ways the professor could better achieve his or her goal by making the change you have in mind. Write the opening sentences of a letter or e-mail you might send to the professor with your suggestion.

Guideline 2 | State your main point

You can usually boost your communication's usability and persuasiveness by stating your main point in your beginning. In earlier chapters, you learned three major reasons for doing so:

- You help your readers find what they most want or need.
- You increase the likelihood that your readers will actually read your main point instead of putting your communication aside before they get to it.
- You provide your readers with a context for viewing the details that follow.

Reasons for stating your main point at the beginning

Choose Your Main Point Thoughtfully

Choose the main point of your communication in the same way you choose the main point of each segment. If you are responding to a request, your main point will be the answer to the question your reader asked. If you are writing on your own initiative, your main point will be what you want your readers to think or do after reading your communication.

Here are some sample statements:

From the beginning of a memo written in a manufacturing company:

Each of these beginnings states the writer's main point.

We should immediately suspend all purchases from Cleves Manufacturing until it can guarantee us that the parts it supplies will meet our specifications.

From the beginning of a memo written to a department head in a food services company:

I request $1,200 in travel funds to send one of our account executives to the client's Atlanta headquarters.

From a research report:

The test results show that the walls of the submersible room will not be strong enough to withstand the high pressures of a deep dive.

Guideline 3 | Tell your readers what to expect

Learn More

For more on forecasting statements, turn to page 216.

Unless your communication is very short, readers will usually be grateful if you tell them at the beginning what to expect in the rest of it. This guideline echoes the advice given in Chapter 8: "Use forecasting statements." A forecasting statement positioned at the beginning of an entire communication can preview its organization and scope.

Preview Your Communication's Organization

By telling your readers about your communication's organization in your beginning, you provide them with a framework for understanding the connections among the various pieces of information you convey. This framework substantially increases your communication's usability by helping your readers to see immediately how each new point you make relates to the points they have already read. It also helps skimming readers to navigate quickly to the information they are seeking.

You can tell your readers about the organization of your communication in various ways:

The writers use sentences to forecast the organization of their report.

In this report, we state the objectives of the project, compare the three major technical alternatives, and present our recommendation. The final sections include a budget and a proposed project schedule.

This booklet covers the following topics:

- Principles of Sound Reproduction

- Types of Speakers

- Choosing the Speakers That Are Right for You

- Installing the Speakers

These writers use a list to forecast their booklet's organization.

Indicate Your Communication's Scope

Readers often want to know from the beginning what a communication does and does not contain. Even if they are persuaded that you are addressing a subject relevant to them, they may still wonder whether you will discuss the specific aspects of the subject they want to know about.

Sometimes, you will tell your readers about the scope of your communication when you tell them about its organization: When you list the topics it addresses, you indicate its scope.

There will be times, however, when you will need to include additional information. That happens when you want your readers to understand that you are not addressing your subject comprehensively or that you are addressing it from a particular point of view. For instance, you may be writing a troubleshooting manual to help factory workers solve a certain set of problems that often arise with the manufacturing robots they monitor. Other problems—ones your manual doesn't address—might require the assistance of someone else. In that case, you should tell your readers explicitly about the scope of your manual:

This manual treats problems you can correct by using tools and equipment normally available to you. It does not cover problems that require work by computer programmers or electrical engineers.

In the first sentence, the writer describes the manual's contents. To avoid possible misunderstanding, the writer adds the second sentence to tell what it does not contain.

Guideline 4 | Encourage openness to your message

Other chapters in this book have emphasized that readers can respond in a variety of ways as they read a communication. For example, when they read a set of recommendations you are making, they can try to understand your arguments or search for flaws. When they read a set of instructions you have prepared, they can follow your directions in every detail or attempt the procedure on their own, consulting your instructions only if they get stumped.

Because the way you begin a communication has a strong effect on your readers' response, you should always pay attention to the persuasive dimension of your beginnings. Always begin in a way that encourages your readers to be open and receptive to the rest of your communication.

Readers' Initial Reactions Can Vary

Ordinarily, you will have no trouble eliciting a receptive response because you will be communicating with fellow employees, customers, and others who want the information you are providing. In certain circumstances, however, your readers may have a more negative attitude toward your message. In such situations, you will need to take special care in drafting the beginning of your communication if you are to win a fair hearing for your message.

Your readers' initial attitude toward your message will be negative if the answer to any of the following questions is "yes." If that is the case, try to pinpoint the attitudes that are likely to shape your readers' reactions to your communication, and then devise your beginning accordingly.

Questions for determining whether readers might resist your message

- Does your message contain bad news for your readers?
- Does your message contain ideas or recommendations that will be unwelcome to your readers?
- Do your readers feel distrust, resentment, or competitiveness toward you, your department, or your company?
- Are your readers likely to be skeptical of your knowledge of your subject or situation?
- Are your readers likely to be suspicious of your motives?

The strategy that is most likely to promote a positive initial reaction or to counteract a negative one differs from situation to situation. However, here are three strategies that often work.

STRATEGIES FOR ENCOURAGING OPENNESS

- **Present yourself as a partner, not as a critic or a competitor.** Suggest that you are working with your readers to help solve a problem they want to solve or to achieve a goal they want to achieve. (See Guideline 1.)

Learn More

See also Chapter 5's discussion of the indirect pattern of organization (page 131).

- **Delay the presentation of your main point.** An initial negative reaction may prompt your readers to aggressively devise counterarguments to each point that follows. Therefore, if you believe that your readers may react negatively to your main point, consider making an exception to Guideline 2, which tells you to state your main point in your beginning. If you delay the presentation of your main point, your readers may consider at least some of your other points objectively before discovering your main point and reacting against it.

Learn More

See also Chapter 5's suggestions for building credibility (page 134).

- **Establish your credibility.** As Chapter 5 suggests, people are more likely to respond favorably to a message if they have confidence in the person delivering it. Consequently, you can promote openness to your message if you begin by convincing your readers that you are an expert in your subject and knowledgeable about the situation. This does not mean, however, that you should announce your credentials in the beginning of *every* communication. If you needlessly present your credentials, you merely burden your readers with unnecessary information. Avoid discussing your qualifications when writing to people, such as your coworkers, who have already formed a favorable opinion of your expertise.

Tell Yourself a Story

Although these strategies will often encourage openness, don't employ them mechanically. Always keep in mind the particular attitudes, experiences, and expectations of your readers as you craft the beginning of a communication.

You might do this by telling yourself a story about your readers. The central figure in your story should be your reader—an individual if you are writing to one person, or a typical member of your audience if you are writing to a group. Begin your story a few minutes before this person picks up your communication and continue it until the moment he or she reads your first words. Although you would not actually include the story in your communication, creating it can help you decide how to begin.

Here is a sample story, written by Jolene, a manager in an insurance company. Jolene wrote it to help herself understand the readers of an instruction manual she is preparing that will teach new insurance agents how to use the company's computer system.

It's Monday afternoon. After half a day of orientation meetings and tours, Bob, the new trainee, sits down at the computer terminal for the first time to try to learn this system. He was a French major who has never used a data entry program. Now, in two hours, he is supposed to work his way through this manual and then enter some sample policy information. He feels rushed, confused, and quite nervous. He knows that the information is critical, and he does not want to make an error.

Jolene predicts her readers' attitudes by imagining a story about one of them.

Despite his insecurity, Bob will not ask questions of the experienced agent in the next office because (being new to the company) he doesn't want to make a bad impression by asking dumb questions.

Bob picks up the instruction manual for the SPRR program that I am writing; he hopes it will tell him quickly what he needs to know. He wants it to help him learn the system in the time allotted without his making any mistakes and without his having to ask embarrassing questions.

This story helped Jolene to focus on several important facts: The reader will be anxious, hurried, and uncertain. Those insights helped her write an effective opening for her manual:

This manual tells you how to enter policy information into our SPRR system. It covers the steps for opening a file for a new policy, entering the relevant information, revising the file, and printing a paper copy for your permanent records.

Jolene adopts a helpful tone.

By following these instructions carefully, you can avoid making time-consuming errors. In addition, the SPRR system is designed to detect and flag possible errors so you can double-check them.

Jolene reassures her readers.

By identifying her readers' probable feelings, Jolene was able to reduce their anxiety and encourage them to be more open to her instructions.

Guideline 5 | **Provide necessary background information**

As you draft the beginning of a communication, ask yourself whether your readers will need any background information to understand what you are going to tell them.

Here are some examples of situations that might require such information at the beginning:

Signs that your readers need background information

- **Your readers need to grasp certain general principles in order to understand your specific points.** For instance, your discussion of the feasibility of locating a new plant in a particular city may depend on a particular analytical technique that you will need to explain to your readers.
- **Your readers are unfamiliar with technical terms you will be using.** For example, as a specialist in international trade, you may need to explain certain technical terms to the board of directors before you present your strategies for opening up foreign markets.
- **Your readers are unfamiliar with the situation you are discussing.** For example, imagine that you are reporting to the executive directors of a large corporation about labor problems at one of the plants it recently acquired in a takeover. To understand and weigh the choices that face them, the directors will need an introduction to the plant and its labor history.

Not all background information belongs at the beginning of your communication. Information that pertains only to certain segments should appear at the beginning of those segments. In the beginning of your communication, include only background information that will help your readers understand your overall message.

Guideline 6 | Include a summary unless your communication is very short

At work, it's quite common for communications as short as a page or two to start with a brief summary of the entire message. These summaries help busy managers learn the main points without reading the entire document, and they give these readers an overview of the communication's content and organization.

In short communications, such as memos and letters, opening summaries often consist of only a few sentences in the opening paragraph or paragraphs.

Here is the opening paragraph of a one-page test report.

In this opening, the writer summarizes a one-page report by briefly identifying the purpose of the test, test method, test results, and future action. Each topic is described in a paragraph in the rest of the report.

On June 29 and 30, we used a computer simulation to test the prototype design for a robotic welder. In the simulation, we required the robot to spot weld front and back panels on an assembly line. We also tested the robot's ability to weld accurately throughout the full range of distances and angles in the technical specifications. The robot performed front welds accurately at all distances, but could not make back welds at certain angles. Consequently, the mountings for four actuators are being redesigned.

In the paragraphs that follow this summary, the writer describes the major features of the robotic welder, capabilities of the simulation program as well as the welding tasks given the robot, test results, and changes to be made in the design.

For longer communications, especially those that are long enough to have covers and tables of contents, the summaries are longer and often printed on a separate page.

These longer summaries are described in Chapter 12, "Writing Reader-Centered Front and Back Matter."

Guideline 7 | Adjust the length of your beginning to your readers' needs

There is no rule of thumb that tells how long the beginning should be. A good, reader-centered beginning may require only a phrase or may take several pages. You need to give your readers only the information they don't already know. Just be sure they know the following:

- The reason they should read the communication (Guideline 1)
- The main point of the communication (Guideline 2)
- The organization and scope of the communication (Guideline 3)
- The background information they need in order to understand and use the communication (Guideline 5)

What your readers need to know

If you have given your readers all this information—and have encouraged them to receive your message openly (Guideline 4)—then you have written a good beginning, regardless of how long or short it is.

Here is an opening prepared by a writer who followed all the guidelines given in this chapter.

> In response to your memo dated November 17, I have called Goodyear, Goodrich, and Firestone for information about the ways they forecast their needs for synthetic rubber. The following paragraphs summarize each of those phone calls.

Brief beginning

The following opening, from a two-paragraph memo, is even briefer:

> We are instituting a new policy for calculating the amount that employees are paid for overtime work.

Briefer beginning

At first glance, this single sentence may seem to violate all the guidelines. It does not. It identifies the topic of the memo (overtime pay), and the people to whom the memo is addressed will immediately understand its relevance to them. It also declares the main point of the memo (a new policy is being instituted). Moreover, because the memo itself is only two paragraphs long, its scope is readily apparent. The brevity of the memo also suggests its organization—namely, a brief explanation of the new policy and nothing else. The writer has correctly judged that his readers need no background information.

Figure 10.1 shows a relatively long beginning from a report written by a consulting firm hired to recommend ways to improve the food service at a hospital. Like the brief beginnings given above, it is carefully adapted to its readers and to the situation.

Examples of longer beginnings are shown in Figures 10.1 and 10.2.

Figure 10.2 shows the long beginning of a 500-page service manual for the Detroit Diesel Series 53 engine manufactured by General Motors.

FIGURE 10.1

Beginning of a Recommendation Report

The writers open their summary by stating the problem they will address. It is one that's important to their readers.

The writers detail the subparts of the problem.

The writers present themselves as the reader's problem-solving partners.

• They tell what they have done to help solve the problem.

• They tell how the report will assist their readers in doing their part of the problem-solving effort.

The writers tell how they have organized their report.

To indicate the scope of this report, the writers highlight a topic it will not cover.

They assist their readers by summarizing their main points.

INTRODUCTION

Wilton Hospital has added 200 patient beds through construction of the new West Wing. Since the wing opened, the food-service department has had difficulty meeting this extra demand. The director of the hospital has also reported the following additional problems:

1. Difficulties operating at full capacity. The equipment, some of it thirty years old, breaks down frequently. Absenteeism has risen dramatically.

2. Costs of operation that are well above average for the hospital industry nationally and in this region.

3. Frequent complaints about the quality of the food from both the patients and the hospital staff who eat in the cafeteria.

To study these problems, we have monitored the operation of the food-service department and interviewed patients, food-service employees, and staff who eat in the cafeteria. In addition, we have compared all aspects of the department's facilities and operations with those at other hospitals of roughly the same size.

In this report, we discuss our findings concerning the food-service department's kitchen facilities. We briefly describe the history and nature of these facilities, suggest two alternative ways of improving them, and provide a budget for each. In the final section of this report, we propose a renovation schedule and discuss ways of providing food service while the renovation work is being done. (Our recommendations about staffing and procedures will be presented in another report in thirty days.)

The first alternative costs about $730,000 and would take four months to accomplish. The second costs about $1,100,000 and would take five months. Both will meet the minimum needs of the hospital; the latter can also provide cooking for the proposed program of delivering hot meals to housebound persons in the city.

FIGURE 10.2

General Information   **DETROIT DIESEL 53**

• SCOPE AND USE OF THE MANUAL

This manual covers the basic Series 53 Diesel Engines built by the Detroit Diesel Corporation. Complete instructions on operation, adjustment (tune-up), preventive maintenance and lubrication, and repair (including complete overhaul) are covered. The manual was written primarily for persons servicing and overhauling the engine and, in addition, contains all of the instructions essential to the operators and users. Basic maintenance and overhaul procedures are common to all Series 53 engines and, therefore, apply to all Inline and Vee models.

The manual is divided into numbered sections. The first section covers the engine (less major assemblies). The following sections cover a complete system such as the fuel system, lubrication system or air system. Each section is divided into subsections which contain complete maintenance and operating instructions for a specific subassembly on the engine. For example, Section 1, which covers the basic engine, contains subsection 1.1 pertaining to the cylinder block, subsection 1.2 covering the cylinder head, etc. The subjects and sections are listed in the Table of Contents on the preceding page. Pages are numbered consecutively, starting with a new Page 1 at the beginning of each subsection. The illustrations are also numbered consecutively, beginning with a new Fig. 1 at the start of each subsection.

Information regarding a general subject, such as the lubrication system, can best be located by using the Table of Contents. Opposite each subject in the Table of Contents is a section number which registers with a tab printed on the first page of each section throughout the manual. Information on a specific subassembly or accessory can then be found by consulting the list of contents on the first page of the section. For example, the cylinder liner is part of the basic engine. Therefore, it will be found in Section 1. Looking down the list of contents on the first page of Section 1, the cylinder liner is found to be in subsection 1.6.3. An Alphabetical Index at the back of the manual has been provided as an additional aid for locating information.

SERVICE PARTS AVAILABILITY

Genuine Detroit Diesel service parts are available from authorized Detroit Diesel distributors and service dealers throughout the world. A complete list of all distributors and dealers is available in the Worldwide Distributor and Dealer Directory, 6SE280. This publication can be ordered from any authorized distributor.

CLEARANCES AND TORQUE SPECIFICATIONS

Clearances of new parts and wear limits on used parts are listed in tabular form at the end of each section throughout the manual. It should be specifically noted that the "New Parts" clearances apply only when all new parts are used at the point where the various specifications apply. This also applies to references within the text of the manual. The column entitled "Limits" lists the amount of wear or increase in clearance which can be tolerated in used engine parts and still assure satisfactory performance. It should be emphasized that the figures given as "Limits" must be qualified by the judgment of personnel responsible for installing new parts. These wear limits are, in general, listed only for the parts more frequently replaced in engine overhaul work. For additional information, refer to the paragraph entitled *Inspection* under *General Procedures* in this section.

Bolt, nut and stud torque specifications are also listed in tabular form at the end of each section.

PARTS REPLACEMENT

Before installing a new or used part, check it thoroughly to make sure it is the proper part for the job. The quality of the replacement part must be equivalent to the quality of the original Detroit Diesel component being replaced and must meet DDC specifications for new or reusable parts.

Parts must also be clean and not physically damaged or defective. For example, bolts and bolt hole threads must not be damaged or distorted. Gasketing must have all holes completely punched with no residual gasket material left clinging to the top or bottom. Flatness and fit specifications in the service manual must be strictly adhered to.

CAUTION: Failure to inspect parts thoroughly before installation, failure to install the proper parts, or failure to install parts properly can result in component or engine malfunction and/or damage and may also result in personal injury.

Page 4   May, 1990   © Copyright 1990 Detroit Diesel Corporation

The writers tell what the manual is about in the first sentence.

In the following sentences, they describe the manual's scope.

The writers indicate the organization of the manual.

They provide a guide to the usability of the manual.

The writers provide background information that is implicit throughout the rest of the manual.

They provide additional background information, including cautions.

Guideline 8 | Global Guideline: Adapt your beginning to your readers' cultural background

Readers' expectations and preferences about the beginning of a communication are shaped by their culture. The suggestions you have just read are suitable for readers in the United States and some other Western countries. However, customs vary widely. For example, the French often open their business correspondence in a more formal way than do people in the United States. In some Spanish-speaking cultures, business letters begin in a flowery manner. In Japan, business letters often begin with a reference to the season. The following example shows one writer's way of adapting to the expectations of a Japanese businessperson (Human Japanese, 2005):

| Beginning of a business letter to a Japanese reader | Here in Seattle, fall has finally arrived, and the days continue to become shorter and shorter. At night it's now cold enough that one needs a sweater to go outside. |

Next, the letter moves to the main topic:

> The reason I'm writing you is that . . . .

Be cautious, though, when writing to a person from another culture. Avoid operating solely on generalizations or stereotypes. Individuals from the same culture can differ from one another in many ways. Also, don't confuse cultural background with race or national origin. People of a particular race or national origin live in many different cultures, and cultures may have members who from different races or national origins. When writing to readers from another culture, as when writing to readers from your own, learn about the specific persons you are addressing.

Guideline 9 | Ethics Guideline: Begin to address unethical practices promptly—and strategically

So far, the guidelines in this chapter have focused on the ways to begin a communication. In contrast, this last guideline concerns situations in which people at work sometimes hesitate to begin writing or speaking at all. Suppose you learn that your employer is engaged in an action you consider to be unethical. Or, suppose you are asked to write something that violates your sense of what is ethical. Should you speak up or express your concerns in writing? New employees are sometimes advised to wait until they have achieved security and status before trying to bring about change. But that means you could spend years before addressing a practice you regard as unethical. Ignoring an unethical act could be seen as unethical in itself.

When determining how to draw attention to something you consider to be unethical, you face a challenge similar to that of figuring out how to begin a memo in which you will recommend a course of action with which you believe your readers will disagree (see pages 293–294). Here are three strategies that could enable you to open a discussion based on your values without jeopardizing your future with your employer's organization.

Readers from different cultures may or may not have different expectations about the ways a work-related communication will begin.

- **Plant the seeds of change.** Instead of trying to alter the situation immediately, plant the seeds of change. For instance, if your employer is thinking of modifying working conditions in a factory, you might inquire, "How would this modification affect workers on the assembly line?" Or, you might ask, "Would this modification be fair to the assembly line workers?" By posing such questions, you extend the range of issues being considered, and you can subtly let others know the values that shape your understanding of the situation. In addition to asking questions, there are many other ways you can introduce your values into the discussion. Think creatively. Taking even a small step toward improving a situation is an ethical act. Moreover, in many circumstances, it's only through a long series of small steps that large improvements are achieved.

- **Use reason rather than accusation.** Your values are much more likely to prevail if you engage people with other views in a reasoned discussion than if you accuse and condemn them. It's not usually possible to persuade people by attacking them. They merely become defensive. Instead, ask them to share their sense of the values that apply to the situation. Appeal to their sense of fairness and of what is right and wrong.

- **Remain open to others' views.** One reason to avoid taking a rigid stand is that it impairs your ability to understand others' views of the situation. People regularly differ on ethical matters, and your own view is not necessarily shared by others. Strive for solutions that will satisfy both you and others. One possibility is to employ a strategy similar to that described in this chapter's discussion of Guideline 1 (see page 264): Identify the values that the other people hold that would lead them to endorse the same course of action that your values lead you to advocate.

You may someday witness a practice that is so outrageous that you will be willing to risk future promotions and even your job in order to stop it. If you find yourself in that situation, seek the aid of influential people inside your company. If the practice you object to violates the law or a government regulation, alert the appropriate agency. This is called *whistleblowing*. Federal law and some state laws are intended to protect whistleblowers, and some laws even reward whistleblowers by giving them a portion of any financial settlement that is made. Still, many whistleblowers do lose their jobs or continue to work under hostile conditions. If you are thinking of whistleblowing, consider the possibility of first attempting a nonconfrontational approach to the problem.

CONCLUSION

The beginning is probably the most important segment of a communication. That's because it can influence the ideas and attitudes your readers derive from the rest of your communication, and it can even determine whether they will read further.

This chapter has suggested that in writing a beginning you start by trying to identify your readers' attitudes toward your message and then determining what you can say to help them understand and use what follows. This reader-centered approach will enable you to create beginnings that prompt your readers to pay careful attention, encourage them to treat your information and ideas with an open mind, and help them read efficiently.

USE WHAT YOU'VE LEARNED

For additional exercises, go to www.cengage.com/english/ anderson7e. *Instructors*: The book's website includes suggestions for teaching the exercises.

EXERCISE YOUR EXPERTISE

1. Select a communication written to professionals in your field. This might be a letter, memo, manual, report, or article in a professional journal. (Do not choose a textbook.) Identify the guidelines from this chapter that the writer applied when drafting the communication's beginning. Is the beginning effective in achieving the writer's usability and persuasive goals? If so, why? If not, how could the beginning be improved?

2. The instructions for many consumer products contain no beginning section at all. For instance, the instructions for some lawnmowers, cake mixes, and detergents simply provide a heading that says "Instructions" and then start right in. Find such a set of instructions and—in terms of the guidelines given in this chapter—evaluate the writer's decision to omit a beginning.

EXPLORE ONLINE

The home page of a website typically looks very different from the opening of a printed document. Nonetheless, the home page has many of the same reader-centered goals as the printed opening. Examine the ways that two websites follow this chapter's guidelines. Consider such things as the communication functions performed by the images, layout, text (if any), and words used for buttons and links. One of the home pages you study should be for a company that sells consumer products such as cars. The other should be for a nonprofit organization or government agency. Present your analysis in the way your instructor requests.

COLLABORATE WITH YOUR CLASSMATES

1. The following paragraphs are from the beginning of a report in which the manager of a purchasing department asks for better-quality products from the department in the company that provides abrasives. Is this an effective beginning for what is essentially a complaint? Why or why not? Working with another student, analyze this beginning in terms of this chapter's guidelines.

> I am sure you have heard that the new forging process is working well. Our customers have expressed pleasure with our castings. Thanks again for all your help in making this new process possible.
>
> We are having one problem, however, with which I have to ask once more for your assistance. During the seven weeks since we began using the new process, the production line has been idle 28 percent of the time. Also, many castings have had to be remade. Some of the evidence suggests that these problems are caused by the steel abrasive supplies we get from your department. If we can figure out how to improve the abrasive, we may be able to run the line at 100 percent of capacity.
>
> I would be most grateful for help from you and your people in improving the abrasive. To help you devise ways of improving the abrasive, I have compiled this report, which describes the difficulties we have encountered and some of our thinking about possible remedies.

2. Form a team of two or three students. Exchange the opening sections (paragraph or longer) of a project you are now preparing for your instructor. First, suggest specific revisions that would make your partner's opening more effective at achieving his or her usability and persuasive goals. Next, suggest an alternative opening that might also be effective.

APPLY YOUR ETHICS

It can be particularly challenging to write an effective opening for a communication in which you are advocating for change based on ethical grounds. Think of some organization's practice, procedure, or policy that your values lead you to believe should be changed. Identify the person or group within the organization that has the authority to make the change that you feel is needed. Then, following the advice given in this chapter, draft the opening paragraph of a letter, memo, e-mail, report, or other communication on the topic that you could send to this person or group.

11 | Ending a Communication

GUIDELINES

DEFINING
OBJECTIVES

PLANNING

CONDUCTING
RESEARCH

DRAFTING

REVISING

WWW

To read additional information, see more examples,
and access links related to
this chapter's guidelines,
go to Chapter 11 at **www
.cengage.com/english/
anderson7e.**

This chapter describes alternative ways of ending a communication that you may use
alone or in combination.

How to choose among
possible strategies for
endings

The ending of a communication can substantially increase its
effectiveness. Researchers have found that readers are better
able to remember what is said at the end of a communication
than information presented in the middle. Consequently, a skillfully written ending can increase a communication's usability by
stating or reviewing points that readers will want to recall later.

In addition, an ending is a transition. It leads readers out of
the communication and into the larger stream of their activities.
Therefore, it provides an excellent opportunity to help readers
by answering the question "What should I do now?"

An ending is also a place of emphasis and a communication's final chance to
influence readers' impressions of its subject matter. Consequently, it's an excellent
place to highlight a communication's most persuasive points.

This chapter's nine reader-centered guidelines suggest specific strategies you can
use to create endings that enhance your communication's usability and persuasiveness. Despite the important role that endings play, however, your best strategy sometimes will be to say what you have to say and then stop, without providing a separate
ending (Guideline 1). Other times, you can use one or a combination of strategies
described in Guidelines 2 through 9.

To decide which strategies to choose, focus (as always) on your readers.

1. **Decide what you want your readers to think, feel, and do as they finish your
   communication.** Then select the strategy or strategies for ending that are most
   likely to create this response.
2. **Determine what kind of ending your readers might be expecting.** Look at
   what other people in your organization and your field have done in similar
   situations. This investigation can tell you what your readers are accustomed to
   finding at the end of the type of communication you are preparing.

Guideline 1 | **After you've made your last point, stop**

As mentioned above, you should end some communications without adding any
words after your last point. That happens when you use a pattern of organization that
brings you to a natural stopping place. Here are some examples:

Communications that you
may want to end after the
last point

- **Proposals.** You will usually end a proposal with a detailed description of what
  you will do and how you will do it. Because that's where your readers expect
  proposals to end, they will enjoy a sense of completion if you simply stop after
  presenting your last recommendation. Furthermore, by ending after your recommendations, you will have given them the emphasis they require.
- **Formal reports.** When you prepare a formal report (a report with a cover, title
  page, and binding), the convention is to end either with your conclusions or
  your recommendations—both appropriate subjects for emphasis.
- **Instructions.** You will usually end instructions by describing the last step.

If your analysis of your purpose, readers, and situation convinces you that you should add something after your last point, review this chapter's other guidelines to select the strategies that will best help you meet your objectives.

Guideline 2 | **Repeat your main point**

Because the end of a communication is a point of emphasis, you can use it to focus your readers' attention on the points you want to be foremost in their minds as they finish reading.

Consider, for instance, the ending of a classic article written to dispel a misunderstanding by family physicians about the proper way to prevent infections in wounds. The writers Mancusi-Ugaro and Rapport (1986) made their main point in the abstract at the beginning of the article, again in the fourth paragraph (where they supported it with a table), and in several other places. Nevertheless, they considered this point to be so important that they restated it in the final paragraph.

> Perhaps the most important concept to be gleaned from a review of the principles of wound management is that good surgical technique strives to maintain the balance between the host and the bacteria in favor of the host. The importance of understanding that infection is an absolute quantitative number of bacteria within the tissues cannot be overemphasized. Limiting, rather than eliminating, bacteria allows for normal wound healing.

Ending that repeats the communication's main point

The strategy of repeating your main point in your ending can also work well when you want to help your readers make a decision or persuade them to take a certain action. The following paragraph is the ending of a memo urging new safety measures:

> I cannot stress too much the need for immediate action. The exposed wires present a significant hazard to any employee who opens the control box to make routine adjustments.

Another example

Guideline 3 | **Summarize your key points**

The strategy suggested by this guideline is closely related to the preceding one. In repeating your main point (Guideline 2), you emphasize *only* the information you consider to be of paramount importance. In *summarizing,* you are concerned that your audience has understood the general thrust of your *entire* communication.

Here, for example, is the ending of a 115-page book entitled *Understanding Radioactive Waste,* which is intended to help the general public understand the impact of the nuclear power industry's plans to open new plants (Murray, 2003):

> It may be useful to the reader for us to now select some highlights, key ideas, and important conclusions for this discussion of nuclear wastes. The following list is not complete—the reader is encouraged to add items.

Key points are summarized.

Try This

Find a printed or online communication that ends in a way that is not described in this chapter. How would you describe the strategy? How effective is it, given the communication's purpose and audience?

1. Radioactivity is both natural and manmade. The decay process gives radiations such as alpha particles, beta particles, and gamma rays. Natural background radiation comes mainly from cosmic rays and minerals in the ground.
2. Radiation can be harmful to the body and to genes, but the low-level radiation effect cannot be proved. Many methods of protection are available.
3. The fission process gives useful energy in the form of electricity from nuclear plants, but it also produces wastes in the form of highly radioactive fission products. . . .

This list continues for thirteen more items, but this sample should give you an idea of how this author ended with a summary of key points.

Notice that a summary at the end of a communication differs significantly from a summary at the beginning. Because a summary at the beginning is meant for readers who have not yet read the communication, it must include some information that will be of little concern at the end. For example, the beginning summary of a report on a quality-control study will describe the background of the study. In contrast, the ending summary would focus sharply on conclusions and recommendations.

Guideline 4 | Refer to a goal stated earlier in your communication

Many communications begin by stating a goal and then describe or propose ways to achieve it. If you end a communication by referring to that goal, you remind your readers of the goal and sharpen the focus of your communication. The following example comes from a seventeen-page proposal prepared by operations analysts in a company that builds customized, computer-controlled equipment used in print shops and printing plants. Notice how the ending refers to the beginning.

Beginning states a goal.

To maintain our competitive edge, we must develop a way of supplying replacement parts more rapidly to our service technicians without increasing our shipping costs or tying up more money in inventory.

Ending refers to the goal.

The proposed reform of our distribution network will help us meet the needs of our service technicians for rapidly delivered spare parts. Furthermore, it does so without raising either our shipping expenses or our investment in inventory.

The next example illustrates the way scientific articles often end by stating the answer to a question asked at the article's beginning. In this case, the researchers want to learn whether an electronic nose could be used to detect spoiled chicken in poultry factories and supermarkets (Rajamäki et al., 2006). The spelling follows conventions in the United Kingdom.

The electronic nose is an instrument which comprises an array of electronic chemical sensors with partial specificity and an appropriate pattern-recognition system, capable of recognizing simple or complex odours. . . . The aim of this study was to determine whether it is possible to distinguish differently stored, MA-packaged, unmarinated broiler chicken cuts with an electronic nose.

Beginning states a goal.

The results were very promising when considering the possible applications of an electronic nose as a screening tool for quality control and inspection purposes. . . . An electronic nose [developed specifically for the chicken industry] would include only a few sensors reacting to spoilage parameters of poultry meat and would therefore cost less than larger electronic nose devices developed for detecting many kinds of volatile compounds in various applications.

Ending refers to the goal.

Guideline 5 | **Focus on a key feeling**

When ending some communications, it may be more important to focus your readers' attention on a feeling rather than a fact. For instance, if you are writing instructions for a product manufactured by your employer, you may want your ending to encourage your readers' goodwill toward the product. Consider this ending of an owner's manual for a clothes dryer. Though the last sentence provides no additional information, it seeks to shape the readers' attitude toward the company.

The GE Answer Center™ consumer information service is open 24 hours a day, seven days a week. Our staff of experts stands ready to assist you anytime.

Ending designed to build goodwill

The following ending is from a booklet published by the National Cancer Institute for people who have apparently been successfully treated for cancer but do not know how long the disease will remain in remission. It, too, seeks to shape the readers' feelings.

Cancer is not something anyone forgets. Anxieties remain as active treatment ceases and the waiting stage begins. A cold or cramp may be cause for panic. As 6-month or annual check-ups approach, you swing between hope and anxiety. As you wait for the mystical 5-year or 10-year point, you might feel more anxious rather than more secure.

These are feelings that we all share. No one expects you to forget you have had cancer or that it might recur. Each must seek individual ways of coping with the underlying insecurity of not knowing the true state of his or her health. The best prescription seems to lie in a combination of one part challenging responsibilities that require a full range of skills, a dose of activities that seek to fill the needs of others, and a generous dash of frivolity and laughter.

You still might have moments when you feel as if you live perched on the edge of a cliff. They will sneak up unbidden. But they will be fewer and farther between if you have filled your mind with thoughts having nothing to do with cancer.

Ending designed to shape complex attitudes

Cancer might rob you of that blissful ignorance that once led you to believe that tomorrow stretched on forever. In exchange, you are granted the vision to see each today as precious, a gift to be used wisely and richly. No one can take that away.

Guideline 6 | Tell your readers how to get assistance or more information

At work, a common strategy for ending a communication is to tell your readers how to get assistance or more information. These two examples are from a letter and a memo:

Endings that offer help

If you have questions about this matter, call me at 523-5221.

If you want any additional information about the proposed project, let me know. I'll answer your questions as best I can.

By ending in this way, you not only provide your readers with useful information, you also encourage them to see you as a helpful, concerned individual.

Guideline 7 | Tell your readers what to do next

In some situations, you can help your readers most by telling them what you think they should do next. If more than one course of action is available, tell your readers how to follow up on each of them.

Ending that tells readers exactly what to do

To buy this equipment at the reduced price, we must mail the purchase orders by Friday the 11th. If you have any qualms about this purchase, let's discuss them. If not, please forward the attached materials, together with your approval, to the Controller's Office as soon as possible.

Guideline 8 | Identify any further study that is needed

Much of the work that is done on the job is completed in stages. For example, one study might answer preliminary questions and, if the answers look promising, an additional study might then be undertaken. Consequently, one common way of ending is to tell readers what must be found out next.

Ending that identifies next question needing study

This experiment indicates that we can use compound deposition to create microcircuits in the laboratory. We are now ready to explore the feasibility of using this technique to produce microcircuits in commercial quantities.

Such endings are often combined with summaries, as in the following example:

Another example

In summary, over the past several months our Monroe plant has ordered several hundred electric motors from a supplier whose products are inferior to those we require in the heating and air conditioning systems we build. Not only must this practice stop immediately, but also *we* should investigate the situation to determine why this flagrant violation of our quality-control policies has occurred.

Guideline 9 | Follow applicable social conventions

All the strategies mentioned so far focus on the subject matter of your communications. When writing the endings of your communications, it is also important to observe the social conventions that apply in each situation.

Some of those conventions involve customary ways of closing particular kinds of communication. For example, letters usually end with an expression of thanks, a statement that it has been enjoyable working with the reader, or an offer to be of further help if needed. In contrast, formal reports and proposals rarely end with such gestures.

Other conventions are peculiar to specific organizations. For example, in some organizations writers rarely end their memos with the kind of social gesture commonly provided at the end of a letter. In other organizations, memos often end with such a gesture, and people who ignore that convention risk seeming abrupt and cold.

Consider, too, the social conventions that apply to your relationship with your readers. Have they done you a favor? Thank them. Are you going to see them soon? Let them know that you look forward to the meeting.

CONCLUSION

This chapter has described nine reader-centered strategies for ending a communication. By considering your readers, your objectives, and the social conventions that apply in your situation, you will be able to choose the strategy or combination of strategies most likely to succeed in the particular communication you are writing.

USE WHAT YOU'VE LEARNED

For additional exercises, go to www.cengage.com/english/ anderson7e. *Instructors:* The book's website includes suggestions for teaching the exercises.

EXERCISE YOUR EXPERTISE

Select a communication written to professionals in your field. This might be a letter, memo, manual, technical report, or article in a professional journal. (Do not choose a textbook.) Identify the guidelines from this chapter that the writer applied when drafting that communication's ending. Is the ending effective in achieving the writer's usability and persuasive goals? If so, why? If not, how could it be improved?

EXPLORE ONLINE

For many websites, there is no single ending point. Visitors may leave the site at any point. In fact, links in many sites lead visitors directly to other sites. Visit three different kinds of sites, such as one that provides explanations or instructions, one that enables people to purchase products, and one created by an advocacy group. Within each site, identify the places where one or more of this chapter's guidelines are employed effectively.

COLLABORATE WITH YOUR CLASSMATES

1. The following paragraphs constitute the ending of a report to the U.S. Department of Energy concerning the

economic and technical feasibility of generating electric power with a special type of windmill (Foreman). The windmills are called diffuser augmented wind turbines (DAWT). In this ending, the writer has used several of the strategies described in this chapter. Working with another student, identify these strategies.

Section 6.0

Concluding Remarks

We have provided a preliminary cost assessment for the DAWT approach to wind energy conversion in unit systems to 150 kw power rating. The results demonstrate economic viability of the DAWT with no further design and manufacturing know-how than already exists. Further economic benefits of this form of solar energy are likely through:

- Future refinements in product design and production techniques
- Economies of larger quantity production lots
- Special tax incentives

Continued cost escalation on nonrenewable energy sources and public concern for safeguarding the biosphere environment will surely make wind energy conversion by DAWT-like systems even more attractive to our society. Promotional actions by national policy makers and planners as well as industrialists and entrepreneurs can aid the emergence of the DAWT from its research phase to a practical and commercial product.

2. Working with another student, describe the strategies for ending used in the following figures in this text:

FIGURE	PAGE	FIGURE	PAGE
1.4	14	8.2	219
2.6	52	19.1	479
2.7	53	20.3	514
5.6	135	27.7	669

APPLY YOUR ETHICS

Examine the student conduct code or values statement of your school. Which of the guidelines in this chapter does its ending follow? Why? Also, identify some of the specific values that underline the code's or values statement's provisions.

12 | Writing Reader-Centered Front and Back Matter

GUIDELINES

| DEFINING OBJECTIVES |
| PLANNING |
| CONDUCTING RESEARCH |
| DRAFTING |
| REVISING |

In the workplace, many communications longer than a page are held together with a paper clip or staple. Usually because of their importance or length, others have many of the features found in the books you buy for your classes. Some of these features are called *front matter* because they precede the opening chapter or section. Front matter includes the following items.

- Title Page
- Executive Summary or Abstract
- Table of Contents
- List of Figures and Tables

The other book-like features are called *back matter,* for an obvious reason.

- Appendixes
- Reference List, End Notes, or Bibliography
- Glossary or List of Symbols
- Index

Communications with front and back matter usually have covers as well. When delivered to their readers, these communications are often accompanied by transmittal letters.

In the workplace, reports and proposals with these features are often *called formal reports* and *formal proposals,* names that will be used in this chapter. Instructions with these features are usually called *instruction manuals.*

HOW TRANSMITTAL LETTERS, COVERS, AND FRONT AND BACK MATTER INCREASE USABILITY AND PERSUASIVENESS

Skillfully created front matter, back matter, covers, and transmittal letters increase a communication's usability and persuasiveness in several ways.

- **They help readers find what they need.** A table of contents in the front and an index in the back guide readers to specific information they want. Even the title helps readers locate the right report or manual from a list of reports or manuals.
- **They provide an abbreviated version of the communication's main points.** The title and abstract or executive summary convey the communication's main points to executives and other busy persons who don't have time to read the entire report.

- **They create a favorable initial impression.** A reader-centered and attractive cover and other front matter foster an initial, positive impression among readers.
- **They help future readers locate the communication efficiently.** Communications with front and back matter are usually kept on hand for future use. When designed well, front matter can assist readers in the future as they look for communications that will help them perform their jobs.
- **They protect the communication from wear.** Communications with front and back matter often encounter frequent use. Their covers and bindings help them remain usable.

To assist you in creating these book-like elements, this chapter begins with four reader-centered guidelines, followed by detailed advice about writing transmittal letters, covers, front matter, and back matter.

Guideline 1 | Review the ways your readers will use the communication

Front and back matter can make especially important contributions to a communication's usability for its readers. The way you write your title and design your table of contents can increase the efficiency with which readers can locate the information they want. To help readers in this way, you need to know what your readers will be looking for and how they will use your communication. As you begin work on the front and back matter, review the information you developed while defining your communication's usability objectives (see Chapter 3).

Guideline 2 | Review your communication's persuasive goals

By thoroughly understanding your readers' goals, values, and attitudes, you can determine what should be included and emphasized in the executive summary to lead readers to make the decision or take the actions you advocate. Reviewing the information you developed while defining your communication's persuasive objectives, you can enhance the effectiveness of the executive summary and other elements of your front and back matter.

Guideline 3 | Find out what's required

Many organizations distribute instructions that tell employees in detail how to prepare front and back matter. These instructions, sometimes called *style guides,* can describe everything from the maximum number of words permitted in an executive summary to the size and color of the title on a report's cover. If you are writing to one of your employer's clients, the *client's* requirements may be the ones you need to follow. Whatever the source, obtain the relevant style guide and follow its requirements precisely.

Guideline 4 | Find out what's expected

Even if nothing is explicitly required, your employer and your readers probably have expectations about the way you draft front and back matter. Also, beyond what's required, there are often unspoken expectations about how the required elements will be prepared. Look at communications similar to the one you are preparing to learn what expectations you should meet.

Guideline 5 | Evaluate and revise your front and back matter

Learn More

See Chapters 15 and 16 for guidelines on revising and testing.

Front and back matter are not mere formalities. Because they impact a communication's usability and persuasiveness, you should treat them seriously. Among other actions, this means including them in drafts you evaluate yourself, give to reviewers, present to readers in user testing, and carefully revise and polish.

CONVENTIONS AND LOCAL PRACTICE

As mentioned above, many employers issue style guides that tell how they want their employees to write front and back matter. Of course, the instructions vary somewhat from organization to organization. For instance, one employer may want the title of any report to be placed in a certain spot on the cover, and another employer may want the title in another spot. Despite these variations, formal reports written on the job look very much alike. The formats described in this chapter reflect common practices in the workplace. You will have little trouble adapting them to other, slightly different versions when the need arises.

As you prepare front and back matter, remember that even if you are following a very comprehensive style manual, you increase your chances of achieving your communication goals by thinking continuously of your readers and applying this chapter's five guidelines.

WRITING A READER-CENTERED TRANSMITTAL LETTER

When you prepare formal reports and proposals, you will often send them (rather than hand them) to your readers. In such cases, you will want to accompany them with a letter (or memo) of transmittal, which may be placed on top of your communication or bound into it. Transmittal letters typically contain the following elements:

- **Introduction**   In the introduction to your letter, introduce the accompanying communication and explain or remind readers of its topic. Begin on an upbeat note. Here is an example written to a client:

    > We are pleased to submit the enclosed report on evaluating three possible sites for your company's new manufacturing plant.

- **Body**   The body of a transmittal letter usually highlights the communication's major features. For example, researchers often state their main findings and the consequences of what they've learned, repeating information in their communication's abstract or summary.
- **Closing**   Transmittal letters commonly end with a short paragraph (often one sentence) that states the writer's willingness to work further with the reader or promises to answer any questions the reader may have.

    Figure 12.1 shows a memo of transmittal.

WWW

To view other examples of front and back matter, link to Chapter 12 at **www.cengage.com/english/anderson7e**.

WRITING A READER-CENTERED COVER

A cover typically includes the communication's title, the name or logo of the organization that created it, and the date it was issued. In some organizations, covers also include the writer's name and a document number used for filing. To write a reader-centered title, use precise, specific terms that tell your current readers exactly what the communication is about and help your future readers determine whether the communication will be useful to them.

Figure 12.2 shows the cover of a research report.

WRITING READER-CENTERED FRONT MATTER

Depending on your readers' needs and your employer's expectations, a formal report, formal proposal, or instruction manual may include some or all of the following elements: title page, summary or abstract, table of contents, and list of figures and tables.

Title Page

A title page repeats the information on the cover and usually adds more. For instance, your employer may want you to include contract or project numbers and such cross-referencing information as the name of the contract under which the work was performed. A title page may also provide copyright and trademark notices. Note that some employers require such information on the cover instead.

Figure 12.3 shows the title page of the report whose cover is shown in Figure 12.2.

FIGURE 12.1
Memo of Transmittal
Written at Work

This letter of transmittal accompanied the empirical research report shown in Figure 24.1 (page 591).

Margaret opens with an introductory sentence that mentions the accompanying report and tells its topic.

In the body of her brief memo, Margaret summarizes the report's findings that will be of most interest to her reader. To do this, she includes very precise information, using exact percentages in the first sentence, for instance. Note that she has not included the broader range of information included in the report's executive summary (page 593).

Again tailoring her memo to her reader, an executive decision maker, Margaret describes in specific detail the next step recommended by the research team.

Margaret closes on an upbeat note and indicates that she would welcome questions.

ELECTRONICS CORPORATION OF AMERICA
MEMORANDUM

To	Myron Bronski, Vice-President, Research
From	MCB Margaret C. Barnett, Satellite Products Laboratory
Date	September 29, 2010
Re	REPORT ON TRUCK-TO-SATELLITE TEST

On behalf of the entire research team. I am pleased to submit the attached copy of the operational test of our truck-to-satellite communication system.

The test shows that our system works fine. More than 91% of our data transmissions were successful, and more than 91% of our voice transmissions were of commercial quality. The test helped us identify some sources of bad transmissions, including primarily movement of a truck outside the "footprint" of the satellite's strongest broadcast and the presence of objects (such as trees) in the direct line between a truck and the satellite.

The research team believes that our next steps should be to develop a new antenna for use on the trucks and to develop a configuration of satellites that will place them at least 25° above the horizon for trucks anywhere in our coverage area.

We're ready to begin work on these tasks as soon as we get the okay to do so. Let me know if you have any questions.

Encl: Report (2 copies)

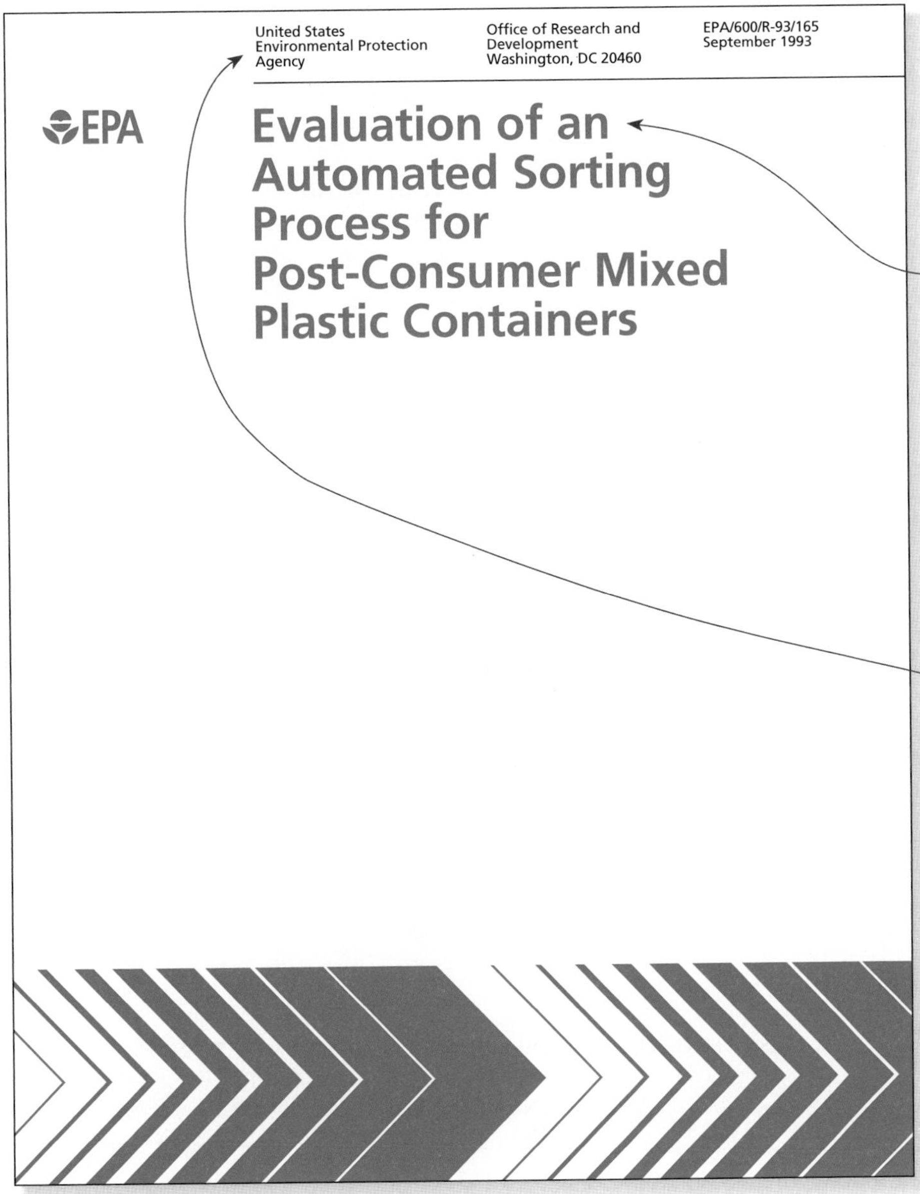

FIGURE 12.2
Cover of a Formal Report

United States
Environmental Protection
Agency

Office of Research and
Development
Washington, DC 20460

EPA/600/R-93/165
September 1993

EPA

Evaluation of an Automated Sorting Process for Post-Consumer Mixed Plastic Containers

The authors give greatest visual prominence to the title of the report. Second-level prominence goes to the logo for the Environmental Protection Agency.

The title conveys very specific information about the report's contents:

- It says that the document contains an "evaluation of" rather than using the more general phrase "report on."
- It specifies that the sorting process is "automated."
- It indicates that the waste being sorted was "containers."
- It describes three specific characteristics of the containers: "post-consumer," "mixed," and "plastic."

In the small type across the top of the cover, the writers provide the following:

- Full name of the EPA.
- Name of the EPA office that sponsored the study, along with its address (because this document is available to members of the general public, who might want to contact the office).
- The report's filing number.
- The report's date.

FIGURE 12.3

Title Page of the Report Whose Cover Is Shown in Figure 12.2

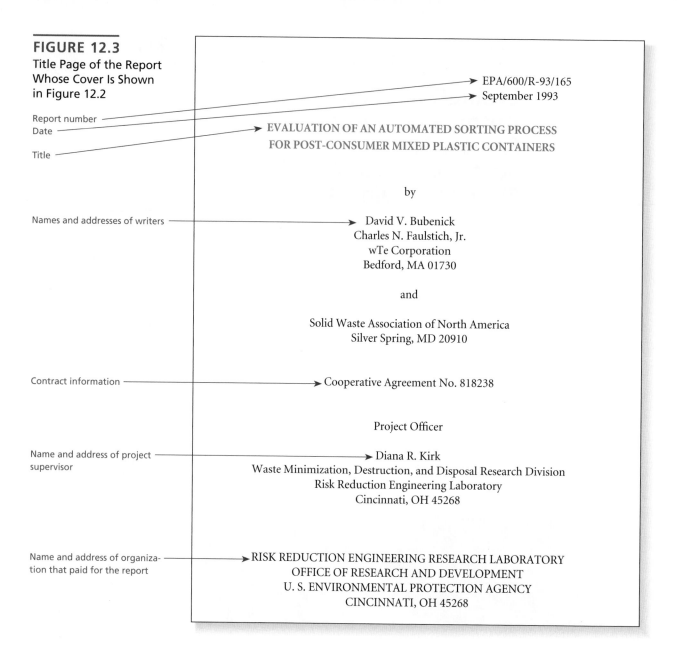

Report number

Date

Title

Names and addresses of writers

Contract information

Name and address of project supervisor

Name and address of organization that paid for the report

EPA/600/R-93/165

September 1993

EVALUATION OF AN AUTOMATED SORTING PROCESS
FOR POST-CONSUMER MIXED PLASTIC CONTAINERS

by

David V. Bubenick
Charles N. Faulstich, Jr.
wTe Corporation
Bedford, MA 01730

and

Solid Waste Association of North America
Silver Spring, MD 20910

Cooperative Agreement No. 818238

Project Officer

Diana R. Kirk
Waste Minimization, Destruction, and Disposal Research Division
Risk Reduction Engineering Laboratory
Cincinnati, OH 45268

RISK REDUCTION ENGINEERING RESEARCH LABORATORY
OFFICE OF RESEARCH AND DEVELOPMENT
U. S. ENVIRONMENTAL PROTECTION AGENCY
CINCINNATI, OH 45268

Summary or Abstract

Most reports and proposals that have front matter include a summary, sometimes called an *abstract*. These summaries serve three purposes:

- They help busy managers learn the main points without reading the entire document.
- They help all readers build a mental framework for organizing and understanding the detailed information they will encounter as they read on.
- They help readers determine whether they should read the full communication.

Purposes of summaries

Writing Reader-Centered Summaries

There are three types of summaries. For each communication you write, choose the one that will give your readers the most help as they perform their tasks.

- **Descriptive summaries** resemble a prose table of contents. They help readers decide whether to read the communication by identifying major topics covered. They do not tell the main points made about each topic. Often, descriptive summaries are used for research reports that will be placed in widely available resources, such as government databases or libraries. To aid in computerized bibliographic searches, they often contain key words that would be used by persons seeking the kind of information the document contains. Figure 12.4 shows a descriptive summary.

Three kinds of summaries

- **Informative summaries** distill the main points from the full document, usually in a page or less, together with enough background information about purpose and method to help readers understand the context in which the key information was developed. Informative summaries are ideally suited to readers who want to make decisions or take other action based on the findings, conclusions, and recommendations reported in the document.

SUMMARY

This handbook is intended to assist personnel involved with the design, construction, and installation of wells drilled for the purpose of monitoring groundwater for pollutants. It presents state-of-the-art technology that may be applied in diverse hydrogeologic situations and focuses on solutions to practical problems in well construction rather than on idealized practice. The information in the handbook is presented in both matrix and text form. The matrices use a numerical rating scheme to guide the reader toward appropriate drilling technologies for particular monitoring situations. The text provides an overview of the criteria that influence design and construction of groundwater monitoring in various hydrogeologic settings.

FIGURE 12.4
Descriptive Summary

In a descriptive summary, writers tell readers the topics discussed in a communication without telling what it says about each topic.

In this example, the writers do the following:
- Identify the intended audience and the way the handbook will assist them.
- Describe its contents.
- Tell how the information is presented.

Figure 12.5 shows an informative summary (named an abstract in this case) from the EPA report whose cover is shown in Figure 12.2.

- **Executive summaries** are a form of informative summaries tailored to the needs of executives and other decision makers who, pressed for time, want to extract the main points of a communication without reading all of it. Typically, executive summaries open by focusing on organizational questions and issues and briefly describe the investigative or research methods used by the writer. Often half or even more of the summary is devoted to precisely summarizing major conclusions and recommended actions—the type of information most helpful to decision makers. Figure 12.6 shows an executive summary.

Summaries that appear in printed and online bibliographic resources are usually called *abstracts*. In many scientific and engineering fields, the term *abstract* is also used for the summary that appears at the front of long reports and proposals.

Whether you are writing a descriptive, informative, or executive summary, follow these reader-centered guidelines.

WRITING READER-CENTERED SUMMARIES

- **Make it 100 percent redundant with the communication.** This purposeful redundancy provides a complete and understandable message to the reader who reads nothing else in the communication. It also means that the summary can't serve as the introduction, even though the introduction that follows it may seem somewhat repetitious.

- **Mirror the structure of the overall communication.** Include information from each major part of the communication, presented in the order of the parts in the overall communication.

- **Meet the needs of your readers.** For instance, if you know that your readers will be primarily interested in a novel method you used, provide more detail about the method than you would for readers who are primarily interested in your results. Likewise, in deciding what to include from any part of the full document, pick the information your specific readers will find most useful.

- **Be specific.** Replace general terms with precise ones. Instead of saying that it was "hot," say that it was "150°." Rather than saying the less expensive alternative "saved money," say that it saved "$43,000 per month." The more specific your abstract, the more useful it will be to your readers.

- **Keep it short.** Summaries are typically only 2 to 5 percent of the length of the body of the communication (not counting attachments and appendixes). That's between half a page and a whole page for every twenty pages in the body of your communication.

- **Write concisely.** Abstracts need to be lean but highly informative. Keep them short by eliminating unnecessary words, not by leaving out information important to your readers.

Learn More

For more strategies for writing concisely, see Guideline 1 for constructing sentences in Chapter 9, page 264.

FIGURE 12.5

Abstract from the Report
Whose Cover Is Shown in
Figure 12.2

ABSTRACT

This project evaluates a proof-of-concept, pilot-scale, automated sorting system for mixed post-consumer plastic containers developed by the Rutgers University Center for Plastics Recycling Research (CPRR). The study evaluates the system's ability to identify and separately recover five types of plastic containers representative of those found in plastics recycling programs. It also addresses the system's potential for full-scale commercial application.

Three series of tests were performed: single composition, short-term tests; mixed composition, short-term tests; and mixed composition, extended tests. A total of 82 test runs were performed during which 66,632 bottles were processed.

The five bottle types considered were natural HDPE, PVC, clear PET, green PET, and opaque HDPE. The containers recovered at each product collection station were counted and the results compared to pre-established recovery goals. Bottle counts were then converted to weight recoveries using average bottle weights. The resulting product purity/contamination weight percents were compared to allowable product contamination limits representative of industry practice. From a detailed videotape analysis of a representative test, an exact profile of bottle feed timing and sequence was reconstructed. This analysis provided valuable insight into system feed and transport dynamic as well as an understanding of how product contamination occurs.

The system produced statistically reproducible results and proved to be mechanically reliable. However, it failed to achieve all of the commercial-level container recovery and product contamination limit goals. It was concluded that bottle singulation and spacing greatly influenced the effectiveness of the identification/separation equipment.

This work was submitted by wTe Corporation in fulfillment of Contract No. 850-1291-4. The contract was administered by the Solid Waste Association of North America and sponsored by the U.S. Environmental Protection Agency. This report covers from May 1992 to July 1993.

i

In this abstract, the researchers included the key information from each of their report's sections.

Because the research reported in this document is for the purpose of testing a general approach to sorting waste, neither the report nor the abstract included a recommendation. (Compare with the executive summary in Figure 12.6.)

In the first sentence, the researchers announce the general **objective** of their study. Later in the paragraph, they identify its specific goals.

The researchers provide detailed information about their **method,** which is important because the method is what they were testing. They provide precise details, such as the exact number of test runs and precise number of bottles processed.

The researchers present their **results,** relating them directly to the primary research questions: Would the method work and, if so, would it work well enough to be commercially successful?

As is often required in government-sponsored projects, the researchers provide information about the sponsors and administrators of the study.

The researchers state the **conclusion** they drew from their analysis of the results.

FIGURE 12.6
Executive Summary

In this executive summary, the writers include the key information from each of the report's sections, condensing a 28-page report into fewer than 250 words.

Because an executive summary is addressed to decision makers, the writers present the information on which their readers would decide what action to take. (Compare with the abstract in Figure 12.5.)

In their first sentence, the writers state the **main point** they make in the body of their report: The airport should purchase a new system.

Because their readers would not be familiar with the details of the Accounting Department's computers, the writers provide the **background information** these readers need in order to understand the rest of the executive summary.

The writers pinpoint the **problem.**

They briefly describe the three **possible solutions** they investigated. Their language echoes the statements they made in the preceding list of the problem's sources.

The writers conclude their summary with their **recommendation and the reason for it.**

EXECUTIVE SUMMARY

The Accounting Department recommends that Columbus International Airport purchase a new operating system for its InfoMaxx Minicomputer. The airport purchased the InfoMaxx Minicomputer in 2005 to replace an obsolete and failing Hutchins computer system. However, the new InfoMaxx computer has never successfully performed one of its key tasks: generating weekly accounting reports based on the expense and revenue data fed to it. When airport personnel attempt to run the computer program that should generate the reports, the computer issues a message stating that it does not have enough internal memory for the job.

Our department's analysis of this problem revealed that the InfoMaxx would have enough internal memory if the software used that space efficiently. Problems with the software are as follows:

1. The operating system, BT/Q-91, uses the computer's internal memory wastefully.
2. The SuperReport program, which is used to generate the accounting reports, is much too cumbersome to create reports this complex with the memory space available on the InfoMaxx computer.

Consequently, we evaluated three possible solutions:

1. Buying a new operating system (BT/Q-101) at a cost of $3500. It would double the amount of usable space and also speed calculations.
2. Writing a more compact program in LINUX, at a cost of $5000 in labor.
3. Revising SuperReport to prepare the overall report in small chunks, at a cost of $4000. SuperReport now successfully runs small reports.

We recommend the first alternative, buying a new operating system, because it will solve the problem for the least cost. The minor advantages of writing a new program in LINUX or of revising SuperReport are not sufficient to justify their cost.

Table of Contents

By providing a table of contents, you help readers who want to find a specific part of your communication without reading all of it. Your table of contents also assists readers who want an overview of the communication's scope and contents before they begin reading it in its entirety.

Tables of contents are constructed from chapter or section titles and from headings. They are most useful to readers if they have two (perhaps three) levels. The highest level consists of the chapters or section titles. Often they are generic: Introduction, Method, Results, and so on. The second level consists of the major headings within each chapter or section. These second-level entries in the table of contents guide readers more directly to the information they are seeking. Using features of Microsoft Word, you can create a table of contents automatically from your communication's headings.

Creating and reviewing your table of contents can help you polish the body of your communication. If you find that your table of contents wouldn't provide as much guidance as your readers would like, that's a sign that you may need to rephrase some of your headings or add additional ones.

Figure 12.7 shows the table of contents from the report whose cover is shown in Figure 12.2. Note that the words *Table of* do not appear.

Lists of Figures and Tables

Some readers search for specific figures and tables. If your communication is more than about fifteen pages long, you can assist readers by preparing a list of figures and tables. Figure 12.8 shows the list of figures from the EPA report whose cover is shown in Figure 12.2. That report also contains a list of tables with the same format.

WRITING READER-CENTERED BACK MATTER

The back matter of a document may include one or more of the following elements: appendixes; reference list, works cited, or bibliography; glossary; and index.

Appendixes

Appendixes enable you to present information you want to make available to your readers even though you know it won't interest all of them. For instance, in a research report you might create an appendix for a two-page account of the calculations you used for data analysis. By placing this account in an appendix, you help readers who will want to use the same calculations in another experiment, but you save readers

FIGURE 12.7
Table of Contents of the Report Whose Cover Is Shown in Figure 12.2

Front matter. Note that lowercase roman numerals are used for the front matter's page numbers.

Body of report. In addition to the chapter titles, the writers included the first-level headings.

Back matter

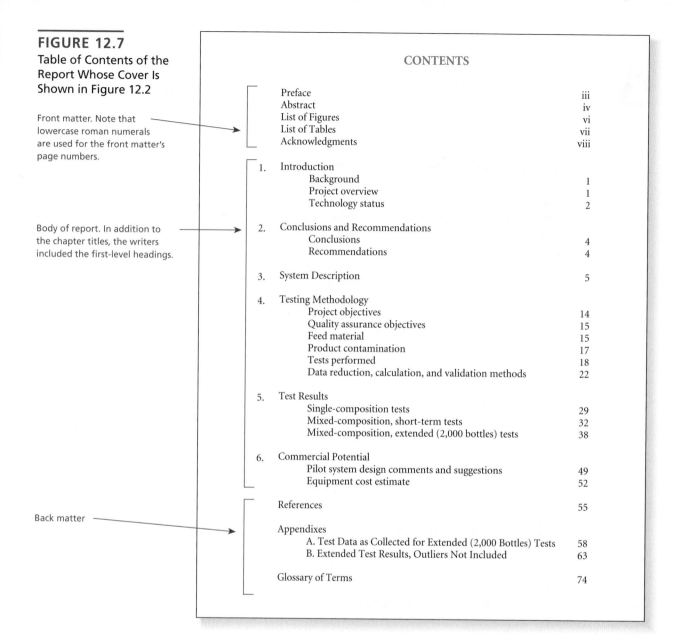

CONTENTS

The writers provided a list of
figures to assist readers interested in finding or reviewing the
information provided in them.

FIGURES

who aren't interested from the necessity of wading through the calculations while reading the body of your report. Appendixes can also assist readers by providing important reference information in a readily accessed place. That is a function of Appendix A in the back of this text.

When you create appendixes, list them in your table of contents (see Figure 12.7) and give each an informative title that indicates clearly what it contains. Also, mention each appendix in the body of your report at the point where your readers might want to refer to it: "Printouts from the electrocardiogram appear in Appendix II."

Begin each appendix on its own page. Arrange and label the appendixes in the same order in which they are mentioned in the body of the communication. If you have only one appendix, label it simply "Appendix." If you have more than one, you may use Roman numerals, Arabic numerals, or capital letters to distinguish them.

References List, Endnotes, or Bibliography

Learn More

For additional information about documenting your sources, including links to other documentation systems, go to Appendix A and to Appendix A at **www.cengage .com/english/anderson7e**.

Place your reference list, endnotes, or bibliography immediately after the body of your communication or after the appendixes. Appendix A explains which sources to cite and how to construct reference lists, endnotes, and bibliographies.

Glossary and List of Symbols

When writing at work, you will sometimes use terms or symbols that are unfamiliar to some of your readers. If you use each term or symbol only once or in only one small segment of your communication, you can explain its meaning in the text. Imagine, however, that you are writing a report in which a special term appears on pages 3, 39, and 72. If you define the term every time it occurs, your report will be repetitious. On the other hand, if you define the term only the first time it appears, your readers may have forgotten the definition by the time they encounter the term again. To solve this problem, create a glossary. It will provide the definition where readers can easily locate it if they need to but where it won't encumber their reading if they don't.

Figure 12.9 shows the first page of the glossary for the EPA report previously mentioned. The author has the terms to be defined in boldface and color so readers can find them easily. Figure 12.10 shows a list of symbols.

Index

If your communication is too long for your readers to thumb through quickly, give them a quick path to specific pieces of information by creating an index. Identify the kinds of information they might want to locate without reading the rest of the communication. If several index topics can be gathered under a single word, indent them under the main word to create second-level entries. See the example index in Figure 12.11.

Some desktop publishing programs can help you create an index by generating an alphabetized list of the words used in your communication. From this list, you can index those that will help your readers find the information they desire.

GLOSSARY OF TERMS

Availability — The probability that equipment will be capable of performing its specified function when called upon at any random point in time. Calculated as the ratio of run time to the sum of run time and downtime.

Bottle — A single plastic container, also referred to as a container.

Cascade — To sequentially increase the belt speed in a series of conveyors.

Contamination — Bottles improperly removed at a station.

Effective Bottle Feed Rate — The average rate at which a specific type of bottle is fed during a mixed composition test. Calculated as the total number of bottles of a specific type fed divided by the total run time for a specific test.

Extended Test — A test series consisting of many replicates at pre-selected conditions.

FIGURE 12.9
First Page of the Glossary of the Report Whose Cover Is Shown in Figure 12.2

In this glossary, the writers provide precise definitions for key terms in their report.

They use bold type and color to make the terms being defined stand out.

List of Symbols

a_k	allpass filter coefficients
a	vector of allpass filter coefficients
A	cross-sectional area
$A(z)$	allpass transfer function
B	area ratio
c	speed of sound
c	vector of transfer functions or of cosine functions
C	scaling coefficient
$C(z)$	transfer function of a subfilter in the Farrow structure
d	fractional part of the total delay D
D	total delay to be approximated
$D(z)$	denominator of $A(z)$

FIGURE 12.10
List of Symbols

This list of symbols, from a study concerning computer modeling of music, is several pages long.

Knowing that readers will search for symbols they have seen in the text, the writer has placed the symbols on left side—the side seen first by people who read from left to right.

To aid the readers' search, the writer has arranged the symbols in alphabetical order.

FIGURE 12.11
Index

This index is from the service manual for a diesel engine used in farm machinery.

The writers use second-level headings to organize related topics.

To help readers scan through the index, the writers used white space and capital letters to break the index into sections.

DETROIT DIESEL 53

ALPHABETICAL INDEX

*General Information and Cautions Section

© Copyright 1990 Detroit Diesel Corporation   May, 1990   Page 1

■ CHAPTER 12 **Writing Reader-Centered Front and Back Matter**

PART VI
DRAFTING VISUAL ELEMENTS

13 | Creating Reader-Centered Graphics

GUIDELINES

| DEFINING OBJECTIVES |
| PLANNING |
| CONDUCTING RESEARCH |
| DRAFTING |
| REVISING |

Graphics are photographs, drawings, flowcharts, tables, and other visual representations. As research shows, they play a critically important role in on-the-job writing.

- Readers look for and want graphics (Day & Gastel, 2005; Schriver, 1997).

- Readers gain more knowledge from communications with graphics (Levie & Lentz,1982).

- Readers remember more from communications with graphics (Simonson, Albright, & Zvacek, 2003).

Other benefits include the following.

- Graphics enhance a communication's visual appeal, thereby increasing the readers' concentration on its message.

- Graphics convey some kinds of information much more efficiently than prose.

FIGURE 13.1
Wordless Instructions for Leaving Plane in an Emergency

Wordless instructions are often used when the readers do not share a common language, as is the case with airline passengers.

In these instructions, arrows show passengers how to remove the cover from the latch, pull the latch, remove the door, and throw the door out of the plane.

Try This

In addition to using arrows, what other design strategies help readers of the airline instructions understand what they should do?
Alternative: Find a set of wordless instructions for something you own. What strategies did the designer use to make them easy to follow?

- Graphics enable writers to convey information to readers who do not share a common language with the writers—or with each other.

Graphics communicate information so effectively that they sometimes convey the entire message. The wordless instructions shown in Figure 13.1 provide an example.

A READER-CENTERED APPROACH TO CREATING GRAPHICS

Although you may never create a wordless communication, you will certainly include many graphics in your on-the-job writing. As you plan and create each one, take the same reader-centered approach that you follow when drafting your prose. Each has its own specific objectives, shaped by the particular information your readers want to gain from it and by the ways your readers want to use that information. Likewise, you may have a distinct persuasive goal for each of the graphics in your communication. For these reasons it's just as important to imagine your readers' moment-by-moment reading process when you are planning and creating your graphics as it is when you are planning and drafting your prose.

This chapter's first eight guidelines guide you through the process of planning, designing, and integrating reader-centered graphics into your communications. A ninth discusses several ethical considerations that apply specifically to graphics. Finally, the Writer's Reference Guide that follows this chapter provides advice for constructing eleven types of widely used graphics.

> **WWW**
>
> For additional information, examples, and exercises about graphics, visit **www.cengage.com/english/anderson7e** and click on Chapter 13.

Guideline 1 | **Look for places where graphics can increase your communication's usefulness and persuasiveness**

When planning, reviewing, and revising your communication, search actively for places where graphics can help you achieve your communication objectives. Contemporary society has become increasingly visual. More and more, people gain information through images rather than words. Paradoxically, though, most of us still think primarily of words when we want to convey information to others. Consequently, unless you devote time to looking for them, you may miss opportunities to increase your communication's effectiveness by replacing, supplementing, or reinforcing your words with visual presentations (Schriver, 1997).

Using Graphics to Increase Usefulness

Begin by making a reader-centered search for spots where graphics can increase your communication's usefulness to your readers—and perhaps save space as well. Look, for example, for places where graphics could help you achieve any of the following objectives. Pages 334 and 335 show a sample graphic suited to each of these objectives.

Writer's Tutorial

Graphics Help Readers Understand and Use Information

SHOW YOUR READERS HOW SOMETHING LOOKS

NASA uses this drawing to show the public the appearance of a space probe sent to Saturn.

NASA artists chose to draw the probe from an angle of view that displays its key parts.

By including images of Saturn in the background, NASA helps readers imagine the satellite performing its mission.

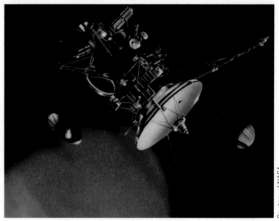

Courtesy of NASA

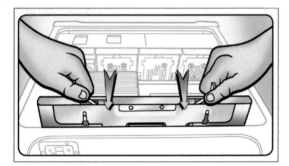

SHOW YOUR READERS HOW TO DO SOMETHING

Graphics can explain operations that would be difficult to describe—and understand—in prose. Writers used this drawing to help readers learn how to perform one step when replacing a computer's memory.

By showing the action from the same angle of view that a person performing this task would have, the writers prepared an easy-to-follow, reader-centered drawing.

EXPLAIN A PROCESS

To help readers understand how seawater picks up chemicals as it flows through hot-spring systems, a writer used this drawing.

This drawing illustrates one way to integrate graphics with text to clarify meaning for readers.

- The text incorporated into the drawing explains the four major steps in the process.
- The caption below the figure describes the overall process.

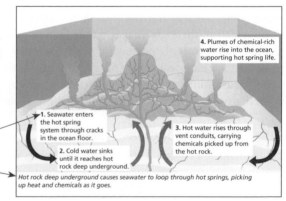

4. Plumes of chemical-rich water rise into the ocean, supporting hot spring life.

1. Seawater enters the hot spring system through cracks in the ocean floor.

2. Cold water sinks until it reaches hot rock deep underground.

3. Hot water rises through vent conduits, carrying chemicals picked up from the hot rock.

Hot rock deep underground causes seawater to loop through hot springs, picking up heat and chemicals as it goes.

SHOW HOW SOMETHING IS CONSTRUCTED

Using this drawing, writers described the construction of lasers used to make computer chips.

By coloring the relevant parts blue, the writers highlighted the path of the laser beams.

To help readers identify the parts of the system, the writers not only placed each label close to the part it names but also drew arrows from each label to its part.

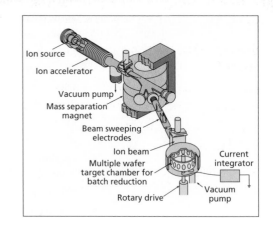

Ion source
Ion accelerator
Vacuum pump
Mass separation magnet
Beam sweeping electrodes
Ion beam
Multiple wafer target chamber for batch reduction
Rotary drive
Current integrator
Vacuum pump

DISPLAY INFORMATION IN A USEFUL WAY

Kodak uses this table to help professional and amateur photographers find the optimum temperature and time for developing film.

By clearly distinguishing information for small and large tanks, Kodak has helped its readers focus on the part of the table that applies to the size of development tank they are using.

Kodak uses bold to highlight the columns for the temperature it recommends.

Kodak Developer	Developing Time (in Minutes)									
	SMALL TANK (Agitation at 30-Second Intervals)					LARGE TANK (Agitation at 1-Minute Intervals)				
	65°F (18° C)	68°F (20° C)	70°F (21° C)	72°F (22° C)	75°F (24° C)	65°F (18° C)	68°F (20° C)	70°F (21° C)	72°F (22° C)	75°F (24° C)
HC-110 (Dil B)	8½	7½	6½	6	5	9½	8½	8	7½	
D-76	9	8	7½	6½	5½	10	9	8	7	
D-76 (1:1)	11	10	9½	9	8	13	12	11	10	
MICRODOL-X	11	10	9½	9	8	13	12	11	10	
MICRODOL-X (1:3)*	—	—	15	14	13			17	16	
DK-50 (1:1)	7	6	5½	5	4½	7½	6½	6	5½	
HC-110 (Dil A)	4½†	3¾†	3¼†	3†	2½†	4¾†	4¼†	4†	3¾†	

* Gives greater sharpness than other developers shown in table.
† Avoid development times of less than 5 minutes if possible, because poor uniformity may result.

Note: Do not use developers containing silver halide solvents.

SHOW TRENDS OR OTHER NUMERICAL RELATIONSHIPS

A team of biological researchers used this graph to help readers see the relationship between the temperature of river water and the abundance of shellfish in the river.

This sophisticated graph has two *X*-axes, one for shellfish larvae (left side) and one for water temperature (right side).

To help their readers see which graphed line is for shellfish and which for water temperature, the researchers provided a label for each line.

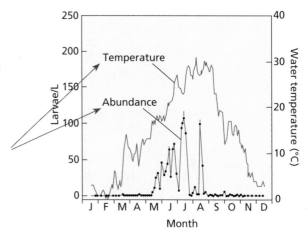

Temperature
Abundance
Larvae/L
Water temperature (°C)
Month
J F M A M J J A S O N D

- **Show your readers how something looks.** Drawings and photographs can often show the appearance of objects with greater clarity than words could achieve. For example, scientists use photographs to show the results of earthquakes, and NASA uses drawings to help readers visualize its space probes.
- **Show how something is constructed.** Drawings and photographs can show the structure of an object, such as the anatomy of a horse or the design of lasers used to make computer chips.
- **Show your readers how to do something.** Ideal for instructions, graphics explain procedures that would be much more difficult to describe—and understand—in prose.
- **Explain a process.** Many processes are best understood when visualized. For example, scientists use diagrams to help readers quickly understand how seawater picks up chemicals as it loops through hot-spring systems under the ocean.
- **Make particular facts easy to find and use.** From bus schedules to the nutrition tables on boxes and jars of food, tables and other graphics help readers quickly locate information they want to use.
- **Show trends and other numerical relationships.** Various graphics such as line graphs, bar graphs, and pie charts enable readers to grasp trends and other numerical relationships much more quickly than they could from sentences.

Using Graphics to Increase Persuasiveness

Graphics can greatly increase a communication's persuasive impact. Often this opportunity arises when a graphic enables you to support your persuasive points by conveying your data in an especially dramatic way (Kostelnick & Roberts, 1998). For example, the 3M Corporation used the bar graph shown in Figure 13.2 to persuade greenhouse owners that they could greatly reduce their winter heating bills by covering their greenhouses with the company's plastic sheets.

Many other kinds of visual representations can enhance a communication's persuasiveness. A drawing, for instance, can help readers envision the desirable outcomes of projects you propose. A photograph of a polluted stream can forcefully portray a problem you want to motivate your readers to address.

When looking for places where graphics can help your communication achieve its goals, be careful to avoid including ones that don't serve a specific purpose. They will detract from your message. Worse, they may confuse your readers.

Guideline 2 | **Select the type of graphic that will be most effective at achieving your objectives**

Once you've decided where to use a graphic, identify the type that will be most effective at achieving your objectives. Most information can be presented in more than one type.

Numerical data can be presented in a table, bar graph, line graph, or pie chart. The components of an electronic instrument can be represented in a photograph,

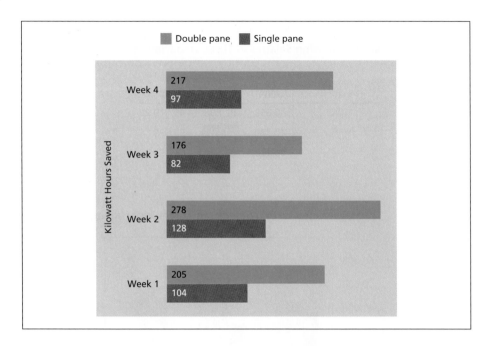

FIGURE 13.2
Graphic That Makes a
Persuasive Point

With this bar graph, a company
that installs replacement windows
for older homes aims to persuade
customers to purchase more expensive double-pane windows.

To emphasize the benefits of
double-pane windows, the company used bright green to color
the bars for the energy saved by
those windows. Green is associated with energy savings and
environmental benefits. For the
single-pane windows, the company used gray, a color associated
with soot.

sketch, block diagram, or schematic. When deciding which type of graphic to use, consider both the tasks you want to help your readers perform and the way you want to affect their attitudes.

Consider Your Readers' Tasks

Different types of graphics support different reading tasks. Consider, for example, Ben's choices.

Ben has surveyed people who graduated over the past three years from three departments in his college. Now he wants to report to the alumni what he has learned about their average starting salaries. He could do this with a table, bar graph, or line graph (see Figure 13.3). Which type of graphic would be best? To decide, Ben must determine the specific task the alumni will want to perform by using the graphic. If they want to learn the average starting salary of people who graduated in their year from their department, a table will help them most. If they want to compare the average starting salaries in their department with the average starting salaries of people who graduated in that same year from the other departments, a bar graph will work best. And if the alumni's task will be to determine how the average starting salary in their department changed over the years and to compare that change with the changes experienced by the other departments, a line graph will be the most useful.

Consider Your Readers' Attitudes

When selecting the type of graphic you will use, think about your readers' attitudes as well as their tasks. Pick the type of graphic that most quickly and dramatically communicates the evidence that supports your persuasive point.

FIGURE 13.3

Three Ways of Showing
Average Starting Salaries
for Three Departments
over Three Years

A table helps a reader who wants
to quickly find a specific piece of
information, such as the average
salary for a particular department
in a specific year.

Average Starting Salaries in Three Departments

Department	Year of Graduation		
	2009	2010	2011
A	30,653	32,898	49,519
B	42,289 →	44,904	52,698
C	46,172	61,538	66,357

A bar graph helps a reader who
wants to make comparisons, for
instance, a comparison among
the average salaries of three
departments in 2010.

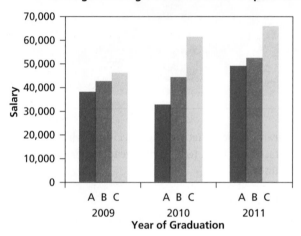

A line graph helps a reader who
wants to see trends in the salaries
for one department or compare
trends in salaries for two or more
departments.

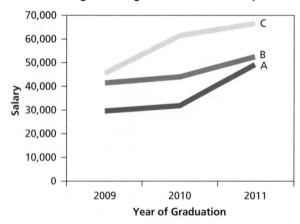

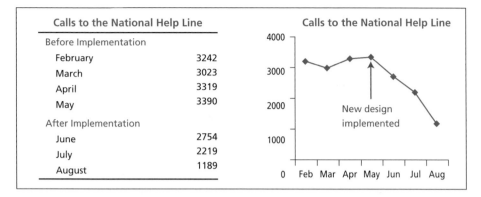

Calls to the National Help Line	
Before Implementation	
February	3242
March	3023
April	3319
May	3390
After Implementation	
June	2754
July	2219
August	1189

FIGURE 13.4

Comparison of the Persuasiveness of Data Presented in a Table and a Line Graph

Consider, for instance, Akiko's choices. Akiko recommended a change in the design of one of her company's products. In order to show how the company has benefited from this change, she has tallied the number of phone calls received by the company's toll-free help line during the months immediately before and after implementation of the new design. As Figure 13.4 shows, if Akiko presents her data in a table, her readers will have to do a lot of subtracting to appreciate the impact of her recommendation. If she presents her data in a line graph, they will be able to recognize her accomplishment at a glance.

At times, you may be able to meet your readers' needs only if you present your data in two ways. For instance, in scientific reports and journals, graphs are used in the text to provide readers with an overall understanding of the data. At the end of the report, the tables are presented for readers who want to examine the data in more detail, perhaps looking for other patterns in them.

Sometimes one type of graphic isn't enough.

Guideline 3 | Make each graphic easy to understand and use

Having chosen the type of graphic you will use, you must then design the item itself. When doing so, focus first on usability. Make your graphics, like your prose, easy for your readers to understand and use. Here are some suggestions.

Design Your Graphics to Support Your Readers' Tasks

First, follow this familiar, reader-centered strategy: Imagine your readers in the act of using your graphic. Then design it to support your readers' efforts. In drawings or photographs for step-by step instructions, for example, show objects from the same angle that your readers will see them when performing the actions you describe.

Likewise, in a table, arrange the columns and rows in an order that will help your readers rapidly find the particular pieces of information they are looking for. Maybe this means you should arrange the columns and rows in alphabetical order, according to a logical pattern, or by some other system. Use whatever arrangement your readers will find most efficient.

Consider Your Readers' Knowledge and Expectations

Of course, your readers will find your graphics useful and persuasive only if they can understand them. Some types are familiar to us all, but other types can be interpreted only by people with specialized knowledge. If you work in a field that employs specialized graphics, use these graphics only when communicating with readers in your field who will understand and expect them. However, when writing to other readers, use an alternative type of graphic—or include explanations these readers need in order to interpret the special-use graphic.

Simplify Your Graphics

You can also make your graphics easy to understand and use by keeping them simple. Simplicity is especially important for graphics that will be read on a computer screen or from a projected image; people have more difficulty reading from these media than from paper. Here are some effective strategies for keeping your graphics simple.

- **Include only a manageable amount of material.** Sometimes, it's better to separate your information into two or more graphics than to cram it all into one.

- **Eliminate unnecessary details.** Like unnecessary words in prose, superfluous details in graphics create extra, unproductive work for readers and obscure the really important information. Figure 13.5 shows how the elimination of extraneous detail can simplify a graph. Figure 13.6 shows two ways to simplify a graphic showing complicated equipment by paring it down to the essentials.

FIGURE 13.5
A Line Graph with and without Grid Lines

Replacing the grid lines with tick marks helps the readers to more easily follow the graphed lines and read the labels.

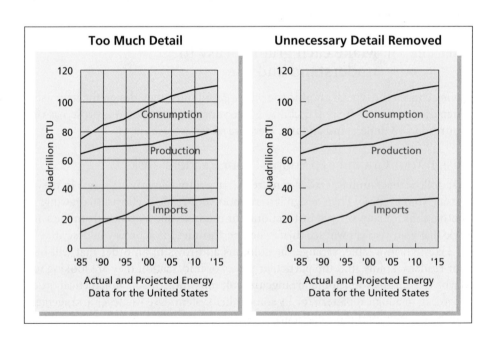

■ CHAPTER 13 **Creating Reader-Centered Graphics**

Less Effective Photograph

The cluttered background makes it difficult to identify the parts of the testing equipment.

More Effective Photograph

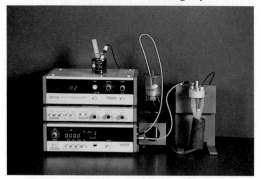

Removal of the clutter allows the parts to be distinguished.

Drawing

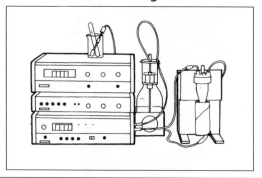

A line drawing shows very clearly some parts that are not obvious even in the uncluttered photograph.

FIGURE 13.6
Two Photographs and a Drawing of the Same Equipment

Label the Important Content Clearly

Labels help readers locate the information in a graphic and understand what it shows. In tables, label every row and column. In diagrams, label every part that is important to your readers. But avoid labeling other features. Unnecessary labels clutter a graphic, making it difficult to understand and use.

For each label, carefully choose the appropriate word or words and place them where they are easy to see. If necessary, draw a line from the label to the item. Avoid placing a label on top of an important part in your graphic. Figure 13.7 shows good placement of labels in a photograph. Note that labels placed in a graphic are much easier than a key for readers to use. See Figure 13.8.

Provide Informative Titles

Titles help readers to find the graphics they are looking for and also to know what the graphics contain once they locate them. Typically, titles include both a number (for example, "Figure 3" or "Table 6") and a description ("Effects of Temperature on the Strength of M312").

Make your titles as brief—and informative—as possible. Use more words if they are needed to give your readers precise information about your graphic. Don't say, "Information on Computer Programs," but say instead, "Comparison of the Speed and Capabilities of Three Database Programs." Be consistent in the placement of your titles. For example, place all titles above your figures or all below your figures.

FIGURE 13.7
Good placement of labels

This photograph shows a "seeing eye" robot developed at Utah State University to assist shoppers who are blind. It directs them to the products they want to purchase.

The creator of this graphic, Jeff D. Allred, placed the labels around the robot, person, and guide dog.

He printed the labels against white rectangles, an effective way of making the text readable when labels are placed inside a photograph. When creating photographs, people sometimes also place the arrows against a white background to make them easier to see.

In the labels, Jeff has printed each component's name in bold, followed by an explanation of the component and what it does.

How it works

Here's how a blind person would make use of the Robotic Guide:

Braille instructions: The visitor would use a Braille piece of paper, hanging on the handle, to find the number of the item he or she needs.

RFID antenna: The diamond-shaped tube is the antenna for a Radio Frequency Identification (RFID) finder that interacts with RFID tags on the store shelves to find the right location.

Keypad: Buyer punches in product number here.

Laptop computer: The brains, located in the middle.

Laser range finder: This box on the robotic base helps the robot maneuver around objects – other shoppers or store displays.

RFID reader: Gray box that sits on the back of the robotic base, behind the laser. It powers the RFID antenna and sends data to the laptop.

Robotic base: The hunk of equipment at the bottom with the wheels.

Source: Vladimir Kulyukin, Utah State University. Photo by Jeff D. Allred for USA TODAY

FIGURE 13.8
Labels Placed in a Graphic
Versus Use of Key

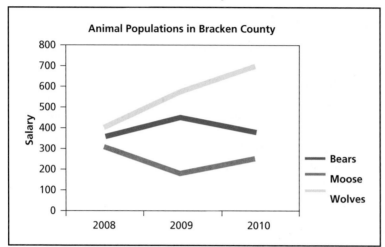

Labels in a Key

Animal Populations in Bracken County

Difficult to Use
When labels are placed in a key,
readers must repeatedly shift
their attention from the graphic
to the key in order to interpret
the information provided. The
same applies to keys for all graphics, including drawings and photos. This extra work is undesirable
from the readers' perspective. The
more labels, the more difficult
keys are to use.

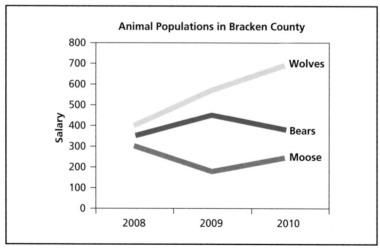

Labels by Their Data

Animal Populations in Bracken County

Easy to Use
When labels are placed next to
the items they identify, readers
can immediately link each part to
its label.

Learn More

For more information on using a list of figures, see page 323.

If readers might seek a specific figure whose location won't be obvious from the regular table of contents, provide a separate list of the figures and the pages where they can be found.

Guideline 4 | Use color to support your message

In recent years, there's been an explosion of color in workplace communications. Web pages and other on-screen communications are saturated with color. Color enlivens and enhances the many oral presentations made by individuals using computer programs such as PowerPoint. In addition, as color printing and copying have become less expensive, even routine printed documents now incorporate multicolor designs.

The widespread availability of color puts a new and powerful aid to communication at your disposal. With color you can clarify your messages, speed your readers' comprehension, and make your information easy for your readers to use. Among other things, color can help you do the following:

Try This

Find a page in a book or magazine where color is used especially well for one of the purposes given in the bulleted list to the right. Can you find a page that uses color well for two of these purposes? Three?

- Highlight a point
- Evoke an emotional response
- Tell the reader where to look first
- Make your communication look more polished and attractive
- Group related items
- Establish hierarchies of importance

Guidelines for Using Color

The impact of color on readers is determined partly by physiology and partly by psychology. Based on what researchers know about these responses, the following suggestions will enable you to use color in a reader-centered way:

1. **Use color primarily for clarity and emphasis, not decoration**. Color attracts the eye. When you use color merely for decoration, you may be drawing your readers' eyes to ornamentation rather than to your communication's important content, thereby reducing the readers' ability to grasp your message.

   Also, attractive colors can actually make reading more difficult, as communication researcher Colin Wheildon (1995) discovered. To a number of people, he showed two versions of the same page, one printed in black and one in blue. The people said the blue version was more attractive, but those who actually read the blue version scored substantially lower on a comprehension test than those who read the black one.

   Deploy colors in a way that promotes rather than hinders comprehension.

2. **Choose color schemes, not just single colors**. Readers see a color in terms of its surroundings. As a result, a color's appearance can change if surrounding colors are changed. To illustrate this point, design expert Jan V. White (1990) uses blocks of color like those shown in Figure 13.9. The top pair of blocks

The same blue square appears lighter against a darker color (such as black) than it does against a lighter color (such as yellow).

The same orange square appears brighter against a complementary color (in this case, green) than it does against a different shade of its own color.

shows that the same shade of blue-gray appears lighter viewed against a dark color than against a pale color. As the bottom pair of blocks demonstrates, the same pure color looks much different when it is surrounded by another shade of the same color than when it is surrounded by a complementary color. The following advice builds on the ways color schemes affect readers.

Use bright colors to focus your readers' attention. For example, to focus his readers' eyes on a central component of a device for creating silicon-germanium crystals, the writer who created Figure 13.10 used bright yellow for the central part and dull colors for the other parts.

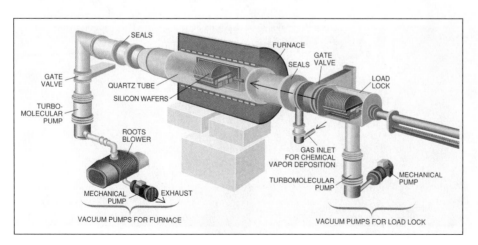

FIGURE 13.10
Bright Color Used to Focus Readers' Attention

The bright color at the center of this figure draws the readers' attention there.

- **Use contrasting colors to create emphasis.** To make something stand out, put it against a much different color rather than a very similar color.

Greater contrast creates more emphasis.

- **Use warm, intense colors to make items "advance" toward the reader.** For example, to make their image of the molecular structure of cytokine "closer" to their readers, the scientists who created Figure 13.11 used warm, intense colors for the molecule and set them against a black background.

3. **To promote easy reading, use a high contrast between text and background.** To comprehend a written message, readers must first pick out the letters from the background on which they are printed. We traditionally use black print on white paper because this color combination creates a very high contrast that makes letter identification easy. However, backgrounds of colors other than

FIGURE 13.11
Warm, Intense Colors Used against a Dark Background to Make Important Material Advance toward the Reader

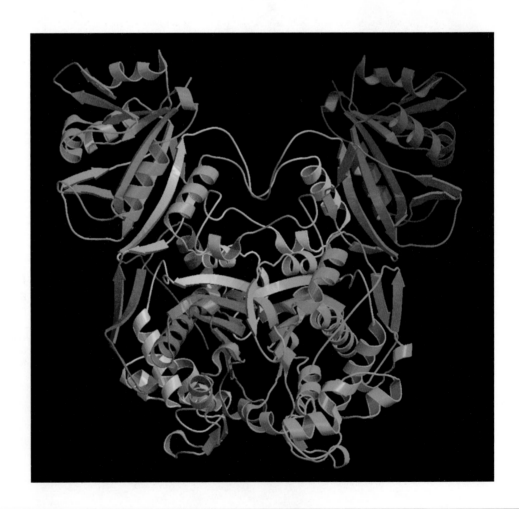

white have become common in on-the-job communications. The following chart shows how much more difficult type becomes to read when contrast with the background is reduced.

Reducing contrast reduces readability.

20%	This type is surprinted in black.	This type is dropped out in white.	20%
40%	This type is surprinted in black.	This type is dropped out in white.	40%
60%	This type is surprinted in black.	This type is dropped out in white.	60%
80%	This type is surprinted in black.	This type is dropped out in white.	80%
100%	This type is surprinted in black.	This type is dropped out in white.	100%

4. Use the associations that colors already have for your readers. In many contexts, colors have specific associations and even symbolic meaning. Driving through a city, we associate red with "Stop" and green with "Go." Use such associations when choosing colors for your graphics. Note the importance, however, of being alert to the fact that different colors have different associations in different contexts (White, 1990). In business, blue is associated with stability, but to a doctor it connotes death. In other contexts, blue suggests sky, water, and cold.

5. Use colors consistently to help your readers establish patterns of meaning. Readers also rely on color patterns to help them understand the information shown in graphics. You can help them out. For example, if you use orange to represent the raw materials and green to represent the products of a manufacturing process, use the same colors for the same purposes in other diagrams in the same communication.

6. Stick to a few colors. While the strategic use of color can increase a communication's effectiveness, an overabundance of color can cause confusion. Colors compete with one another for attention. To help readers see the patterns of meaning you are establishing, limit yourself to as few colors as possible.

Learn More

The same colors also have different associations in different cultures. See page 352.

Guideline 5 | Use graphics software and existing graphics effectively

Software companies offer a variety of powerful programs that can help you create reader-centered graphics. Even standard desktop publishing programs include many capabilities for creating graphics. For example, Microsoft Word and similar programs enable you to create line graphs, bar charts, organizational charts, and drawings, among other graphics.

As handy as such software features are, they sometimes have default settings or limitations that produce results that are not reader-centered. For example, when producing bar graphs and line graphs, the default settings in word processing and spreadsheet programs include grid lines whether they are needed or not (see page 340). Similarly, some have default settings that place labels for graphed lines in a key beside the graph, a location that makes reading difficult (see page 342 and Figure 13.8 on page 343).

However, you shouldn't avoid using such programs because they have default settings. Most allow you to modify the generic graphs they produce in ways that are consistent with this chapter's reader-centered guidelines. The writer's tutorial on pages 350–351 shows how to create and modify a line graph with Microsoft Excel.

The important point is that you should always review graphics you produce with software from your readers' perspective, being sure that it will be as useable and persuasive as possible for them.

The same caution applies when you are incorporating into a communication a graphic previously made by you or by coworkers. A graphic that was extremely well suited for the audience and purpose for which it was originally created is not necessarily designed in a way that will best address your objectives in a new situation or with different readers.

Guideline 6 | Integrate your graphics with your text

To enable your graphics to achieve their potential for usability and persuasiveness, carefully integrate them with a communication's prose. Here are four strategies you can use to create a single, unified message in which your graphics and prose work harmoniously together.

Introduce Your Graphics in Your Text

When people read, they read one sentence and then the next, one paragraph and then the next, and so on. When you want the next element they read to be a table or a chart rather than a sentence or a paragraph, you need to direct their attention from your prose to the graphic and tell them how the graphic relates to the statements they just read. There are various ways of doing this:

Learn More

For advice on integrating graphics into oral presentations and web pages, see Chapters 19 and 20.

Two sentences

> In a market test, we found that Radex was much more appealing than Talon, especially among rural consumers. See Figure 3.

One sentence with an introductory phrase

> As Figure 3 shows, Radex was much more appealing than Talon, especially among rural consumers.

One sentence with the reference in parentheses

> Our market test showed that Radex was much more appealing than Talon, especially among rural consumers (Figure 3).

Sometimes your text reference to a graphic will have to include information your readers need in order to understand or use the graphic. For example, here is the way the writers of an instruction manual explained how to use one of their tables:

Writer tells reader how to use the graphic.

> In order to determine the setpoint for the grinder relay, use Table 1. First, find the grade of steel you will be grinding. Then read down column 2, 3, or 4, depending upon the grinding surface you are using.

Whatever kind of introduction you make, place it at the exact point where you want your readers to focus their attention on the graphic.

Place Your Graphics Near Your References to Them

When your readers come to a statement asking them to look at a graphic, they lift their eyes from your prose and search for the graphic. You want to make that search as short and simple as possible. Ideally, you should place the graphic on the same page as your reference to it. If there isn't enough room, put the graphic on the page facing or the page that follows. If you place the figure farther away than that (for instance, in an appendix), give the number of the page on which the figure can be found.

WWW

To see communications whose creators have integrated graphics effectively with their text, go to Chapter 13 at **www .cengage.com/english/ anderson7e.**

State the Conclusions You Want Your Readers to Draw

One way to integrate your graphics into your text is to state explicitly the conclusions you want your readers to draw from them. Otherwise, readers may draw conclusions that are quite different from the ones you have in mind.

For example, one writer included a graph that showed how many orders she thought her company would receive for its rubber hoses over the next six months. The graph showed that a sharp decline would occur in orders from automobile plants, and the writer feared that her readers might focus on that fact and miss the main point. So she referred to the graph in the following way:

> As Figure 7 indicates, our outlook for the next six months is very good. Although we predict fewer orders from automobile plants, we expect the slack to be taken up by increased demand among auto parts outlets.

Writer tells reader how to use the graphic.

When Appropriate, Include Explanations in Your Figures

Sometimes you can help your readers understand your message by incorporating explanatory statements into your figures. Figure 13.12, which shows how paint powder is applied to cars in an automobile assembly line, provides an example.

Guideline 7 | Get permission and cite the sources for your graphics

Copyright law treats graphics in a way that differs significantly from its treatment of text. Under the "fair use" provision of the law, you may quote a small part of a text without permission. However, you must obtain permission for *every* graphic you wish to use, even if you want to use only one out of hundreds in a particular book. The only exceptions are graphics in the public domain because they belong to a government agency or are owned by your employer. Even for graphics in the public domain, cite the source of your graphics.

You must obtain permission and cite sources for graphics obtained at websites as well as those from print documents.

Learn More

For more information on copyright and fair use, turn to page 158.

Writer's Tutorial

Creating Reader-Centered Graphs with a Spreadsheet Program

This tutorial tells how to create graphs using Excel for Windows 2007. For instructions on using Word forMacIntosh, go to www.cengage.com/english/anderson7e and click on Chapter 13. If you use a different wordprocessor or get stuck, click on your program's Help menu for assistance.

CREATING A GRAPH

PLAN YOUR TABLE

1. Decide which labels will go on your vertical axis.
2. Decide which labels will go on your horizontal axis.

	Oct	Nov	Dec
NYC			
Hou			
Sea			

CREATE A TABLE IN A NEW SPREADSHEET

1. Create a table by arranging the labels for your vertical and horizontal axes as shown.
2. Enter your data into the table.

	Oct	Nov	Dec
Houston	130	300	350
Seattle	100	150	250
New York	250	500	650

MAKE A GRAPH BASED ON YOUR TABLE

1. Highlight your table.
2. Click the **Insert** tab.
3. In the **Charts** group on the ribbon, click **Line**.
4. In the top row of the dropdown menu, choose **Line**.

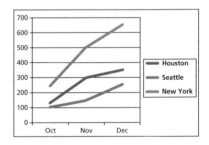

ADD A TITLE

1. Click in your graph.
3. Click the **Layout** tab.
4. In the **Labels** group on the ribbon, click **Chart Title**.
5. In the drop-down menu, click **Above Chart.**
6. In the text box that appears in your graph, enter the title.

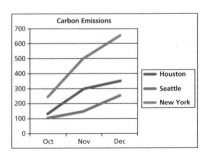

LABEL THE AXES

1. Click in your graph.
2. Click the **Layout** tab.
3. In the **Labels** group on the ribbon, click **Axis Titles**.
4. In the drop-down menu, click **Primary Vertical Axis Title**.
5. In the window that appears, click **Rotated Title**.
6. In the text box that appears in the graph, enter the title.
7. In the **Labels** group on the ribbon, click **Axis Titles**.
8. In the drop-down menu, click **Primary Horizontal Axis Title**.
9. In the text box that appears in your graph, enter the title.

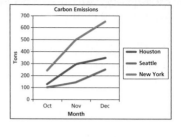

LABEL THE GRAPHED LINES

By replacing the legend with labels placed next to the graphed lines, you make your graph easier to read. See Figure 13.7.

1. In your graph, click the legend.
2. On the keyboard, press **Delete**.

   • The graph will change shape.

3. Click the **Layout** tab.
4. In the **Insert** group on the ribbon, click **Text Box**.
5. Draw a text box near one of the graphed lines.
6. In the text box, enter the label for the graphed line.

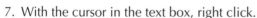

Steps 7 through 11 make your graph easier to read by separating the labels from the grid lines. See Figure 13.6.

7. With the cursor in the text box, right click.
8. In the small toolbar that appears, click **Fill** icon (paint can).
9. From the drop-down menu, click **Automatic**.
10. Click the icon for placing a Line around the text box (pencil).
11. From the drop-down menu, click **Automatic**.
12. Move the text box near the graphed lines it labels.
13. Repeat Steps 7 through 12 for the other graphed lines.

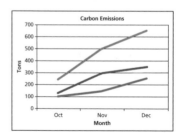

EDIT YOUR GRAPH

If changing the appearance of any element in your graph will increase the graph's readability and persuasiveness, do the following:

1. Click the area to be changed (for instance, plot area, title, or vertical label).
2. Right click.
3. Use the menu that appears to make the changes you want.

Note: Using very similar steps, you can create bar graphs and pie charts.

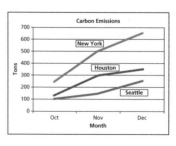

FIGURE 13.12
Explanatory Text
Incorporated in a Figure

The numbered explanations help
readers understand the process by
describing it step by step.

The boxed explanations help
readers identify and understand
the function of two important
parts of the paint facility.

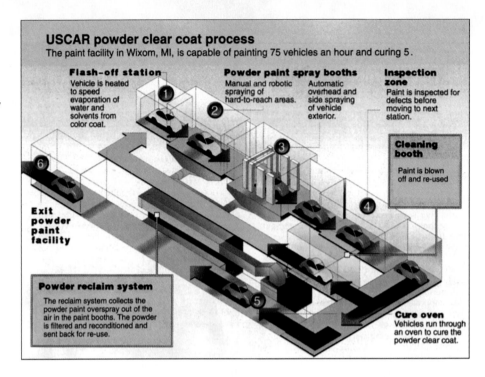

Guideline 8 | Global Guideline: Adapt your graphics when writing to readers in other cultures

Visual language, like spoken and written language, differs from nation to nation. For
example, Dwight W. Stevenson (1983) reports that while people in Western nations
typically read graphics from left to right, the Japanese read them from right to left—
the same way they read prose. Moreover, though many technical symbols are used
worldwide, others are not.

A major U.S. corporation reports that it once encountered a problem with publications intended to market its computer systems abroad because they included photos of telephones used in the United States rather than the much different-looking
telephones used in the other countries (Anonymous). The design of many other
ordinary objects differs from country to country. If you use a picture of an object
that looks odd to people in your target audience, the effect you are striving for may be
lost. Customs vary even in such matters as who stands and who sits in various working situations. Certain hand gestures that are quite innocent in the United States are
regarded as obscene elsewhere in the world.

Colors too have different connotations in different cultures. While yellow suggests caution or cowardice in the United States, it is associated with prosperity in

Egypt, grace in Japan, and femininity in several other parts of the world (Thorell & Smith, 1990).

The point is simple: Whenever you are writing for readers in another country, discuss your plans and review your draft graphics with people familiar with that country's culture. If possible, also test your drafts with these individuals.

Guideline 9 | Ethics Guideline: Avoid graphics that mislead

Graphics can mislead as easily as words can. When representing information visually, you have an ethical obligation to avoid leading your readers to wrong conclusions. This means not only that you should refrain from intentional manipulation of your readers but also that you should guard against accidentally misleading readers. Here are some positive steps you can take when creating several common types of graphs.

WWW

For additional information about ethical issues involved with graphics, go to Chapter 13 at **www.cengage.com/ english/anderson7e.**

Ethical Bar Graphs and Line Graphs

To design bar graphs and line graphs that convey an accurate visual impression, you may need to include zero points on your axes. In Figure 13.13, the left-hand graph makes the difference between the two bars appear misleadingly large because the Y-axis begins at 85 percent instead of at zero. The center graph of Figure 13.13 gives a more accurate impression of the data represented because the Y-axis begins at zero.

If you cannot use the entire scale, indicate that fact to your readers by using hash marks to signal a break in the axis and, if you are creating a bar graph, in the bars themselves. The right-hand graph in Figure 13.13 shows an example.

Note, however, that zero points and hash marks are sometimes unnecessary. For example, in some technical and scientific fields, certain kinds of data are customarily represented without zero points, so readers are not misled by their omission.

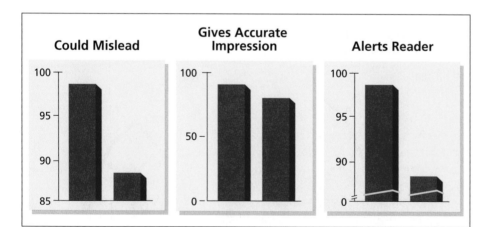

FIGURE 13.13
Creating Ethical Bar Graphs

Because its Y-axis begins above zero, the graph on the left can mislead.

The two graphs on the right illustrate ways to avoid misleading.

FIGURE 13.14
Creating Ethical
Pictographs

The left-hand pictograph makes
the difference between the two
amounts larger than it actually is
because the red apple is larger in
height *and* width, thereby enlarg-
ing its area disproportionately.

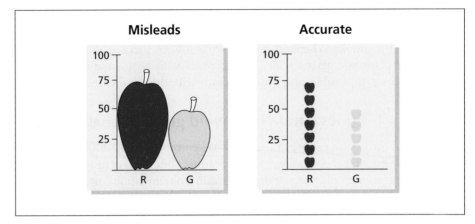

Ethical Pictographs

Pictographs can also mislead readers. For example, the graphs in Figure 13.14 represent
the average percentage of an apple harvest that a grower should expect to be graded
Extra Fancy. This left-hand graph makes the percentage of Red Delicious apples seem
much larger than the percentage for Golden Delicious, even though the actual difference is only 20 percent. That's because the picture of the Red Delicious apple is larger
in height *and* width, so its area is much greater. The right-hand graph shows how to
represent the data accurately: by making columns that differ in height alone.

Ethical Use of Color

Like any other element of a graph, color can be used unethically. For example,
because bright colors attract the eye, they can be used to distract the reader's focus
from the most important point. The left-hand graph in Figure 13.15 uses a bright
line to deemphasize the crucial bad news by emphasizing relatively unimportant
news. The right-hand graph shows how the problem can be corrected.

Try This

Using your favorite search
engine, search for "misleading graphs." Can you find one
that seems to be misleading
through carelessness, one that
the creator appears to have
made to manipulate readers,
and one that is called misleading but wouldn't really
cause anyone to draw an
inaccurate conclusion?

FIGURE 13.15
Using Color Ethically

The left-hand graph misleads
readers if Trend B represents the
crucial information.

The right-hand graph shows how
to use color to avoid misleading.

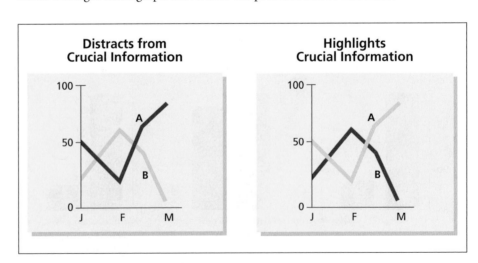

CONCLUSION

Graphics can greatly increase the clarity and impact of your written communications. To use graphics well, you need to follow the same reader-centered strategy that you use when writing your prose: Think about the tasks your readers will perform while reading and think about the ways you want your communication to shape your readers' attitudes. Doing so will enable you to decide where to use graphics, determine the most effective types of graphics to use, make them easy to understand and use, and integrate them successfully with your prose.

Figure 13.16 provides a guide you can use when designing your graphics. The Writer's Reference Guide that follows this chapter supplements this chapter's general advice by providing detailed information about how to construct eleven types of graphics that are often used at work.

USE WHAT YOU'VE LEARNED

EXERCISE YOUR EXPERTISE

1. Create an outline for an assignment in your course. On the finished outline, list each place where a graphic could make your communication more usable and persuasive in your reader's eyes. For each place, identify the type of graphic that will be most effective in helping you achieve your communication objectives.

2. Using a desktop publishing program, make a table that displays the number of hours in a typical week that you spend in each of several major activities, such as attending class, studying, eating, and visiting with friends or family. The total number of hours should equal 168. Be sure to include a brief heading for the columns and rows. Next, use your desktop publishing program to convert your table to a graph or chart. Refine your table so that the graph is as easy to read as possible. For assistance in using these features of your program, use the program's Help feature. Alternatively, do this exercise using a different kind of data.

3. Using a desktop publishing program's drawing feature, create a simple illustration of a piece of equipment used in your major or in one of your hobbies or activities. For assistance in using the drawing feature of your program, use the program's Help feature.

Note: Exercises for developing your expertise in applying this chapter's guidelines when creating specific types of graphics are available at the book's website, www.cengage.com/english/anderson7e.

EXPLORE ONLINE

Browse the web for the sites of companies that provide products or services related to your major. Find sites for several companies in your country and several in a country located in a very different part of the world. For instance, if you are in Asia, look at sites in the Middle East, Africa, Europe, or North America. Using the guidelines in this chapter, compare the graphics in the two sets of websites. Although you will have looked at too few websites to make any broad generalizations about the use of graphics in the two countries, what hypotheses can you form about their similarities and differences?

COLLABORATE WITH YOUR CLASSMATES

1. Team up with another student. First, each of you is to locate two different types of graphics in journals, textbooks, or other sources in your field (including websites). Working together, evaluate each of the graphics from the perspective of the guidelines in this chapter. What features of the designs are effective? How could each graphic be improved?

2. Working with another student, exchange drafts of a project in which you are both using one or more graphics. Using the guidelines in this chapter, review the graphics in the drafts. What features are most effective in achieving the writer's usability and persuasive goals? How could the graphics be made more effective from the target readers' perspective?

APPLY YOUR ETHICS

Graphs and various types of charts appear frequently in the newspapers and popular magazines. Occasionally, they are criticized for presenting data in a misleading fashion. Find a graphic that you feel may mislead readers. Explain the reason for your assessment and tell how the graphic could be redesigned. Also, identify persons who might be adversely affected or who might unfairly benefit if some action were taken on the basis of the misinterpretation. Present your results in the way your instructor requests.

FIGURE 13.16   Writer's Guide for Creating Graphics

 To download a copy of this Writer's Guide, go to www.cengage.com/english/anderson7e and click on Chapter 13.

Writer's Guide
CREATING GRAPHICS

Planning

1. Identify places where graphics will increase your communication's usability.
2. Identify places where graphics will increase your communication's persuasiveness.

Selecting

1. Select the types of graphics that will best support your readers' tasks.
2. Select the types of graphics that will effectively influence your readers' attitudes.

Designing

1. Design graphics that are easy to understand and use.
2. Design them to support your readers' tasks.
3. Design graphics that your readers will find persuasive.
4. Keep your graphics simple enough for easy use.
5. Label content clearly.
6. Provide your graphics with informative titles.

Using Color

1. Use colors to support your message.
2. Use color for emphasis, not decoration.
3. Choose a color scheme, not just individual colors.
4. Provide high contrast between text and background.
5. Select colors with appropriate associations.
6. Limit the number of colors.
7. Use color to unify the overall communication.

Integrating with the Text

1. Introduce each graphic in the text.
2. Tell your readers the conclusions you want them to draw.
3. Provide all explanations your readers will need in order to understand and use each graphic.
4. Locate each graphic near the references to it.

Addressing an International Audience

1. Check your graphics with persons from the other nations.

Using Graphics Ethically

1. Avoid elements that might mislead your readers.
2. Obtain permission from the copyright owner of each image that is not in the public domain.

CREATING ELEVEN TYPES OF READER-CENTERED GRAPHICS

This guide provides detailed reader-centered advice for creating eleven types of graphics that are widely used on the job. The table below will help you choose the type that is best suited to your purpose.

WHEN YOU WANT TO HELP YOUR READERS . . .	BEST GRAPHICS	PAGE
Tables and Graphs		
■ Find and use data, facts, or advice	Table	358
■ Understand the relationships among variables	Line graph	360
■ Compare quantities	Bar graph	362
	Pictograph	364
■ See a trend	Line graphs	360
	Bar graph	362
■ See the relative sizes of the parts that make up a whole	Pie chart	366
Images		
■ See how something looks	Photograph	368
	Drawing	370
	Screen shot	372
■ Understand how something is constructed	Photograph	368
	Drawing	370
■ Understand a process	Flowchart	374
■ Understand how to do something	Photograph	368
	Drawing	370
	Screen shot	372
Management Graphics		
■ Understand the structure of an organization	Organizational chart	376
■ Understand the schedule for completing a project	Schedule chart	377

Tables

Use: To help your readers find data or other information rapidly

Creating a Table

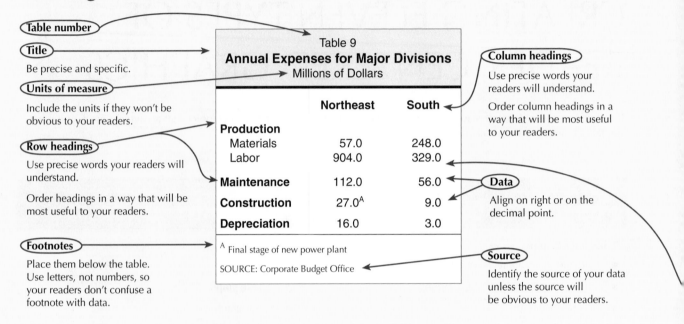

Table number

Title
Be precise and specific.

Units of measure
Include the units if they won't be obvious to your readers.

Row headings
Use precise words your readers will understand.

Order headings in a way that will be most useful to your readers.

Footnotes
Place them below the table. Use letters, not numbers, so your readers don't confuse a footnote with data.

Table 9
Annual Expenses for Major Divisions
Millions of Dollars

	Northeast	South
Production		
Materials	57.0	248.0
Labor	904.0	329.0
Maintenance	112.0	56.0
Construction	27.0[A]	9.0
Depreciation	16.0	3.0

[A] Final stage of new power plant

SOURCE: Corporate Budget Office

Column headings
Use precise words your readers will understand.

Order column headings in a way that will be most useful to your readers.

Data
Align on right or on the decimal point.

Source
Identify the source of your data unless the source will be obvious to your readers.

INFORMAL TABLES

Informal tables flow right into the text. They are useful when the preceding sentence tells what the table is about and when the interpretation of the data is obvious.

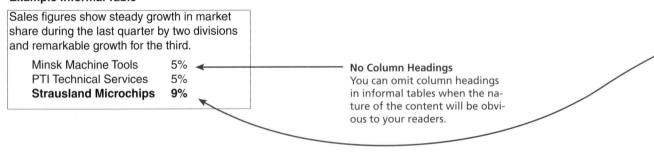

Example Informal Table

Sales figures show steady growth in market share during the last quarter by two divisions and remarkable growth for the third.

Minsk Machine Tools	5%
PTI Technical Services	5%
Strausland Microchips	**9%**

No Column Headings
You can omit column headings in informal tables when the nature of the content will be obvious to your readers.

Learn More at the Website

To see more sample tables, go to www.cengage.com/english/anderson7e and choose "Writer's Reference Guide: Creating Eleven Types of Reader-Centered Graphics." The website also includes tips on creating tables in Microsoft Word.

TABLES WITH TEXT

Compared with paragraphs, tables can present many kinds of textual information in a manner that is much easier for readers to understand and use.

Lawn Mower Troubleshooting		
Problem	**Cause**	**Action**
Engine fails to start	A Blade control handle disengaged	A Engage blade control handle.
Hard starting or loss of power	A Spark plug wire loose B Carburetor improperly adjusted	A Connect and tighten spark plug wire. B Adjust carburetor.

Comparisons of Foods		
Food	**Calories**	**Fat**
Apple	101	0
Apple Pop Tart	191	2.9 g

Aligning text in columns
Align text on the left or center it within the column.

TIPS FOR CREATING READER-CENTERED TABLES

- Use extra space or draw horizontal lines to guide your readers' eyes across rows or groups of rows.
- Make key information stand out with bold, color, highlighting.
- Sort row and column headings to help readers find the information they want.

Unsorted	**Sorted**
Fruits	Nutritious Foods
Sweets	Fruits
Legumes	Grains
Grains	Legumes
Fried products	Nonnutritious foods
	Fried products
	Sweets

Group related items under a common heading.

Arrange headings and sub-headings in an order that will help the readers find what they want. In this example, alphabetizing is used for subheadings. The heading "Nutritious Foods" is placed before "Nonnutritious Foods" for emphasis.

- Avoid tables that are too large for readers to use easily.
  - · Include only what your readers need.
  - · If the table is still large, divide it into two or more separate tables.

Line Graphs

Creating a Line Graph

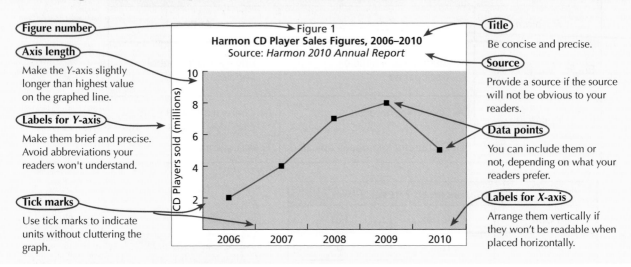

Figure number

Axis length
Make the Y-axis slightly longer than highest value on the graphed line.

Labels for Y-axis
Make them brief and precise. Avoid abbreviations your readers won't understand.

Tick marks
Use tick marks to indicate units without cluttering the graph.

Title
Be concise and precise.

Source
Provide a source if the source will not be obvious to your readers.

Data points
You can include them or not, depending on what your readers prefer.

Labels for X-axis
Arrange them vertically if they won't be readable when placed horizontally.

Figure 1
Harmon CD Player Sales Figures, 2006–2010
Source: *Harmon 2010 Annual Report*

LINE GRAPHS COMPARING TRENDS

By using two or more lines to represent a quantity over time, you enable your readers to compare overall trends.

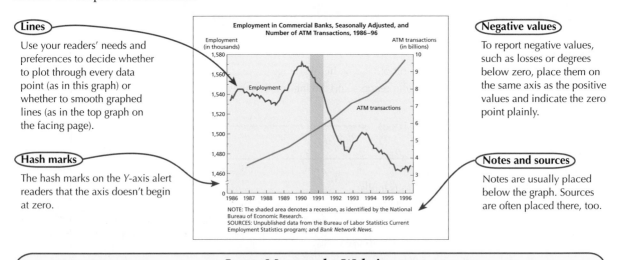

Lines
Use your readers' needs and preferences to decide whether to plot through every data point (as in this graph) or whether to smooth graphed lines (as in the top graph on the facing page).

Hash marks
The hash marks on the Y-axis alert readers that the axis doesn't begin at zero.

Negative values
To report negative values, such as losses or degrees below zero, place them on the same axis as the positive values and indicate the zero point plainly.

Notes and sources
Notes are usually placed below the graph. Sources are often placed there, too.

Employment in Commercial Banks, Seasonally Adjusted, and Number of ATM Transactions, 1986–96

NOTE: The shaded area denotes a recession, as identified by the National Bureau of Economic Research.
SOURCES: Unpublished data from the Bureau of Labor Statistics Current Employment Statistics program; and *Bank Network News*.

Learn More at the Website

For instructions on using Excel to create line graphs, go to www.cengage.com/english/anderson7e and choose "Writer's Reference Guide: Creating Eleven Types of Reader-Centered Graphics."

LINE GRAPHS SHOWING INTERACTIONS AMONG VARIABLES

In technical and scientific communications, your readers may want to see the interactions of several variables.

Labels for axes

Make the labels precise. In this graph, the writers used highly technical labels appropriate for the scientists who are their readers.

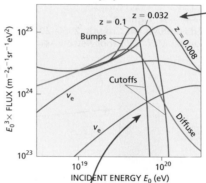

Expected Effects of GZK Cutoff on the Cosmic-Ray Spectrum

Placement of labels

By carefully placing the labels, you help your readers identify the lines in a complex graph.

Color of lines

Use colors that contrast enough with the background to be seen easily.

GRAPHS SHOWING DATA POINTS ONLY

In some situations, readers want to see the individual data points rather than graphed lines. Or, they may want both.

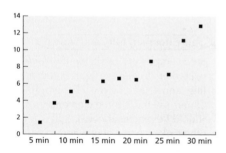

TIPS FOR CREATING READER-CENTERED LINE GRAPHS

■ Use different colors to enable readers to distinguish readily among the lines. If you can't use color, use different styles for the lines—dashes for one, dots for another, and so on.

■ If possible, begin lines at zero to avoid misleading readers.

■ If it isn't practical to start the lines at zero, use hash marks to signal this fact to your readers.

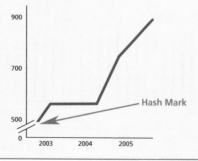

Bar Graphs

Creating a Bar Graph

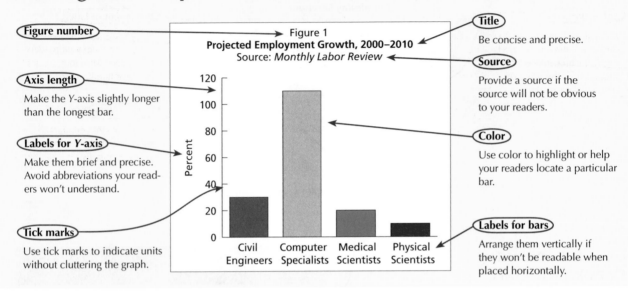

Figure number

Axis length

Make the *Y*-axis slightly longer than the longest bar.

Labels for *Y*-axis

Make them brief and precise. Avoid abbreviations your readers won't understand.

Tick marks

Use tick marks to indicate units without cluttering the graph.

Figure 1
Projected Employment Growth, 2000–2010
Source: *Monthly Labor Review*

Title

Be concise and precise.

Source

Provide a source if the source will not be obvious to your readers.

Color

Use color to highlight or help your readers locate a particular bar.

Labels for bars

Arrange them vertically if they won't be readable when placed horizontally.

BAR GRAPHS SHOWING TRENDS

By using a series of bars to represent a quantity over time, you enable your readers to see an overall trend.

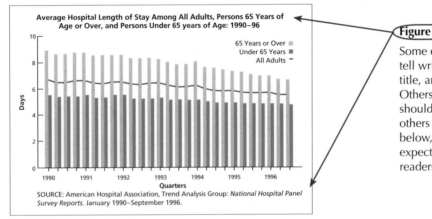

Figure number, title, and source

Some employers and scientific fields tell writers to place a figure number, title, and source below a bar graph. Others say these three elements should be placed above it. Still others place some above and some below, as in this example. Learn the expectations of your employers and readers.

Learn More at the Website

For instructions on using Excel to create bar graphs, go to www.cengage.com /english /anderson7e and choose "Writer's Reference Guide: Creating Eleven Types of Reader-Centered Graphics."

MULTIBAR GRAPHS SHOWING SEVERAL COMPARISONS

With a multibar graph, you help your readers compare the same groups on many topics (e.g., male and female participants in each of several sports activities).

Label for bars

When you are using horizontal bars, you can line up their labels using the last letter of each word rather than the first letter.

Key to bars

When the bars are repeated, use a key to label them. Otherwise, place labels next to the bars, as in the top graph on the opposite page.

Grid lines

Use grid lines when tick marks aren't enough to enable readers to readily gauge each bar's length.

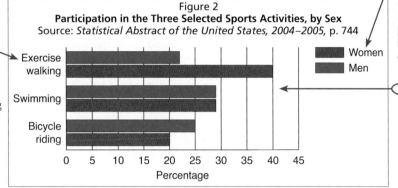

Figure 2
Participation in the Three Selected Sports Activities, by Sex
Source: *Statistical Abstract of the United States, 2004–2005*, p. 744

TIPS FOR CREATING READER-CENTERED BAR GRAPHS

- Arrange bars in the order that your readers will find most helpful. Alternatives include arranging them alphabetically, chronologically, or from longest to shortest.
- If possible, begin bars at zero to avoid misleading your readers.
- If it isn't practical to start the bars at zero, use hash marks to signal this fact to your readers.

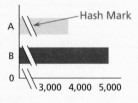

MORE TYPES OF BAR GRAPHS

Bar graphs come in many other varieties.

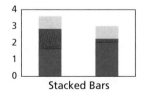

Stacked Bars

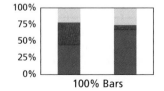

100% Bars

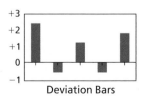

Deviation Bars

Pictographs

Creating a Pictograph

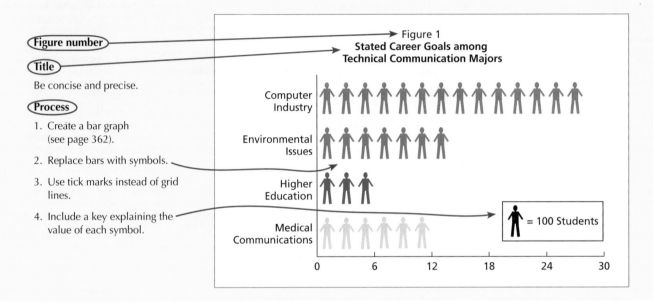

Figure number

Title

Be concise and precise.

Process

1. Create a bar graph (see page 362).

2. Replace bars with symbols.

3. Use tick marks instead of grid lines.

4. Include a key explaining the value of each symbol.

Figure 1
Stated Career Goals among Technical Communication Majors

- Computer Industry
- Environmental Issues
- Higher Education
- Medical Communications

0   6   12   18   24   30

= 100 Students

WHEN TO USE PICTOGRAPHS

Pictographs help you accomplish the following goals:

- Emphasize the practical impact of the data you represent. For example, using silhouettes of people to represent the workers who will be hired at a new factory emphasizes a benefit the factory will bring to the community.

- Make your data visually interesting and memorable. Visual interest is especially important when you are addressing the general public.

In some situations, readers expect a more abstract representation of information and would consider pictographs to be inappropriate. For example, pictographs are rarely used in technical and scientific reports. Be sure to check your readers' expectations.

Learn More at the Website

You can obtain symbols free from online clipart sites. Go to www.cengage.com /english /anderson7e and choose "Writer's Reference Guide: Creating Eleven Types of Reader-Centered Graphics."

AVOIDING DISTORTION

To avoid distortion, represent larger quantities with more symbols, not larger ones.

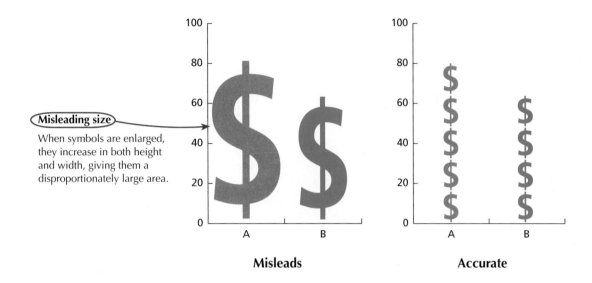

Misleading size

When symbols are enlarged, they increase in both height and width, giving them a disproportionately large area.

Misleads

Accurate

TIPS FOR CREATING READER-CENTERED PICTOGRAPHS

- Choose symbols that your readers will readily associate with the topic of your pictograph.
- Keep the symbols simple. Unneeded details distract readers and clutter the pictograph.

Correct

Too much detail

Pie Charts

Creating a Pie Chart

Process

1. Draw a circle.

2. Divide the circle into wedges proportional to each item's percentage of the whole.

3. Label each wedge; include the percentage in each label.

4. Use a different color or shading for each wedge.

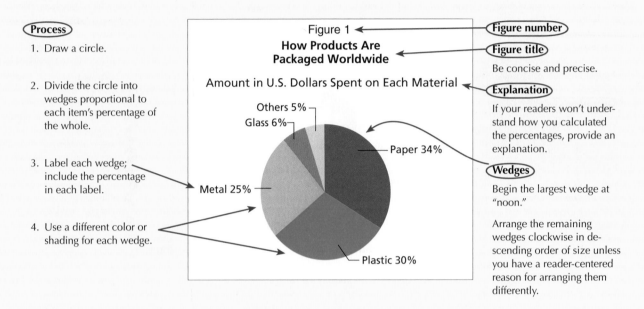

Figure 1

How Products Are Packaged Worldwide

Amount in U.S. Dollars Spent on Each Material

Others 5%
Glass 6%
Paper 34%
Metal 25%
Plastic 30%

Figure number

Figure title

Be concise and precise.

Explanation

If your readers won't understand how you calculated the percentages, provide an explanation.

Wedges

Begin the largest wedge at "noon."

Arrange the remaining wedges clockwise in descending order of size unless you have a reader-centered reason for arranging them differently.

AVOIDING DISTORTION

Stick with two-dimensional pie charts. Three-dimensional ones can create a visual distortion.

Visual distortion

The 30% wedge looks largest because the reader sees both the top and the side of the slice. However, the 35% slice represents a larger portion of the chart.

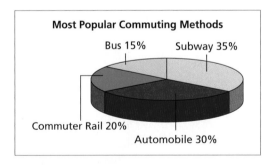

Most Popular Commuting Methods

Bus 15%   Subway 35%

Commuter Rail 20%

Automobile 30%

Learn More at the Website

For instructions on using Excel to create pie charts, go to www.cengage.com /english /anderson7e and choose "Writer's Reference Guide: Creating Eleven Types of Reader-Centered Graphics."

EMPHASIZING A PARTICULAR WEDGE

You can emphasize a particular wedge in two ways:

• Use a contrasting color that is much lighter or darker than the other colors.

• Pull a wedge out of the pie.

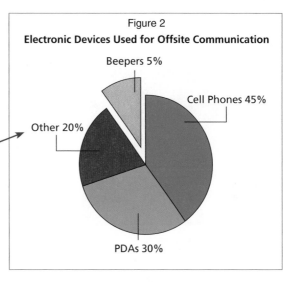

Figure 2
Electronic Devices Used for Offsite Communication

Beepers 5%

Cell Phones 45%

Other 20%

PDAs 30%

TIPS FOR CREATING READER-CENTERED PIE CHARTS

■ Be sure that your wedges add up to 100%.
■ Limit the number of wedges to eight or fewer.
■ Create an "Other" wedge if you have several small quantities whose sizes are similar. If your readers will wonder what is included in the "Other" category, explain in a footnote.

Photographs

Uses: To show readers how to perform a task, locate an object,
or see how something looks

Creating a Photograph

HELPING YOUR READERS DO SOMETHING

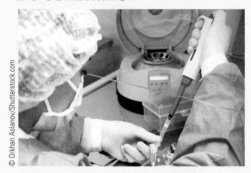

© Orkhan Aslanov/Shutterstock.com

Take your readers' point of view
When photographing for instructions, include all the
details your readers will find helpful—and trim away
the rest. Take photos from the same angle of view that
your readers will have.

HELPING YOUR READERS LOCATE SOMETHING

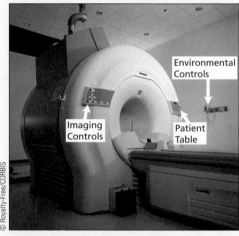

Environmental Controls

Imaging Controls

Patient Table

© Royalty-Free/CORBIS

Provide context and pointers
In addition to the item your readers want to find, show enough
of the surrounding area so they know the general place
in which to search. Arrows and labels are very helpful
to readers.

HELPING YOUR READERS SEE HOW SOMETHING LOOKS

Display condition
Photographs often surpass words
in showing readers the condition
of an object.

Show identifying marks
Photographs help readers identify
animals, plants, and other objects.

Focus on what matters most
Fill your image with the details most
useful or persuasive to your readers.

© Sascha Burkard/Shutterstock.com

Red-Eyed Leaf Frog

© tkachuk/Shutterstock.com

Condition of *Eastern Pegasus*

Learn More at the Website

For instructions on using desktop publishing software to crop, label, and sharpen photos, go to www.cengage.com/english/
anderson7e and choose "Writers Reference Guide: Creating Eleven Types of Reader-Centered Graphics."

USING SOFTWARE TO CREATE READER-CENTERED PHOTOGRAPHS

Using ordinary desktop publishing software, you can increase the effectiveness of the photographs you have scanned or taken with a digital camera. The following examples use features available in Microsoft Word.

CROPPING

By cropping (or trimming) a photograph, you eliminate unnecessary detail and enlarge the items you want your readers to see.

The Crop **tool**

To crop a photograph, click on it and then on the **Crop** tool.

- In Word 2007, the **Crop** tool is at the right-hand end of the ribbon.

- In Word 2008 for Mac, the **Crop** tool is under **Picture** in the **Formatting Palette**.

Drag the thick black lines from the corners.

LABELING

Label the items that are significant to your readers.

The Text **Box tool**
To create a label, use the **Text Box** and **Line** tools. They are under the **Insert** tab in Word 2007 and on the **Drawing** toolbar in Word 2008 for Mac.

Typhoons Mindulle and Tingting spin
side-by-side off the coast of China
(Courtesy of NASA)

Label placement
Labels can be placed inside or outside a photograph.

Position them so they do not cover important parts of the image.

Typhoons Mindulle and Tingting spin
side-by-side off the coast of China
(Courtesy of NASA)

TIPS FOR CREATING READER-CENTERED PHOTOGRAPHS

- Plan your photographs carefully in advance, considering the angle of view and removing items that would clutter the picture.
- Include a person or familiar object in a photo if this will help your readers understand the size of the object you are photographing.

Drawings

Uses: To show how to do something or how something looks or is constructed

Creating a Drawing

Process

1. Identify the features that are most important for your readers to see.

2. Choose the angle of view that will help your readers most.

3. Focus your readers' attention on the key features by using color, heavier lines, and labels.

4. Eliminate unnecessary and distracting details.

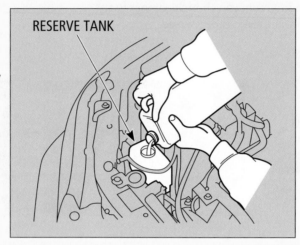

RESERVE TANK

About this drawing

This drawing helps car owners locate the reserve tank for radiator fluid.

- A label identifies the tank.

- The tank, container, and arms are highlighted by the use of heavy lines. Lighter lines are used for everything else.

- The shading of everything else also highlights the tank, container, and arms.

- The angle of view is close to the one that owners would have when filling the tank.

DRAWINGS THAT SHOW HOW TO DO SOMETHING

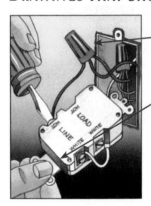

Instructional drawings

This drawing shows how the wires will look when a reader has pigtailed them as described in the caption.

It shows the reader exactly how to hold the switch, using the same angle of vision that the reader would have when performing this step.

The caption provides additional detail.

> ### TIPS FOR CREATING READER-CENTERED DRAWINGS
>
> - Use a perspective that shows three sides of an object (e.g., front, top, side) unless you have a reader-centered reason for using another view. In most cases, the three-sided perspective gives readers the most information about the item being drawn.
>
> - Labels can be placed in the drawing or next to it.

2. Pigtail all the block wires together and connect them to the terminal marked HOT LINE on the GFCI.

Learn More at the Website

To see more sample drawings, go to www.cengage.com /english /anderson7e and choose "Writer's Reference Guide: Creating Eleven Types of Reader-Centered Graphics." The website also includes tips on creating drawings in Microsoft Word.

DRAWINGS SHOWING APPEARANCE AND STRUCTURE

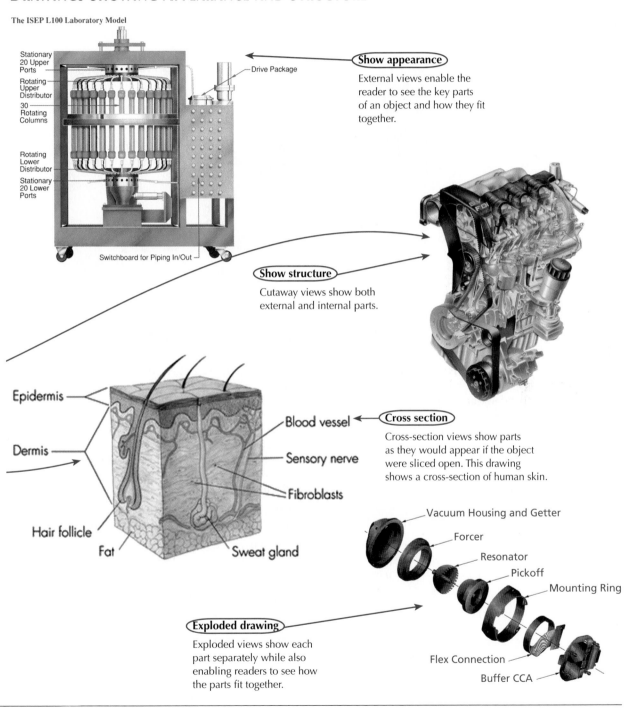

The ISEP L100 Laboratory Model

Stationary 20 Upper Ports

Rotating Upper Distributor

30 Rotating Columns

Rotating Lower Distributor

Stationary 20 Lower Ports

Drive Package

Switchboard for Piping In/Out

Show appearance
External views enable the reader to see the key parts of an object and how they fit together.

Show structure
Cutaway views show both external and internal parts.

Epidermis

Dermis

Hair follicle

Fat

Blood vessel

Sensory nerve

Fibroblasts

Sweat gland

Cross section
Cross-section views show parts as they would appear if the object were sliced open. This drawing shows a cross-section of human skin.

Vacuum Housing and Getter

Forcer

Resonator

Pickoff

Mounting Ring

Flex Connection

Buffer CCA

Exploded drawing
Exploded views show each part separately while also enabling readers to see how the parts fit together.

Screen Shots

Creating a Screen Shot

Using programs included on most operating systems, you can easily make the following types of screen shots:

1. Full screen

2. A window

3. Part of a window or screen

CREATING A SCREEN SHOT IN WINDOWS:
1. To take a full-screen shot, press **Print Screen**.
2. To capture a window's image, press **Print Screen** and **Alt** simultaneously.
3. Paste the image into your document.
4. Click on the image.
5. Click on the **Format** tab.
6. Click on **Crop**.
7. Drag the black lines from the corners.

CREATING A SCREEN SHOT IN MAC OS:
1. To take a full-screen shot, press these keys simultaneously: **Command**, **Shift**, and **3**.
2. To take a cropped screen shot
   - Press these keys simultaneously: **Command**, **Shift**, and **4**.
   - Drag the cursor around the **desired area.**
3. Place the image in your document by pulling down the **Insert** Menu and choosing **Picture>From File**.

TIPS FOR CREATING READER-CENTERED SCREEN SHOTS

- Be sure that your screen shots are large enough to enable readers to read the words and recognize the icons—unless reading these items is not important for their purpose.
- Use arrows, colored borders, or other visual cues to help readers locate items.
- If your readers are supposed to fill in fields in a window, fill in the fields in your screen shot (instead of leaving the fields blank).
- To present directions compactly, you can sometimes overlap screen shots.
- See page 349 for guidelines concerning copyright and screen shots.

HELPING READERS LOCATE AN ON-SCREEN ITEM

Show a sufficient area of a screen or window to let the reader know where to look for the item.

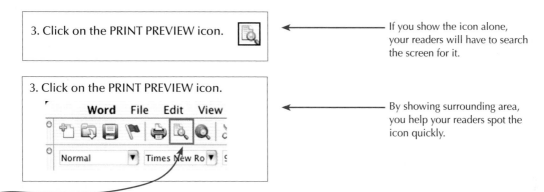

3. Click on the PRINT PREVIEW icon.

⟵ If you show the icon alone, your readers will have to search the screen for it.

3. Click on the PRINT PREVIEW icon.

Word File Edit View

Normal ▼ Times New Ro ▼

⟵ By showing surrounding area, you help your readers spot the icon quickly.

GUIDING YOUR READERS THROUGH A SEQUENCE OF STEPS

In some cases, you can overlap screen shots to guide your readers through a series of steps, as this example shows.

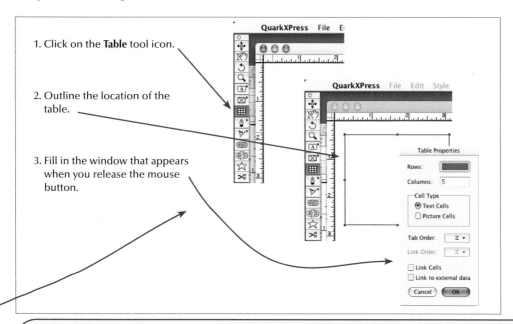

1. Click on the **Table** tool icon.

2. Outline the location of the table.

3. Fill in the window that appears when you release the mouse button.

QuarkXPress File E

QuarkXPress File Edit Style

Table Properties

Rows:

Columns: 5

Cell Type
◉ Text Cells
○ Picture Cells

Tab Order: Z ▾

Link Order: Z ▾

☐ Link Cells
☐ Link to external data

Cancel OK

Learn More at the Website

To learn more about taking screen shots, go to www.cengage.com /english /anderson7e and choose "Writer's Reference Guide: Creating Eleven Types of Reader-Centered Graphics."

Flowcharts

Use: To help your readers understand the steps in a process or procedure

Creating a Flowchart

Process

1. List the steps in the process.
2. Create a symbol or drawing for each type of step.
3. Label each step.
4. Arrange symbols from left to right, top to bottom, or in a circle.

Symbols

The symbols in the left-hand chart carry a specific meaning to readers in many fields:

☐ = process or activity

◇ = decision

Arrows

Arrows show the direction of activity.

Title

The title may be at the bottom or top.

Drawings

In the right-hand chart, each drawing is an image readers will associate with the activity represented.

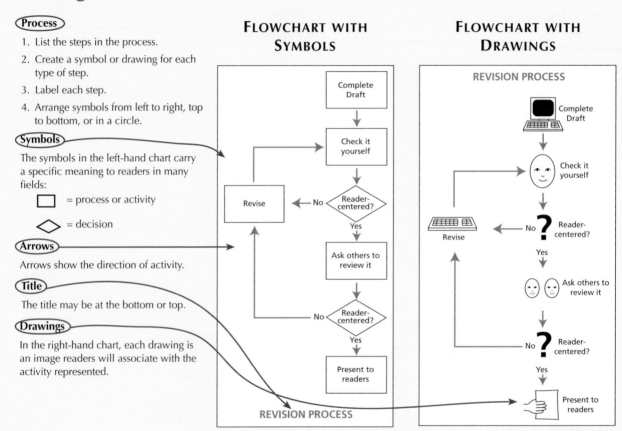

FLOWCHART WITH SYMBOLS

Complete Draft → Check it yourself → Reader-centered? —No→ Revise; Yes ↓ Ask others to review it → Reader-centered? —No→ ; Yes ↓ Present to readers

REVISION PROCESS

FLOWCHART WITH DRAWINGS

REVISION PROCESS

Complete Draft → Check it yourself → Reader-centered? —No→ Revise; Yes ↓ Ask others to review it → Reader-centered? —No→ ; Yes ↓ Present to readers

TIPS FOR CREATING READER-CENTERED FLOWCHARTS

- Choose between symbols and drawings by considering the knowledge and expectations of your readers.
- Make symbols and labels large enough for your readers to see plainly.
- When writing to readers who work in a field that uses specialized flowchart symbols, use the symbols in your flowchart.

Learn More at the Website

To learn how to create flowcharts using Word, go to www.cengage.com /english /anderson7e and choose "Writer's Reference Guide: Creating Eleven Types of Reader-Centered Graphics." The website also includes more sample flowcharts.

FLOWCHART WITH DRAWINGS

In this example, the flowchart and caption work together to help readers understand the carbon cycle.

Circular layout

Because the diagram portrays a cycle that has no beginning or end, the writers arranged the steps in a circle.

Caption

The caption helps readers understand the process illustrated in the drawing.

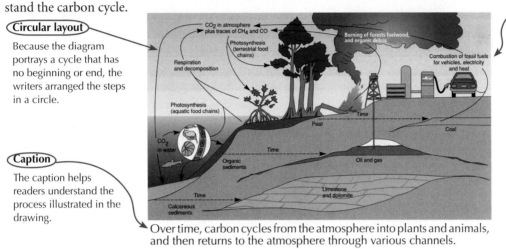

Labels

The writers used a variety of drawings to symbolize the various elements in the carbon cycle.

Over time, carbon cycles from the atmosphere into plants and animals, and then returns to the atmosphere through various channels.

FLOWCHART WITH SPECIALIZED SYMBOLS

Using this flowchart, a computer consulting company described the general structure of a computer system it proposed to create for a buying club. Members pay a fee to join and then order products online or by phone.

Symbol shapes

The writers used different shapes for different types of elements in the system: rectangles for servers and squares with rounded corners for other elements.

Colors

The writers used different colors to group related elements visually.

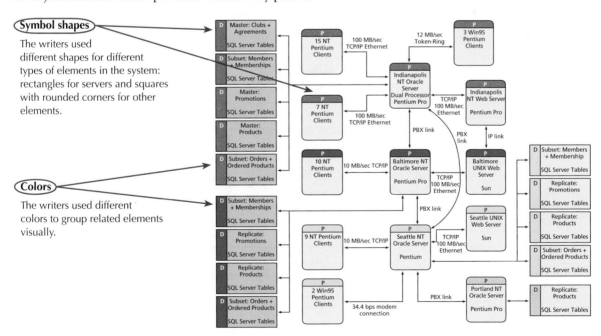

Organizational Charts

Uses: To help readers understand the scope and arrangement of an organization

To help readers understand the formal lines of authority and responsibility in an organization

Purpose

This chart's purpose is to help readers understand the structure of the units in this organization. The chart at the bottom of this page has a different purpose.

Arrangement of boxes

All boxes at the same level have the same size, shape, and color.

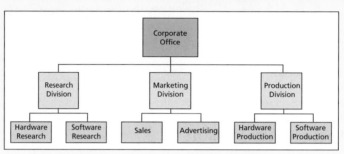

Additional Information

Depending on how readers will use the chart, the boxes may include additional information, such as the name of the person in charge of each unit, the number of employees in the unit, or the unit's central phone number.

ORGANIZATIONAL CHART SHOWING REPORTING LINES

Purpose

This chart's purpose is to help readers understand the lines of authority and responsibility in the organization. The chart at the top of this page has a different purpose.

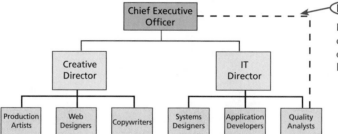

Dotted lines

Dotted lines indicate consulting relationships or secondary reporting lines.

Learn More at the Website

To learn how to make an organizational chart quickly in Word, go to www.cengage.com /english /anderson7e and choose "Writer's Reference Guide: Creating Eleven Types of Reader-Centered Graphics."

Schedule Charts

Creating a Schedule Chart

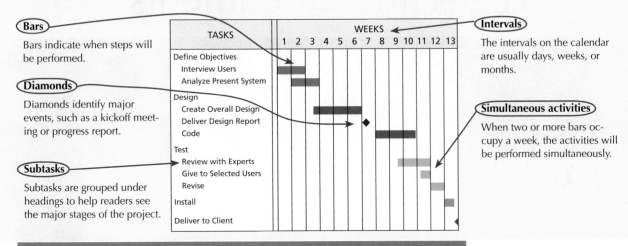

Bars

Bars indicate when steps will be performed.

Diamonds

Diamonds identify major events, such as a kickoff meeting or progress report.

Subtasks

Subtasks are grouped under headings to help readers see the major stages of the project.

Intervals

The intervals on the calendar are usually days, weeks, or months.

Simultaneous activities

When two or more bars occupy a week, the activities will be performed simultaneously.

TIPS FOR CREATING READER-CENTERED SCHEDULE CHARTS

- Adjust the amount of detail to the needs and expectations of your readers.
- Use color to group the bars for related steps, a different color for each group.

GANTT CHARTS

Gantt charts are detailed schedule charts used to plan and monitor progress on complex projects. Usually created using special computer programs, they may involve hundreds or thousands of tasks. However, you can adopt some of their conventions in simpler schedule charts. Gantt charts are named for Henry Gantt, the early twentieth-century engineer who invented them.

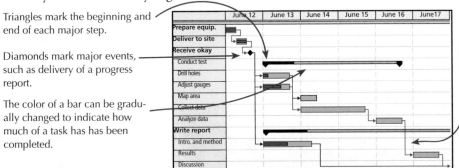

Triangles mark the beginning and end of each major step.

Diamonds mark major events, such as delivery of a progress report.

The color of a bar can be gradually changed to indicate how much of a task has has been completed.

Bars linked by red lines indicate the "critical path" in a project, which is the sequence of steps that must be completed in the time specified for the entire project to be finished by its deadline.

Learn More at the Website

To learn how to make a schedule chart quickly in Word, go to www.cengage.com /english /anderson7e and choose "Writer's Reference Guide: Creating Eleven Types of Reader-Centered Graphics."

14 | Designing Reader-Centered Pages and Documents

GUIDELINES

Y ou build your communications out of *visual* elements: the dark marks of your words, sentences, and paragraphs against the light background of the page, as well as your drawings and graphs and tables. Your readers *see* the visual design of these elements before they read and understand your message. And what they see has a powerful effect on the success of your communications, on its usability and persuasiveness.

DEFINING OBJECTIVES	
PLANNING	
CONDUCTING RESEARCH	
DRAFTING	
REVISING	

CHAPTER **14**

Here, for example, are some of the ways that good design enhances usability.

- **Good design helps readers understand your information.** For example, you can use visual design to signal the hierarchy of ideas and information in a report. This helps your readers understand what you are saying and what its significance is to them. Similarly, when you write instructions, you can place a direction and a figure next to each other to indicate that the two work together to explain a step.

- **Good page design helps readers locate information.** At work, readers often want to find part of a communication without reading all of it. With headings and other design elements, you can help them do that quickly.

- **Good design helps readers notice highly important content.** With good design, you can emphasize for readers the content that is especially important to them, such as a warning in a set of instructions or a list of actions to take in a recommendation report.

Good design increases usability.

Here are some of the ways good design affects readers' attitudes, thereby increasing a communication's persuasiveness.

- **Good design encourages readers to feel good about the communication itself.** You've surely seen pages—perhaps in a textbook, set of instructions, or website—that struck you as uninviting, even ugly. As a result, you may have been reluctant to read them. And undoubtedly you've seen other printed or online pages that you found attractive, so that you approached them eagerly and receptively. Good design has the same impact on readers of work-related communications. It increases readers' willingness to read reports, proposals, and similar documents carefully, and it extends their willingness to read instructions and websites at all.

- **Good design encourages readers to feel good about the communication's subject matter.** The impact of design on readers' attitudes toward subject matter was dramatically demonstrated by researcher Karen Schriver (1997), who asked people to comment on two sets of instructions for a microwave oven. The sets differed only in their page design. When people commented on the design they preferred, they also said that the oven—which they had never seen—was easy to use. No matter what you are writing about, your page design can influence—for better or worse—your readers' attitudes toward your subject.

Good design increases persuasiveness.

© Gary Conner/Index Stock Imagery/PhotoLibrary

A READER-CENTERED APPROACH TO DESIGN

WWW

To read additional information, see more examples, and access linksrelated to this chapter's guidelines, go to Chapter 14 at **www.cengage. com/english/anderson7e**.

Because page design can have such a significant impact on your communication's usability and persuasiveness, you should approach design in the same reader-centered manner that you use when drafting text and graphics: Think continuously about your readers, including who they are, what they want from your communication, and the context in which they will be reading it.

This chapter's guidelines will lead you through a step-by-step process for creating effective, reader-centered page designs. The chapter also includes instructions for applying these guidelines using a word processing program.

DESIGN ELEMENTS OF A COMMUNICATION

As you read and work with this chapter's guidelines, it will be helpful to think about the building blocks of a page design in the way that professional graphic designers do. When they look at a page, they see six basic elements.

The six design elements of a page

- **Text.** Paragraphs and sentences.
- **Headings and titles.** Labels for sections of your communication.
- **Graphics.** Drawings, tables, photographs, and so on—including their captions.
- **White space.** Blank areas.
- **Headers and footers.** The items, such as page numbers, that occur at the top or bottom of each page in a multipage document.
- **Physical features.** These include paper, which may take many shapes and sizes, and bindings, which come in many forms.

Learn More

For information on the visual design of web pages and websites, go to Chapter 20, page 501.

The Writer's Tutorial on pages 382–383 displays a small sample of the numerous ways you can arrange these six elements as you create designs that will help you achieve your communication objectives.

This chapter's guidelines for working with the six visual elements are just as valid for pages on the web as for pages on paper. Chapter 20 explains their application to screens and also provides additional advice about designing web pages and websites.

Guideline 1 | **Begin by considering your readers and purpose**

Start your design work in the same way you begin plans for any other aspect of your communication. Review your communication's objectives so that you have the following considerations clearly in mind.

- **Who your readers are.** Review their needs, attitudes, and expectations. All should influence your design decisions.

- **What tasks your communication should enable readers to perform.** You need to understand your readers' tasks if you are going to create a highly usable design that will increase the ease with which they can perform these tasks. For example, if your readers are going to use your communication as a reference work from which they will seek only specific pieces of information, you will need to design it differently than if they will read it straight through.
- **How you want to influence readers' attitudes.** To design pages that will increase your communication's persuasiveness, you need to define the ways you want to affect your readers' attitudes and you need to understand the goals, values, preferences, and other factors that will shape their responses to your communication.

Guideline 2 | Create a grid to serve as the visual framework for your pages

When designing a page, your aim is to create simple, consistent, attractive, and meaningful relationships among the text, graphics, white space, headings and titles, and headers and footers.

The first step is to decide where each of these elements will go on the page. For this purpose, graphic designers draw a grid of vertical and horizontal lines that provides the framework for their pages. They place text, graphics, and other visual elements in the rectangles, called *cells,* formed by the grid.

The simplest grid is one you know well: the grid used for most college papers, as illustrated here.

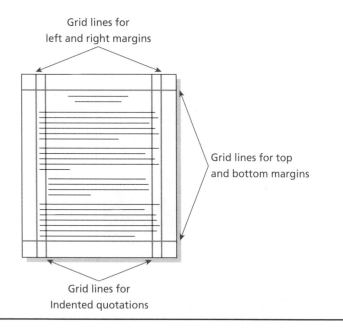

Grid lines for left and right margins

Grid lines for top and bottom margins

Grid lines for Indented quotations

The simplest grid design

Writer's Tutorial

Designing Grid Patterns for Print

BASIC DESIGNS

RESEARCH JOURNAL

This page from the *Journal of Waste Environment Research* uses a basic two-column design. Both columns have the same width.

The title and author are centered in an area that goes from the left-hand margin of the left column to the right-hand margin of the right column.

In the footer, the page number is flush with the left margin. Other information is flush against the right margin.

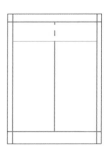
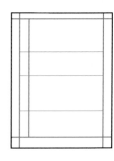

COMPUTER MANUAL

Apple Computer uses a two-column design for this manual.

- The wide column is for the text and figures. The chapter title and most of the information in the footer align with the left-hand margin of this column.
- The narrow column is used only for the major headings. By extending headings into this column, the writers made them easy for readers to spot.

The page number aligns with the wide column's right margin.

ENGINEERING JOURNAL

This page from *Machine Design* uses a four-column design.

- The title and subtitle extend across three columns and are placed flush left against the left-hand marginv of the left column.
- The photograph also extends across three columns and is placed flush right against the right-hand margin of the right column.

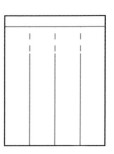

Learn More at the Website

To learn more, go to www.cengage.com /english /anderson7e and choose Chapter 14.

VARIATIONS ON BASIC DESIGNS

ENGINEERING JOURNAL

On this page from *Machine Design,* most elements align within a basic three-column grid. However, the journal created a dynamic appearance by allowing the photograph to intrude into the right and left columns. The text wraps around the photograph's angular edges.

The design also includes a minor grid line within the left-hand column. Aligned with it are the following:

- The author's title (Vice President).
- The word "Machining" in the title.
- Indentations for the first line of a paragraph, and the left notation in the footer.

Pages in which elements intrude into columns are usually found in journals that include advertising, magazines addressed to the general public, annual reports, and instructions for consumer products. They are uncommon in technical proposals and reports.

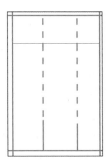

RESEARCH LABORATORY ANNUAL REPORT

In this page from one of its annual reports, the Lawrence Livermore National Laboratory uses a three-column grid.

To create a dynamic design, the laboratory extended some visual elements across the outside margins so that they reach the edges of the page:

- The banner at the top of page goes to both edges.
- The color behind the left side of the title also goes to the edge.
- The photo at the bottom reached the right edge of the page.

The title, text, and footer remain strictly within the grid columns.

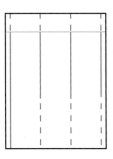

Using grid patterns creatively, you can build an unlimited array of functional and persuasive designs. See Figure 14.1 for some samples.

With such variety being possible, how can you explore and assess grids you might use for a particular communication? Graphic designers accomplish these tasks by sketching various possibilities for the same page. These sketches are called *thumbnails*, because they are miniature versions of possible end results. Take a look at Figure 14.1 again. The nine sketches shown there are thumbnails used by a graphic

FIGURE 14.1
Nine Thumbnail Page Designs for the Same Material

Each of these nine thumbnails describes a different way to present the same content.

Notice how different designs give different levels of emphasis to the title, text, figure, and headings. The choice among these designs would depend on the readers and purpose of this communication.

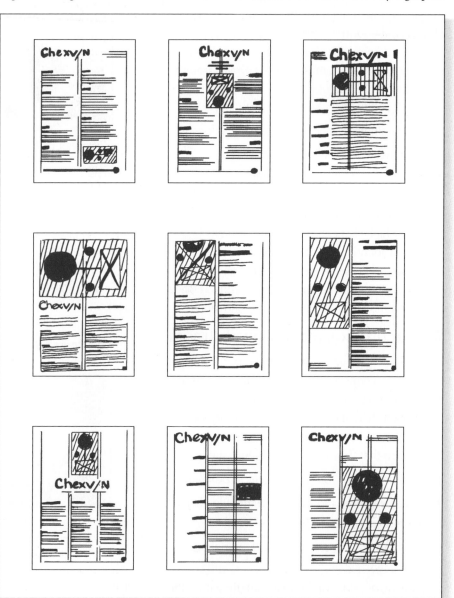

artist to try out and evaluate possible designs for the *same* communication. Notice that the different designs place different levels of emphasis on the figure, title, text, and headings. Using them, the designer was able to select the design most likely to achieve her communication objectives.

As you create thumbnails and choose among them, keep three aspects of your communication in mind.

- **Usability.** Will your readers want to read straight through your report from start to finish? The best design may be the one used for college reports: lots of paragraphs grouped under headings and subheadings. Are you preparing instructions for a process that requires readers to read one step, look away to perform the step, and then look back at the directions to find the next step? In this case, a page design that relied on paragraphs would make reading difficult. Readers will be able to read a step, do a step, and locate the next step more easily if your page design separates the steps visually and numbers the steps prominently.

- **Persuasiveness.** The ways you want to affect your readers' attitudes should also influence your page design. Is it likely that your readers will be indifferent to the topic of your report unless you attract their attention? You may need to create a design that features appealing or intriguing photos or illustrations. Many technical magazines do this. Do you need to persuade your readers that the process described in your instructions is easy? Then you may need to use larger photos and more white space than you otherwise would.

- **Content.** Finally, you should also consider the amount and kinds of content you have. If you are writing instructions for which there is one graphic for every step, you may want to use a two-column design that places all the steps in the left-hand column and the corresponding figures in the right-hand column. However, if you have a graphic for every eight or ten steps, devoting one column to graphics would create an unattractive and not especially useful design.

> Aspects of your communication to consider as you explore and assess thumbnails

The rest of this chapter's guidelines provide details that will help you construct thumbnails and also build your communication's actual pages.

INTRODUCTION TO GUIDELINES 3 THROUGH 6

Whether you are making thumbnails or creating a draft of your communication, you need to place your communication's contents into the grid with which you are working. For thumbnails, you use sketchy lines like those in Figure 14.2 to show where you will position your text, headings and titles, graphics, white space, and headers and footers. For your drafts, you place the contents in your desktop publishing pages. Based on the following design principles described by Robin Williams (1994), Guidelines 3 through 6 provide advice about how to do that.

DESIGN PRINCIPLES	
WILLIAMS'S PRINCIPLES	**THIS CHAPTER'S GUIDELINES**
Alignment	Guideline 3, page 386
Grouping	Guideline 4, page 388
Contrast	Guideline 5, page 391
Repetition	Guideline 6, page 395

Guideline 3 | **Align related elements with one another**

One goal of visual design is to help your readers see how your information is organized. The first principle of visual design—*alignment*—helps you do this. By comparing the following two designs, you can see how readily alignment establishes connections among related items.

In the right-hand design, the writer has used alignment to establish relationships that are not apparent in the left-hand design.

Arbitrary Placement

Alignment Establishes Relationships

In this demonstration, the right-hand design connects related items with one another by aligning them along the invisible grid lines at the right and left margins. To coordinate the more complex array of visual elements included in many on-the-job communications, writers add more vertical and horizontal lines, all arranged with the readers' needs and attitudes in mind.

Sample multicolumn grid designs

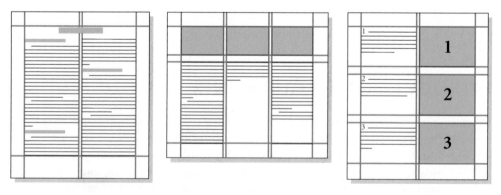

In the left-hand example, a pair of vertical grid lines is used to create a simple, two-column page. The white space between the columns is called a *gutter*.

The center example represents a page in which the writer presents three figures, each accompanied by an associated block of text. To connect each figure to its text, the writer aligns each text-and-figure pair vertically in a column of its own, leaving varying amounts of white space at the bottoms of the columns.

The right-hand example represents a set of instructions in which the writer places all the directions in the left-hand column and all the illustrations in the right-hand column. To link each direction visually with its corresponding illustration, the writer aligns their tops along the same horizontal grid line. Horizontal gutters separate the direction-and-figure pairs from one another.

Figure 14.2 shows sample grid designs for several types of communication.

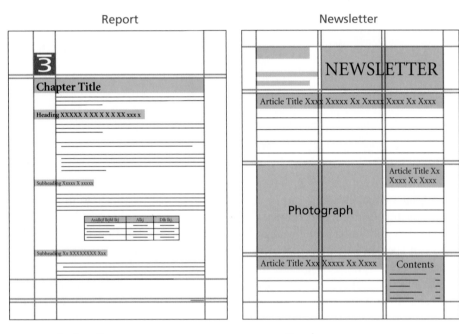

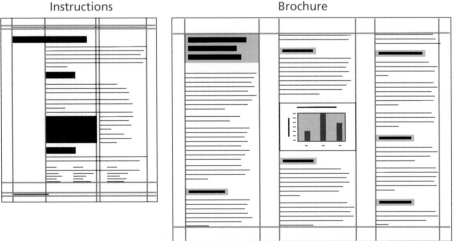

FIGURE 14.2
Sample Grids for Several Types of Communication

Grid lines create the visual structure of a page.

- Visual elements are aligned with one another by being placed against the same grid line.

- When visual elements extend across vertical grid lines, they usually extend to the far sides of the adjacent areas.

Sometimes writers break out of the basic grid pattern to emphasize one particular graphic element.

Try This

Can you find a grid pattern significantly different from the ones at the right in the books, magazines, or other printed documents near you while you are reading this chapter? Can you find two? Do the grid patterns make the communications more usable for the target readers? More persuasive?

Ways to Align Visual Elements with Grid Lines

All visual elements can be arranged in one of three ways with respect to the vertical grid lines: flush left, flush right, or centered.

Ways visual elements can align with vertical grid lines

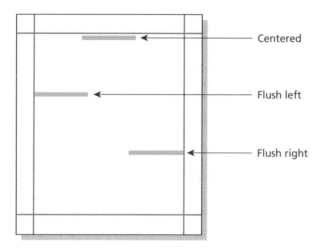

Centered

Flush left

Flush right

A page's visual elements can be aligned with horizontal grid lines in much the same way: top, bottom, or centered.

Headers and footers usually align flush right or left. A heading, figure, or other item can span two or more columns as long as it aligns within the grid system. Extending items across columns is one way of emphasizing them.

To place a drawing or other item with irregular outlines within a grid, try enclosing it in a rectangle whose sides, top, and bottom run along grid lines. Figure 14.3 shows how such rectangles can anchor the figures in the framework of the page.

Guideline 4 | **Group related items visually**

The second design principle is *grouping* (sometimes called *chunking*). It emphasizes that readers judge the relationship between adjacent items by interpreting the distance, or amount of white space, between them. Less distance (less white space) means that two items are closely related to one another; more distance signals that they are not closely related.

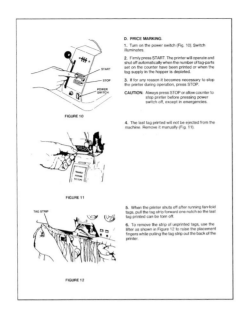

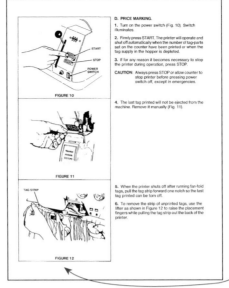

Because the three illustrations in the left-hand page have an irregular shape, readers can't see how they fit into the page's grid.

The problem is solved in the right-hand page by putting rectangular borders around the illustrations.

Not Grouped

> **Martina Alverez**
>
> AAA Consultants, Inc.
>
> 4357 Evington Street
>
> Minneapolis, MN 50517
>
> (416) 232-9999

Proximity Establishes Groupings

> **Martina Alverez**
> AAA Consultants, Inc.
>
>
> 4357 Evington Street
> Minneapolis, MN 50517
> (416) 232-9999

In the right-hand design, the writer has used grouping (adjustment of the white space) to establish relationships that are not apparent in the left-hand design.

Here are some ways in which you can use white space to group elements in your page designs.

- Use less white space below headings than above to place the heading in closer proximity to the text it labels than to the preceding section.
- Use less white space between titles and the figures they label than you use to separate the title and figure from adjacent items.
- In lists that have subgroups, use less white space between items within a subgroup than between one subgroup and another subgroup.

In Figure 14.4, notice how grouping helps to convey the organization of the page.

Ways to use grouping

FIGURE 14.4
Grouping Used to Show
the Organization of the
Information on a Page

This page illustrates some ways
that writers use proximity to help
readers see how the information
on a page is organized.

The writers used less space
above these captions than below;
this helps readers see that the
captions are associated with the
screens.

To create a visual link between
this heading and the paragraph
below, the writers put only one-
half as much space below the
heading as they put above it.

By placing these marginal notes
close to the screens, the writers
enabled readers to see that
the captions supply information
about the screens.

Here, too, the writers placed the
captions closer to the screens
than to the material that follows.

Formatting Text

First, select (block) the text you want to format. To format a single word, just
click in the word. Then specify the formats you want; see the table on the next
page for ways to specify the formats. To select text, hold down the mouse button
as you drag the I-beam pointer over the text. Or hold down SHIFT as you press an
arrow key.

Select the text you want to format. | Then specify a format. For example, click the Bold button on the Formatting toolbar. | Word changes only the text that you selected.

You can also format text as you type. For example, to type bold text, first press
the shortcut keys for bold (or click the Bold button on the Formatting toolbar) and
then type the text. Press the shortcut keys again or click the button again to return
to regular text.

Copying Formatting

Once you've formatted text to look the way you want, use the Format Painter
button on the Standard toolbar to copy the formatting to other text you select.

(Windows)

(Macintosh)

Format Painter pointer

Select text with the formats you want to copy. Then double-click the Format Painter button. | Select the text to format. You can copy the formatting to several selections. | To return to the normal mouse pointer, click the Format Painter button again.

Changing the Preset (Default) Character Formatting

You can change the font, font size, and other character formats that are preset for
Word. From the Format menu, choose the Font command. Select the formats that
you want Word to use in the current document and all new documents based on
the current template. Then choose the Default button.

Guideline 5 | Use contrast to establish hierarchy and focus

Usually, you will want some items to stand out more than others. For instance, you may want to indicate that some items are at a higher level than others within your organizational hierarchy. Or you may want to focus attention on certain items, such as a warning in a set of instructions. However, nothing will stand out if everything on the page looks the same. To make some things stand out, you must use *contrast*, the third of Williams's design principles.

No Contrast	Contrast Establishes Emphasis
Martina Alverez AAA Consultants, Inc. 4357 Evington Street Minneapolis, MN 50517 (416) 232-9999	**Martina Alverez** **AAA Consultants, Inc.** 4357 Evington Street Minneapolis, MN 50517 (416) 232-9999

Color, type size, and bold are used in the right-hand design to establish a focus and a visual hierarchy.

When deciding how to establish the visual hierarchy of your page, pay special attention to text items: paragraphs, headings, figure titles, and headers and footers. To create distinctions among these items, you can control four variables.

- **Size.** Making an element larger increases its visual prominence. Using this principle, most books (including this one) use larger type for major headings than for smaller ones.

    When working with headings, remember that type size is measured in *points*, which refers to the distance between the top of *ascenders* of the tallest letters (e.g., *b* and *l*) and the bottom of *descenders* (e.g., *g* and *y*). There are 72 points to an inch.

M M M M M M M M M M
8 10 12 14 18 24 36 48 60 72

A variety of type sizes, measured in points

When designing a page, avoid making some elements so large that they look out of proportion or pull your readers' eyes away from the rest of the page. As a rule of thumb, keep a larger type to no more than one and one-half times the size of the next smaller size.

- **Type treatment.** Use plain type (called *roman*) for most type. Create contrast for items that you want to stand out by printing them in **bold** or *italics*.

- **Color.** Choose a different color for words or other elements you want to emphasize.
- **Typeface.** To help readers distinguish one kind of element on your page from another, you can use different typefaces for each. At work, this is often done to help headings stand out from paragraphs.

Of course, using different typefaces to distinguish different kinds of content will work only if there is enough contrast between the two typefaces. Consider the following six examples.

Times Roman	Helvetica
Baskerville	**Arial**
Palatino	**Folio**

Try This

Some advertisements ignore some or all of the suggestions made in Guidelines 1 through 5. Find an example and identify the suggestions it follows and the ones it doesn't.
How does ignoring the suggestions improve the ad's effectiveness?
How would ignoring the same suggestions affect a communication you are preparing for your technical communication course?

As you can see, the typefaces in left-hand column look very similar to one another. Those on the right also closely resemble each other. Choosing two typefaces from the left-hand column would not provide enough contrast for most readers to notice immediately. The same is true for choosing two from the right-hand column.

However, readers can perceive immediately the difference between any typeface on the left and any on the right. Here's why. The typefaces on the left all have lines, called *serifs*, across the ends of their strokes. They are called *serif* typefaces. In contrast, the typefaces on the right have no serifs. They are called *sans serif* typefaces (in French, *sans* means "without"). To make their headings stand out, many writers use a sans serif type for their headings and a serif type for their paragraphs.

To create even greater contrast among the visual elements on the page, you can use the variables of size, type treatment, typeface, and color to reinforce one another. By coordinating these variables with one another, you can create distinctive appearances for all of the visual elements on a page. The following table illustrates this principle. Figure 14.5 shows a page created using this table.

ELEMENT	TYPE CATEGORY	SIZE	TREATMENT	COLOR
Paragraph	serif (e.g., Times)	12 point	roman	black
Headings				
First level	sans serif (e.g., Arial)	16 point	bold roman	green
Second level	sans serif	14 point	bold roman	green
Figure caption	sans serif	10 point	bold roman	black
Footers	sans serif	10 point	italic	black

You can use the principle of contrast to distinguish other elements as well. For example, in a set of instructions, you can set off warnings in a box that might have a

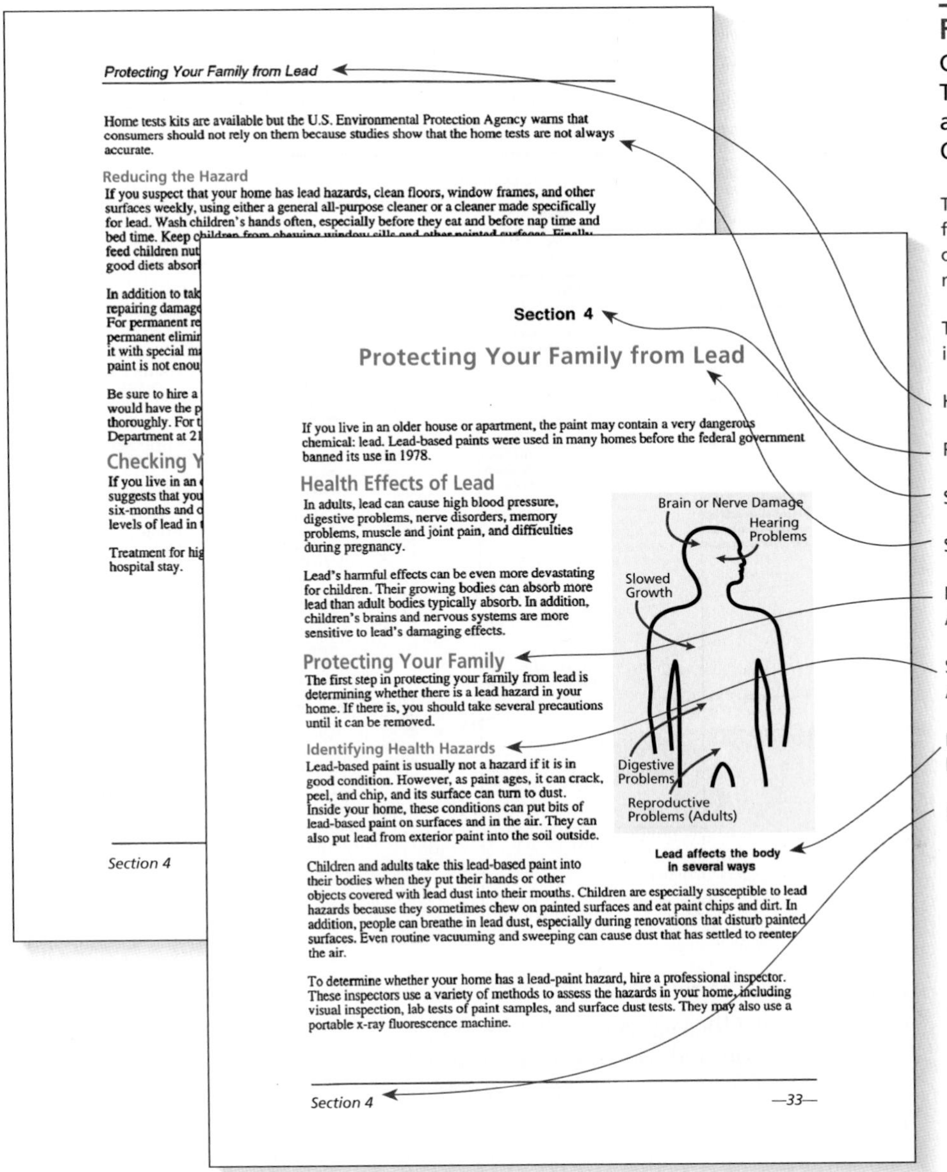

Protecting Your Family from Lead

Home tests kits are available but the U.S. Environmental Protection Agency warns that consumers should not rely on them because studies show that the home tests are not always accurate.

Reducing the Hazard

If you suspect that your home has lead hazards, clean floors, window frames, and other surfaces weekly, using either a general all-purpose cleaner or a cleaner made specifically for lead. Wash children's hands often, especially before they eat and before nap time and bed time. Keep c

feed children nut

good diets absor

In addition to tak

repairing damag

For permanent re

permanent elimi

it with special m

paint is not enou

Be sure to hire a

would have the p

thoroughly. For t

Department at 2

Checking Y

If you live in an

suggests that yo

six-months and c

levels of lead in

Treatment for hi

hospital stay.

Section 4

Section 4

Protecting Your Family from Lead

If you live in an older house or apartment, the paint may contain a very dangerous chemical: lead. Lead-based paints were used in many homes before the federal government banned its use in 1978.

Health Effects of Lead

In adults, lead can cause high blood pressure, digestive problems, nerve disorders, memory problems, muscle and joint pain, and difficulties during pregnancy.

Lead's harmful effects can be even more devastating for children. Their growing bodies can absorb more lead than adult bodies typically absorb. In addition, children's brains and nervous systems are more sensitive to lead's damaging effects.

Protecting Your Family

The first step in protecting your family from lead is determining whether there is a lead hazard in your home. If there is, you should take several precautions until it can be removed.

Identifying Health Hazards

Lead-based paint is usually not a hazard if it is in good condition. However, as paint ages, it can crack, peel, and chip, and its surface can turn to dust. Inside your home, these conditions can put bits of lead-based paint on surfaces and in the air. They can also put lead from exterior paint into the soil outside.

Brain or Nerve Damage
Hearing Problems
Slowed Growth
Digestive Problems
Reproductive Problems (Adults)

Lead affects the body in several ways

Children and adults take this lead-based paint into their bodies when they put their hands or other objects covered with lead dust into their mouths. Children are especially susceptible to lead hazards because they sometimes chew on painted surfaces and eat paint chips and dirt. In addition, people can breathe in lead dust, especially during renovations that disturb painted surfaces. Even routine vacuuming and sweeping can cause dust that has settled to reenter the air.

To determine whether your home has a lead-paint hazard, hire a professional inspector. These inspectors use a variety of methods to assess the hazards in your home, including visual inspection, lab tests of paint samples, and surface dust tests. They may also use a portable x-ray fluorescence machine.

Section 4 —33—

FIGURE 14.5
Coordination of Size, Typeface, Type Treatment, and Color to Reinforce One Another

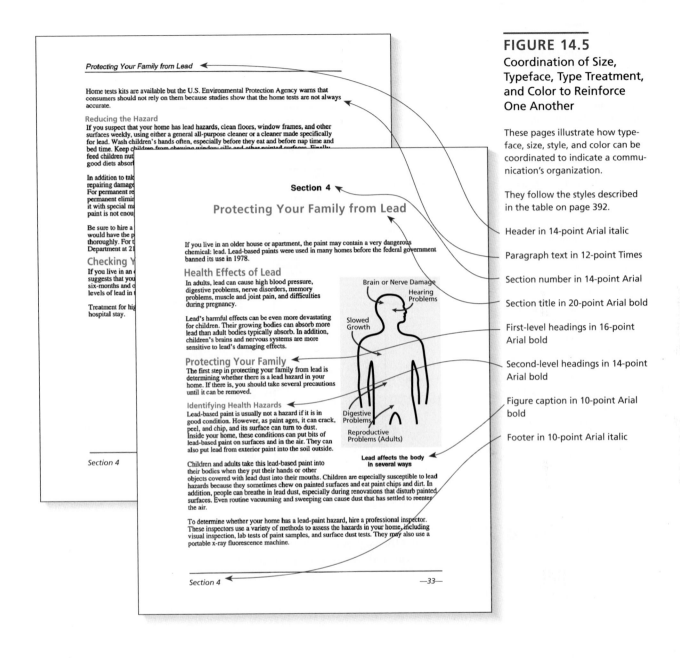

These pages illustrate how typeface, size, style, and color can be coordinated to indicate a communication's organization.

They follow the styles described in the table on page 392.

Header in 14-point Arial italic
Paragraph text in 12-point Times
Section number in 14-point Arial
Section title in 20-point Arial bold
First-level headings in 16-point Arial bold
Second-level headings in 14-point Arial bold
Figure caption in 10-point Arial bold
Footer in 10-point Arial italic

colored background. In a report, you might set off your tables from your paragraphs by using the same typeface for your tables that you use for your headings.

When working with page design variables, remember that your reason for creating visually contrasting elements is to simplify the page for your readers so they

FIGURE 14.6

Rules and Icons Used to Signal Organization

The first sentence on page xxvii explains that the icons' purpose is to make "finding your way around" the guide easier.

The icons help readers distinguish special kinds of information from the rest of a page's contents:

- "Warnings" and "Cautions" that require attention.
- "Notes" that might be helpful but could be skipped.

Note how the rules (horizontal lines) also help to organize the information visually.

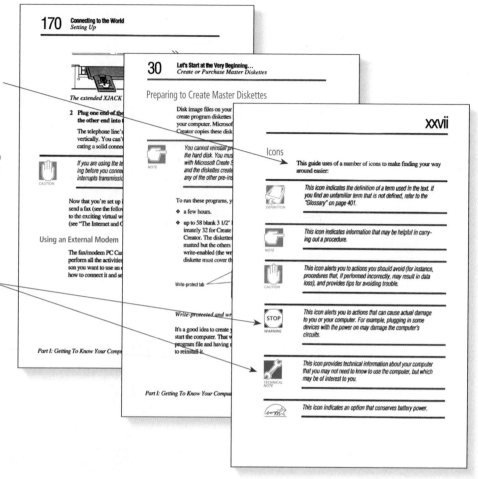

WWW

For additional information on using Microsoft Word to design pages, go to Chapter 14 at **www.cengage.com/english/anderson7e.**

can readily see the role played by each element on the page. Don't create so many distinctions that the page seems chaotic to readers. In most cases, for example, stick to only two colors and two typefaces for the text.

Other tools you can use to help readers see the organization of a page include *rules,* which are lines that mark the boundaries between parts, and *icons,* which highlight information for readers or help them locate a certain kind of content. See Figure 14.6.

USING WORD PROCESSORS TO CREATE PAGE DESIGNS

A Writer's Tutorial available at www.thomsonedu.com/english/anderson7e shows how to use a word-processing program to create page designs that employ Guidelines 1 through 5.

Guideline 6 | Use repetition to unify your communication visually

In communications that are more than a page long, you should think not only about creating well-designed single pages but also about creating a well-designed *set* of pages.

When you are creating a set of pages, the chief goal is to make the pages harmonize visually in a way that both supports your readers' use of them and creates an aesthetically pleasing design. To achieve this goal, use the fourth of Williams's design principles: *repetition*.

For instance, use the same grid pattern throughout. Similarly, use the same type treatment (typeface, size, color, etc.) for all major headings, for all second-level headings, all headers and footers, and so on. Such repetition enables your readers to "learn" the structure of your pages so that they immediately understand the organization of each new page as they turn to it. Repetition also creates visual harmony that is more pleasing to the eye than a set of inconsistent designs.

Of course, it's not possible to have exactly the same design for every page in some communications. In long reports, for instance, each chapter may begin with a page that looks different from every other page in the chapter. Instruction manuals often have troubleshooting sections that present different kinds of information than the manuals' other pages do. For such communications, follow these two strategies:

- **Use the same design for all pages that present the same kind of information.** For instance, use one design for all pages that begin new chapters and another for all pages that don't.

- **Even for pages that have different overall designs, use the same treatment for all elements they share.** For example, use the same style for headings wherever they appear. On all pages, stick with the same typefaces and the same style for headers and footers.

Figure 14.7 shows pages from a communication that uses these strategies.

Using Word Processing Programs to Achieve Consistency

The *styles* feature included with word-processing programs enables you to create consistent pages very efficiently. For example, you can name your first-level headings "Heading 1" and then use a series of menus and checkboxes to define the typeface, size, color, margins, indentations, and other features for these headings. To apply all the details of this style to each first-level heading in your communication, you simply highlight the item and select the corresponding style from a menu. This process is much more efficient than formatting each heading individually.

The styles feature is also helpful to writing teams. They can define a single set of styles for all team members to use when drafting, thereby saving the time it would take to achieve consistency by reformatting everyone's draft.

The styles feature has the added advantage of letting you adjust styles easily. For example, to change the size and color of *all* of the first-level headings, you just redefine the style itself.

FIGURE 14.7
Page Design Used to Create Visual Unity in a Communication

By consistently using various design elements, this car owner's manual achieves overall visual unity.

All pages have the same design at the top.

All pages use the same two-column grid, regardless of the type of content.

Wide illustrations exactly fill both columns, thereby keeping the same outside margins.

The same styles are used throughout for the headings.

Color is applied according to the same pattern throughout.

All right-hand pages have blocks of color that give the section number.

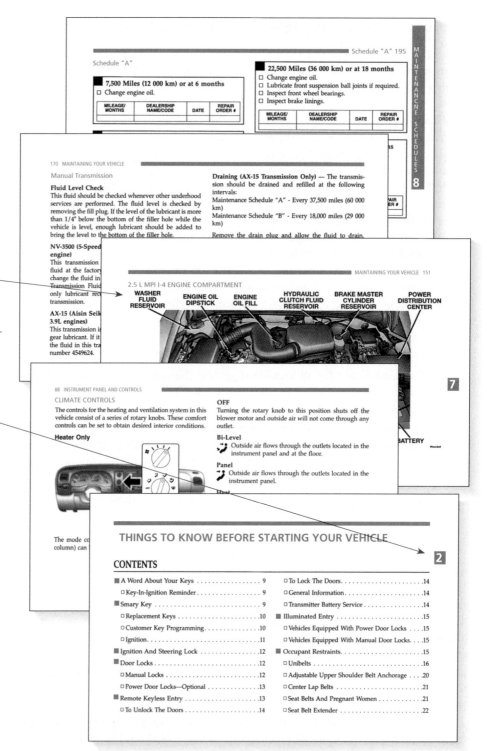

Guideline 7 | **Select type that is easy to read**

We recognize letters by seeing their main lines. These lines are more obvious with some typefaces than with others, as you can see by comparing the two typefaces on the left with the one on the right:

Times Roman	**Ultra**	*Script*
All major lines of all the letters are clearly defined.	The second leg of the *u* is very thin, as is the cross-stroke of the *t*.	The shape of the *s* is hidden by the ornate lines, and the circle of the *p* is not closed.

Comparison of three typefaces

When we are reading only a few words, as when looking at a magazine ad, these differences are inconsequential. However, when we are reading an entire page, a full paragraph, or even a few sentences, type with the more distinct main lines is easier to read. When choosing typefaces for your text, apply the following principles.

PRINCIPLES OF TYPE SELECTION

- **Use a typeface with strong, distinct main lines.**
- **Avoid using italics for more than a sentence at a time.** Although they are great for emphasis, italics are difficult to read in long passages because they represent a "distorted" or "ornate" version of the alphabet that obscures the main lines.
- **Avoid using all capital letters for more than a few words at a time.** Research has shown that capital letters are difficult to read in long stretches (Tinker, 1969). Why? Many capital letters have very similar shapes. For instance, capitals "D" and "P" are the same height and face the same direction. In contrast, lowercase "d" and "p" have different heights and face in different directions. Such differences make lowercase letters easier to identify and promote more rapid reading.

After studying the extensive literature on the readability of different typefaces for printed communications, Karen Schriver (1997) concludes that in extended text serif type may be somewhat easier to read than sans serif, although the two categories can be read equally quickly in small amounts. Schriver notes, however, that several factors may come into play, including cultural preferences and familiarity with the typeface. In the United States, serif type is most common in published books, but in Europe publishers often use sans serif type. She concludes that if you choose a typeface with strong, distinct lines, your text will be readable whether you select a serif or sans serif typeface.

Of course, no type will be legible unless it is large enough. Research has found that people find it easiest to read passages printed in type between 8 and 12 points

Learn More

To review the difference between serif and sans serif type, go to page 392.

high (Tinker, 1969). At work, writers usually avoid type smaller than 10 points and reserve type larger than 12 points for headings, titles, and the like. They sometimes use type as small as 9 points in headers, footers, and captions for figures.

Guideline 8 | **Design your overall document for ease of use and attractiveness**

For some communications, you will need to think about how to design not only the elements on the page but also the overall product that you will provide to your readers. Here, too, consider your communication from your readers' perspective.

- **Size.** Choose a size that your readers will find convenient in the situations in which they will use your communication. For example, if you are creating instructions that readers will use at a desk already crowded by a computer and a keyboard, imitate software manufacturers who print manuals that are smaller than 8½ by 11 inches.
- **Shape.** Choose a shape that will make it convenient for your readers to carry and store. If appropriate, for example, make it a shape that will fit into a shirt pocket or into a special case (such as the case for a notepad computer or a blood glucose monitor).
- **Binding.** Some bindings stay open more easily than others. If your readers will want your communication to lie flat because they need to use both hands for something else, consider a spiral binding or three-ring binder.
- **Paper.** If you are creating a shop manual that might be spilled on or a document that might be used outdoors, consider a coated paper that resists liquids and soil.

CONCLUSION

Every page has some sort of page design. This chapter has explained how you can look at your pages in the way that graphic designers do, thinking about each graphic element in terms of the way it affects your readers. By following this chapter's advice, you will be able to design pages and communications that help your readers read efficiently, emphasize the important contents of your communication, and create a favorable impression.

USE WHAT YOU'VE LEARNED

For additional exercises, go to www.cengage.com/english/anderson7e. *Instructors*: The book's website includes suggestions for teaching the exercises.

EXERCISE YOUR EXPERTISE

1. Figure 14.8 shows four package inserts for a prescription medication. All contain the same information, but each was prepared by a different graphic designer. Describe the page designs used in each. What purpose do you think each designer had in mind? Which design do you think works best? Worst? Why?

2. Find three different page designs. (Choose no more than one from a popular magazine; look at instructions, insurance policies, leases, company brochures, technical reports, and the like.) If your instructor asks, find all the samples in documents related to your major. Photocopy one page illustrating each design. Then, for each page, describe what you think the purpose of the document is and discuss the specific features of the page design that help or hinder the document from achieving that purpose.

3. Do the Informational Page project in Appendix B.

4. Experiment with the page design of a draft or finished communication of yours that has headings and at least one list. Using your desktop publishing program's styles feature, define styles for the communication's design elements. Instead of using the program's default styles, create your own. After you have applied your styles throughout your communication, change the size, color, and typeface for the headings. Also try different designs for the other elements. (To learn how to use the program's styles feature, use its Help feature.)

EXPLORE ONLINE

Search for "Page Design" with your web browser. Find two suggestions that you consider to be good supplements to the ones given in this chapter. Would each of them be an additional subpoint under one of the chapter's guidelines or would one or both of them be new guidelines? Provide a reader-centered reason for following each of them.

COLLABORATE WITH YOUR CLASSMATES

Work with another student on this exercise, which will provide you with practice at evaluating and improving a page design.

1. List the ways in which the design of the page shown in Figure 14.9 (page 404) could be improved.

2. Redesign the page using the strategies described in the Writer's Tutorial for Creating a Multicolumn Page Design available at www.cengage.com/english/anderson7e (click on Chapter 14). Specifically, do the following:
   a. Create three thumbnail sketches for the page.
   b. Create a full-size mockup of the best design.

3. Display your mockup on the wall of your classroom along with the mockups prepared by the other students in your class. Decide which mockups work best. Discuss the reasons.

APPLY YOUR ETHICS

This exercise invites you to consider the role that page and screen design can play in your ability to communicate effectively on issues that are important to you for ethical reasons. Visit three or more websites whose creators are advocating a position on an ethical basis. Sites maintained by groups taking various positions on environmental policy or education policy are examples. Evaluate the ways that the screen design of each site promotes or hinders the group's ability to achieve the usability and persuasive goals it has for the site.

FIGURE 14.8

Package Inserts for a Prescription Medication

HYGROTON®
chlorthalidone usp

50 mg. Tablets, 100 mg. Tablets

Oral Antihypertensive-Diuretic

DESCRIPTION
HYGROTON (chlorthalidone) is a monosulfamyl diuretic which differs chemically from thiazide diuretics in that a double-ring system is incorporated in its structure. It is 2-Chlor-5-(1-hydroxy-3-oxo-1-isoindolinyl) benzenesulfonamide, with the following structural formula:

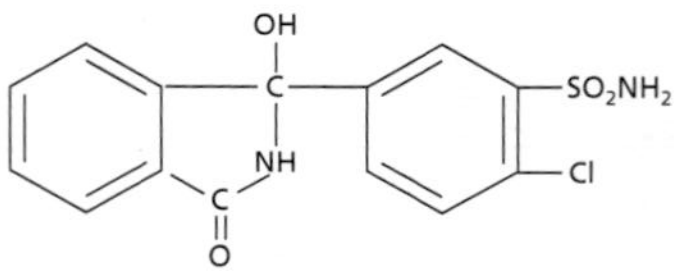

ACTIONS
HYGROTON is an oral diuretic with prolonged action (48-72 hours) and low toxicity. The diuretic effect of the drug occurs within two hours of an oral dose and continues for up to 72 hours.
INDICATIONS
Diuretics such as HYGROTON are indicated in the management of hypertension either as the sole therapeutic agent or to enhance the effect of other antihypertensive drugs in the more severe forms of hypertension and in the control of hypertension of pregnancy.
CONTRAINDICATIONS
Anuria. Hypertensitivity to chlorthalidone. The routine use of diuretics in an otherwise healthy pregnant woman with or without mild edema is contraindicated and possibly hazardous.
WARNINGS
Should be used with caution in severe renal disease. In patients with renal disease, chlorathalidone or related drugs may precipitate azotemia. Cumulative effects of the drug may develop in patients with impaired renal function.

USAGE IN PREGNANCY: Reproduction studies in various animal species at multiples of the human dose showed no significant level of teratogenicity; no fetal or congenital abnormalities were observed.
NURSING MOTHERS: Thiazides cross the placental barrier and appear in cord blood and breast milk.
PRECAUTIONS
Periodic determination of serum electrolytes to detect possible electrolyte imbalance should be performed at appropriate intervals.
Chlorthalidone and related drugs may decrease serum PBI levels without signs of thyroid disturbance.

ADVERSE REACTIONS
Gastrointestinal System Reactions:
anorexia constipation
gastric irritation jaundice
nausea (intrahepatic
vomiting cholestatic
cramping jaundice)
diarrhea pancreatitis
Central Nervous System Reactions:
dizziness headache
vertigo xanthopsia
paresthesias
Hematologic Reactions:
leukopenia thrombocytopenia
agranulocytosis aplastic anemia
Other Adverse Reactions:
hyperglycemia muscle spasm
glycosuria weakness
hyperuricemia restlessness
impotence

Whenever adverse reactions are moderate or severe, chlorthalidone dosage should be reduced or therapy withdrawn.

DOSAGE AND ADMINISTRATION
Therapy should be individualized according to patient response. This therapy should be titrated to gain maximal therapeutic response as well as the minimal dose possible to maintain that therapeutic response.
Initiation: Preferably, therapy should be initiated with 50 mg. or 100 mg. daily. Due to the long action of the drug, therapy may also be initiated in most cases with a dose of 100 mg. on alternate days or three times weekly (Monday, Wednesday, Friday). Some patients may require 150 or 200 mg. at these intervals.
Maintenance: Maintenance doses may often be lower than initial doses and should be adjusted according to the individual patient. Effectiveness is well sustained during continued use.
OVERDOSAGE
Symptoms of overdosage include nausea, weakness, dizziness and disturbances of electrolyte balance.
HOW SUPPLIED
HYGROTON (chlorthalidone). White, single-scored tablets of 100 mg. and aqua tablets of 50 mg. in bottles of 100 and 1000; single-dose blister packs, boxes of 500; Paks of 28 tablets, boxes of 6.
CAUTION: Federal law prohibits dispensing without prescription.
ANIMAL PHARMACOLOGY
Biochemical studies in animals have suggested reasons for the prolonged effect of chlorthalidone. Absorption from the gastrointestinal tract is slow, due to its low solubility. After passage to the liver, some of the drug enters the general circulation, while some is excreted in the bile, to be reabsorbed later.

USV PHARMACEUTICAL MFG. CORP.
Manati, P.R. 00701

A

FIGURE 14.8

(continued)

hygroton* chlorthalidone usp

oral
antihypertensive-
diuretic

50 mg. tablets
100 mg. tablets

Description: *HYGROTON* (chlorthalidone) is a monosulfamyl diuretic which differs chemically from thiazide diuretics in that a double-ring system is incorporated in its structure. It is 2-Chloro-5-(1-hydroxy-3-oxo-1-isoindolinyl) benzenesulfonamide, with the following structural formula:

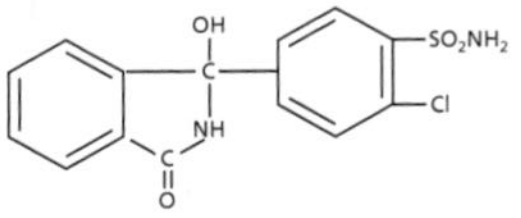

Actions: *HYGROTON* is an oral diuretic with prolonged action (48-72 hours) and low toxicity. The diuretic effect of the drug occurs within two hours of an oral dose and continues for up to 72 hours.

Indications: Diuretics such as HYGROTON are indicated in the management of hypertension either as the sole therapeutic agent or to enhance the effect of other antihypertensive drugs in the more severe forms of hypertension and in the control of hypertension of pregnancy.

Contraindications: Anuria. Hypersensitivity to chlorthalidone.

The routine use of diuretics in an otherwise healthy pregnant woman with or without mild edema is contraindicated and possibly hazardous.

Warnings: Should be used with caution in severe renal disease. In patients with renal disease, chlorathalidone or related drugs may precipitate azotemia. Cumulative effects of the drug may develop in patients with impaired

Usage in pregnancy: Reproduction studies in various animal species at multiples of the human dose showed no significant level of teratogenicity; no fetal or congenital abnormalities were observed.

Nursing mothers: Thiazides cross the placental barrier and appear in cord blood and breast milk.

Precautions: Periodic determination of serum electrolytes to detect possible electrolyte imbalance should be performed at appropriate intervals.

Chlorthalidone and related drugs may decrease serum PBI levels without signs of thyroid disturbance.

Adverse reactions:
Gastrointestinal System Reactions:
anorexia
gastric irritation
nausea
vomiting
cramping
diarrhea
pancreatitis

constipation
jaundice
    (intrahepatic
    cholestatic
    jaundice)

Central Nervous System Reactions:
dizziness
vertigo
paresthesias

headache
xanthopsia

Hematologic Reactions:
leukopenia
agranulocytosis

thrombocytopenia
aplastic anemia

Other Adverse Reactions:
hyperglycemia
glycosuria
hyperuricemia
impotence

muscle spasm
weakness
restlessness

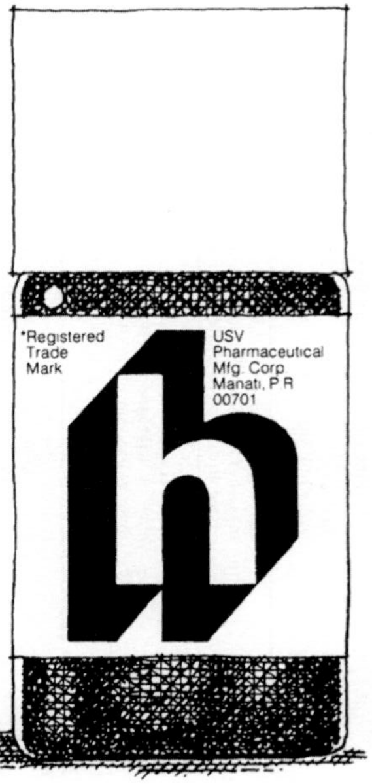

Whenever adverse reaction are moderate or severe, chlorthalidone dosage should be reduced or therapy withdrawn.

Dosage and administration: Therapy should be individualized according to patient response. This therapy should be titrated to gain maximal therapeutic response as well as the minimal dose possible to maintain that therapeutic response.

Initiation: Preferably, therapy should be initiated with 50 mg. or 100 mg. daily. Due to the long action of the drug, therapy may also be initiated in most cases with a dose of 100 mg. on alternate days or three times weekly (Monday, Wednesday, Friday). Some patients may require 150 or 200 mg. at these intervals.

Maintenance: Maintenance doses may often be lower than initial doses and should be adjusted according to the individual patient. Effectiveness is well sustained during continued use.

Overdosage: Symptoms of overdosage include nausea, weakness, dizziness and disturbances of electrolyte balance.

How supplied: HYGROTON (chlorthalidone). White, single-scored tablets of 100 mg. and aqua tablets of 50 mg. in bottles of 100 and 1000; single-dose blister packs, boxes of 500; Paks of 28 tablets, boxes of 6.

Caution: Federal law prohibits dispensing without prescription.

Animal pharmacology: *Biochemical studies* in animals have suggested reasons for the prolonged effect of chlorthalidone. Absorption from the gastrointestinal tract is slow, due to its low solubility. After passage to the liver, some of the drug enters the general circulation, while some is excreted in the bile, to be reabsorbed later.

USV
Pharmaceutical
Mfg. Corp.
Manati, P.R.
00701.

*Registered
Trade
Mark

B

FIGURE 14.8
(continued)

Hygroton®
chlorthalidone USP

50 mg. Tablets 100 mg. Tablets	Oral Antihypertensive-Diuretic		
Description	**Hygroton** (chlorthalidone) is a monosulfamyl diuretic which differs chemically from thiazide diuretics in that a double-ring system is incorporated in its structure. It is 2-Chlor-5-(1-hydroxy-3-oxo-1-isoindolinyl) benzenesulfonamide, with the following structural formula		
Actions	**Hygroton** is an oral diuretic with prolonged action (48-72 hours) and low toxicity. The diuretic effect of the drug occurs within two hours of an oral dose and continues for up to 72 hours		
Indications	Diuretics such as **Hygroton** are indicated in the management of hypertension either as the sole therapeutic agent or to enhance the effect of other	antihypertensive drugs in the more severe forms of hypertension and in the control of hypertension of pregnancy	
Contraindications	Anuria. Hypertensitivity to chlorthalidone	The routine use of diuretics in an otherwise healthy pregnant woman with or without mild edema is contraindicated and possibly hazardous	
Warnings	Should be used with caution in severe renal disease. In patients with renal disease, chlorathalidone or related drugs may precipitate azotemia	Cumulative effects of the drug may develop in patients with impaired renal function	
Usage in Pregnancy:	Reproduction studies in various animal species at multiples of the human dose showed no significant level of teratogenicity; no fetal or congential abnormalities were observed		
Nursing Mothers:	Thiazides cross the placental barrier and appear in cord blood and breast milk		
Precautions	Periodic determination of serum selectrolytes to detect possible electrolyte imbalance should be performed at appropriate intervals	Chlorthalidone and related drugs may decrease serum PBI levels without signs of thyroid disturbance	
Adverse reactions:			
Gastrointestinal System Reactions:	anorexia gastric irritation nausea	vomiting cramping diarrhea	constipation pancreatitis jaundice (intrahepatic cholestatic jaundice)
Central Nervous System Reactions:	dizziness vertigo paresthesias	headache xanthopsia	
Hematologic Reactions:	leukopenia agranulocytosis	thrombocytopenia aplastic anemia	
Other Adverse Reactions:	hyperglycemia glycosuria hyperuricemia muscle spasm	weakness restlessness impotence	Whenever adverse reaction are moderate or severe, chlorthalidone dosage should be reduced or therapy withdrawn
Dosage and Administration	Therapy should be individualized according to patient response. This therapy should be titrated to gain maximal therapeutic response as well as	the minimal dose possible to maintain that therapeutic response	
Initiation:	Preferably, therapy should be initiated with 50 mg. or 100 mg. daily. Due to the long action of the drug, therapy may also be initiated in most cases with a dose of 100 mg. on alternate days or	three times weekly (Monday, Wednesday, Friday). Some patients may require 150 or 200 mg. at these intervals	
Maintenance:	Maintenance doses may often be lower than initial doses and should be adjusted according to	the individual patient. Effectiveness is well sustained during continued use	
Overdosage	Symptoms of overdosage include nausea, weakness, dizziness and disturbances of electrolyte balance		
How Supplied	**Hygroton** (chlorthalidone). White, single-scored tablets of 100 mg. and aqua tablets of 50 mg in	bottles of 100 and 1000; single-dose blister packs, boxes of 500; Paks of 28 tablets, boxes of 6	
Caution:	Federal law prohibits dispensing without prescription		

C

FIGURE 14.8
(continued)

Hygroton®

Chlorthalidone USP

50 mg. Tablets
100 mg. Tablets
Oral Antihypertensive-Diuretic

DESCRIPTION

HYGROTON (chlorthalidone) is a monosulfamyl diuretic which differs chemically from thiazide diuretics in that a double-ring system is incorporated in its structure. It is 2-Chlor-5-(1-hydroxy-3-oxo-1-isoindolinyl) benzenesulfonamide, with the following structural formula:

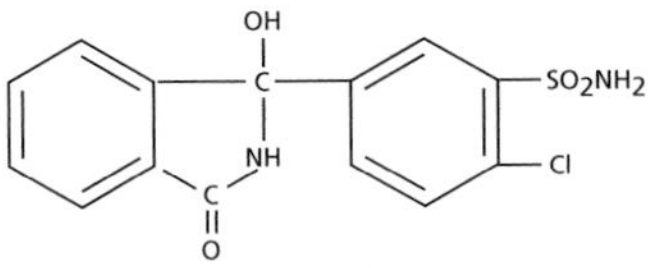

ACTIONS

HYGROTON is an oral diuretic with prolonged action (48-72 hours) and low toxicity. The diuretic effect of the drug occurs within two hours of an oral dose and continues for up to 72 hours.

INDICATIONS

Diuretics such as HYGROTON are indicated in the management of hypertension either as the sole therapeutic agent or to enhance the effect of other antihypertensive drugs in the more severe forms of hypertension and in the control of hypertension of pregnancy.

CONTRAINDICATIONS

Anuria.

Hypersensitivity to chlorthalidone. The routine use of diuretics in an otherwise healthy pregnant woman with or without mild edema is contraindicated and possibly hazardous.

WARNINGS

Should be used with caution in severe renal disease. In patients with renal disease, chlorathalidone or related drugs may precipitate azotemia. Cumulative effects of the drug may develop in patients with impaired renal function.

Usage in Pregnancy: Reproduction studies in various animal species at multiples of the human dose showed no dignificant level of teratogenicity; no fetal or congenital abnormalities were observed.

Nursing Mothers: Thiazides cross the placental barrier and appear in cord blood and breast milk.

PRECAUTIONS

Periodic determination of serum selectrolytes to detect possible electrolyte imbalance should be performed at appropriate intervals. Chlorthalidone and related drugs may decrease serum PBI levels without signs of thyroid disturbance.

ADVERSE REACTIONS

Gastrointestinal Systems Reactions:
  anorexia
  gastric irritation
  nausea
  vomiting
  cramping
  diarrhea
  constipation
  jaundice (intrahepatic cholestatic jaundice)
  pancreatitis
Central Nervous System Reactions:
                dizziness
  vertigo
  paresthesias
  headache
  xanthopsia Hematologic Reactions:
                leukopenia
  agranulocytosis
  thrombocytopenia
  aplastic anemia
Other Adverse Reactions:
  hyperglycemia
  glycosuria
  hyperuricemia
  impotence
  muscle spasm
  weakness
  restlessness
Whenever adverse reaction are moderate or severe, chlorthalidone dosage should be reduced or therapy withdrawn.

DOSAGE AND ADMINISTRATION

Therapy should be individualized according to patient response. This therapy should be titrated to gain maximal therapeutic response as well as the minimal dose possible to maintain that therapeutic response.

Initiation: Preferabley, therapy should be initiated with 50 mg. or 100 mg. daily. Due to the long action of the drug, therapy may also be initiated in most cases with a dose of 100 mg. on alternate days or three times weekly (Monday, Wednesday, Friday). Some patients may require 150 or 200 mg. at these intervals.

Maintenance: Maintenance doses may often be lower than initial doses and should be adjusted according to the individual patient. Effectiveness is well sustained during continued use.

OVERDOSAGE

Symptoms of overdosage include nausea, weakness, dizziness and disturbances of electrolyte balance.

HOW SUPPLIED

HYGROTON (chlorthalidone). White, single-scored tablets of 100 mg. and aqua tablets of 50 mg. in bottles of 100 and 1000; single-dose blister packs, boxes of 500; Paks of 28 tablets, boxes of 6.

CAUTION: Federal law prohibits dispensing without prescription.

ANIMAL PHARMACOLOGY

Biochemical studies in animals have suggested reasons for the prolonged effect of chlorthalidone. Absorption from the gastrointestinal tract is slow, due to its low solubility. After passage to the liver, some of the drug enters the general circulation, while some is excreted in the bile, to be reabsorbed later.

USV PHARMACEUTICAL
MFG. CORP.
Manati, P.R. 00701

D

FIGURE 14.9
Sample Page for Use with
Collaboration Exercise

4) Flatten the clay by pounding it into the table.
Remember, work the clay thoroughly!!

Slab Roller

The slab roller is a simple but very
efficient mechanical device. A uniform
thickness of clay is guaranteed. The
ease of its operation saves countless
hours of labor over hand rolling techniques.

1) The thickness of the clay is determined by the number of masonite®
boards used. There are three $1/4$" boards and one $1/8$" board. They can be
used in any combination to reduce or increase the thickness of the clay.
If no boards are used, the clay will be 1"
thick. If all the boards are used, the clay
will be $1/8$" thick. So, for a $1/2$" thickness of
clay, use two $1/4$" boards.

2) The canvas cloths must be arranged
correctly. They should form a sort of
envelope around the wet clay. This
prevents any clay from getting on the
rollers and masonite® boards.

3) Place the clay flat on the canvas near
the rollers. It may be necessary to trim
some of the clay since it will spread out
as it passes through the rollers.

4.

PART VII

REVISING

15 | Revising Your Drafts

GUIDELINES

CHAPTER 15

DEFINING
OBJECTIVES

PLANNING

CONDUCTING
RESEARCH

DRAFTING

REVISING

Successful writers at work know that even though they follow a thoroughly reader-centered approach when drafting
a communication, they can still increase the usability and persuasiveness of almost any draft they create. Consequently, these
highly skilled writers pause to revise before placing their communications in their readers' hands. Employers, too, recognize
the importance of revision. Many have developed procedures
and policies that require employees to revise.

THE THREE ACTIVITIES OF REVISING

Revising involves three separate activities. Although they are often performed almost
simultaneously, understanding that each requires a separate mental activity can
greatly increase your revising expertise.

1. **Identifying possible improvements you could make in your draft.** In this activity, you evaluate the draft from the perspective of its intended readers. Will
they find it to be helpful? Persuasive?
2. **Deciding which of the possible improvements to make.** At work, you will
rarely have time to make all possible improvements. Deadlines will loom.
Other responsibilities will demand your attention. In this activity, you identify
which of the possible revisions will produce the greatest improvement in the
time available.
3. **Making the selected changes.**

This chapter provides guidelines for the two methods of *identifying possible
improvements* that are most widely used in the workplace.

- **Checking.** You carefully examine your draft yourself. (See the next section of
this chapter.)
- **Reviewing.** You ask for advice, principally about your draft's usability and persuasiveness, from people who are not part of your target audience. (See page
413 in this chapter.)

A final section of this chapter explains how to *perform* the second revising activity:
determining which changes will bring about the greatest improvement (see page 419).

To supplement this chapter's advice, the next chapter explains a testing procedure for identifying possible improvements. This procedure will be particularly useful when you are writing instructions, but it also has many other applications.

You may have deduced that this chapter does not include guidelines for the third
step of revising, which is to make the changes you've selected. The reason is simple.
Whether you are drafting or revising, the principles and guidelines for good communication remain the same. Therefore, if you determine that you need to polish

© Creatas/PhotoLibrary

your writing style, you can find advice for doing so in Chapter 9, "Developing an Effective Style." Similarly, if you discover that you need to reorganize, make your graphics more usable, or design your pages more effectively, consult the appropriate chapters in this book.

GUIDELINES FOR CHECKING YOUR DRAFT YOURSELF

The first thing to know about checking over your own drafts is this: It is extremely difficult to recognize problems in your own writing. Several obstacles hamper your ability to do so. The following guidelines identify these obstacles and suggest ways of overcoming them.

Guideline 1 | Check from your readers' point of view

Writers sometimes create the first obstacle by defining the focus of their checking too narrowly. They concentrate on spelling, punctuation, and grammar, forgetting to consider their communication from the perspective of readers. Consequently, you should always begin your checking by reminding yourself of your communication's objectives. If you wrote them down, pull out your notes. If you recorded them mentally, review them now. What tasks do you want your communication to help your readers perform? How do you want it to alter their attitudes?

Then, with your communication's objectives fresh in mind, read your draft while imagining your readers' moment-by-moment responses. As they proceed, how usable will they find each section, paragraph, and sentence to be? How persuasive will they find each of the writing strategies by which you hope to shape their attitudes? Look especially for places where the readers may not respond in the way you desire. That's where you need to revise.

Guideline 2 | Check from your employer's point of view

In many working situations, you must consider your communication not only from your readers' perspective but also from your employer's. Here are three questions you should ask:

- **How will my communication impact others in the organization?** By anticipating conflicts and objections that your communication might stir up, you may be able to reduce their severity or avoid them altogether.
- **Does my communication promise something on my employer's behalf?** For example, does it make a commitment to a client or customer? If so, ask yourself whether you have the authority to make the commitment, whether the commitment is in your employer's best interest, and whether it is one your employer wishes to make.

WWW

To read additional information, see more examples, and access links related to checking your drafts, go to Chapter 15 at **www.cengage.com/english/anderson7e.**

Things to consider when evaluating from your employer's perspective

- **Does my communication comply with my employer's policies?** First, check your draft against your employer's policies on writing style and format. Second, determine whether it complies with the employer's regulations concerning what you can say. For example, for legal reasons, your employer may want to restrict when and how you reveal information about a certain aspect of your work. This might happen, for instance, if you are working on a patentable project or a project regulated by a government body, such as a state or federal Environmental Protection Agency.

Guideline 3 | Distance yourself from your draft

Another obstacle to effective checking is being too "close" to what we write. Because we know what we meant to say, when we check for errors we often see what we intended to write rather than what is actually there. A word is misspelled, but we see it spelled correctly. A paragraph is cloudy, but we see clearly the meaning we wanted to convey.

To distance yourself from your draft, try the following measures:

Ways to distance yourself from your draft

- **Let time pass.** As time passes, your memory of your intentions fades. You become more able to see what you actually wrote. Set your draft aside, if for only a few minutes, before checking it.
- **Read your draft aloud, even if there is no one to listen.** Where you stumble when saying your words, your readers are likely to trip as well. Where the "voice" you've created in your draft sounds a little off to you, it will probably sound wrong to your readers, too (see Chapter 9).
- **Let your computer read your paper aloud to you.** PC and Mac computers come with free programs that read your writing aloud through your computer's speakers. This simple way of distancing yourself from your draft can help you detect problems. For instructions on using the programs, go to www.cengage.com/english/anderson7e and click on Chapter 15.

Try This

Why not proofread by letting your computer read your writing aloud to you? For instructions, go to **www.cengage.com/english/anderson7e** and click on Chapter 15.

Guideline 4 | Read your draft more than once, changing your focus each time

Yet another obstacle to effective checking is summed up in the adage, "You can't do two things at once." To "do" something, according to researchers who study the way humans think, requires "attention." Some activities (such as walking) require much less attention than others (such as solving calculus problems). We can attend to several more or less automatic activities at once. However, when we are engaged in any activity that requires us to concentrate, we have difficulty doing anything else well at the same time (J. R. Anderson, 1995).

This limitation has important consequences. When you concentrate on one aspect of your draft, such as the spelling or consistency of your headings, you diminish your ability to concentrate on the others, such as the clarity of your prose.

To overcome this limitation on attention, check your draft at least twice, once for *substantive* matters (such as clarity and persuasive impact) and once for *mechanical* ones (such as correct punctuation and grammar). If you have time, you might read through it more than twice, sharpening the focus of each reading.

Guideline 5 | Use computer aids to find (but not to cure) possible problems

Desktop publishing programs offer a variety of aids that can help you check your drafts:

- **Spell checkers.** Spell checkers identify possible misspellings by looking for words in your draft that aren't in their dictionaries.
- **Grammar checkers.** Grammar checkers identify sentences that may have any of a wide variety of problems with grammar or punctuation. See Figure 15.1.
- **Style checkers.** Some computer programs analyze various aspects of your writing that are related to style, such as average paragraph, sentence, and word length. Using special formulas, they also compute scores that attempt to reflect the "readability" of a draft. See Figure 15.2.

All three types of aids can be helpful. But use them with caution. They all have serious shortcomings. For example, spell checkers ignore words that are misspelled but look like other words, such as *forward* for *foreword* or *fake* for *rake*. Grammar

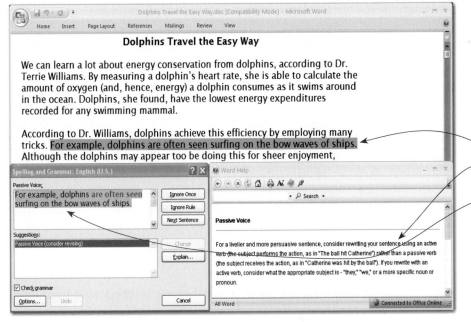

FIGURE 15.1
Grammar Checker

A grammar checker looks for sentences that seem to violate a set of rules it has been given.

If it finds one, it does the following:

- Highlights the sentence.
- Provides a brief explanation of the problem it may have found.
- Indicates the part of the sentence that may have the problem.

Usually, grammar checkers look for misspellings also; note that it missed the use of the wrong "too" in the second line of the second paragraph (see the discussion of spell checkers above).

FIGURE 15.2
Style Checker

Style checkers provide statistics that are intended to help writers determine how difficult their prose is to read.

Typically, they calculate the average lengths of the paragraphs, sentences, and words.

They also employ these statistics to estimate the "readability" of the passage.

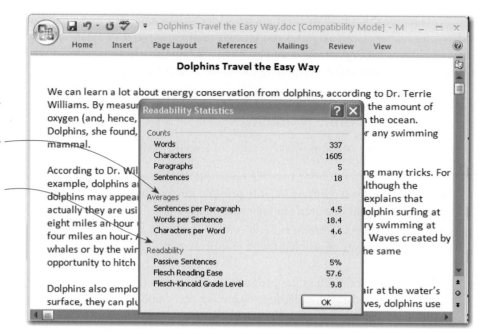

Learn More

To learn how to distinguish effective from ineffective uses of the passive voice, turn to page 272.

Learn More

To learn more about treating your communication's stakeholders ethically, turn to pages 90 and 112.

checkers can't tell a good use of the passive voice from a poor one. And style checkers can't tell a clearly written long sentence from a murky short one. Sometimes, these aids suggest revisions that would make your writing worse. Because of these and similar limitations, review their suggestions carefully. And always read over your drafts yourself.

Guideline 6 | Ethics Guideline: Consider the stakeholders' perspective

If you followed the ethics guidelines provided earlier in this book, you will already have considered your communication from the stakeholders' point of view at least twice. Early in your work on your draft, you identified its stakeholders and learned about the impact your message might have on them. Then, while drafting, you will have kept your stakeholders constantly in mind.

However, there are so many things to think about when drafting that there's a chance you may have overlooked your stakeholders' concerns in some way. Therefore, when checking a draft, review your communication one last time from their perspective.

Checklist for Checking

Figure 15.3 is a checklist you can use to ensure that you are checking your drafts in a reader-centered way.

FIGURE 15.3   **Writer's Guide for Checking Your Own Drafts**

To download a copy of this Writer's Guide, go to Chapter 15 at www.cengage.com/english/anderson7e.

Writer's Guide
CHECKING YOUR OWN DRAFTS

1. Gain enough distance from your draft to read it from other people's perspectives.

2. Read your draft from your readers' perspective.

    Look for ways you can make it more usable for your readers.

    Look for ways you can make it more persuasive to your readers.

3. Read your draft from your employer's perspective.

    How will your draft affect others in your organization?

    Does your draft make promises your employer won't want you to make?

    Does your draft comply with your employer's policies?

4. Read your draft from the perspective of its stakeholders.

    Does it create undesirable consequences for the stakeholders?

    If so, how can these consequences be reduced or eliminated?

5. Read your draft more than once, focusing on different issues each time.

6. Check for errors missed by the computer program you used.

GUIDELINES FOR REVIEWING DRAFTS

In *reviewing*, writers give their drafts to someone else to look over. You may be familiar with this process through the peer reviewing often done in classes. At work, writers are often required to have their drafts reviewed to ensure that their communications are well written, conform to the employer's policies, and say things that serve the organization's interests. For routine communications, reviews are usually conducted by the writer's manager. For sensitive communications, the reviewers may also include company executives, lawyers, public relations specialists, and many others. Commonly, the persons performing required reviews have the authority to demand changes.

Even when a review is not required, employees often ask coworkers or managers to look over their drafts. Through these voluntary reviews, employees learn additional ways to increase the usability and persuasiveness of their communications.

You can influence the quality of the reviews of your drafts. The helpfulness of the suggestions you receive will depend largely on the way you manage your interactions

WWW

To read additional information, see more examples, and access links related to reviewing drafts, go to Chapter 15 at **www.cengage.com/english/anderson7e.**

with your reviewers. The following six guidelines will help you develop expertise at eliciting supportive and helpful reviews of your drafts.

Several of these guidelines also include suggestions for reviewing other people's drafts effectively. If you are promoted to a supervisory or managerial position, the ability to review other people's drafts will be essential. Even if you are not a manager, you will almost surely be asked to look over someone else's draft, perhaps as early as your first months at work. Moreover, you will be evaluated on the value of the advice you offer.

Note: The guidelines are as useful for peer reviewing done in class as for reviewing conducted in the workplace.

Guideline 1 | Discuss the objectives of the communication and the review

Reviewing needs to be as reader-centered as any other writing activity. Don't just hand your draft to reviewers with a request that they "please look it over." Share the detailed information you developed while defining your communication's objectives. Describe the tasks you want to help your readers perform, the ways you want to influence your readers' attitudes, and the facts you have about your readers and their situation that will affect the way they respond. Only if they have this information can reviewers suggest effective strategies for increasing your communication's usability and persuasiveness.

Also discuss the scope and focus of the review. As the writer, you may have a good sense of areas that most need attention—whether organization, selection of material, accuracy, tone, page design, or something else. Without discouraging your reviewers from noting other kinds of improvement, guide them to examining what you think is most critical.

Similarly, if you are the reviewer and your writers don't tell you their communications' objectives or their desires for the reviews, ask for this information. You need it in order to be able to give the best possible assistance.

At work, drafts are sometimes reviewed in person with a supervisor or a group of coworkers

Guideline 2 | Build a positive interpersonal relationship with your reviewers or writer

The relationship between writers and their reviewers is not only intellectual but also interpersonal and emotional. The quality of this human relationship can greatly affect the outcome of the review process. When writers feel criticized and judged rather than helped and supported, they become defensive and closed to suggestions. When reviewers feel their early suggestions have been rebuffed rather than welcomed, they cease giving the advice the writer needs. The following strategies will help you build positive, productive relationships with your reviewers and writers.

When You Are the Writer

As a writer, interact with your reviewers in ways that encourage them to be thoughtful, supportive, and generous. Reviewers are more helpful if they believe you welcome their comments. Throughout your discussions with them, project a positive attitude, treating your reviewers as people who are on your side, not as obstacles to the completion of your work. Begin meetings by thanking them for their efforts. Paraphrase their comments to show that you are listening attentively, and express gratitude for their suggestions.

Ways to encourage your reviewers to help as much as they can

When your reviewers offer comments, stifle any temptation to react defensively, for instance by explaining why your original is superior to a suggested revision. Even if a particular suggestion is wrongheaded, listen to it without argument. On the other hand, don't avoid dialogue with your reviewers. When they misunderstand what you are trying to accomplish, explain your aims while also indicating that you are still open to suggestions. If your reviewers misunderstood what you were attempting, it's probably a sign that you aren't being clear and would benefit from their advice.

When You Are a Reviewer

When you are the reviewer, use these strategies to build a good relationship with the writer:

Ways to encourage openness to your suggestions

- **Begin with praise.** By doing so, you show the writer that you recognize the good things he or she has done, and you indicate your sensitivity to the writer's feelings. In addition, some writers don't know their own strengths any better than they know their weaknesses. By praising what they have done well, you encourage them to continue doing it and you reduce the chances they will weaken strong parts of their communication by revising them.

- **Focus your suggestions on goals, not shortcomings.** For example, instead of saying, "I think you have a problem in the third paragraph," say, "I have a suggestion about how you can make your third paragraph more understandable or persuasive to your readers."

- **Use positive examples from the writer's own draft to explain suggestions.** For example, if the writer includes a topic sentence in one paragraph but omits the topic sentence in another, cite the first paragraph as an example of a way to improve the second. By doing this, you indicate that you know the writer understands the principle but has slipped this one time in applying it.

- **Project a positive attitude toward the writer.** Even if you have the authority to demand changes, think of the writer as a person you want to help, not judge. If the writer feels you are judgmental, he or she will almost surely resent and resist the revisions you suggest.

Guideline 3 | Rank suggested revisions—and distinguish matters of substance from matters of taste

When you review a writer's draft, one of the most helpful things you can do is rank or prioritize your suggestions. By doing so, you help the writer decide which revisions will bring the greatest improvement in the least amount of time. Also, many writers can face only a limited number of suggestions before feeling overwhelmed and defeated. As a reviewer, look over your suggestions to determine which will make the greatest difference. Convey those to the writer. Keep the rest to yourself, or make it clear that they are less important. Ranking also helps *you* work productively. Often, you will not have time to review an entire draft in detail. Begin by scanning the draft to identify the issues that most need attention, then focus your effort on them.

When ranking suggestions, distinguish those based on substantial principles of writing from those based on your personal taste. If you suggest changes that would merely replace the writer's preferred way of saying something with your preferred way, you will have done nothing to improve the writing—and you will almost certainly spark the writer's resentment. Unfortunately, it can be difficult to determine whether we like a certain way of saying something because it is good in itself or because it matches our own style. When you are faced with this uncertainty, try to formulate a reason for the change. If all you can say is, "My way sounds better," you are probably dealing with a matter of taste. If you can offer a more objective reason—for instance, one based on a guideline in this book—you are dealing with a matter of substance.

One caution, however. Sometimes your sense that something doesn't sound right is a clue that a problem exists. For example, you may stumble over a sentence that doesn't sound right and discover, after closer examination, that it contains a grammatical error. If something sounds wrong, see whether you can find an objective reason for your dissatisfaction. If so, tell the writer. If not, let the matter drop.

So far, the discussion of this guideline has focused on the reviewers' perspective. When you are the writer, you can be helped greatly by your reviewers' rankings of their suggestions. If your reviewers don't volunteer a ranking, request one.

Guideline 4 | Explore fully the reasons for all suggestions

The more fully writers and reviewers explore the reasons for the reviewers' suggestions, the more helpful the review will be. Understanding these reasons can help you recognize and avoid similar problems in the future. Even when a suggestion seems useless, knowing the reviewers' reasons for making it can help you identify a problem you previously overlooked so you can devise your own way of solving it.

When you are acting as a reviewer, it's also in your interest to provide a full explanation of your suggestions. If you don't, the writer may think that you are simply expressing a personal preference and dismiss your suggestions, leaving the draft weaker than it might otherwise have been.

Full explanations also help the writer learn to write better. For example, imagine that you suggest rephrasing a sentence so that an echo word is at the beginning rather than the end. The writer may agree that your version is better but may not know the reason why unless you explain the *principle* that you applied.

Some reviewers withhold such explanations because they think the reasons are obvious. However, reasons that are obvious to the reviewer are not necessarily obvious to the writer. Otherwise, the writer would have avoided the problem in the first place.

Whenever possible, phrase your suggestions from the perspective of the intended readers. Instead of saying, "I think you should say it like this," say, "I think your intended readers will be able to understand your point more clearly if you phrase it like this." Such a strategy will help the writer take a reader-centered approach to revising the draft. It will also help you present yourself not as a judge of the writer's writing, but as a person who wants to help the writer achieve his or her communication objectives.

Guideline 5 | Present your suggestions in the way that will be most helpful to the writer

Whenever possible, discuss suggestions in person with your writer or reviewer. Conversation enables the writer to ask for clarification and the reviewer to provide full explanations. If you can't meet face to face, talk on the phone or through a free program such as Skype (www.Skype.com) or iChat (if you both have Mac computers). Or use e-mail.

Whether you are the reviewer or writer, talk with your partner about the best way to convey written suggestions. In particular, think about the method that will make it easiest for the writer to incorporate suggested revisions into the draft being reviewed. If you are the writer, consider providing your reviewer with a computer copy of your draft. Your reviewer can then insert suggestions and return the draft to you. You will then be able to use the desktop publishing program to accept or decline one by one, thereby eliminating the need to retype suggested text that you want to incorporate into your communication (see Figure 15.4).

If more than one person will review your draft, however, it can be inefficient to have each reviewer use a desktop publishing program to enter suggestions into your draft. Each reviewer will return his or her comments to you in a separate file. You will need to read back and forth among drafts to see if different reviewers make different suggestions about the same issue. You will also need to go repeatedly to the different files to extract suggestions you want to accept so you can create a single new draft that incorporates them all. To avoid this work, use a program like GoogleDocs (http://docs.google.com) or Buzzword (www.buzzword.com) that enables you to upload your draft to a central location on the Internet. All reviewers can use their browsers to access and insert comments into this single copy, greatly simplifying your work. As a bonus, because the reviewers can read each other's comments they can supplement rather than repeat one another's advice. Figure 15.5 shows an example.

Learn More

To learn about the use of echo words to smooth the flow of thought through a communication, turn to pages 226 and 227.

Learn More

Programs like Google Docs and Buzzword also help teams write collaboratively. See Guideline 7, Chapter 18 (page 466) for more information.

Try This

Use Google Docs or Buzzword to invite a friend or family member to comment on something you've written for a course or on your own.

FIGURE 15.4

Reviewer's Suggestions Inserted into the Writer's Draft Using a Desktop Publishing Program

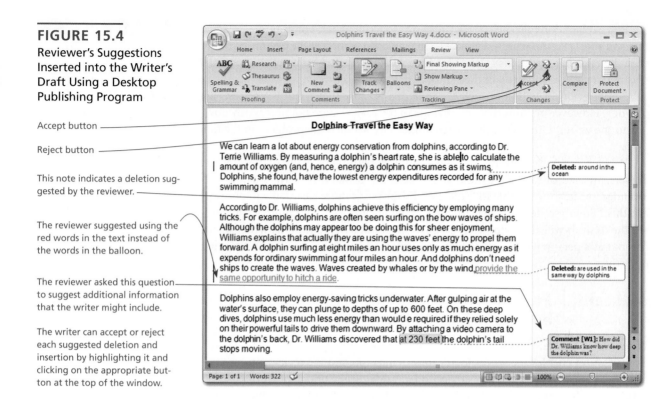

Accept button

Reject button

This note indicates a deletion suggested by the reviewer.

The reviewer suggested using the red words in the text instead of the words in the balloon.

The reviewer asked this question to suggest additional information that the writer might include.

The writer can accept or reject each suggested deletion and insertion by highlighting it and clicking on the appropriate button at the top of the window.

Guideline 6 | ## Ethics Guideline: Review from the stakeholders' perspective

Reviewers can be as helpful in evaluating stakeholder impacts as in identifying ways to increase a communication's effectiveness for its target readers. Be sure to ask your reviewers to assist you in this way. Tell them whom you've identified as stakeholders and how you think these individuals might be affected by your communication. Ask them to help you discover other stakeholders or impacts that you might have missed.

Similarly, when you review a draft written by someone else, ask about his or her assessment of its stakeholder impacts. Then, as you read, make your own judgments. Who will be affected? How? Should the stakeholders be consulted directly about the decisions and actions discussed in this communication? Can potential negative impacts on the stakeholders be reduced or eliminated? Is it ethical to take the proposed action at all?

Ethical problems can be difficult for writers and reviewers to discuss. None of us likes to be accused of acting unethically, so a writer may quickly become defensive. Furthermore, the writer and reviewer may hold quite different ethical principles. Consequently, whether you are the writer or the reviewer, you may find it helpful to refer to the ethical code of your employer or profession, rather than insist on your own ethical views. But remember that despite the difficulties of doing so, you have an obligation to raise ethical questions even if you have not been asked.

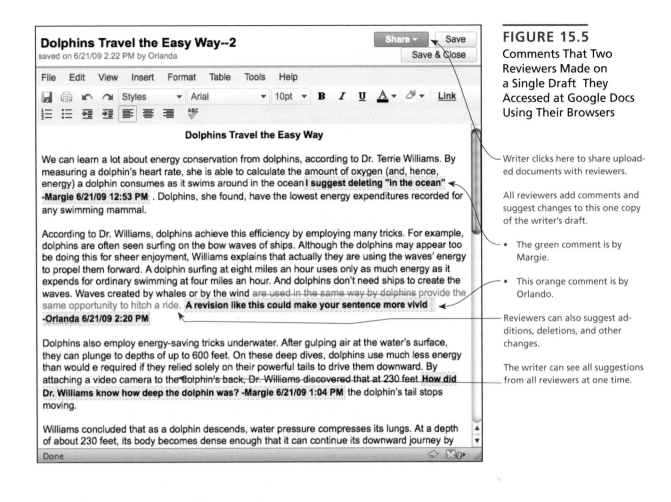

FIGURE 15.5
Comments That Two Reviewers Made on a Single Draft They Accessed at Google Docs Using Their Browsers

Writer clicks here to share uploaded documents with reviewers.

All reviewers add comments and suggest changes to this one copy of the writer's draft.

- The green comment is by Margie.

- This orange comment is by Orlando.

Reviewers can also suggest additions, deletions, and other changes.

The writer can see all suggestions from all reviewers at one time.

Writer's Guide for Reviewing

Figure 15.6 is a Writer's Guide you can use to ensure that you are checking your drafts in a reader-centered way.

GUIDELINES FOR MANAGING YOUR REVISING TIME

The two preceding sets of guidelines focused on ways you can identify improvements you could make in a draft—by checking the draft yourself and by having someone else look over your draft. The following guidelines take you one step farther in the revision process: choosing which of the possible improvements you should actually make.

WWW

To read additional information and see more examples related to revising, go to Chapter 15 at **www.cengage.com/english/anderson7e.**

FIGURE 15.6 **Writer's Guide for Reviewing Drafts**

To download a copy of this Writer's Guide, go to Chapter15 at www.cengage.com/english/anderson7e.

Writer's Guide
REVIEWING DRAFTS

1. When preparing for the review, do the following:

    ☐ Discuss the objectives of the review.

    ☐ Discuss the objectives of the communications.

    ☐ Build a positive relationship with the writer or reviewer.

2. When discussing comments and suggestions, do the following:

    ☐ Identify the revisions that are most important.

    ☐ Distinguish matters of substance from matters of taste.

    ☐ Explore the reasons for all suggestions.

3. Present your suggestions in the way that will be most helpful to the writer.

4. Review the draft from the perspective of its stakeholders.

In some cases, of course, you won't need to make any choices. You may find that your draft (probably not your first one) is already good enough to influence your readers in the way you intend. Or you will have adequate time to make all the changes that have been identified.

Often, however, writers at work face daunting decisions about which changes to make and which to forgo. The need for these decisions can arise for any of the following reasons:

*   When there isn't enough time to make all the revisions that checking, reviewing, or testing suggest.
*   When two or more reviewers give contradictory advice.
*   When a reviewer, such as a boss, suggests changes that the writer believes will actually diminish the communication's effectiveness.

The guidelines that follow will help you develop expertise at dealing effectively with these more complex revising situations.

Guideline 1 | Adjust your effort to the situation

This guideline suggests that you approach revising as if you were an investor. As you begin revising any communication, let your first concern be the amount of time and energy you might need to invest in it rather than in your other duties.

Some students think it is peculiar to ask how good a communication needs to be. "After all," they observe, "shouldn't we try to make everything we write as good as it can be?" On the job, the answer is "No."

Of course, all your communications should present their central points clearly and achieve an appropriate tone. At work, however, different communications need different levels of additional polish. For example, in many organizations, printed messages are expected to be error-free, but e-mail messages sent internally may contain spelling errors and grammatical mistakes as long as their meaning is clear and their tone inoffensive. Even among printed communications, different levels of polish are expected. In a survey of workers in a diverse mix of professions, researchers Barbara Couture and Jone Rymer (1993) found that the proportion of employees who "often" or "very often" make major revisions in communications they consider to be "special" is three times larger than the proportion who "often" or "very often" revise communications they consider to be "routine."

To determine how "good" your communication needs to be, do the following:

Ways to determine how good a communication has to be

- **Think about your communication's usability and persuasive objectives.** How carefully does your communication need to be crafted to help your readers perform their tasks? To affect your readers' attitudes in the way you want?

- **Consider general expectations about quality.** At work you usually need to polish your communications most thoroughly when addressing people at a higher level than you in the organizational hierarchy, when addressing people outside your organization rather than inside, and when trying to gain something rather than give something (for example, when writing proposals rather than reports). These general expectations can be translated into a list that ranks some typical communications according to the level of polish they usually need. Figure 15.7 illustrates a typical ranking. Of course, the ranking in your workplace may be different.

- **Look at similar communications.** Often, the level of quality needed is determined largely by what's customary. By looking at communications similar to the one you're preparing, you can see what level of quality has succeeded in the past.

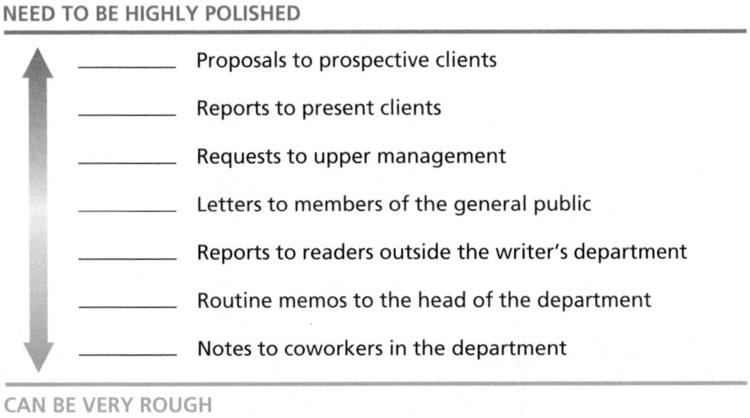

NEED TO BE HIGHLY POLISHED

_____ Proposals to prospective clients

_____ Reports to present clients

_____ Requests to upper management

_____ Letters to members of the general public

_____ Reports to readers outside the writer's department

_____ Routine memos to the head of the department

_____ Notes to coworkers in the department

CAN BE VERY ROUGH

FIGURE 15.7
Amount of Polish Typically Needed by Several Kinds of Communication

- **Ask someone.** Your boss or coworkers can be excellent sources of information about the level of quality needed, as can your intended readers.

Reasons for making your communications better than "good enough"

There are sometimes good reasons for wanting your communication to go beyond the minimum level needed to make your communication "good enough."

First, your writing is a major factor on which your job performance will be evaluated. In fact, people who are not involved with your work from day to day may base their opinion of you solely on your writing. There's a real advantage to having readers say not only, "Your report provided us with lots of useful information," but also, "And you wrote it so well!"

Second, communications prepared at work often have a wider audience than the writer assumes. Notes to coworkers and to your immediate manager can often be fairly rough. Sometimes, however, those people may pass your communications along to upper management or to members of other departments for whom you would want to prepare more polished communications. When determining what level of quality to strive for, consider who all your possible readers may be.

Guideline 2 | Make the most significant revisions first

After you've decided how much effort to invest in revising, identify the revisions that will bring the greatest improvement in the shortest time. If you spend fifteen minutes on revisions that result in small improvements rather than large ones, you have squandered your time.

Because the advice to make the most important revision first sounds so sensible, you may wonder whether anyone ever ignores it. In fact, many people do. They begin by looking at the first paragraph, making revisions there, and then proceed through the rest of the communication in the same sequential way. This procedure works well if the most serious shortcomings occur in the opening paragraphs, but not if they come in later passages the writers might not even reach before the time available for revising runs out. Also, some problems—for instance, inconsistencies and organizational problems—require simultaneous attention to passages scattered throughout a communication. A sequential approach to revising may ignore them altogether.

How to Rank Revisions

No single ranking applies to all communications. The importance of most revisions depends on the situation. For one thing, the relative importance of a certain revision is determined by the other revisions that are needed or desirable. Consider, for instance, the place of "supplying missing topic sentences" in the ranked list (Figure 15.8) that Wayne created after surveying the comments made by four reviewers of a proposal he had drafted. In comparison with the other items in the list, supplying topic sentences is relatively unimportant. However, in a draft with fewer problems in more critical areas, providing topic sentences might bring more improvement than any other revision Wayne could make.

MAKE GREAT IMPROVEMENT

↑

_____ Correcting errors in key statements

_____ Adding essential information that was overlooked

_____ Correcting misspellings

_____ Repairing obvious errors in grammar

_____ Fixing major organizational difficulties

_____ Supplying missing topic sentences

_____ Revising sentences that are tangled but still understandable

_____ Correcting less obvious problems in grammar

↓

MAKE SMALL IMPROVEMENT

FIGURE 15.8
Amount of Improvement That Various Revisions Would Make in Wayne's Proposal

Another factor affecting the relative importance of a revision is its location. Improvements in an opening summary would increase the usability and persuasiveness of a long report more than would a similar revision to an appendix. Revisions in passages that convey key points have greater impact than similar revisions to passages that provide background information and explanations.

A third factor can come into play when you are ranking revisions requested by reviewers: the importance of the person making the request. For obvious and practical reasons, you may want to give first attention to revisions requested by people in positions of authority, even if you would not otherwise have given high priority to these same revisions.

When you have limited time for revising, some of the revisions may never be made. By creating a ranked list, you ensure that the revisions you do complete are the ones that contribute most to your communication's effectiveness.

Be Sure to Correct Mechanical Problems

No matter how you rank other possible improvements, place the correction of mechanical problems—those that involve a clear right and wrong—at the top of your list, unless you are writing in a situation where you are sure that errors won't matter. Here are the major mechanical problems you should focus on:

- **Correctness.** Fix all errors in spelling, grammar, and punctuation as well as such things as the numbering of figures and the accuracy of cross-references.

- **Consistency.** Where two or more items should have the same form, be sure they do. For instance, revise parallel tables and parallel headings that should look the same but don't.

- **Conformity.** Correct all deviation from your employer's policies on such matters as the width of margins, the use of abbreviations, and the contents of title

Mechanical issues to focus on

pages. Even if your employer hasn't developed written policies on these things, remember that your communication must still conform to your employer's informal policies and expectations.

- **Attractiveness.** Be sure your communication looks neat and professional.

Problems in any of these four areas diminish your communication's persuasiveness by undermining your credibility as a person capable of careful work. They can also reduce usability by distracting readers from your main points, slowing comprehension, and even causing misunderstanding. Often, it's best to postpone revising mechanical matters until the end of your work on a communication unless you are going to give your draft to someone for review. There is no point in investing time in correcting passages that may later be deleted or changed. On the other hand, mechanical problems can distract reviewers from concentrating on more substantive issues.

Guideline 3 | **Be diplomatic**

When your drafts are reviewed, you may confront situations requiring considerable diplomacy. For example, managers or other influential reviewers may suggest revisions you know will weaken your communication. Without giving offense, you must persuade the reviewers to change their minds. One diplomatic approach is to refer to relevant reader-centered guidelines in this book and explain the reasons for them.

Another awkward situation arises when two reviewers give contradictory advice or make incompatible demands. Sometimes such conflicts are over writing strategy. In other cases, they are about something entirely different. While studying a small software company, researcher Stephen Doheny-Farina (1992) witnessed a months-long dispute over the draft of a business plan designed to raise money from investors. The disagreement, he discovered, was not over the writing but rather the company's mission, organization, and future plans.

If your draft is caught in a crossfire between contending parties, ask the reviewers to explain the reasons for their advice. You may be able to fashion a statement that is acceptable to all parties. If not, ask the reviewers to talk directly with one another and then tell you what they have decided.

Guideline 4 | **To revise well, follow the guidelines for writing well**

This guideline echoes an observation at the beginning of this chapter: When making changes, follow the same guidelines that you follow when drafting. Whether you are drafting or revising, the principles and guidelines for good communication remain the same. Whether you've found that you need to adjust your organization, polish your style, or recast a graphic, refer back to the relevant advice in this book. And always abide by the central advice of this book: Whether drafting or polishing any aspect of a communication, keep your readers' needs, expectations, preferences, and attitudes chiefly in mind.

Guideline 5 | **Revise to learn**

The first four guidelines for managing your revising time focus on ways in which revising can improve the effectiveness of your communications. Revising can also help you learn to write better.

When revising, you will often be applying writing skills that you have yet to master fully. By consciously practicing those skills, you can strengthen them for use when drafting future communications. Because you can learn so much from revising, it can often be worthwhile to continue polishing a draft after it is good enough to satisfy the needs and expectations of your readers and employer.

CONCLUSION

The complexities of working situations often require you to play several roles while revising. As an investor, you decide how much time to devote to revising a draft. As a diplomat, you negotiate with reviewers about changes they have requested or demanded. As a "student," you revise for the sake of what you can learn. The primary strategies underlying all the guidelines in this chapter are the same ones you should employ when you are planning and drafting any communication on the job: Consider the communication from the perspective of its readers, the contexts in which they will read and you are writing, and the communication's ethical impact on its stakeholders.

USE WHAT YOU'VE LEARNED

For additional exercises, go to **www.cengage.com/english/ anderson7e**. *Instructors:* The book's website includes suggestions for teaching the exercises.

EXERCISE YOUR EXPERTISE

1. Following the advice given in this chapter, carefully check a draft you are preparing for class. Then give your draft to one or more of your classmates to review. Make a list of the problems they find that you overlooked. For three of those problems, explain why you missed them. If possible, pick problems that have different explanations.

2. The memo shown in Figure 15.9 contains eighteen misspelled words. Find as many as you can. Unless you found all the words on your first reading, explain why you missed each of the words you overlooked. Based on your performance in this exercise, state in a sentence or two how you could improve your reading for misspellings and typographical errors.

3. The memo shown in Figure 15.9 has several other problems, such as inconsistencies and missing punctuation. Find as many of those problems as you can.

EXPLORE ONLINE

1. Exchange electronic files of the drafts for your current assignment with a classmate. Using your desktop publishing program, insert comments in your partner's file, then return it to him or her. Evaluate the comments your partner provided for their clarity and ease of use, while he or she does the same for your comments. Share praise for comments that were presented effectively, and share suggestions concerning comments that could be presented more effectively.

2. Open an account at Google Docs (http://docs.google.com) or Buzzword (www.buzzword.com). Upload a short communication you've prepared for this course. Share the uploaded documents with a classmate. When you receive an e-mail saying that a classmate has shared

FIGURE 15.9
Memo for Use in Expertise
Exercises 2 and 3

MARTIMUS CORPORATION
Interoffice Memorandum

February 19, 2011

From	T. J. Mueller, Vice President for Developmnet
To	All Staff
RE	PROOFREADING

Its absolutely critical that all members of the staff carefully proofread all communications they write. Last month we learned that a proposal we had submited to the U.S. Department of Transportation was turned down largely because it was full of careless errors. One of the referees at the Department commented that, "We could scarcely trust an important contract to a company that cann't proofread it's proposals any better than this. Errors abound.

We received similar comments on final report of the telephone technology project we preformed last year for Boise General.

In response to this widespread problem in Martinus, we are taking the three important steps decsribed below.

I. TRAINING COURSE

We have hired a private consulting form to conduct a 3-hour training course in editing and proofreading for all staff members. The course will be given 15 times so that class size will be held to twelve participants. Nest week you will be asked to so indicates times you can attend. Every effort will be made to accommodate your schehule.

II. ADDITIONAL REVIEWING.

To assure that we never again send a letter, report, memorandum, or or other communication outside the company that will embarrass us with it's carelessness, we are establishing an additional step in our review proceedures. For each communication that must pass throug the regular review process, an additional step will be required. In this step, an appropriate person in each deparmtent will scour the communication for errors of expression, consistancy, and correctness.

III. WRITING PERFORMANCE TO BE EVALUATED

In addition, we are creating new personnel evaluation forms to be used at annual salary reviews. They include a place for evaluating each employees' writing.

III. Conclusion

We at Martimus can overcome this problem only with the full cooperation of every employe. Please help.

a document with you, go to the document and enter reviewer suggestions. Return to the document you uploaded to see the comments your classmate made in it. For instructions on using Google Docs and Buzzword, go to www.cengage.com/english/anderson7e and click on Chapter 15.

COLLABORATE WITH YOUR CLASSMATES

This exercise will help you strengthen your ability to make constructive suggestions to a writer. (See this chapter's advice on reviewing other people's writing.)

1. Exchange drafts with a classmate, together with information about the purpose and audience for the drafts.

2. Carefully read your partner's draft, playing the role of a reviewer. Ask your partner to read your draft in the same way.

3. Offer your comments to your partner.

4. Evaluate your success in delivering your comments in a way that makes the writer feel comfortable while still providing substantive, understandable advice. Do this by writing down three specific points you think you handled well and three ways in which you think you could improve. At the same time, your partner should be making a similar list of observations about your delivery. Both of you should focus on such matters as how you opened the discussion, how you phrased your comments, and how you explained them.

5. Talk over with your partner the observations that each of you made.

6. Repeat steps 3 through 5, but have your partner give you his or her comments on your draft.

APPLY YOUR ETHICS

Find a print or online communication that you believe to be unethical or that you believe might possibly be unethical. Advertising is one possible source of such communications. Imagine that you are seeing it not as a finished communication but in draft form because its creator has asked you to review it. Plan your approach to raising and discussing the ethical issue with the person who drafted the communication. Because this person is free to accept or ignore your suggestions, you will need to convey your recommendation in a way that this person finds persuasive. Present your plan in the manner your instructor requests.

16 | Testing Drafts for Usability and Persuasiveness

CHAPTER GUIDELINES

By following the reader-centered strategies described elsewhere in this book, you try your best to draft a communication that your readers will find highly usable and persuasive. However, you can't know for sure how your readers will respond until you actually place your draft in their hands. This chapter describes ways you can *test* a draft so that you can identify places that don't work the way you intended. With this knowledge, you can make improvements that increase the effectiveness of the final version you will deliver to your readers.

Companies such as Microsoft and IBM use testing extensively for the online and print instructions that accompany their products. By identifying problems and making improvements before products are launched, these companies reduce customer frustration and also save the significant amounts of money it would take to staff help lines or otherwise help customers out of problems that could have been avoided with clearer communication. In addition to instructions, drafts of websites, informational documents, and a wide variety of other communications are tested by both commercial organizations and government agencies. Testing is such an important activity in some organizations that they employ testing specialists and build special test facilities like the ones shown in Figures 16.1 and 16.2. Even without such facilities, however, you can increase the effectiveness of your communications by using the same basic strategies that testing experts employ.

The goal of this chapter is to guide you through the process of planning, conducting, and interpreting a test, whether you are testing a project for your technical communication course or a document you are writing on the job.

CHAPTER 16

DEFINING OBJECTIVES

PLANNING

CONDUCTING RESEARCH

DRAFTING

REVISING

WWW

To read additional information, see more examples, and access links related to this chapter's guidelines, go to Chapter 16 at **www.cengage.com/english/anderson7e.**

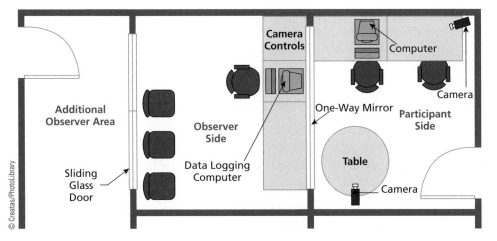

FIGURE 16.1
Layout of a Facility for Testing Computer Manuals

Test readers can use a draft manual or website in the room on the Participant side.

On the Observer side, writers and testing specialists watch through a one-way mirror. They can also record all of the participant's actions and keystrokes.

© Creatas/PhotoLibrary

FIGURE 16.2
Professional Testing Facility

The test session is recorded on a video camera.

In the Test Room, the person with his back to the Observation Room asks readers about communications they read or view simultaneously.

The Test Room can be set up to test communications used in large spaces, such as instructions for assembling, servicing, or operating large equipment.

The observers are looking through a one-way mirror.

© Spencer Grant/PhotoEdit

THE LOGIC OF TESTING

The basic logic of testing is simple. You ask one or more people, called *test readers*, to read and use your draft in the way your intended readers will read and use your finished communication. Then, by various means, you gather information from your test readers that will enable you to predict how other readers will respond. If your test is constructed effectively, the parts that worked well for your test readers will probably work well for your target readers. Where the test readers encounter problems, your target audience probably will too.

This chapter's first eight guidelines tell how to get the most benefit from your test. The ninth guideline explains how to test communications addressed to readers from other cultures. The tenth will help you treat your test readers ethically.

Guideline 1 | Establish your test objectives

At a general level, all your tests will aim to answer two questions:

- How can I improve my draft? In the workplace, most tests emphasize the diagnostic goal of learning ways the draft can be improved.
- Is my communication good enough? It is always helpful to learn also how close your draft is to a "final" draft you can deliver to your target audience.

The first step in creating a test for answering these questions is to define its objectives. Then you can tailor your test to provide the specific information you need.

1. **Review your communication's usability and persuasiveness objectives.** These objectives, which you developed by following the guidelines in Chapter 3, tell exactly how your communication is intended to affect your readers while they are reading it. Your test should help you determine where your draft succeeds in achieving these objectives and where it needs improvement.

2. **Identify the features you hope will contribute most to your communication's usability and persuasiveness.** For longer communications, you may not be able to test everything. Therefore, you need to focus on what you believe will have the greatest impact on your target readers' responses.

3. **Identify any features about which you feel unsure.** You can design your test to determine their effectiveness.

You can encapsulate your test's objectives in a set of questions you want your test to answer. How well does my draft support my readers as they perform their first task? Their second one? How effective is it in influencing their attitudes in the way I want? How well do the headings work? Are the technical terms I used in the introduction understandable to them?

If you wish to determine whether your draft is "good enough," you may also want to establish *measurable criteria* for your test. For example, if you are drafting an instruction manual, you might decide that the manual is good enough if it enables test readers to perform their tasks in less than two minutes or with only one error that is easily corrected.

However, in many cases a subjective judgment will be fully adequate: Your draft is good enough if it enables your test readers to perform their tasks without much difficulty and if they say your draft had a positive impact on their attitudes.

WWW

For more information about establishing measurable criteria for testing, go to Chapter 16 at **www.cengage.com/ english/anderson7e.**

Guideline 2 | Pick test readers who truly represent your target readers

To construct a test that accurately predicts how your target audience will respond to your finished communication, you must pick test readers who truly represent your target readers. If you are writing instructions for plumbers or computer programmers, choose plumbers or computer programmers for your test readers. If you are creating a website for adults who suffer from asthma, choose adult asthma sufferers. You could actually make your communication *less* effective for your intended readers if you made revisions based on responses from people who are not in your target audience.

If it is impossible to locate test readers from your target audience, find people as similar as possible to that audience. Begin by reviewing the knowledge of your readers that you developed by following the guidelines in Chapter 3. Pay special attention to your target readers' familiarity with your subject. If you select test readers who already know what your communication is intended to convey, they will probably

The selection of test readers is a critical part of test design.

Try This

Using the search term "usability lab" or "usability test," find a company whose website enables people to sign up to participate in user tests. Can you find a site that conducts user tests on products or services that interest you?

understand passages that could baffle your target readers. Consequently, they won't help you detect needed improvements.

On the other hand, if your test readers know less than your target readers about your subject, their puzzlement could lead you to include background information and explanations your target readers neither need nor want. Imagine, for instance, that in one of your classes you have drafted a report to engineers who design robots. If you can't enlist such engineers as test readers, ask engineering students (rather than history majors) at your school—and ask engineering students who've studied robotics rather than those who haven't.

How many test readers are enough? Because there is generally little variation in the ways readers respond to step-by-step instructions, two or three test readers may be sufficient for them. In contrast, people vary considerably in the ways they interpret and respond to a report or informational website. Common practice is to use about a dozen test readers for such communications. However, testing with even a single reader is infinitely better than no testing at all.

Even a single test reader can provide valuable insights.

Guideline 3 | Focus on usability: Ask your test readers to use your draft the same way your target readers will

Whenever you test a draft, one of your primary goals will be to evaluate its usability for its target readers. To learn where you need to make revisions, you must construct a test in which the test readers use your draft to perform the same tasks your target readers will want to perform with the finished communication.

At work, most readers' tasks usually fall into three major categories.

- Perform a procedure, as in reading instructions.
- Locate information, as in looking for a certain fact in a reference manual or website.
- Understand and remember content, as when trying to learn about something through reading.

For many communications, all three types of tasks are important. For example, when people consult the owner's manuals for a desktop publishing program, they usually want to *locate* directions for a particular procedure, *understand* the directions, and then use the directions as a guide while they *perform* the procedure. Although the three types of tasks are often bound together, each can be tested in a different way. Consequently, the following sections discuss separately the design of performance, location, and understandability tests.

Performance Tests

To test a draft's effectiveness at helping its readers perform a procedure, you could simply give the draft to your test readers and watch them use it. To get the best results, however, you must construct several elements in your test so that your test readers' reading situation resembles as closely as possible the situation of your target readers.

Try This

Go to the usability website created by the government of any country (e.g., www .usability.gov for the United States). Find its advice about usability testing. On what points does its advice agree with the advice in this chapter? On what points does its advice disagree?

STRATEGIES FOR PERFORMANCE TESTS

Tasks	Ask your test readers to perform the same tasks your target readers will perform.
Location	Conduct the test in the same setting as the target readers would use.
Resources	Provide the test readers with the same tools, equipment, reference materials, and other resources that the target readers would have—but not additional ones.
Information gathering	Gather information in ways that enable you to observe the details of the test readers' efforts without interfering in their use of your draft.

The above strategies will help you plan the four major elements of *performance tests.*

To see how you might apply these guidelines, consider the way Imogene tested instructions that would enable ordinary consumers to install a car radio and CD player manufactured by her employer.

First, Imogene recruited two friends as her test readers. Both owned cars and neither had made a similar installation before, so both represented her target readers. Imogene asked Rob to install the equipment in his garage at home and Janice to install hers in the parking lot of her apartment building. These are work areas that would be used by Imogene's target readers. Rob already had the necessary tools. Imogene supplied Janice with a power drill and Phillips head screwdriver, but nothing else because these might be the only tools an ordinary consumer would have.

While Rob and Janice worked, Imogene took detailed notes, using the form shown in Figure 16.3. During the test, she focused only on the left-hand column, writing down everything that seemed to indicate that Rob or Janice was having difficulty. So that she could pay full attention to the details of Rob's and Janice's efforts, she waited to fill in the other columns until after the test had been completed. To obtain a full sense of what Rob and Janice were thinking as they used her instructions, she asked them to "think aloud," verbalizing their thoughts throughout the process.

Courtesy of Adage

Courtesy of Central and East European Center for Cognitive Science

Some usability labs have equipment that mirrors the test user's computer monitor (top) or detects where on the screen a test user is looking at each moment (bottom).

As he worked on the installation, Rob asked Imogene several questions. Instead of answering, she urged him to do his best without her help. If she had begun to provide oral instructions, she would no longer have been testing her written ones.

FIGURE 16.3
Test Observation Form

After the test, you can ask the test reader to tell you what caused the problems you observed.

Observation Sheet for User Tests

Problem (What difficulty did the reader have?)	Interpretation (What might have caused the difficulty?)	Solution (What might prevent this difficulty?)
After completing a step, Rob sometimes hunted around the page for his place.	1. The steps don't stand out plainly enough.	1. Enlarge the step numbers. 2. Print the steps in bold so they are easier to distinguish from the notes and cautions.
At Step 7, Rob couldn't find the hole for the mounting bracket.	1. Rob couldn't understand the figure.	1. Simplify the figure by eliminating unnecessary details.

However, she did assist at a point where Rob was completely stumped. Without her help, he would have had to stop work altogether, and Imogene wouldn't have been able to find out how well the rest of her instructions worked.

For some communications, you may need to create *scenarios*, or stories, that tell your test readers the tasks you want them to perform. For example, Shoon has drafted a website for a company that sells sporting goods on the Internet. In his performance test, Shoon asked his test users to imagine that they want to purchase a product for some particular activity, such as day hikes along easy trails or a weeklong ascent of a glacier-wrapped mountain peak. He then requested that they use his draft of the website to determine the appropriate products and order them.

Several other features of the test Shoon conducted illustrate variations that are possible when applying the strategies for performance tests. Because Shoon's website offered so many different kinds of products, it was not feasible for Shoon to ask his test readers to use every part of the site. Instead, he asked them to use only certain carefully selected parts of the site. Also, instead of using a worksheet to record information, Shoon placed a video camera behind his test users to record the entire test for later study. To learn what his test users were thinking as they used his website, Shoon engaged them in a conversation in which he continuously asked them what they were trying to do and how they were attempting to do it. He did not, however, tell them how to overcome problems they encountered. He chose this technique, developed by researchers M. Ted Boren and Judith Ramey (2000), because it is especially useful for testing online communications, where users often work so quickly that it is difficult to follow their actions.

Location Tests

Location tests closely resemble performance tests: You give your draft to your test readers, asking them to find specific pieces of information or the answer to a particular question as rapidly as possible. This procedure enables you to evaluate the effectiveness of the headings, topic sentences, table of contents, and other guideposts that reveal the organization of print communications (see Chapter 8). For websites and other online communications, location tests can help you assess the organization of your site and the navigational aids it contains. With location tests, as with performance tests, you must simulate your target readers' reading situation as closely as possible in order to achieve the most helpful results.

Understandability Tests

In an *understandability test,* you ask your test readers to read your draft and then you ask them questions designed to determine whether they understood it accurately. Understandability tests often are combined with other kinds of tests. For example, because readers cannot use information unless they understand it, understandability is an element in every performance test, such as those designed by Imogene and Shoon. Similarly, an understandability test can be combined with a location test by asking test readers to use the information they find.

If you want to test understandability in isolation, ask your test readers to read your draft and then pose questions about what they've read. Several types of questions are suggested below. Most were used by Norman, an insurance company employee, to test the following paragraph from a draft version of a new automobile insurance policy.

> In return for your insurance payments, we will pay damages for bodily injuries or property loss for which you or any other person covered by this policy becomes legally responsible because of an automobile accident. If we think it appropriate, we may defend you in court at our expense against a claim for damages rather than pay the person making the claim. Our duty to pay or defend ends when we have given the person making a claim against you the full amount of money for which this policy insures you.

A passage Norman tested

- **Ask your test readers to recognize a correct paraphrase.** By testing your test readers' ability to recognize a paraphrase, you can determine whether they have understood your communication well enough to identify the same meaning expressed in different words.

  > **True or false:** We will defend you in court against a claim when we believe that is the best thing to do.
  >
  > When will we defend you in court? (a) When you ask us to. (b) When the claim against you exceeds $20,000. (c) When we believe that is the best thing to do.

Questions to determine whether test readers accurately recognize a paraphrase

- **Ask your test readers to create a correct paraphrase.** By asking them to paraphrase, you can find out whether they understood well enough to explain your message in their own words.

A request that test readers create a paraphrase

> In your own words, tell when we will defend you in court.

- **Ask your test readers to apply your information.** To test your readers' ability to apply information, create a fictional situation in which the information must be used. Norman created the following question to test the understandability of one section of his draft (not the section quoted above).

A scenario used to determine whether test readers understood Norman's draft well enough to apply its information.

> You own a sports car. Your best friend asks to borrow it so he can attend his sister's wedding in another city. You agree. Before leaving for the wedding, he takes his girlfriend on a ride through a park, where he loses control of the car and hits a hot dog stand. No one is injured, but the stand is damaged. Is that damage covered by your insurance? Explain why or why not.

Guideline 4 | Focus on persuasiveness: Learn how your draft affects your readers' attitudes

To test the persuasive impact of a draft, compare your test readers' attitudes before and after reading. At work, user tests usually focus on the readers' attitudes toward one or both of the following:

- **The *subject matter* of the communication.** Among other things, Imogene would look for positive changes in the readers' attitudes toward the radio/CD player and the job of installing it by themselves. Among other things, Shoon would want to determine his website's ability to prompt positive changes in his test readers' attitude toward products sold there.
- **The *quality* of the communication.** Imogene would be concerned with test readers' attitudes toward the instructions she drafted and Shoon toward the website he created.

There are several ways to obtain information about your test readers' attitudes before and after they read your draft. Once they've finished reading, you can ask how their attitudes have changed. This approach may yield flawed results, however, because readers sometimes misremember their earlier attitudes.

To obtain more reliable results, ask the same question before and after their reading. Avoid questions that can be answered with "Yes" or "No." Attitude changes are usually a matter of degree: "I felt a little positive, but now I feel very positive." Here are two effective types of questions:

- **Open-ended questions.** "How do you feel about buying sporting goods online?" "Describe your level of confidence that you could easily install a radio/CD player in your car using a set of printed instructions."
- **Questions constructed around a scale.** "On a scale of 1 (very appealing) to 5 (very unappealing), how do you feel about buying sporting goods online?"

Guideline 5 | Interview your test readers after they have read and used your draft

By interviewing your test readers after they've used your draft, you can often obtain a wealth of insights you couldn't otherwise have discovered. Here are five especially productive questions.

- **What were you thinking when you encountered each problem?** Even when test readers try to speak their thoughts aloud as they read your draft, their account may be incomplete. For example, some test readers become so engrossed in their task that they forget to speak. Or, their thoughts fly faster than they can report them.

- **How did you try to overcome each problem?** Test readers' strategies may suggest ways you can improve the presentation of your information.

- **How do you respond to specific elements of my communication?** Ask about any element about which you'd like feedback. Imogene might ask, "Do you find the page design appealing?" Shoon could inquire, "Do you like the website's colors?"

- **What do you suggest?** Readers often have excellent ideas about ways a communication can better meet their needs.

- **How do you feel about the communication overall?** When responding to such an open-ended question about their feelings, test readers may reveal some very helpful information about your communication's usability or persuasiveness that your other questions didn't give them an opportunity to express.

Interviewing can be informative and enjoyable.

Guideline 6 | Avoid biasing your test results

The value of your test depends on how closely your test readers' responses to your draft match those of your target audience. Anything that diminishes the closeness of this match can cause you to miss needed changes or to revise in counterproductive ways. Here are some strategies for avoiding bias.

AVOIDING BIAS

- **Remain unobtrusive.** Let the test readers focus on their tasks. If they are overly conscious of being observed, they may have difficulty concentrating.

- **Refrain from intervening unless absolutely necessary.** Even if your test readers ask for help, request that they rely only on the draft to solve problems. Intervene only if you must do so to enable them to continue with other parts of the test or if they are about to injure themselves or others.

- **Phrase questions in an unbiased way.** When interviewing test readers, avoid phrasing questions in ways that seem to steer them toward a certain answer. Your test readers may comply, thereby depriving you of accurate information needed as the basis for solid, reader-centered revisions.

- **Ask someone else to conduct your test—if you can watch.** Research indicates that test readers are reluctant to criticize a draft to the writer. By asking someone else to conduct your test, you increase your chances of obtaining unbiased responses. Note, however, that someone else's written or oral summary of a test cannot provide you with as many detailed, helpful insights as actually watching the test. Have someone else conduct your test only if you can watch, whether by sitting nearby, viewing the test through a one-way mirror, or watching a videotape of it.

- **Deemphasize your relationship to the draft.** If you conduct the test yourself, avoid emphasizing that you are the writer. When interviewing your test readers, welcome criticisms. If you seem disappointed or unreceptive to criticisms, test readers will stop making them. Assure the readers that all comments are helpful to you.

Guideline 7 | Interpret your test results thoughtfully

As explained at the beginning of this chapter, testing involves prediction. Based on the responses of your test readers, you predict how members of your target audience will respond. However, every reader is an individual. Different persons respond in different ways to any communication.

Your challenge when interpreting test results is to determine whether the responses to a particular paragraph, graphic, or design element by your test readers is likely to be typical of the responses that your larger audience will have. There's no surefire way to answer that question. However, answering the following three can be helpful.

- **How many test readers did you have?** The more you have, the more likely that responses they share represent your target audience's likely responses.

- **In what ways do your test readers differ from your target readers?** Do they have a different level of prior knowledge about your subject? Different expectations and needs? You will need to make such comparisons in order to determine whether your target readers would respond differently than your test readers did.

- **What did you observe during the test that can help you interpret the results?** Sometimes your observations of test readers can help you interpret results. For example, if you notice that some test readers rushed through the instructions you are testing, you might conclude that their hurry, not a deficiency in your draft, may account for some of the difficulties they encountered.

Guideline 8 | Test early and often

You can often increase your efficiency by conducting many small user tests as you are working on your draft rather than waiting to conduct one large one when your draft is complete.

In fact, Shoon conducted his first user test while he was still planning his website. On a sheet of paper, he sketched a site map showing the major elements he planned to include in his site and the structure in which he planned to organize them. Shoon also sketched a homepage and several second-level pages. He then showed these pages to test users. For the site map, he asked what was missing and what seemed to be misplaced. For the sample web pages, he asked what his test users thought would happen if they clicked on each of the navigational buttons. Using the insights he gained, Shoon was able to refine his site's organization and its chief navigational features before he began the time-consuming work of building the site itself.

Additionally, when Shoon completed the first major segment of his site, he conducted another user test on it. Through this test, he not only identified changes he could make in this segment but also enabled himself to design other segments so they were done right the first time. If he'd waited to complete the entire site before testing, he would have had to make the changes in many places, instead of just one.

Guideline 9 | Global Guideline: With communications for readers in other cultures, choose test readers from the culture

Testing communications that will be used in other cultures can present special challenges. For reasons explained throughout this book, the language, images, and other aspects of a communication that seem perfectly clear and appropriate in one's own culture can be incomprehensible, laughable, or offensive in another. When preparing communications for audiences outside your own culture, make a special effort to use test readers that represent those audiences. Readers from different cultures can approach communications from different perspectives and expectations. For example, researcher Daniel D. Ding (2003) found significant differences between instruction manuals for installing a water heater that were written in China and in the United States. The Chinese manual describes a series of ideal relationships among the parts but does not describe the specific actions the installer would take to achieve these relationships. Installers must bring their experience to the instructions in order to perform the task. Ding explains that this understanding of the way instructions should be written emerges from Chinese culture, in which context and individual objects are seen as a unity. In contrast, the U.S. instructions consist of lists of actions, described in detail, so that the installation can be accomplished independent of the installer's previous knowledge. These substantial differences illustrate how important it is for writers in one culture who are addressing readers in another culture to use test readers from that other culture.

If it is impossible for you to arrange for such test readers, it will be particularly important for you to have your communication reviewed by persons who are knowledgeable about those audiences.

Guideline 10 | **Ethics Guideline: Obtain informed consent from your test readers**

An ethical principle developed for research in medicine, psychology, and many other fields also applies to the testing of the drafts you write at work or in class. This principle states that the people involved should be volunteers who have agreed to participate after being fully informed about what you are asking them to do.

For certain kinds of research, this is not only an ethical requirement but also a legal one. Researchers must tell potential volunteers about the study in writing and obtain in writing the volunteers' consent to participate.

Although such legal requirements probably will not apply to the testing you do of draft communications you prepare at work, the ethical principle still does. When asking others to try out something you have drafted, let them know the following:

WWW

For information on the legal requirements that apply to certain tests, go to **www .cengage.com/english/ anderson7e.**

- The test's purpose
- The things you will ask them to do during the test
- The time required for the test
- Where the test will take place
- Any potential risks to them (some lab procedures, for instance, do involve risk if not done properly)
- Their right to decline to participate
- Their right, if they do volunteer, to stop participating at any time

Figure 16.4 shows a sample informed consent form.

WWW

To see a proposal for testing and a report on the results of a test, go to Chapter 16 at **www.cengage.com/english/ anderson7e.**

CONCLUSION

Figure 16.5 shows a planning guide that will help you learn how your readers will respond to a communication you are drafting. With the insights your tests produce, you will have the surest possible guide to revising your drafts so you can create final versions that work for your readers in exactly the ways you want them to.

Informed Consent

I am seeking volunteers to test a _____ that I am creating as part of my work for_____.

If you are interested in volunteering, please read the following information carefully. If you decide to participate, please sign below.

1. In the test, you will use a draft of _____.

2. You will not encounter any risk of harm in this process.

3. The purpose of this test is to enable me to identify ways to make my draft more effective. It is not a test of your abilities.

4. I will observe you as you read and use my draft. Afterward, I will ask you questions about the draft.

5. The entire test will take between _____ and _____ minutes.

6. You may stop participating in this test at any time.

7. I will conduct the test at a time that is convenient for you.

8. The test will take place at _____.

9. I will be glad to answer any questions you have about the test.

_____

I have read and understood the description given above concerning the test. I volunteer to participate.

Name _____

Signature _____

Date _____

FIGURE 16.4
Sample Informed Consent Form

Using this informed consent form, a student efficiently communicated the information needed to treat his test readers ethically.

- The writer tells potential volunteers what they would be doing during the test.
- The writer describes the risks involved. (There are none in this case. For tests that involve risk, the statement would need to be changed.)
- The writer explains the test's purpose.
- He lets potential volunteers know that they would also be asked to answer questions.
- The writer explains how much time he is asking potential volunteers to give.
- He explains that volunteers can stop participating at any time.
- The writer tells when and where the test will take place.
- The writer provides a place for volunteers to confirm that they are participating voluntarily.

Learn More

For a copy of this form that you can download and edit, go to Chapter 16 at **www .cengage.com/english/ anderson7e**.

For usability tests that are subject to federal regulation, a different informed consent form is used. For information go to Chapter 16 at www.cengage.com/english/ anderson7e.

FIGURE 16.5   **Writer's Guide for Testing Drafts**

To download a copy of this Writer's Guide, go to Chapter 16 at
www.cengage.com/english/anderson7e.

Writer's Guide
TESTING DRAFTS

Objectives of the Test

1. What are your communication's usability objectives?

2. What are its persuasive objectives?

3. What features or parts do you need to test for the following reasons?

   ☐ Because they are so crucial to your communication's success

   ☐ Because you are unsure how well they will work

Test Readers

1. Who are your target readers?

2. Can you find test readers from this group? If you cannot, what test readers would closely resemble your target readers? What are the important differences between your test readers and the target readers?

Test Readers' Tasks

1. What will you ask your test readers to do so you can evaluate your draft's usability?

2. What will you ask your test readers to do so you can evaluate its persuasiveness?

Resources

1. What resources will your target readers have that you should provide to your test readers?

Location

1. Can you conduct your test where your target readers will use your communication?

2. If not, how can you simulate that location?

Information Gathering

1. How will you observe what your test readers do while using your draft?

2. How will you learn what your test readers are thinking as they use your draft?

3. What questions will you ask your test readers when you interview them after they've finished using your draft?

Avoiding Bias

1. How will you avoid biasing your results?

Ethics

1. What do you need to say in your informed consent form in order to treat your test readers ethically?

USE WHAT YOU'VE LEARNED

For additional exercises, go to **www.cengage.com/english/anderson7e.** *Instructors:* The book's website includes suggestions for teaching the exercises.

EXERCISE YOUR EXPERTISE

1. Explain how you would test each of the following communications:
   a. A display in a national park that is intended to explain to the public how the park's extensive limestone caves were formed.
   b. Instructions that tell homeowners how to design and construct a patio. Assume that you must test the instructions without having your test readers actually build a patio.

2. Following the guidelines in this chapter, design a test for a communication that you are now drafting.

3. Complete the assignment on user testing given in Appendix B.

4. Design a user test for your school's website. Imagine that the target audience is high school students who are deciding what college to attend. Who would you select as test readers? How would you define the objectives of the test? What would you ask them to do, and how would you gather information from them concerning the site's usability and persuasiveness?

EXPLORE ONLINE

At one of its websites, the United State Government provides instructions for conducting usability testing (www.usability.gov). Some of its agencies also publish instructions for usability testing, for example the National Aeronautics and Space Administration (http://www.hq.nasa.gov/pao/portal/usability/overview/index.htm). After visiting one of these sites, write a memo to your instructor explaining why the government feels that usability testing is so important that it publishes instructions for performing the tests.

COLLABORATE WITH YOUR CLASSMATES

Interview a classmate about a project he or she is drafting. Then design a test that will help your partner identify ways that he or she could revise the draft to make it more usable and persuasive for its target audience.

APPLY YOUR ETHICS

Following the advice given in Guideline 10, develop an informed consent form that you could give to people you ask to volunteer as test readers of a draft you are preparing for your class. How could you recruit volunteers so that they won't feel pressured to agree to be test readers even though they would prefer not to? Are there any aspects of your test that could conceivably harm your test readers physically, emotionally, or socially? If so, how can you alter your test to eliminate this risk?

PART VIII

APPLICATIONS OF THE READER-CENTERED APPROACH

445

17 | Communicating and Collaborating in the Globally Networked World

We live and work in a globally networked world. An ever-growing variety of Internet resources link us with people almost anywhere in the world, enabling us to instantaneously correspond with, talk to, see, and collaborate with them. At work, many employees routinely communicate via the Internet with coworkers, clients, and suppliers on other continents with the same ease as when sending a message to someone on the next floor of their own building. Some employees now write collaboratively with colleagues who live thousands of miles away in different countries and cultures. In some international corporations, work on a large project follows the sun around the earth. After finishing a day's work, employees in Europe send it to the United States, where another group works on it until the end of their day, when they forward it to Asia. Besides increasing the range of people with whom you will interact in your career, the Internet has also introduced new forms of communication and new ways for writing teams to create communications. For instance, using their web browsers, employees (or students) who are in different locations can simultaneously edit a single copy of a report or proposal they are preparing together.

In this chapter, you learn how to use Internet services to do the following in the globally networked working world:

- Correspond digitally
- Write collaboratively
- Meet virtually

WWW

For additional information, examples, and exercises related to this chapter, visit this book's website at **www.cengage.com/english/anderson7e** and click on Chapter 17.

All of the Internet resources described in this chapter are available for free. Consequently, you can put the chapter's information and advice to immediate use in your classes, extracurricular and service organizations, and personal life.

CORRESPONDING DIGITALLY

A variety of digital technologies are replacing paper as the way employees send messages to coworkers, clients, and customers. E-mail is the most widely used. Because of its speed and efficiency, it has replaced letters and memos for much workplace and personal correspondence. Ironically, e-mail is now seen as too slow and cumbersome in many situations, where it is being replaced by instant messaging, blogs, and miniblogs. For both work and personal use, all technologies for digital correspondence have the added convenience that they enable a person to send and receive messsages on cell phones and other handheld devices.

E-mail

In addition to speeding your message to its readers, e-mail has the additional appeal of looking familiar. Like a memo, it has spaces at the top for the writer's

where reader
science, knitt
pictures, link
lists of messa
appearing at t
at once, enab
to the origina
with e-mail,

Although
originally cre
neers who w
online locatic
respondence,
common goa
ways for wor
information i
and General
only very sho
which limits

Like all th
for you to set

Guidelin

Despite the d
in the globall

1. **Keep it**
2. **Use a si**
   informa
3. **Stick to**
   message
4. **Follow**
   and org
   respond
   Adapt t
5. **Use list**
   some e
6. **Don't i**
   digital

name, reader's name, date, and subject of the message. E-mail also has several limitations. When you and another person exchange several messages on a topic, the relevant information may be scattered in multiple e-mails. When you and three or more people are corresponding, the exchanges can become very confusing. In addition, as we all have the misfortune of knowing, e-mail accounts are susceptible to spam.

Although e-mail still holds a preeminent place worldwide, its popularity varies by region. It is the primary online communication tool for 87 percent of people in North America, 74 percent in Europe, and only 58 percent in the Asian-Pacific (Epsilon, 2009). Especially when corresponding internationally, learn the preferences of the other persons.

The Writer's Tutorial on page 450 provides advice about using e-mail effectively on the job. Its most important suggestion is to provide a specific, informative subject line. Many people use subject lines to determine whether to even open messages in their overstuffed inboxes. The subject line also helps readers find e-mails they want to review after reading them the first time. The Guidelines for Online Correspondence (page 451) also provide important advice for writing e-mail at work.

Instant Messaging

As fast as e-mail can be, instant messaging can be even faster. To read an e-mail, people must open their e-mail programs, pick the e-mail out from the other messages in their inboxes, and then open the message itself. This process can create a long delay between sending a message and readers reading it. In contrast, instant messages appear instantly on the computer, cell phone, or other screen of the intended reader. Consequently, they are noticed immediately and may receive an immediate reply. The exchange of instant messages can be so rapid that IM communication is called a *chat*.

In the Asian-Pacific region, instant messaging is the preferred tool for 28 percent of the people who communicate online (Epsilon, 2009).

AOL, Google, and many other Internet companies offer free instant messaging accounts. With these same services, you can also conduct audio and video chats if your computer has a microphone and a webcam. Text message services offered by telephone companies send instant messages over telephone lines rather than the Internet.

Blogs

A blog is a website that can be updated continuously. Its main feature is a list of messages organized with the most recent one at the top. Blogs are most familiar as personal online journals, social networking sites such as Facebook, or websites

Try This

If you are not already using instant messaging, sign up for an IM service at www.google.com, www.aol.com, or another service listed under Chapter 17 at this book's website at **www.cengage.com/english/anderson7e**. Try audio and video chats as well as text chats.

Provide an informati
specific subject line.
Readers use this line
decide whether to op
e-mail and to find e-
they want to review a
reading them the first

Highlight main point:
and make reading ea:
by using bullets.

Use short paragraphs
and place blank lines
between them to
promote reading ease

Tell readers how to
contact you by using
your e-mail system's
"signature" feature.

To

Learn More

For advice about choosing the technologies best suited to a particular project, see Chapter 18's Guideline 7 (page 466).

Try This

Try collaborating with a classmate or friend on the creation of a homework assignment or course project. To use a wiki, go to www.wikispaces.com. To collaborate in the cloud, go to www.google.com and click on Documents in the More pull-down menu or go to www.acrobat.com and click on File Sharing. To share your desktop, go to www .acrobat.com and click on Web Conferencing. For links to other sites that provide these services, visit this book's website at **www.cengage .com/english/anderson7e** and click on Chapter 17.

employer's computer, your employer can open it without violating your privacy (Berghel, 1997).

7. **Don't write something about a person that you wouldn't say directly to him or her.**

8. **Don't include confidential or proprietary information.** A major concern of employers is the ease with which employees can accidentally share confidential information in online correspondence. Some companies and government agencies have strict guidelines for using online correspondence.

Note to Instructors: You can enable your students to experience any of the kinds of online correspondence just mentioned. Course management programs such as Blackboard and WebCT enable you easily to create discussion boards, which are blogs, and wikis for your course. You can also set up course blogs and wikis at the sites mentioned in the margins. Links to those and other free resources are provided at the book's website.

WRITING COLLABORATIVELY ONLINE

Internet technology has opened up new opportunities for employees and students to write collaboratively without being in the same location and without shuffling drafts back and forth as e-mail attachments. At easily accessible websites, members of your writing team can assemble their resources, identify goals, debate ways to proceed, create new documents, and revise and edit the evolving texts. Here are three kinds of Internet services that provide this kind of support.

Wikis

A wiki is a website where members of your writing team can work collaboratively on a communication using web browsers. The best-known example is the Wikipedia. For writing projects at work or school, wikis support your writing team by enabling members to brainstorm, discuss ideas, make outlines, establish project schedules, and create, read, and edit drafts of the communication they are creating together. Members can also upload data, graphics, and other resources for the team's use. Thus, a wiki becomes not only a place where a document can be drafted but also a storeroom where relevant materials can be placed for easy access by all team members. Each page in a wiki is a web page, but you can download the content and format it as a word-processed document.

Documents Created in "the Cloud"

Another type of Internet service enables your writing team to use web browsers to create communications that don't need to be revised with a word processer when they are downloaded. When you create and edit communications with one of these services, you use a word-processing program that the service provides and you access through your browser. The program exists at the service's computers rather than on your computer, which is where Microsoft Word is located. The communications your team creates with these services are also stored on the service's computers. Because the programs and communications are remote from your computer, they are said to be "in the cloud." Using them is called "cloud computing." Examples of cloud-computing services are Google Docs and Buzzword. Both services enable writers to download the documents they've created in a variety of formats, including Microsoft Word.

Desktop Sharing

Members of your writing team can also work together on a document by sharing a draft that is stored on the computer of only one team member. Using ConnectNow, for example, team members can hold an online meeting in which they all view one member's computer desktop. This person can turn control of his or her desktop over to another team member who can make additions and revisions using the programs on the "host" member's computer. Meanwhile, all team members can watch and discuss the work in progress through text, audio, or video chat.

GUIDELINES FOR WRITING COLLABORATIVELY ONLINE

1. **Keep to the topic.** Avoid digressions and sidelights.
2. **Insert comments and questions.** Often the best way to advance work on a collaborative piece is to pose questions to other team members: What do you mean? What do you think of this? Can anyone solve this problem?
3. **Clean up the content and organization.** The changes, comments, and questions entered into a collaborative text can eventually overwhelm the text itself so that no one can see what's been produced so far. Take the time to eliminate old comments and responses that are no longer needed as the communication develops.
4. **Develop sections of long communications in separate files.** The longer a file is, the more difficult it will be for team members to find their locations when adding or editing material, especially if there is reorganization. The separate files can be joined together toward the end of the process.

MEETING VIRTUALLY

Try This

Hold a virtual meeting with two classmates or friends by going to www.skype .com or to www.acrobat .com and clicking on Web Conferencing. If you own an Apple computer, try iChat. For links to other sites that support virtual meetings, visit this book's website at **www .cengage.com/english/ anderson7e** and click on Chapter 17.

Several Internet services enable two or more members of your team to hold meetings on their computers using an audio or video chat. Skype and iChat are examples. Such live meetings can be much more efficient than the long e-mail or IM exchanges that teams might otherwise use to discuss the same issues.

When your team is discussing a draft, it can be especially helpful for each of you to see the draft as you are all talking. One way to achieve this goal is for each team member to open one window for a chat and another for viewing the document with Google Docs or a similar service. Services like ConnectNow enable team members to chat and view their draft in a single browser window.

GUIDELINES FOR VIRTUAL MEETINGS

1. **Make and distribute an agenda in advance.** An agenda helps to focus the discussion and pace the conversation so that all items can be addressed in the time available.

2. **Assure that before the meeting all members have materials to be discussed.** Don't take meeting time to share these items.

3. **Ask all members to work in a location without distractions.** When the other team members aren't physically present, it's easy to let your attention drift.

4. **Give everyone a chance to speak.**

5. **Let each speaker finish.**

6. **Ask for clarification and elaboration.**

7. **Be sensitive to possible cultural and gender differences.** They can be more difficult to detect than when you are meeting face to face.

Learn More

For additional advice about working effectively in team meetings, see Chapter 18's Guidelines 5, 6, and 8.

USE WHAT YOU'VE LEARNED

For additional exercises, go to **www.cengage.com/english/ anderson7e**. *Instructors:* The book's website includes suggestions for teaching the exercises.

EXERCISE YOUR EXPERTISE

1. Evaluate a social networking blog or miniblog such as Facebook or Twitter in terms of its usefulness for teams on which you might work in your career. Present your evaluation in a memo written to a real company for which you might work in the future.

2. As part of their marketing strategy, many organizations have pages at social networking blogs such as Facebook. Find the page for one company or nonprofit group. In a letter to the group, state the goals you think it has for the page. Based on your reading of messages posted at

the site, evaluate the site's effectiveness at achieving those goals. Recommend ways that the site's effectiveness could be increased.

EXPLORE ONLINE

1. Using your ingenuity, find two blogs on a topic related to your major or a particular academic interest of yours (not a blog about celebrities, etc.). Write a memo to your classmates explaining the strategy you used to find the blogs. Include your assessment of the quality of the information people were posting on the blog.

2. Find two online discussions of cloud computing. Using information in them, write a memo to your classmates about the ways your future employer might ask you or other employees to use cloud computing.

COLLABORATE WITH YOUR CLASSMATES

Write a 200-word summary of this chapter. Partner with another student in your class who has done the same.

Upload summaries to a wiki or a service like Google Docs or Buzzword. Suggest changes to your partner's draft by inserting comments in it while he or she does the same for your draft. Return to your draft and edit it in light of your partner's suggestions. If your instructor suggests that you use a different draft (such as an assignment you are preparing), use it instead of a summary of the chapter.

APPLY YOUR ETHICS

On the Internet, examine the privacy statement by Facebook or another blog for social networking. Then search for articles describing privacy concerns that some people have about such sites. Write a memo to your classmates defending the privacy policy of the blog from an ethical perspective or arguing from an ethical perspective that the policy should be changed.

18 | Creating Communications with a Team

GUIDELINES

Teamwork skills are highly valued by employers hiring and promoting employees. Prominent among team skills are those required to write collaboratively.

Employers place a great emphasis on team skills not only because team projects are very common in the workplace but also because teamwork can be difficult. And the difference in the quality of work produced by effective and ineffective teams can be tremendous.

This chapter's nine guidelines will help you successfully contribute to–and lead–writing teams, whether they include two persons or two hundred. Moreover, the strategies this chapter teaches for collaborative writing will assist you in all other aspects of team projects. Likewise, the technologies for collaborative writing discussed here are also used in other ways by project teams. Thus, this chapter is as much about effective teamwork in general as it is about writing on teams.

WWW

To read additional information and access links related to this chapter's guidelines, go to Chapter 18 at **www.cengage .com/english/anderson7e.**

VARIETIES OF TEAM STRUCTURES

Team structures vary in two important ways: the way leadership is assigned and the way tasks are distributed.

Leadership

On some teams, especially large ones, a single person is designated as the leader or manager. In other cases, team members are collectively responsible for all aspects of their work. Both arrangements have variations. For example, the person assigned as team manager either may personally make all critical decisions or may facilitate discussions that lead to consensus among team members.

Distribution of Tasks

While working on a communication, teams perform the same activities as an individual writer does: defining objectives, planning, drafting, and revising. A team may distribute work on these activities in a variety of ways.

- **All team members may work together on all tasks.** Because all team members talk together throughout the process, this structure can produce a very thorough and original result. On the other hand, it works only for small teams and it can be inefficient.
- **A team may divide the work, with each person working independently on his or her own task.** For instance, the team leader may define the communication's objectives, and a second person may plan, a third may draft, and a fourth (perhaps a technical editor)

© Dmitry Yashin 2009/Shutterstock.com

© Yuri Arcurs 2009/Shutterstock.com

Members of a writing team can share leadership responsibilities (previous page) or one person can be designated as the team leader (above).

may revise. For larger projects, two or more people may draft, each working on a section related to his or her specialty. This pattern can be efficient but may not produce the creativity coherence that can result from greater dialog among team members.

● **Team members may share some tasks and work independently on others.** For example, the whole team may define the communication's objectives and plan its overall strategy. Individuals then research and draft sections related to their specialties. The team then reviews all drafts, and one person edits the communication based on the group's suggestions.

No pattern is best for all situations. Your teams' organizational plans should respond to the nature of the communications being prepared and the talents and areas of expertise of the team members. For many projects, a hybrid pattern produces the best result by combining the benefits of team discussions at some points with the efficiency of individual work at other points.

This chapter's guidelines are strategies teams can use to work together efficiently and effectively whether one person serves as leader or all team members share leadership responsibilities.

Guideline 1 | Develop a shared understanding of the communication's objectives

The first steps in creating a reader-centered communication as a team are the same as the first steps you take when writing reader-centered communications by yourself:

● Learn about your readers' characteristics, situation, needs, and expectations.
● Define your communication's usability objectives by learning about the tasks the communication will help your readers perform.
● Define your communication's persuasive objectives by describing the ways you want your communication to influence your readers' attitudes and actions.

The following suggestions will help team members achieve a common understanding of the communication they are working together to create.

Learn More

To review the reader-centered approach to defining a communication's objectives, go to Chapter 3.

WWW

The Worksheet for Defining Objectives can be used for team projects as well as for individual ones. To download a copy, go to Chapter 3 at **www.cengage.com/english/anderson7e**. Also, see Chapter 3's guidelines for defining objectives.

DEFINING OBJECTIVES AS A TEAM

■ Take time to explore the diverse views of all team members regarding the communication's objectives. Ask questions yourself and invite questions from others.

■ Keep the team talking until it is clear that all team members agree on the objectives.

■ Record the decisions and provide all team members with copies.

■ Remain open to new insights about your readers and purposes as your team's work progresses.

Guideline 2 | **Make and share detailed plans**

When planning a communication you will write by yourself, you can begin with fairly general plans that you will fill out as you proceed. With team projects, however, general or vague plans can create many problems. When team members begin their individual work based on a vague plan, they often find that they have invested a significant amount of time and creativity in producing drafts that have to be significantly modified or even discarded because they don't match the more detailed plan that the team eventually develops.

Here are four actions that can help your teams avoid the frayed nerves and wasted time that such misunderstandings can cause.

PLANNING TEAM PROJECTS

- **Discuss plans as a team.** When teams discuss plans, the individual members gain a fuller understanding of what they are trying to accomplish. In addition, the plans themselves often improve as a result of the discussion.
- **Discuss the communication's outline together.** By discussing the outline, teammates can develop a common, detailed understanding of the communication's logic and structure. The team may create the outline from scratch during the discussion or start with an outline developed in advance by one team member.
- **Create storyboards.** After your team has discussed the outline, each team member can complete a storyboard form for the sections he or she will write. As you can see in Figure 18.1, the form asks for the section's main point, supporting points, and graphics. The team can then assemble and review the storyboards, adjusting each one so it fits together tightly with the others. At the end of this storyboard review, each writer should have a clear sense of what he or she should write.
- **Establish rules for naming files.** For example, all team members might agree to name files with the titles of the sections they are writing. When files are all named in the same manner, teams can easily review and discuss drafts that are shared electronically and they can quickly assemble the drafts to form the final document.

Learn More

For information about a different variation of storyboarding used with oral presentations, see page 483.

Guideline 3 | **Make a project schedule**

As part of your team's planning, make a schedule. A schedule lets each person know exactly when his or her tasks must be completed. It also motivates team members to complete their assignments on time because it shows how missing their deadlines creates problems for later steps in the process.

When you create a schedule, include the following elements.

- **Include time for defining objectives and planning.** The team will have an opportunity to act on Guidelines 1 and 2 only if the necessary time is provided in the schedule.

- **Include milestones.** Milestones indicate the dates by which various stages of the work will be completed. These interim deadlines help teams make

Learn More

For advice about creating schedules for client and service-learning projects, see Chapter 21, page 529.

FIGURE 18.1
Storyboard for Use by Writing Teams

Before anyone begins drafting, team members can use storyboards to decide together what the content for each part of their communication will be.

In this space, team members state the main point to be made in a section or subsection.

Here, team members describe the points that will be made to support or explain the major point.

Team members use this space to sketch or make notes about the graphics that will be integrated into this part of the communication.

After each team member's storyboard has been reviewed by the team or by the leader, team members will have a clear idea about how to develop their sections in a way that fits easily into the communication's overall structure.

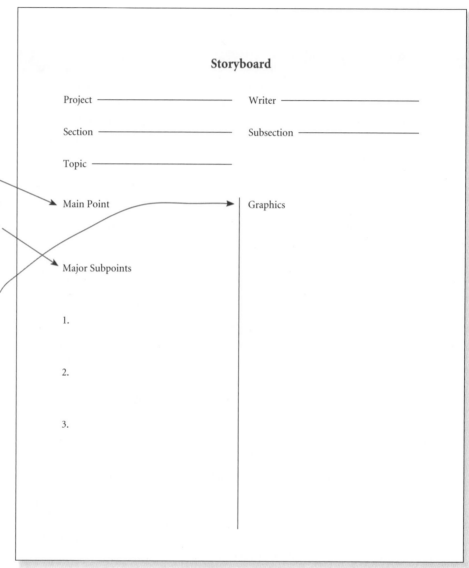

■ CHAPTER 18 **Creating Communications with a Team**

progress at a steady pace. If a milestone is missed, the team knows that it must speed up its remaining work to complete the entire project on time.

- **Create enough milestones to catch potential problems early.** For instance, in addition to a milestone for the completion of final drafts, set a date for completing rough drafts for team review. This deadline allows the team to identify writers who are having difficulty so that assistance can be provided early enough for them to complete the final draft on time.
- **Include time for editing.** No matter how much discussion and planning has happened earlier in the process, almost every team-written communication needs editing before delivery to its readers. The sections may be written in different styles, leave gaps, or repeat some of the same information. Whether the editing is performed by one member or all the members together, time will be needed for this activity.

Learn More

For guidelines for creating schedule charts, see the Writer's Reference Guide for Creating Eleven Types of Reader-Centered Graphics, page 377.

Guideline 4 | Share leadership responsibilities

Communication experts Kenneth D. Benne and Paul Sheats (2000) have identified a wide range of roles that someone must play if the team is going to maximize its productivity. On teams with shared leadership, all team members should assume equal responsibility for performing these roles when the situation requires. On teams that have a designated leader, this person would usually perform most of these roles. Even then, however, all team members can increase the team's effectiveness by performing one or more of these roles as needed.

Benne and Sheats divide the roles into two groups: task roles, which keep the team moving toward its goal, and group maintenance roles, which assure good working relationships among the team members. The roles most essential for successfully completing team-written communications are described here.

ESSENTIAL ROLES IN TEAM COMMUNICATION	
TASK ROLES	**CONTRIBUTIONS**
Initiators	offer new ideas, propose new solutions, and restate old issues in a novel way. They provide creativity and direction as the team explores its subject matter and communication strategies.
Information seekers	request clarification and additional information. They ensure that the team members understand all relevant factors—including their subject matter, readers, and communication alternatives.

Try This

Identify the task and group maintenance roles at which you excel. Among the other roles, which ones would be the most valuable for you to learn how to perform more expertly?

Information givers	furnish the facts needed by the team, sometimes on their own initiative, sometimes in response to information seekers.
Opinion seekers	ask others to express their judgments, values, and opinions. They also share their views with the team.
Clarifiers	clear up misunderstandings or confusion by explaining points or providing additional information.
Summarizers	consolidate the team's deliberations by stating concisely what has been said or decided. They help team members see what has been accomplished so the team can proceed to the next task.
Energizers	motivate the team to take action, often by communicating a sense of enthusiasm or by emphasizing its commitment to its goals.

GROUP MAINTENANCE ROLES	CONTRIBUTIONS
Encouragers	offer warmth, praise, and recognition during team discussions. They support quieter team members, whom they gently encourage to join in.
Harmonizers	help team members explore differences of opinion without hurting one another's feelings. They detect and reduce friction by helping the team to focus on ideas rather than on personalities.
Feeling expressers	share their own feelings or vocalize those of the team, thereby enabling members to deal with emotions that might interfere with the team's ability to work together productively.
Compromisers	volunteer concessions of their own positions on controversial issues and suggest a middle ground when other team members seem stuck in opposing positions. They help all team members realize that they are contributing even when their ideas are altered.
Gatekeepers	assure that all team members have an opportunity to speak, sometimes by asking more talkative members to be brief and by inviting quieter members for their contributions.

Guideline 5 | **Make meetings efficient**

Nothing will be more precious to your communication teams than time. Because it's usually so hard to find moments when all members are free, every one of these moments should be used well. Moreover, every member usually has so many other responsibilities that meetings must be productive to justify the minutes and hours taken away from other duties. The following strategies help make meetings productive.

Try This

Think of a meeting in which you recently participated. Which of these strategies for conducting efficient meetings would have most improved that meeting?

- **Prepare an agenda.** Before the meeting, have someone list the major issues to be discussed. Open the meeting by having the team review the agenda to be sure everyone agrees on what is to be accomplished. As the meeting proceeds, team members can keep the discussion on track by referring to the agenda.

- **Bring discussions to a close.** Communication teams sometimes keep debating a topic even after everything useful has been said. When a discussion becomes repetitious, focus attention on the decision to be made by saying something like, "We seem to have explored the options pretty thoroughly. Let's make a decision."

   Some teams continue talking even after they have reached consensus. If your team does that, try formalizing the agreement by saying, "I think I hear everyone agreeing that . . . ." If someone objects, the team can clear up any point that still needs to be resolved. Otherwise, the team can move on to the next item of business.

- **Sum up.** After all the topics have been covered to everyone's satisfaction, sum up the results of the meeting. Such a summation consolidates what the team has accomplished and reinforces its decisions.

- **Set goals for the next meeting.** Before the meeting breaks up, make sure everyone knows exactly what he or she is to do before the next meeting. This strategy helps assure that when you meet again, you will have new ideas and material to discuss.

Guideline 6 | **Encourage discussion, debate, and diversity of ideas**

One of the chief benefits of group work is that many people bring their expertise and creativity to a project. To take full advantage of that benefit, all team members must offer their ideas freely—even if the ideas conflict with one another. In fact, debate and disagreement can be very useful if carried out in a courteous and non-threatening way. Debate ensures that the team won't settle for the first or most obvious suggestion. It also enables your team to avoid *groupthink*, a condition in which everyone uncritically agrees at a time when critical thinking is really what's needed.

Encouraging debate and diversity of ideas can be difficult. Some people are naturally shy about speaking,

In group meetings, active listening is as important as it is in one-to-one conversation.

© Anton Vengo/Superstock

and many avoid disagreeing, especially if they fear that their ideas will be treated with hostility rather than openness and politeness. The following sections describe strategies your teams can use to promote productive debate and consideration of a rich diversity of ideas.

Active Listening

Good listening is one key to good discussions. In fact, the quality of your contributions to team discussions can depend at least as much on how you listen as on what you say. By following the strategies for active listening described below, your teams can encourage full and open participation by all members. The strategies will also help team members fully understand one another and build on one another's ideas.

LISTENING ACTIVELY

- **Give everyone a chance to speak.** If someone is quiet, ask for the person's thoughts and ideas.
- **Let each speaker finish.** Hold your thoughts until they've finished theirs.
- **Focus on ideas.** Be sure you understand what the person is saying.
- **Ask for clarification and elaboration.** Before responding, get a full picture of the speaker's thoughts.
- **Show that you are paying attention.** If you agree with the person, say such things as "That's interesting," "I hadn't thought of that," or even simply "Uh-huh."
- **Look at the speaker.** People believe that listeners who maintain lots of eye contact are friendlier than those who maintain less eye contact (Kleinke, Bustos, Meeker, & Staneski, 1973). When speakers think they are receiving more eye contact, they also think their listeners are more attentive (Kleck & Nuessle, 1967).
- **Refrain from distracting gestures.** Researchers have found that speakers feel uncomfortable if their listeners engage in such nervous gestures as cleaning their fingernails, drumming their fingers, or holding their hands over their mouths (Mehrabian, 1972). Replace such gestures with note taking or assume a relaxed but alert posture.

Fostering Productive Debate and Handling Conflict

The goal of team meetings is not simply to get people to share ideas but also to test and develop them through discussion and debate. The following strategies can increase team members' willingness to disagree and work together to resolve differences. They can also provide the basis for handling situations where a debate over ideas turns into a conflict between individuals.

- **Be open to others' ideas.** Don't be so attached to your own thoughts that you can't hear theirs.

- **Respect the other person's ideas.** First show that you understand the other person's position by paraphrasing it. Identify the points on which you agree before explaining why you have a different view on other points.

- **Discuss ideas, not persons.** Refrain from implying that the person is wrong. Express your reasons for thinking that another idea is better.

- **Watch nonverbal communication.** A team member's facial expression or body language may suggest that they are feeling uncomfortable, hurt, or angry about an issue. Respond in a respectful and sensitive way.

- **Look for the "third way."** Sometimes, ideas that seem to be in conflict can both be accommodated.

- **Be prepared to compromise.** Stalemate is unproductive. If another team member continues to disagree with you, your best contribution to the team may be to compromise—even if you are certain you are correct. Being a productive team member can require supporting writing choices you wouldn't make on your own.

- **Paraphrase and vote if team members become deadlocked.** If two team members are unable to find a resolution to their disagreement, suggest that the team should decide. Then paraphrase each position fairly and take a vote.

Discussing Drafts

Teams can increase the usability and persuasiveness of their drafts by sharing suggestions for revision. To benefit fully from these discussions, members must help one another overcome the fear that their suggestions may offend the person who drafted the section under discussion and the tendency of some members to resist changes to parts they drafted. Chapter 15's guidelines for reviewing drafts can help teams avoid these barriers to productive reviews. Especially relevant to team writing are the following six guidelines from Chapter 15, plus two more that apply specifically to team projects.

DISCUSSING DRAFTS

STRATEGIES FROM CHAPTER 15

- Begin comments about a draft with praise for what's strong.
- Explain the reader-centered reasons for all suggestions.

Learn More

For a full discussion of these suggestions, go to Chapter 15's Guidelines 2, 3, and 4 for reviewing, pages 409–411.

- Avoid phrasing suggestions in ways that sound like criticisms of the writer's work.
- Rank suggested revisions, focusing on the ones that will make the greatest improvement.
- Distinguish matters of substance from matters of style.
- Accept all reasonable suggestions for improving your own draft and thank team members for their ideas.

SPECIAL GUIDELINES FOR TEAM PROJECTS

- **Treat drafts as team property, not individual property.** Encourage each other to give up the sense of personal "ownership" of drafts. While still taking pride in your contributions, consider your drafts as something you've created for the team; they have become its property.

  This doesn't mean that you should be silent if someone suggests a change that would weaken rather than strengthen your draft. It does mean that during discussions of your draft, you should join in a reasoned conversation about the best way to write "our" communication rather than struggling to protect "my" writing from criticism by others. Your openness about your draft will encourage others to adopt the same attitude concerning theirs.

- **Swap responsibilities.** The team could agree that, after drafts have been reviewed, the responsibility for revising each section will be given to someone other than the person who drafted it. Members might also agree that when all the drafts are combined, either a single individual or the whole team will be responsible for editing and polishing. With these arrangements, all drafts clearly become team drafts.

Guideline 7 | ## Choose the computer technology best suited to your team's project

On the job, you will have access to a variety of computer programs that support team projects, including team writing. Fortunately, the same kinds of programs are available for you for free right now so you can begin using them in your technical communication and other classes. These programs support team projects in significantly different ways. The following discussion will help you identify the one best suited to your team's specific project.

Criteria For Choosing The Most Supportive Program

When evaluating programs, consider how they support each of the five collaborative tasks involved with team writing.

- Sharing drafts among team members
- Reviewing drafts, making suggestions, asking questions, and so on
- Revising drafts
- Tracking drafts (keep track of which draft of each section is the most recent one)
- Managing team activities by making plans together, discussing progress, and so on

Begin by determining whether your team will want to share drafts by distributing copies of them to each team member or using programs that provide all team members with access to a single copy of each draft for each team member.

Distributing Copies of All Drafts to All Team Members

To distribute drafts, your team can use programs with which you are already familiar. After they've drafted their sections using their favorite word-processing program, team members can send drafts to the rest of the team as e-mail attachments. To convey suggestions for change, they can insert comments into the drafts. If they make revisions in the drafts, they can use Track Changes or a similar feature to enable other team members to see where the changes are. To manage the team's work they can e-mail, make phone calls, or use a free program for audio and video chats, such as Skype and iChat.

Sending copies of each draft to all members can be very effective if your team is small and only one person works on a draft at a time. However, keeping track of documents can become difficult. If you send a draft to the two other persons on your team, three copies exist of the same document, one on each team member's computer. If one team member revises the draft and sends a copy to the other two, two versions exist. If the other team member does the same thing, three versions exist. None of the three versions includes both sets of revisions, so someone must merge the revisions into a fourth version. The larger the team, the bigger the chore of keeping track of the versions.

Learn More

For more on the Comment feature in Word, go to Chapter 15, pages 417–418.

WWW

For instruction on using the Track Changes feature in Word, go to **www.cengage .com/english/anderson7e**, Chapter 18.

Providing Access to a Single Copy for All Team Members

The second distribution method eliminates the difficulties caused by multiple copies of the same document. Using any of several free programs, you upload copy to a central location and then send your team members an e-mail that enables them to

WWW

For instructions on using GoogleDocs, Skype, and ConnectNow, go to **www.cengage.com/english/anderson7e,** Chapter 18.

Try This

Set up a free account at GoogleDocs. Upload a draft of a project you are preparing for one of your courses. Using GoogleDocs' Share feature, send an e-mail to a classmate asking him or her to provide suggestions. Return to your account to review his or her recommendations.

access, comment on, and edit copy through a web browser such as Internet Explorer, Safari, or Firefox. Because all changes are made to this single draft, there's no need to merge changes from separate versions.

Sharing drafts this way helps teams that want to use audio and video chats to discuss drafts. If they are using GoogleDocs (http://docs.google.com), for instance, to share the document, they can open the document in one window and open a Skype audio or video chat in another. ConnectNow (https://www.acrobat.com) enables teams to view a draft and have a video chat in the same window (see Figure 18.2).

Unfortunately, programs like GoogleDocs and ConnectNow have limited word-processing options, so you can't add many of the page design features that increase a communication's usability and persuasiveness until you've downloaded it to your own computer. Likewise, when you upload a draft with sophisticated page design, much of the design may be lost.

Making the Choice

As the preceding discussion indicates, you must consider many factors when choosing the technology that best supports your team's project. The key is to determine the specific ways your team will work together and then look in detail at the ways each alternative supports them.

Guideline 8 | ## Global Guideline: Help your team work across cultural differences

Successful teamwork requires good relationships among team members. Each team member needs to be fully engaged. Each must encourage the other members to contribute their knowledge and talents fully. Such positive team dynamics depend largely on the ways team members interpret one another's behavior. Is that quiet member uninterested in your team's work or full of ideas but shy about speaking up? Your attitudes and actions toward this person will depend on your answer. If you base your attitudes and actions on an incorrect interpretation, you may offend the person, thereby diminishing his or her commitment to the team and decreasing the team's overall effectiveness.

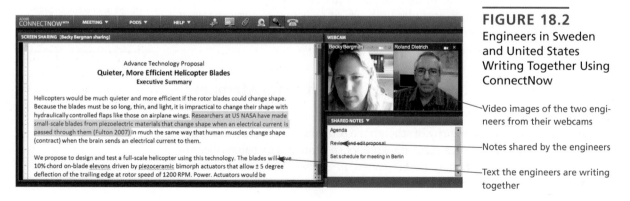

FIGURE 18.2
Engineers in Sweden and United States Writing Together Using ConnectNow

—Video images of the two engineers from their webcams

—Notes shared by the engineers

—Text the engineers are writing together

Two factors reduce our ability to interpret others' behavior accurately. First, we usually interpret so quickly that we aren't aware that other interpretations are possible. Second, we tend to forget that our interpretations are shaped by our cultural background, leading us to draw sometimes wildly erroneous conclusions about team members from other countries. The ability to avoid such mistakes is becoming increasingly valuable as the number of workplace teams with members from different countries grows and as the number of teams with members from the same country but different cultural backgrounds increases.

A first step in learning how to work across cultural difference is to recognize the many ways these differences can impact teams.

Cultural Differences in Behaviors On Teams

Bosley (1993) has identified four important differences in the ways that people from various cultures may interact on workplace teams.

- **Expressing disagreement.** In some cultures, individuals typically express disagreement directly. In some other cultures, people say "no" only indirectly in order to save face for themselves and for the other person.

- **Making suggestions.** People from some cultures offer suggestions freely. To avoid embarrassing others, people from some other cultures avoid saying anything that might be interpreted as disagreement.

- **Requesting clarification.** In some cultures, individuals frequently ask others to explain themselves more clearly. People from some other cultures feel that it is rude to ask for clarification because doing so would imply that the speaker doesn't know what he or she is talking about or hasn't succeeded in explaining things clearly.

- **Debating ideas.** Whereas members of work teams in some cultures debate ideas vigorously as a way of exploring ideas, people from some other cultures regard such behavior as disloyal and unacceptable.

Recognizing cultural differences

Cultural Differences in Behaviors in Conversation

The ways people interact in everyday conversation show up in the ways they converse in team meetings. In Sweden, listeners signal that they are being attentive by sitting straight and folding their arms in front of their torsos (Rabinowitz & Carr, 2001). However, if you were raised in the United States, you might interpret the Swedish posture as signaling boredom or disagreement. In the U.S., most people signal attentiveness by learning forward and openness by keeping arms spread.

Similarly, cultural differences exist concerning eye contact. People raised in the United States are likely to interpret eye contact as a signal of sincerity and interest in the other person. Eye contact is even more important in Arab cultures, where

WWW

For more on culture and collaboration, go to **www.cengage.com/english/anderson7e**.

people use intense eye contact to read someone's real intentions. However, if you were raised in Korea, you might interpret eye contact that is maintained for a long time as rudeness. In Japan, looking someone in the eye is to invade his or her space (Varner & Beamer, 2005).

Cultural Differences and Gender

WWW

For more on gender and collaboration, go to **www .cengage.com/english/ anderson7e**.

Different cultures also have different expectations about the ways men and women will behave. For example, research shows that many men in the United States present their ideas and opinions as assertions of fact. When exploring ideas, they may argue over them in a competitive manner. In contrast, some women offer their ideas tentatively, introducing them with statements such as, "I think," or, "I'm not sure about this, but . . ." If there is disagreement, they may support part or all of the other person's ideas and seek to reach consensus (Lay, 1989). Differences in gender expectations can lead to misinterpretations even when team members are from the same culture. Interpreting the behaviors of teammates whose gender is different than yours can lead to misinterpretations.

Improving Team Effectiveness Across Cultural Difference

Responding to possible cultural differences

The best tools for assuring team effectiveness across cultural difference are knowledge, self-awareness, and flexibility.

- **Knowledge.** The more team members know about one another's cultural expectations, the greater their ability to avoid counterproductive misinterpretations of one another's behavior. For developing this knowledge, no guidebook can beat the effectiveness of conversations in which team members talk about these differences. These conversations can occur during team meetings or outside of them. One of the great pleasures of working with persons with different cultural backgrounds is learning about the cultures and the people who reside in them.

- **Self-awareness.** Also important is your own awareness of the extent to which your own cultural background influences your interpretations of your teammates' behaviors. You need to keep in mind continuously the possibility that your assumptions about and evaluations of others may be inaccurate because you are interpreting them from the wrong perspective, your own culture's. Talking with other team members about your understanding of them is the best way to test your interpretations.

- **Flexibility.** Through conversation, you will likely learn that some of your ways of interacting are being interpreted incorrectly by others. In addition to discussing the "meaning" of these behaviors in your culture and that of one or more teammates, think of ways you can modify the behavior to reduce the chance of being misunderstood.

Ultimately, each multicultural team must develop its own ways of working as productively as possible. A team can't succeed in doing that if the members merely share generalizations about their cultures. Instead, they must develop a mutual understanding that sees each member's ways of interacting as *partly* an expression of his or culture. This understanding is an important element in the personal relationships (still within a workplace setting) that team members develop.

Guideline 9 | **For virtual teams, foster personal relationships and conversational interchanges**

As you learned in Chapter 1, some teams, called *virtual teams*, work entirely online. Research indicates that along with benefits of enabling collaboration among people in different locations, this arrangement creates some special challenges. Of particular concern for employers are the risks of low motivation for the team's project and misunderstandings that slow and even degrade project results.

Research also suggests that these difficulties are caused by the absence of two kinds of interaction that are almost always part of face-to-face teamwork: personal connections and conversations among team members.

Even when teams meet only briefly and infrequently in person, the individuals develop a direct connection with the others. This may occur as they introduce themselves, as they chat while waiting for a meeting to start or during a break or meal. However much e-mail, telephone calls, and video chats may seem to overcome distance, they don't seem to create the same personal connections unless there's a specific effort to use them that way. Researchers now recommend that virtual teams now include time devoted specifically to getting to know one another in their project plans.

Research also suggests that teams need to be aware of the importance of helping one another interpret their ways of interacting. The sorts of cultural differences described in Guideline 8 can also affect virtual teams. For instance, team members who are culturally attuned to exploring ideas through competitive debate are likely to do that in digital exchanges as much as in direct conversation. Likewise a person who is culturally attuned to thinking it is best to make suggestions indirectly in face-to-face meetings will probably use the same strategy online. Though the speaker/writer is likely to communicate in the same way in both circumstances, the other members of a team that meet virtually are much less likely to follow up than members of a team that meet face to face in ways that open discussion of the cultural expectations involved. You can help your team members compensate by encouraging them to communicate in a nonconfrontational way whenever one of them feels puzzled or irritated by the way interactions are proceeding.

CONCLUSION

Team writing is very common in the workplace. Although all the guidelines in this book apply to team writing as well as to individual writing, team projects require some additional skills. The nine guidelines presented in this chapter will help you create team efforts in which the team members work together productively and enjoy their mutual effort. Figure 18.3 shows a feedback and evaluation form that team members can use to help one another develop their collaboration skills.

FIGURE 18.3   Collaboration Feedback and Evaluation Form

For a copy of this form that you can download, go to Chapter 18 at www.cengage.com/english/anderson7e.

COLLABORATION FEEDBACK AND EVALUATION FORM

Team Member _____   Date _____

	Poor			Excellent
Organization				
Helps team to:				
Formulate objectives	1	2	3	4
Plan strategies for achieving objectives	1	2	3	4
Divide tasks effectively and fairly	1	2	3	4
Set schedule	1	2	3	4

Comments:

	Poor			Excellent
Interpersonal Relations				
Makes all team members feel welcome and appreciated	1	2	3	4
Helps others express and address their concerns	1	2	3	4
Understands and responds to others' feelings	1	2	3	4
Shows sensitivity to others' style of interacting	1	2	3	4
Motivates others	1	2	3	4

Comments:

	Poor			Excellent
Meetings				
Helps team to:				
Set agenda	1	2	3	4
Maintain focus	1	2	3	4
Bring discussion to a close	1	2	3	4
Sum up	1	2	3	4
Set goals for next meeting	1	2	3	4

Comments:

(*continued*)

FIGURE 18.3
(continued)

COLLABORATION FEEDBACK AND EVALUATION FORM
(Continued)

Team Member _____   Date _____

	Poor			Excellent

Discussion, Debate, and Development of Ideas

	Poor			Excellent
Encourages all to contribute their ideas and judgments	1	2	3	4
Requests information and explanations when needed	1	2	3	4
Supplies needed information he or she has	1	2	3	4
Presents ideas and information fully and clearly	1	2	3	4
Listens with attention and respect	1	2	3	4
Understands others' ideas and values	1	2	3	4
Makes suggestions diplomatically	1	2	3	4
Helps team explore differences without hurting feelings	1	2	3	4
Focuses on ideas, not persons, when conflict arises	1	2	3	4
Makes concessions	1	2	3	4

Comments:

Individual Contributions

Takes initiative	1	2	3	4
Offers new ideas and suggestions	1	2	3	4
Assumes responsibility	1	2	3	4
Completes assignments thoroughly	1	2	3	4
Meets deadlines	1	2	3	4
Produces high-quality results	1	2	3	4

Comments:

USE WHAT YOU'VE LEARNED

For additional exercises, go to www.cengage.com/english/anderson7e. *Instructors:* The book's website includes suggestions for teaching the exercises.

APPLY YOUR EXPERTISE

1. Write a memo to your instructor in which you describe difficulties encountered by a team on which you have worked. Describe strategies presented in this chapter that you employed in an attempt to address these problems. Also reflect on things that you could have done differently that might have overcome the problems.

2. Write a memo to your instructor describing the style and strategies you have brought to team projects in the past. First, identify the strengths of your approach, referring to specific advice given in this chapter. Second, pinpoint ways in which you can achieve an even higher level of effectiveness.

3. If you are working on a team project in your technical communication course or another class, write a memo to your teammates suggesting three ways the team could improve its productivity. Persuasively explain the reasons for each suggestion.

EXPLORE ONLINE

1. Compare two services that enable teams to post a single draft that all team members access to add comments and revise. If you will be working on a team project in your technical communication course or another class, which one would be most helpful to your team? Why? If you wish, your two services may be GoogleDocs (http://docs.google.com) and ConnectNow (https://www.acrobat.com).

2. Services like GoogleDocs and ConnectNow are considered to be examples of "cloud computing." Using a search engine for your research, develop a brief definition of cloud computing and explain to other students how it is affecting or might affect professionals in your field of study.

3. Find and describe a college course in which students at different schools (possibly in different countries) collaborated on one or more course projects.

COLLABORATE WITH YOUR CLASSMATES

1. If you are currently involved in a team project, fill out the Collaboration Feedback and Evaluation Form shown in Figure 18.3 for each of the other team members.

2. If you are about to begin a group project, write a memo to your instructor detailing the plans you have made in accordance with the advice given in Guidelines 3 and 4.

APPLY YOUR ETHICS

Various team members may bring different talents, commitments, and values to collaborative projects. What are your ethical obligations if one team member doesn't contribute, persistently completes work late, or resists compromise? What if one member is plagiarizing? Who are the stakeholders in these cases? What are your options? What is an ethical course of action? What are your obligations if one or more members of your team ask you to contribute more or in a manner that others consider more productive? Would any of your responses be different if the project were assigned at work rather than in school? Present your results in the way your instructor requests.

19 | Creating and Delivering Listener-Centered Oral Presentations

GUIDELINES

In your career, you will almost certainly make many oral presentations. In some, you may briefly and informally address individuals with whom you work closely each day. At other times, you may speak more formally to upper-level managers, clients, professional organizations, or the general public. Usually you will convey your entire message orally, but on other occasions your talk may supplement a report or proposal you have prepared on the same subject.

Compared to written communications, oral presentations are more personal and more interactive. You'll be able to see your audience's reactions as you are speaking. Your listeners will usually have an opportunity to make comments and ask questions when you have finished, perhaps even before.

This chapter will lead you through the process of planning, preparing, rehearsing, and delivering oral presentations at work. Its eleven guidelines teach a listener-centered approach to speaking that parallels the reader-centered approach to writing.

Guideline 1 | Define your presentation's objectives

Begin work on an oral presentation in the same way you begin work on a written communication—by figuring out exactly what you want to accomplish and studying the situation in which you will communicate:

1. **Think about your listeners and your communication goals.** Think about who your listeners are and how you want to affect them. To review the procedures for defining objectives, turn to Chapter 3.

2. **Think about what your listeners expect.** Consider their expectations about such things as the topics you will cover and the way you will treat them. Different groups have different ideas about what makes an appropriate presentation.

3. **Find out how much time you will have for your presentation.** Even when you aren't given a time limit in advance, your listeners are likely to have firm expectations about your presentation's length. Don't ruin an otherwise successful presentation by talking too long.

Although their audiences, topics, and goals vary considerably, oral presentations are everyday occurrences in many careers.

4. **Assess the scene of your talk.** Three aspects of the scene are especially important:
   - *Size of audience.* The smaller your audience, the smaller your graphics can be, the more likely your audience will expect you to be informal, and the more likely your listeners will interrupt you with questions.
   - *Location of your talk.* If the room has fixed seats, you will have to plan to display your graphics in a place where they are visible to everyone. If you can move the seats, you will be able to choose the seating arrangement best suited to your presentation.
   - *Equipment available.* The kinds of equipment that are available determine the types of graphics you can use. You can't show overheads or PowerPoint slides unless the necessary projector is available.

Guideline 2 | Plan the verbal and visual parts of your presentation as a single package

When people read, they can focus on only one item at a time, either your paragraphs or your graphics. In contrast, when people listen to a presentation, they can simultaneously hear your words and look at a graph, table, diagram, or other visual item you show them. By "speaking" to their eyes as well as to their ears, you can greatly increase the effectiveness of your oral presentations. A study conducted at the Wharton Applied Research Center found that speakers who use graphics are judged by their listeners to be

- Better prepared
- More professional
- More persuasive
- More credible
- More interesting

than speakers who don't use graphics (Andrews & Andrews, 1992; Jewett & Margolis, 1987).

To gain the full advantage of the combined communicative power of the verbal and visual dimensions of your presentations, you must integrate them fully. First choose the type of oral delivery and the visual medium you will use. Then plan how you will weave your words and images together.

Choosing the Type of Oral Delivery

At work, people generally use three forms of oral delivery: the scripted talk, the outlined talk, or the impromptu talk. Sometimes the situation or the profile of your listeners will dictate the type of talk you will give. At other times, you will be free to choose.

Scripted Talk   A scripted talk is a speech that you write out word for word in advance and deliver by reading the script or by reciting it from memory.

Because the scripted talk lets you work out your exact phrasing ahead of time, it is ideal for presenting complex information and speaking when a small slip in phrasing

Scripted talks offer security but can be rigid.

could be embarrassing or damaging. A scripted talk enables you to be sure you will make all your major points and, if you timed your rehearsals, keep within your time limit.

The scripted talk is also a good choice when you expect to be extremely nervous, as might happen when you are addressing a large or unfamiliar audience. Even if you become too unsettled to think straight, you will have all your words in front of you.

On the other hand, scripted talks take a long time to prepare, and you cannot alter them in response to audience reactions. Even more importantly, they are difficult to compose and deliver in a natural speaking style that maintains the interest of your audience. Rehearsals are a must.

Outlined talks offer flexibility within a general framework.

Outlined Talk   To prepare an outlined talk, you do what the name implies: Prepare an outline—perhaps very detailed—of what you plan to say.

At work, outlined talks are more common than scripted talks. You can create them more quickly, and you can deliver them in a "speaking voice," which helps increase listener interest and appeal. Furthermore, outlined talks are very flexible. In response to your listeners' reactions, you can speed up, slow down, eliminate material, or add something that you discover is needed. Outlined talks are ideal for situations in which you speak on familiar topics to small groups, as in a meeting with other people in your department.

The chief weakness of the outlined talk is that it is so flexible that unskilled speakers can easily run over their time limit, leave out crucial information, or have difficulty finding the phrasing that will make their meaning clear. For these reasons, you may want to avoid giving outlined talks on unfamiliar material or in situations where you might be so nervous that you become tongue-tied or forget your message.

Impromptu Talk   An impromptu talk is one you give on the spur of the moment with little or no preparation. At most, you might jot down a few notes beforehand about the points you want to cover.

Impromptu talks are ideal for topics you know well.

The impromptu talk is well suited to situations in which you are speaking on a subject so familiar that you can express yourself clearly and forcefully with little or no forethought.

The chief disadvantage of the impromptu talk is that you prepare so little that you risk treating your subject in a disorganized, unclear, or incomplete manner. You may even miss the mark entirely by failing to address your audience's concerns. For these reasons, the impromptu talks given at work are usually short, and they are usually used in informal meetings where listeners can interrupt to ask for additional information and clarification.

Choosing the Visual Medium You Will Use

The visual medium you will employ will depend partly on what's available and what your listeners expect.

Learn More

There is a Writer's Tutorial for using PowerPoint on pages 480–482.

In the workplace, people use many different media for the visual part of their oral presentations. The four most common are computer projections made with programs such as PowerPoint, overhead transparencies, chalkboards or dryboards, and handouts. Figure 19.1 summarizes the chief advantages and disadvantages of each. A Writer's Tutorial for using PowerPoint is on pages 480–482.

MEDIUM	ADVANTAGES AND DISADVANTAGES

FIGURE 19.1
Advantages and Disadvantages of Several Media

Computer projections (for example, PowerPoint)

Advantages

Enable you to create very polished-looking slides without the aid of a graphic artist

Enable you to provide visual harmony to your presentation by using a single colorful design for all your slides

Allow you to prepare slides quickly by typing words and inserting other content directly into them

Enable you to expand your media by incorporating tables, graphs, photos, sound, animation, and movies

Allow you to use special aids such as a remote control for changing slides and a laser pointer for highlighting elements on the screen

Disadvantages

Special equipment is required for projecting the slides for your audience

Must be delivered in a dark room, making it difficult for your listeners to take notes or see you

Slides cannot be altered or reordered during a presentation to accommodate audience response

Preparation can be time-consuming

Ability to create special effects leads some speakers into making overly elaborate presentations that detract from their content

Overhead transparencies

Advantages

Can be made simply by copying a word-processed page onto an acetate sheet with an ordinary copy machine

Projectors are available in almost every organization

Can reorder the slides and draw on them with markers even as you are giving your talk

Disadvantages

Require some preparation

Look plain, especially compared to computerized presentations

Can't include motion or sound

Chalkboard and dryboard

Advantages

Require no preparation

Very flexible

Can be used to record contributions from audience

Work well for small meetings and discussion sessions

Disadvantages

Can be used only for words and line drawings

Can delay presentation while you do your writing

Can leave you speaking to the board rather than your audience

Handouts

Advantages

Give readers something to take away that provides key information from your talk

Aid listeners with notetaking

Disadvantages

Require preparation

May tempt audience to read ahead rather than listen to what you are saying

Writer's Tutorial

Creating a Listener-Centered Presentation

MAKE YOUR PRESENTATION

Follow this tutorial by typing your own text rather than the text shown in the sample slides. This tutorial was created for Microsoft PowerPoint 2007. If you are using another program, you may need to use its online help for assistance.

DEFINE YOUR OBJECTIVES

- Who are your listeners?
- What are their characteristics?
- What are your communication's usability and persuasive goals? (See Chapter 3.)
- How long should your presentation be?

PLAN

- Pick three or four main points.
- Pick your graphics.
- Plan a reader-centered introduction and conclusion.
- Assign a time to each part of your presentation.

Tip

Assign the least time to your introduction and conclusion.

Introduction	1 minute
Problem	2 minutes
Method	2 minutes
Results	3 minutes
Discussion	4 minutes
Recommendation	2 minutes
Conclusion	1 minute

DRAFT

Choose a "theme" or visual style.

1. Open a new PowerPoint file.
2. Click on the **Design** tab.
3. Roll the cursor over the options in the **Themes** area, watching the demonstrations in your slide.
4. To see more options, click on the arrows to the right of the samples.
5. Click on the design you've chosen.

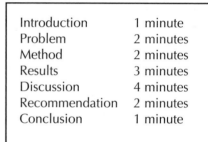

Tip

Choose a simple design. At work, fancy designs can distract readers and may seem inappropriate for a business setting.

Learn More at the Website

To learn advanced features of PowerPoint, go to Chapter 19 at www.cengage.com /english /anderson7e.

CREATE A TITLE SLIDE

1. In the top rectangle, type a short, informative title.
2. In the bottom rectangle, type a subtitle.

Tip

Use the subtitle to explain the topic's relevance to your listeners or to identify your approach to the topic.

CREATE A BULLET SLIDE

1. Click on the **Home** tab.
2. Click on **New Slide**.
3. Type the slide's title in the space provided.
4. In the box marked "Click to add text," type your first bullet item.
5. To create another bullet item, press **Enter**.
   - Stick to six or fewer bullet items per slide.
   - Use parallel constructions (see page 225).
6. To use bold, color, and other choices for the text, use the options under the **Home** tab.

Tip: Write slides that present an outline, not a script. Your listeners will not want you to read your PowerPoint slides to them.

CHOOSE A SPECIAL LAYOUT

1. Click on the Home tab.
2. Next to the New Slide icon, click on the small Layout icon (top icon).
3. Click on a layout appropriate for your needs. (For this example, choose Two Content.)

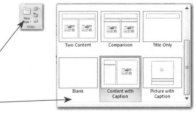

INSERT A GRAPHIC

1. In the left-hand area, type your text.
2. In the right-hand area, click on the icon for **Chart**.
3. Click on the image of the type of graphic you want to insert. (For this example, choose Pie.)
4. Replace the data in the spreadsheet that appears with your data.
5. Close the spreadsheet.

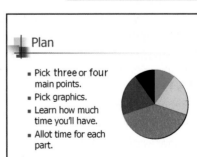

(*continued*)

Writer's Tutorial (*continued*)

Creating a Listener-Centered Presentation

MAKE POWERPOINT NOTES

The notes you create can be printed with your slides or viewed by you (but not your audience) when you give your presentation.

1. Click in the space at the bottom of the PowerPoint window that says, "Click to Add Notes."
2. Type your notes.

Tip

You can increase the size of the notes to make them easier to read when you are giving your presentation.

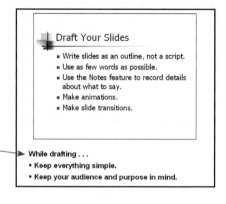

CREATE ANIMATIONS

Animations define the ways items on a slide come into view and exit.

1. Click on the **Animations** tab.
2. Highlight the items you want to animate, such as a bullet list.
3. Click on the arrowhead to the right of **No Animation**.
4. Watch the demonstrations as you roll over the choices in the dropdown menu.
5. Click on the animation you want.

Tip

Animations can slow your presentation and distract listeners. Use them only where they serve a purpose.

CREATE TRANSITIONS

A transition is the visual effect as you switch from slide to slide.

1. Click on the slide for which you want a special transition.
2. Click on the **Animation** tab.
3. Roll the cursor over the options, watching the demonstrations.
   - If your window is wide, the options will appear in the ribbon.
   - If your window is narrow, click on **Transition Scheme**.
4. Click on the transition you choose.
5. To use the transition for all slides, click on **Apply to All**.

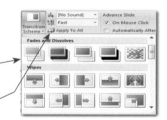

REVISE

1. Rehearse by yourself.
   - Time your presentation.
   - Practice making your points without reading slides to your listeners.
2. Rehearse with reviewers.
3. Based on what you learn, polish your presentation.

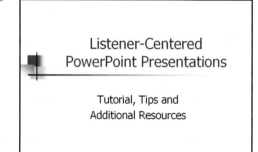

When choosing the medium that is most suitable to your presentation, consider the following criteria:

- **Your purpose and audience.** A computerized presentation is appropriate when you need to impress, as with a formal presentation or in a sales meeting. At a small meeting in your own department, overhead transparencies or a handout may be much more suitable.

Criteria for selecting media for graphic presentations

- **Your listeners' expectations.** If you provide glitz when your listeners expect simplicity, your credibility will diminish, just as it will if you keep pausing to draw diagrams on the board for listeners who expect you to have prepared your graphics in advance.

- **Your resources.** The amount of time you have, the size of your budget, and the unavailability of equipment may limit your choices.

In some situations, there's an advantage to using more than one medium in a presentation. For instance, when speaking to a group of managers, a chemical engineer used slides to show photographs of the crystals grown in an experiment, overhead transparencies to explain the process used, and a handout containing detailed data that her audience could study later.

Planning the Way You Will Integrate the Verbal and Visual

Having selected the form of oral delivery and the visual medium you will use, plan the ways you will integrate the two. Storyboards provide an excellent tool for this purpose. Devised by people who write movie scripts, storyboards are a series of pages that are divided into two parts. One part shows the words and other sounds that moviegoers will hear and the other describes (in words or sketches) what they will see at each moment of the film.

Figure 19.2 shows how to create a storyboard using PowerPoint. Chapter 18 explains how to create a storyboard on paper.

When creating a storyboard, define the way that your words and graphics will supplement each other during each part of your presentation. Avoid having them simply repeat one another. The most frequent complaint about PowerPoint presentations is that some presenters simply read their slides to their audiences.

Learn More

For an explanation of a storyboard you can create on paper, turn to page 459.

Here are three commonly used relationships between the verbal and visual parts of a presentation.

- The slides can focus on key words, while you speak the full sentences.
- The slides can state key points, while you explain.
- The slides can show a graphic that you discuss.

Guideline 3 | **Focus on a few main points**

For a variety of reasons, people find it more difficult to understand what they hear than what they read.

- Listeners have more difficulty than readers in concentrating for extended periods. People can read for hours at a time, but many listeners have trouble concentrating for more than twenty minutes.

Difficulties listeners face

FIGURE 19.2
Two Types of Storyboards
Using PowerPoint

By pulling down the **View** menu
and selecting **Notes Page,** you
can prepare storyboards for your
presentations.

At the top of the page, one of
your slides appears.

At the bottom is an area in which
you can write notes or even an
entire script associated with the
slide.

During a presentation, you can
show your notes on your computer screen as an aid to speaking.
See page 490.

In its Slide Sorter View,
PowerPoint also enables you
to see thumbnails of all your
slides so that you can review
and adjust their order.

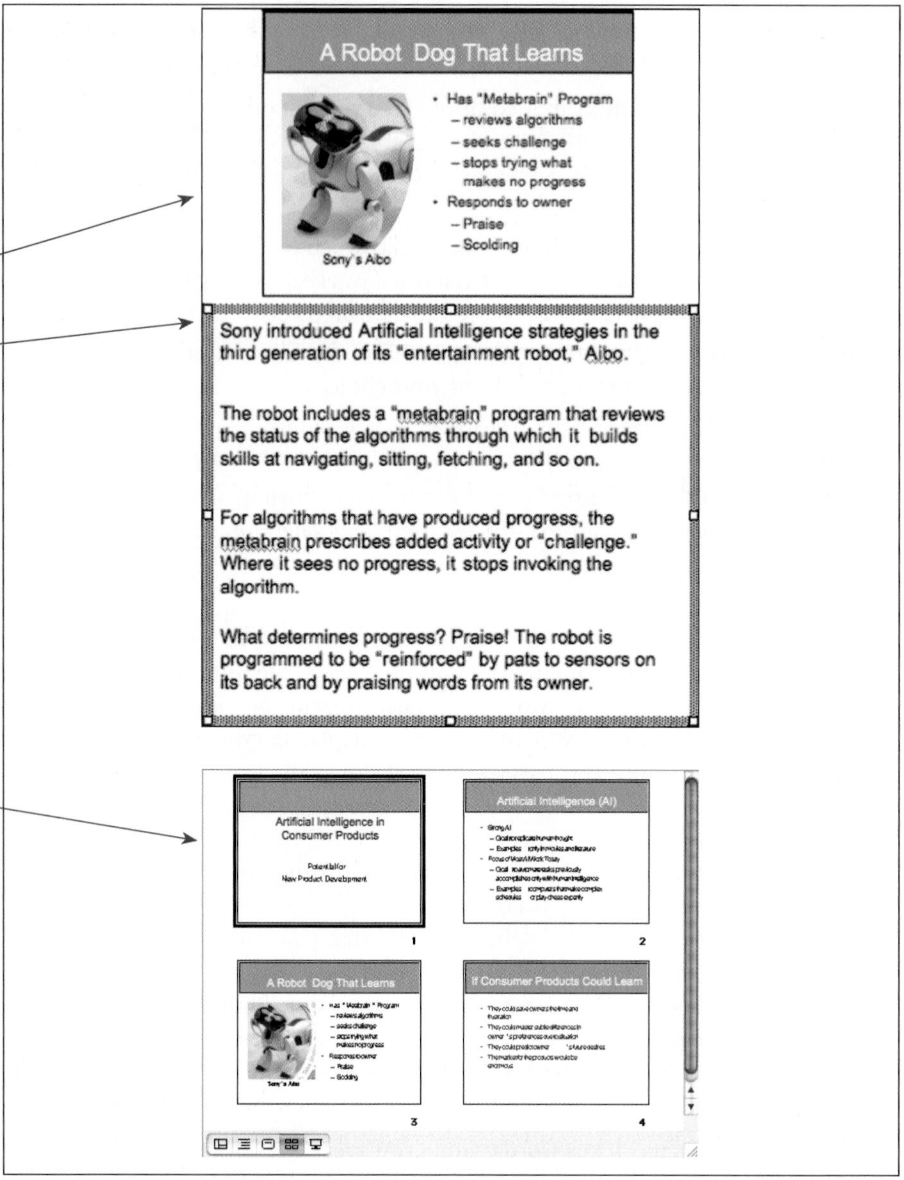

- A talk proceeds at a steady pace, so listeners have no chance to pause to figure
  out a difficult point.
- If listeners fail to understand a point or let their attention wander, they cannot
  flip the pages back and reread the passage. The talk goes forward regardless.

One way to make your oral presentations easy to understand is to concentrate on only
a few major points. Many experienced speakers limit themselves to three or, at most,
four. Of course, major points may have subpoints, but their number should also be

limited. And it is always important for you to select points that are directly relevant to your listeners' interests and needs. For instance, when making a recommendation for overcoming a problem in your employer's organization, your first point might be that the problem has two major features, your second might be that there are three principal causes of the problem, and your third might be that you have three recommendations for overcoming it. To help your listeners focus on and remember your major points and subpoints, present them visually as well as orally.

Guideline 4 | Use a simple structure—and help your listeners follow it

You can also increase the understandability and memorability of your presentations by employing a structure that's easy to follow. A widely used structure consists of three major parts: introduction, body, and conclusion. Here are typical contents for each.

TYPICAL STRUCTURE OF AN ORAL PRESENTATION

Introduction	■ Introduce your topic.
	■ Explain its relevance to your listeners.
	■ Forecast the organization of your presentation.
	■ Depending on your purpose and audience, you may also use the introduction to do the following:
	■ Provide background information.
	■ State your main point.
Body	■ Present your three or four main points, elaborating and explaining each one.
Conclusion	■ Sum up your main points.
	■ Identify next steps listeners can take.
	■ Invite questions.

Simple structure for oral presentations

Even when you employ a very simple organization, your listeners may need help discerning it. Help them by forecasting the structure in the introduction, clearly signaling transitions, and summarizing the structure (and main points) in the conclusion. The following list describes techniques for guiding your listeners in these ways.

SIGNALING THE STRUCTURE OF ORAL PRESENTATIONS

GENERAL STRATEGIES	TECHNIQUES
Forecast the structure	■ In the introduction, tell what the structure will be: "In the rest of my talk, I will take up the following three topics . . . ."
	■ Show a graphic that outlines the major parts of your talk.

Signal Transitions	■ Announce transitions explicitly, "Now I would like to turn to my second topic, which is . . . ."
	■ Show a graphic that announces the next topic and, perhaps, lists its subtopics.
	■ Highlight your main points. Your discussion of each main point is a major section of your talk. "I'm going to shift now to the second cause of our problem, a cause that is particularly important to understand."
	■ Pause before beginning the next topic. This pause will signal to your listeners that you have completed one part of your presentation.
	■ Slow your pace and speak more emphatically when announcing your major points, just as you would when shifting to a new topic in conversation.
	■ Move about. If you are speaking in a setting where you can move around, signal a shift from one topic to another by moving from one spot to another.
Review	■ In your conclusion, remind listeners explicitly about what you've covered: "In conclusion, I've done three main things during this talk: . . . ."

Guideline 5 | Speak in a conversational style

For most talks, a conversational style works best. It helps you express yourself clearly and directly, and it helps you build rapport with your listeners. A more abstract, impersonal style can lead you to make convoluted statements that are difficult to understand and leave your listeners feeling that you are talking *at* them rather than *with* them. Although people often associate a conversational style with informality, the two are not synonymous. The two keys to a conversational style are (1) speaking directly to your listeners, and (2) expressing yourself in the simple, natural, direct way you do in conversation. Here are suggestions for mastering this style.

STRATEGIES FOR DEVELOPING A CONVERSATIONAL STYLE

■ **Create your talk with your audience "present."** While preparing your talk, imagine that members of your audience are right there listening to the words you are planning to speak.

■ **Use the word *you* or *your* in the first sentence.** Thank your listeners for coming to hear you, praise something you know about them (preferably something related to the subject of your talk), state why they asked you to address them, or talk about the particular goals of theirs that you want to help them achieve.

- **Continue using personal pronouns throughout.** Let the use of you or your in the first sentence establish a pattern of using personal pronouns (*I, we, our, you, your*) throughout.
- **Use shorter, simpler sentences than you might use when writing.**
- **Choose words your listeners will understand immediately.** If your audience stops to figure out the meaning of a word you have used, they will stop listening to the next point you are making.

As you strive to create a conversational style, remember that the way you use your voice can be as important as the words and phrasing you select. Listen to yourself and your friends converse. Your voices are lively and animated. To emphasize points, you change the pace, the rhythm, and the volume of your speech. You draw out words. You pause at key points. Your voice rises and falls in the cadence of natural speech.

Using your voice in the same way during your oral presentations can help you keep your listeners' interest, clarify the connections between ideas, identify the transitions and shifts that reveal the structure of your talk, and distinguish major points from minor ones. It can even enhance your listeners' estimate of your abilities. Researcher George B. Ray (1986) found that listeners are more likely to believe that speakers are competent in their subject matter if the speakers vary the volume of their voices than if the speakers talk in a monotone.

Exhibit enthusiasm for your subject. In conversation, we let people know how we feel as well as what we think. Do the same in your oral presentations, especially when you are advocating ideas, making recommendations, or promoting your employer's products and services. If you express enthusiasm about your topic, you increase the chances that your listeners will share your feelings.

Finally, use gestures. In conversation, you naturally make many movements—pointing to an object, holding out your arms to show the size of something, and so on. Similarly, when making oral presentations, avoid standing stiffly and unnaturally. Use natural gestures to help hold your listeners' attention and to make your meaning and feelings clear.

Guideline 6 | Create easy-to-read, understandable graphics

To contribute fully to a presentation's effectiveness, graphics must be easy to read and understand. These two qualities depend on the way you design your graphics.

Learn More

For additional advice about creating graphics, see Chapter 13 and also the "Writer's Guide for Creating Eleven Types of Reader-Centered Graphics".

Designing Effective Graphics

All of Chapter 13's guidelines for designing pages apply also to the design of graphics for oral presentations. The following list highlights particularly important points from Chapter 13 and adds some that apply specifically when you are addressing an audience in person. Figure 19.3 shows how these guidelines can be applied to slides in a computerized presentation.

WWW

To see examples of effective and ineffective PowerPoint presentations, go to Chapter 19 at **www.cengage.com/ english/anderson7e.**

DESIGNING GRAPHICS THAT ARE EASY TO READ AND UNDERSTAND

- Use type large enough to be read throughout the room.
- Use an easy-to-read typeface.
- Use a light background and dark letters—or vice versa—for high contrast.
- Use color to highlight key points and focus attention.
- Put text in bullet lists.
- Use key words and phrases rather than whole sentences.
- Avoid overcrowding.
  - Limit each slide to five or fewer bullet items.
  - Break larger topics into several graphics.
  - Leave plenty of white space (blank area) between items and around margins.
  - Keep tables, graphs, and drawings simple.
- Provide a brief title for every graphic.
- Use a consistent design for all your graphics

So that the verbal and visual elements of your presentation can support one another harmoniously, distill your verbal statements into key words and phrases for the bullet lists in your graphics.

Statement you will make verbally

In this report, I will discuss three major causes of the extinction of animal and bird species in Asia: loss of habitat due to development and the harvesting of natural resources, increasing pollution from automobiles and factories, and poaching.

Content of the corresponding visual

Causes of extinction
- Loss of habitat
- Pollution
- Poaching

Like any other element of a communication, graphics should be tested beforehand to make certain they will be as understandable and persuasive as possible. Once you've completed the set, show them to other people, preferably members of your target audience. Do this early enough to allow time to polish them, if that seems advisable, before your presentation.

Slide 1

CRI — Crandon Research Institute

Developing
Hydrogen Fuels Cells for
Commercial Boats

Feasibility of Conducting a
Prototype Research Project

1

Slide 2

CRI — Overview

- Fuel cell technology
- Benefits for commercial boats
- Current research
- Potential funding

2

Slide 3

CRI — A proven technology

- Provided power on NASA space shuttles since 1970s
- Used in prototype cars and buses worldwide

3

Slide 4

CRI — How fuel cells work

Electricity (Electrons) Out

Water Out

H_2 H^+ O_2 H_2O

Hydrogen In   Oxygen In

Anode Electrolyte Cathode

4

Slide 5

CRI — Benefits for boats

Gas/Diesel	Fuel Cell
1. Pollution	1. No pollution
2. Fuel and engine heavy	2. Cells and motors light
3. Can create dangerous fire	3. No danger from fire
4. Prices will rise	4. Prices will fall

5

Slide 6

CRI — Extensive research already

- Car companies
  - BMW, Ford, GM, Honda
- Oil companies
  - Shell, Exxon/Mobil
- Research companies
  - GE, H-Tec

6

Slide 7

CRI — Federal spending grows

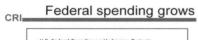
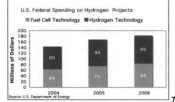

U.S. Federal Spending on Hydrogen Projects

■ Fuel Cell Technology ■ Hydrogen Technology

Millions of Dollars

	2004	2005	2006
Fuel Cell Technology	80	94	99
Hydrogen Technology	64	75	84

Source: U.S. Department of Energy

7

Slide 8

CRI — Recommendation

- Assemble research team
- Contact potential funding sources
- Prepare proposal

8

FIGURE 19.3
Presentation Slides Made with PowerPoint

Katerine created this presentation for a meeting about possible new projects for her employer, a research company.

Her title slide clearly indicated the topic and focus of her presentation.

Katerine's second slide provided an overview of what she would say.

Her slides provided only key words or phrases, not her script.

Katerine used large letters and high contrast to make her slides very readable.

She used graphics to highlight major points.

Katerine used a consistent overall design for all her slides while also adding graphics and varying other elements in some slides to maintain visual interest.

Displaying Graphics Effectively

During your talk, use your graphics in ways that support and reinforce your presentation rather than detract from it:

Try This

Presentation programs such as PowerPoint enable you to make "kiosk" presentations that run continuously on their own on a computer. Try making a presentation that, by itself, informs viewers about something important to you. Play with your program's features, including those that let you add music or record a script.

- **Display a graphic only when you are talking about it.** If you talk about one thing but display a graphic about something else, each person in your audience must choose whether to listen to your words or read your graphic. Either way, your audience may miss part of your message.

- **Leave each graphic up long enough for your listeners to digest its contents.** Sometimes this pacing will require you to stop speaking while your listeners study your graphic.

- **Explain the key points in your graphic.** If you want your readers to notice a particular trend, compare certain figures, or focus on a particular feature in a drawing, say so explicitly.

- **Do not read from your graphics.** Remember: Your visuals should provide key words and concepts on which you elaborate. Your listeners will become very restless if you merely read your graphics.

- **Stand beside projected graphics, not in front of them.** Your listeners can't read what they can't see.

Guideline 7 | Involve your audience in your presentation

In an oral presentation, you can interact in person with your audience, something you can't do through a written communication. By taking advantage of this opportunity, you can ensure that your listeners pay close attention and that you are meeting their needs.

Look At Your Listeners

Learn More

For more on cultural conventions concerning eye contact, turn to page 469.

In North America and much of Europe, the most basic way to involve listeners in your talk is make them feel that that are in a direct interchange with you—something you can accomplish simply by making eye contact with them. Through eye contact, you convey that you are interested in your listeners as individuals, both personally and professionally.

Eye contact has the additional benefit of giving your audience a more favorable impression of you. Researcher S. A. Beebe (1974) first demonstrated this effect when he asked two groups of speakers to deliver the same seven-minute talk to various audiences. One group was instructed to look often at their listeners, whereas the other group was told to look rarely. Beebe found that the speakers who looked more often at their listeners were judged to be better informed, more experienced, more honest, and friendlier than those who used less eye contact.

Benefits of maintaining eye contact with your listeners

Eye contact also enables you to judge how things are going. You can see the eyes fastened on you with interest, the nods of approval, the smiles of appreciation, or the puzzled looks and the wandering attention. These signals enable you to adjust your talk, if necessary.

There are situations when it is appropriate for you to look briefly away from your listeners. For example, if you are using an outline or script, you may want to refer occasionally to your notes. Just don't rivet your gaze on your papers. Look back at your audience. The same is true if you look at graphics that are behind you, for instance on a chalkboard or a screen for an overhead or data projector.

If you are using a presentation program such as PowerPoint, place your computer screen so that it faces you as you look at your audience. By looking at the screen, you will see the same graphics that your audience sees projected behind you. Also, if your computer supports two monitors (most laptops do), you can use the Presenter View (PC) or Presenter Tools (Mac) to display helpful items on your screen that your audience won't see (Figure 19.4). Consult PowerPoint Help for instructions.

If you have difficulty looking at your listeners, try the following strategies.

To maintain a connection with your audience when using projected visual aids, face your listeners and look at your slides on your computer screen rather than on the large screen behind you.

STRATEGIES FOR LOOKING AT YOUR LISTENERS

- **Look around at your audience before you start to speak.** This will give you an opportunity to make initial eye contact with your listeners when you aren't also concentrating on what to say.

- **Follow a plan for looking.** For instance, at the beginning of each paragraph of your talk look at a particular section of your audience—to the right for the first paragraph, to the left for the second, and so on.

- **Target a particular feature of your listeners' faces.** You might look at their eyes, but you could use their foreheads or noses instead. Unless they are very close, they won't notice the difference.

- **When rehearsing, practice looking at your audience.** For instance, develop a rhythm of looking down at your notes, then up at your audience—down, then up. Establishing this rhythm in rehearsal will help you avoid keeping your head down throughout your talk.

- **Avoid skimming over the faces in your audience.** To make someone feel that you are paying attention to him or her, you must focus on an individual. Try setting the goal of looking at a person for four or five seconds—long enough for you to state one sentence or idea.

Ask Your Listeners to Say or Do Something

In some oral presentations, you can ask your audience to contribute information. For example, if you are making a presentation to fellow employees about ways to improve customer satisfaction, you might ask what kinds of complaints they have heard. If your purpose is to train your listeners to perform a procedure, you may be able to ask them to perform the steps as you talk and demonstrate.

FIGURE 19.4
Presenter Tools from PowerPoint

By pulling down the **View** menu and choosing **Presenter Tools**, you can show a screen like this on your computer during your presentation.

It shows the slide being projected and your notes about it.

On the left side, it shows the current slide between the one before it and also the next one.

In the top left-hand corner, a clock tells how many minutes have passed since you began speaking.

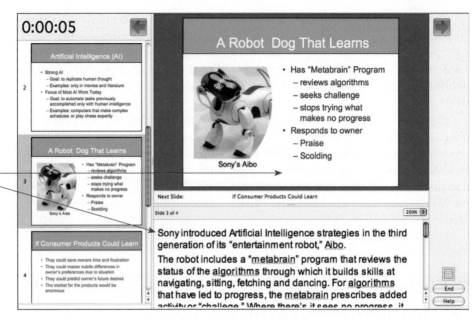

This figure shows the Presenter Tools in Microsoft PowerPoint 2008 for Mac. PowerPoint 2007 for PC has a different layout but similar content.

WWW

For instructions on printing slides for a PowerPoint presentation, go to Chapter 19 at **www.cengage.com/ english/anderson7e**.

Help Your Listeners Take Notes

Note taking is another way in which listeners actively participate in a presentation. You can encourage note taking by distributing an outline at the beginning of your talk. If you are using a program such as PowerPoint, you can print out copies of the slides you will project so that members of your audience can jot notes on them.

Invite Questions

In almost every situation, it is appropriate to invite listeners to ask questions either during your presentation or after it. Guideline 8 provides advice about ways to prepare for and respond to questions.

Give Your Listeners Something to Take Away

You may want to encourage your listeners to continue their interaction with you after your presentation. For instance, your listeners may include people who might become customers or clients of your employers' organization. If so, provide them with something to take away that includes the main points you made or supplemental information, along with your name, phone number, and e-mail address. If you distributed a handout for note taking, that would be perfect.

Guideline 8 | Prepare for interruptions and questions— and respond courteously

Audiences at work often ask questions and make comments. In fact, most of the talks you give there will be followed by discussion periods during which members of your audience will ask you for more information, discuss the implications of your talk, and even argue with you about points you have made. This give-and-take helps explain the popularity of oral presentations at work: They permit speaker and audience to engage in a discussion of matters of common interest. Part of preparing to deliver a talk is preparing yourself for questions and discussions. In a sense, you do this when you plan the presentation itself, at least if you follow Chapter 4's advice that you begin planning your communication by thinking about the various questions that your readers will want it to answer. Usually, however, you will not have time in your talk to answer all the questions you expect your listeners to ask. The questions you can't answer in your talk are ones your listeners may raise in a question period. Prepare for them by planning your responses.

When a member of the audience asks a question—even an antagonistic one— remember that you want to maintain good relations with all your listeners. If you are speaking in a large room, be sure that everyone hears the question. Either ask the questioner to speak loudly or repeat the question yourself. If people hear only your answer, they may have no idea what you are talking about. Respond to all questions courteously. Remember that the questions and comments are important to the people who ask them, even if you don't see why. If you don't know how much detail the questioner wants, offer some and then ask the questioner if he or she wants more. If you don't know the answer to a question, say so.

Some speakers ask that questions and comments be held until after they have finished. Others begin by inviting their listeners to interrupt when they have a question. By doing so, they are offering to help listeners understand what is being said and relate it to their own concerns and interests. Sometimes listeners will interrupt without being invited to do so. Such interruptions require special care. Speak to the person immediately. If you are planning to address the matter later in your talk, you may want to ask the questioner to wait for your response. If not, you may want to respond right away. After you do so and resume your talk, be sure to remind your listeners of where you broke off: "Well, now I'll return to my discussion of the second of my three recommendations."

Guideline 9 | Global Guideline: Adapt to your audience's cultural background

An audience's cultural background affects its expectations about an oral presentation in much the same way that it influences readers' expectations about written communications.

- **An audience's expectations may be shaped by several cultures.** These include the culture of their region and the national or ethnic group to which they belong, the culture of their employer's organization, and the culture of their profession, among others.

- **The members of an audience are individuals, not cultural stereotypes.** Although it can be useful to learn about your audience's culture, also find out as much as you can about the specific people you will be addressing.

Several aspects of oral presentations vary from culture to culture. Investigate your audience's expectations about each of them as you plan and prepare your presentation. You may learn that you can create a truly listener-centered presentation only if you do something different from what this chapter's other guidelines suggest.

ADAPTING TO YOUR AUDIENCE'S CULTURAL BACKGROUND

- **Opening.** In some cultures, presentations typically begin with the main point. In others, they begin with formal greetings. In the United States, presentations often begin with an effort to build rapport with an audience— for instance by telling a joke or personal anecdote. Beginning in a way that differs from what the audience expects can get a talk off to a weak start.

- **Organization.** Different cultures have developed different patterns for organizing presentations. For example, Wolvin and Oakley (1985) report that in China a presentation may have the following structure:.

Typical structure of a Chinese talk

KI	An introduction offering an observation of a concrete reality
SHO	A story
TEN	A shift or change in which a new topic is brought into the message
KETSU	A gathering of loose ends, a "conclusion"
YO-IN	A last point to think about, which does not necessarily relate to the rest of the talk

In this style of Chinese presentation, speakers do not state the main points explicitly nor do they preview the structure of the talk as is customary in some European countries and the United States.

- **Directness.** As the example of the traditional form of a Chinese presentation suggests, some cultures prefer to make their points indirectly, while others, including the general U.S. culture, prefer to make them explicitly.

- **Tone.** To illustrate the way expectations about tone vary from one culture to another, Carté and Fox (2004) describe the stereotypes that two English-speaking cultures have of each other's presentations. To the British, U.S. presentations seem overly optimistic, boastful, and superficial. To U.S. audiences, British presentations seem gloomy, pessimistic, and preoccupied with problems, not solutions. Both believe that the other's presentations are badly prepared.

- **Eye contact.** Whereas eye contact is highly valued in some cultures such as in the United States, it can seem aggressive and unwelcome in some other cultures.

- **Gestures.** Gestures that have a positive meaning in some cultures can be incomprehensible, rude, or obscene in others.
- **Visuals.** Images of people and gestures that seem natural in one culture can offend audiences in another. Carté and Fox (2004) describe a presentation in which a large European company used an image of a lion tamer and lion to symbolize the way their computer systems would tame potential clients' problems. Although the presentation worked in several countries, it failed in an African country whose symbol was a lion. To this audience, the image suggested a colonial power subduing their state.

When speaking to an audience from another culture, you may also be addressing persons who are not fluent in English. In those situations, the following suggestions should help you succeed.

SPEAKING TO LISTENERS WHO ARE NOT FLUENT IN ENGLISH

- **Use words your audience can understand.** Learning the audience's level of proficiency in English may require research. Many technical and scientific terms are the same in many languages, so that you may be able to use them even when you need to simplify your language in other ways.
- **Use graphics.** Although you should be cautious about graphics, people who don't understand one another's languages can still understand the same drawings, graphs, flowcharts, and other visuals.
- **Provide a handout.** If you are speaking from a script, the script itself is the best handout, along with copies of your key graphics. Otherwise, create a detailed outline. Consider using full sentences—even in an outline—rather than words or brief phrases. Beyond helping your listeners follow your talk, a handout provides a resource they can study later to understand points they missed.

Guideline 10 | Rehearse

All of your other good preparations can go for naught if you are unable to deliver your message in a clear and convincing manner. Whether you are delivering a scripted or outlined talk, consider the following advice.

GUIDELINES FOR REHEARSING

- **Rehearse in front of other people.** They can help you identify weak spots and make suggestions for improvement.
- **Pay special attention to your delivery of the key points.** These are the points where stumbling can cause the greatest problems.

- **Rehearse with your graphics.** You need to practice coordinating your graphics with your talk.
- **Time your rehearsal.** But be sure that you speak at the same pace you will use in your actual presentation so you have an accurate sense of how long your talk will require.

Guideline 11 | Accept your nervousness—and work with it

This guideline is difficult for many novice speakers to follow. But it is very important. Not only is nervousness unpleasant to experience, it can lead to distracting and unproductive behaviors that greatly impair the effectiveness of your talk. These include the following:

- Looking away from your listeners instead of looking into their eyes
- Speaking in an unnatural or forced manner
- Exhibiting a tense or blank facial expression
- Fidgeting, rocking, and pacing

How should you deal with nervousness? First, accept it. It's natural. Even practiced speakers with decades of experience sometimes feel nervous when they face an audience. If you fret about being nervous, you merely heighten the emotional tension. Also keep in mind that your nervousness is not nearly so obvious to your listeners as it is to you. Even if they do notice that you are nervous, they are more likely to be sympathetic than displeased. Furthermore, a certain amount of nervousness can help you. The adrenaline it pumps into your system will make you more alert and more energetic as you speak.

Here are some strategies for reducing your nervousness and controlling the pacing, fidgeting, and other undesirable mannerisms it fosters.

STRATEGIES FOR CONTROLLING NERVOUSNESS

- **Arrive early.** Avoid rushing from a previous activity to your talk. Give yourself plenty of time to set up and look around before you begin.
- **Devote a few minutes before your talk to relaxing.** Take a walk or spend a quiet moment alone.
- **Speak with audience members before your presentation begins.** Doing this enables you to make a personal connection with at least some of your listeners before you begin.
- **Remind yourself that your listeners are there to learn from you, not to judge you.**
- **When it's time to begin, pause before you start your talk.** Look at your audience, say "Hello," and take out your outline or notes as you accustom yourself to standing before your listeners.

MAKING TEAM PRESENTATIONS

At work, team presentations are as common as team writing. They occur, for example, when a project team reports on its progress or results, when various departments work together on a joint proposal, or when a company is selling a technical product (such as customized software) that requires the participation of employees from several units of the company.

Team presentations are so common because it is usually more effective to have several members make a presentation than to have one person speak for all team members. In a team presentation, each topic can be discussed by the person who is most expert in it. Moreover, the use of a variety of speakers can help retain the audience's attention.

The following sections provide advice for creating and delivering effective team presentations.

Learn More

The guidelines in Chapter 18, "Creating Communications with a Team", are as helpful to teams creating oral presentations as they are to teams creating written communications. Turn to page 456.

Plan Thoroughly

When developing a team presentation, plan as carefully as you would if preparing a team-written document. Devoting a team meeting to making plans can be very helpful. Decide which topic each team member is to discuss, which points are to be made, and how each part fits with the others. Set a time limit for each part so that the total presentation doesn't run too long. Also, decide whether the team is going to ask the audience to hold questions until the end or invite the audience to interrupt the speakers with questions. In either case, provide time for the interchange between the team and its listeners.

Allow for Individual Differences

In a team-written project, the goal is usually to produce a document with a single voice for the entire document. In team presentations, however, each speaker can speak in his or her own style and voice, provided that the general tone of the presentation is relatively consistent.

Make Effective Transitions between Speakers

By carefully planning the transitions from one speaker to the next, a team can substantially increase the effectiveness of its presentation. Switch speakers where you are making a major shift in topic. This will help the audience discern the overall structure of the presentation. Also, have both speakers explain how the two parts of the presentation fit together. For example, the speaker who is finishing might say, "Now, Ursula will explain how we propose to solve the downtime problem I have just identified." The next speaker might then say, "In the next few minutes, I'll outline our three recommendations for dealing with the downtime problem Jefferson has described."

Show Respect for One Another

Your team can increase the effectiveness of your presentation if each member shows, perhaps implicitly, respect for the others. If team members seem confident of one another's contributions and capabilities, your audience is more likely to adopt the same attitudes.

Rehearse Together

Rehearsals are crucial to the success of team presentations. Team members can help one another polish their individual contributions and ensure that all the parts are coordinated in a way that the audience can easily understand. Moreover, they can see to it that the entire presentation can be completed in the time allotted. Running overtime is a common and serious problem for groups that have not worked together previously.

CONCLUSION

Making oral presentations can be among your most challenging—and rewarding—experiences at work. By taking a listener-centered approach that is analogous to the reader-centered approach described elsewhere in this book and by striving to communicate simply and directly with your audience, you will prepare talks that your listeners will find helpful, informative, interesting, and enjoyable. The Writer's Guide shown in Figure 19.5 will help you succeed in doing so.

USE WHAT YOU'VE LEARNED

For additional exercises, go to **www.cengage.com/english/ anderson7e**. *Instructors:* The book's website includes suggestions for teaching the exercises.

EXERCISE YOUR EXPERTISE

1. Outline a talk to accompany a written communication that you have prepared or are preparing in one of your classes. The audience for your talk will be the same as for your written communication. The time limit for the talk will be ten minutes. Be sure that your outline indicates the following:
   - The way you will open your talk
   - The overall structure of your talk
   - The main points from your written communication that you will emphasize
   - The graphics you will use
   - Be ready to explain your outline in class.

2. Imagine that you must prepare a five- to ten-minute talk on some equipment, process, or procedure. Identify your purpose and listeners. Then write a script or an outline for your talk (whichever your instructor assigns). Be sure to plan what graphics you will use and when you will display each of them.

3. Do one of the Oral Briefing projects in Appendix B.

4. Using PowerPoint or a similar program, prepare a brief set of slides for the talk you will give in one of the preceding exercises. Follow the advice given in Guideline 2.

EXPLORE ONLINE

1. Using a search engine, find an online source for advice about oral presentations. Compare the advice given there with the advice you have found in this chapter.

2. Locate two PowerPoint presentations online that are about a topic of interest to you. View them in html and

then download them for viewing on your computer. Which of the two is most effective when viewed in html? Why? Which of the two is most effective when viewed on your computer? Why? Write a memo to your classmates explaining what you've learned about the Internet and about PowerPoint. To find PowerPoint presentations using Google, click on Advanced Search. On the page that opens, enter your topic. From the dropdown menu for Filetype, choose Microsoft PowerPoint.

COLLABORATE WITH YOUR CLASSMATES

At work, you will sometimes be asked to contribute to discussions about ways to make improvements. For this exercise, you and a classmate are to deliver a five-minute impromptu talk describing some improvement that might be made in some organization you are familiar with. Topics you might choose include ways of improving efficiency at a company that employed you for a summer job, ways of improving the operation of some club you belong to, and ways that a campus office can provide better service to students.

Plan your talk together, but divide your content so that you both talk. In your talk, clearly explain the problem and your solution to it. Your instructor will tell you which of the following audiences you should address in your talk:

- Your classmates in their role as students. You must try to persuade them of the need for, and the reasonableness of, your suggested action.
- Your classmates, playing the role of the people who actually have the authority to take the action you are suggesting. You should take one additional minute at the beginning of your talk to describe these people to your classmates.

APPLY YOUR ETHICS

Prepare a brief (three- to five-minute) presentation concerning an ethical issue that you or someone you know has confronted on the job. Describe the situation, identify the stakeholders, tell how the stakeholders would be affected, and tell what was done by you or someone else. If nothing was done, tell why. Be sure to keep within your time limit. According to your instructor's assignment, give your report to the entire class or to two or three other students.

FIGURE 19.5 **Writer's Guide for Creating and Delivering Oral Presentations**

To download a copy of this Writer's Guide, go to Chapter 19 at www.cengage.com/english/anderson7e.

Writer's Guide
CREATING AND DELIVERING ORAL PRESENTATIONS

Planning Your Presentation

1. Define your presentation's objectives.
2. Select the form of oral delivery best suited to your audience and purpose.
3. Identify ways to assure that you look at your audience.
4. Anticipate the questions your listeners are most likely to ask.
5. Identify ways to work with your nervousness.

Organizing

1. Choose a simple structure.
2. Select a few main points as your primary focus.
3. Help your listeners follow your structure by forecasting it, signaling transitions, and reviewing your main points.

Creating Graphics

1. Identify places where graphics will help you explain your subject matter, describe the structure of your presentation, or highlight your main points.
2. Avoid the temptation to use too many graphics or to use graphics that are too fancy.
3. Select the medium best suited to your purpose, audience, and situation.
4. Create graphics that are easy to read and understand.

Rehearsing

1. Rehearse with your graphics.
2. Time your rehearsal to stay within your time limit.

20 | Creating Reader-Centered Websites

GUIDELINES

I n this chapter, you will learn how to design and construct effective websites, as well as how to help others evaluate and devise strategies for improving their websites. This knowledge will increase your value to your employer. Traditional websites have long been important to business, nonprofit, and government organizations to shape public perceptions, sell products and services, and provide information to employees. More recently, various types of websites associated with the personal world, such as blogs and social networking sites like Facebook, are also gaining use in the workplace for internal as well as external uses. Your knowledge of what makes a website succeed in achieving its objectives will enable you to design new sites and evaluate existing ones, devise ways to make them better, and help your employer find novel, productive uses for them.

The best way to acquire this knowledge is to create a website yourself. Fortunately, this knowledge is inexpensive to obtain. You probably have the necessary tools on the computer you use right now. They are included for free with Microsoft Word and most other word-processing programs. You can also use similar tools for free at Google and other Internet sources. This chapter will teach you to make reader-centered websites using these free tools.

Learn More

Chapter 17, "Communicating and Collaborating in the Globally Networked World," describes several uses of websites designed by employers for internal use.

WWW

For more information about creating websites, links to helpful resources, example sites, and additional exercises, link to Chapter 20 at **www .cengage.com/english/ anderson7e.**

Reader-centered process for creating websites

READER-CENTERED PROCESS FOR CREATING A WEBSITE

To create an effective website, follow the same reader-centered process that you use to create usable, persuasive print communications—with some adjustments in the details.

Define Your Communication's Objectives
  1. Learn about your website's readers and define its purpose.

Plan
  2. Create a map for the website that includes what your readers want and enables them to locate it quickly.

Conduct Research
  3. Gather the information your readers need.
  4. Ethics Guideline: Respect intellectual property and provide valid information.

Draft
  5. Design pages that are easy to use and attractive.
  6. Provide navigational aids that help your readers move quickly through your website to the information they want.
  7. Unify your website verbally and visually.
  8. Ethics Guideline: Construct your website so readers with disabilities can use it.
  9. Global Guideline: Design your website for international and multicultural readers.

Revise
10. Test your website before launching it.

DIGITAL PORTFOLIO WEBSITES

This chapter guides you through the process of creating a website by showing you how to construct a special kind that can be a powerful aid in your job search. Called a *digital portfolio*, this type of website displays projects you've completed. The contents may include anything that can be shown on a computer, such as reports you've written, computer programs you've designed, videos of presentations you made, and photos of projects you've constructed, to name a few possibilities. Your digital portfolio provides *evidence* that you really have the abilities you *claim* to have in your résumés and application letters. You can post it on the web for employers to view or put it on CDs that you hand to them at interviews or mail to them with your applications. Figure 20.1 shows two pages of a digital portfolio.

Information about creating this special kind of website is incorporated throughout the chapter. Step-by-step directions for making one with Microsoft Word are provided in the Writer's Tutorial that begins on page 504. If you work along on a computer while reading this chapter, you will have completed a draft digital portfolio website by the time you reach the last page.

Learn More

For an explanation of the relationship between claims and evidence, see Chapter 5's Guideline 4 (page 127).

The Writer's Tutorial on pages 504–506 gives step-by-step directions for creating a website using Microsoft Word.

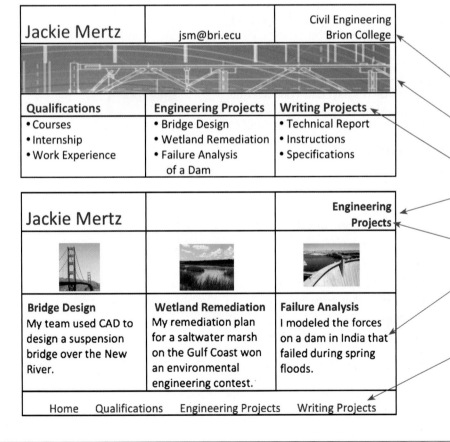

FIGURE 20.1
Digital Portfolio Main Page and Section Page

Jackie includes her name and contact information.

Jackie uses a graphic related to the job she wants.

These headings are links to her project, portfolio's major sections.

Jackie uses the same three-column design here as in her main page.

She identifies the subject of this section. The photos are links to her projects.

Jackie briefly describes each project, giving employers background information that will help her impress them with her talents.

These links provide easy navigation throughout the portfolio.

Jackie prepared a similar section page for her writing projects.

Writer's Tutorial

Creating a Website Using Tables

This tutorial shows how to create a website using Word for Windows 2007. If you use a different word processor or get stuck, click on your program's **Help** menu for assistance. To illustrate the process, the tutorial shows how to create a digital portfolio website, but you can create a website on any other topic using these instructions.

DEFINE YOUR WEBSITE'S OBJECTIVES (GUIDELINE 1)

1. Pick a topic or use "My Digital Portfolio."
2. Identify your target readers.
3. Define your website's usability and persuasive objectives.

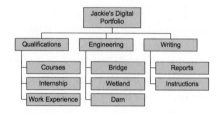

PLAN YOUR WEBSITE (GUIDELINE 2)

1. Determine your website's topics and subtopics, each identified by one word.
   - If you are making a digital portfolio, the major topics might identify your areas of expertise.
2. Draw a site map that organizes the topics and subtopics.

MAKE YOUR WEBSITE FOLDER (GUIDELINES 3 AND 4)

1. Make a new folder.
2. Title it "Website," followed by your initials.
   - All files for your website must be saved in this folder.
3. In the folder, put all files that will be displayed at your website.
   - For a digital portfolio, these might be projects you want employers to admire.

CREATE YOUR HOME PAGE (GUIDELINE 5)

1. Open a new Word document.
2. Under the **Insert** tab, click **Table**.
3. In the box that appears, choose 3 columns and 4 rows.
4. Under the **Microsoft Office Button**, choose **Save As**.
5. Click on **Other Formats**.
6. For **Save In**, navigate to the website folder you created.
7. For **File Name**, type "Index."
   - "Index" is the name browsers look for at websites.
8. For **Save as type**, choose **Web Page**.
9. Click **Save**.

CREATE ONE OTHER PAGE

1. **Copy** the table from Index.
2. **Open** a new Word document.

3. **Paste** the table into the new document.

4. **Save** the page by using Steps 4 to 9 in Create Your Home Page.

OPEN YOUR HOME PAGE FOR EDITING

1. From the **File** menu, choose **Open**.

2. From the pull-down menu by **Enable**, choose **All Readable Documents**.

3. Navigate to "Index.htm."

4. Click **Open**.

ENTER INFORMATION INTO YOUR HOME PAGE (GUIDELINES 5 AND 7)

1. In Row 1, enter heading text for your website.

   • For a digital portfolio, you might enter your name, contact information, major, and college.

2. In Row 3, enter the words that identify your website's topics.

3. In Row 4, enter information related to each topic.

   • For a digital portfolio, you might enter your projects' names.

4. In Row 2, highlight all the cells.

5. Under the **Table Tools** tab for **Layout**, click **Merge Cells**.

6. In Row 2, **Insert** a graphic related to your website's subject.

7. Change the color, type size, and alignment of the items to guide the reader's eye and make the page attractive.

8. **Save** your home page.

Jackie Mertz	jsm@bri.ecu	Civil Engineering Brion College
Qualifications	Engineering Projects	Writing Projects
• Courses	• Bridge Design	• Technical Report
• Internship	• Wetland Remediation	• Instructions
• Work Experience	• Failure Analysis of a Dam	• Specifications

Jackie Mertz	jsm@bri.ecu	Civil Engineering Brion College
Qualifications	Engineering Projects	Writing Projects
• Courses	• Bridge Design	• Technical Report
• Internship	• Wetland Remediation	• Instructions
• Work Experience	• Failure Analysis of a Dam	• Specifications

ENTER INFORMATION INTO YOUR TOPIC PAGES (GUIDELINES 5 AND 7)

1. In Word, open the second page you created.

2. In Row 1 of the table, enter heading information.

3. In Row 3, enter the titles of your website's topics.

4. In Row 3, you may also enter a description of each topic.

5. In Row 2, enter an appropriate image for each topic.

6. In Row 4, highlight all the cells.

7. Under the **Table Tools** tab for **Layout**, click **Merge Cells**.

8. In Row 4, type "Home" and the names of all your topics.

9. Change the color, type size, and alignment of all items to harmonize with those used in your home page.

10. **Save** the page.

Jackie Mertz		Engineering Projects
Bridge Design My team used CAD to design a suspension bridge over the New River.	Wetland Remediation My remediation plan for a saltwater marsh on the Gulf Coast won an environmental engineering contest.	Failure Analysis I modeled the forces on a dam in India that failed during spring floods.

Jackie Mertz		Engineering Projects
Bridge Design My team used CAD to design a suspension bridge over the New River.	Wetland Remediation My remediation plan for a saltwater marsh on the Gulf Coast won an environmental engineering contest.	Failure Analysis I modeled the forces on a dam in India that failed during spring floods.
Home Qualifications Engineering Projects Writing Projects		

(continued)

11. Make one copy of this page for each of your other topics.
12. Create each topic's page by replacing the appropriate items.

FINISH YOUR PAGES

1. On all pages, eliminate the table's grid lines.
   - Highlight the entire table.
   - Under the **Table Tools** tab for **Design**, click **Borders**.
   - In the menu that appears, click **No Border**. ⟶
2. Create links among your pages. (Guideline 6)
   - In the home page's Row 3, highlight the text in one cell.
   - Under the **Insert tab**, click **Hyperlink**.
   - In the window that opens, click on **Existing File or Web Page**.
   - Click on **Current Folder**.
   - Click on the appropriate file.
   - Click **OK**.
   - Repeat for all links on all pages. ⟶
3. **Save** each page when you finish.

TEST YOUR WEBSITE (GUIDELINE 10)

1. Open a web browser.
2. From the **File** menu, select **Open File**.
3. Navigate to Index.
4. Click **Open**.
5. Test all links.
6. Ask other people to review your portfolio.
7. Revise.
8. Test all links again after revising.

Learn More at the Website

To see digital portfolios made with Word, go to www.cengage.com /english /anderson7e.

GUIDELINE FOR DEFINING OBJECTIVES

Guideline 1 focuses on a few special considerations that arise when defining the objectives of a website rather than a printed communication.

Guideline 1 | Learn about your website's readers and define its purpose

Begin creating a website by defining its objectives. Who are its readers? How is it intended to help and influence them? Answer these questions as specifically as you can. Chapter 3's guidelines for defining objectives will help you describe the site's usability and persuasive objectives.

In defining your site's objectives, pay particular attention to usability. Research shows that it is the major concern of website readers (Nielsen, 2000). The overwhelming majority go to a website to find specific information or a particular service. Unless you design your site so they can locate what they want quickly, they will abandon your site and look elsewhere. Facing the challenge posed by these goal-oriented, impatient users, web developers try to create "sticky" sites, ones that give users enough of what they want quickly enough to hold them there. They can achieve this goal only if they have learned exactly what their readers want and how they will look for it.

If you are creating a digital portfolio website, review Chapter 2's discussion of the special objectives of employment materials. Your portfolio combines objectives of a résumé and a job application letter.

> Start by following Chapter 3's guidelines for defining your communication's objectives.

> **Try This**
>
> List the qualifications desired by employers for the kind of job you would like. This list can guide your selection of samples to include in a digital portfolio.

GUIDELINES FOR PLANNING

Guideline 2 | Create a map for the website that includes what your readers want and enables them to locate it quickly

There are many ways to plan the content and organization of any communication. For websites, developing a site map is particularly effective.

First, using your knowledge of your readers and the results of your research about them, list the information and assistance they will want to locate at your site. Add the additional information you realize will help them.

Each piece of information on your list will be either the sole content or one of the contents of a page in your website. From the perspective of users who will be trying to locate this information, each is a *destination*.

Next, plan an easy-to-follow path from your home page to each destination. If your site has only a few pages, readers will be able to locate rapidly what they want, provided that your home page includes a menu with links to every other page. If your site is larger, such a menu will be impractical. In this case, create a reader-centered hierarchy that enables readers to reach their destinations in no more than three clicks. Creating three-click paths is a goal commonly adopted for large sites built in the workplace.

Readers can reach their destinations in three clicks only if they don't make mistakes. Every wrong choice requires one or more clicks to return to the starting point. Consequently, you must create categories that readers will intuitively think of when searching for a topic. For example, if you are creating a site about your college, place rules for running for an office in student government in the area for Students (and in the subcategory for Student Government) rather than the area for General College Policies. Figure 20.2 shows the plan for a hierarchically structured website that uses intuitive categories.

In some websites, including digital portfolios, some ways of grouping your content can have greater persuasive impact than others. A student who has gathered samples under headings for First-Year Projects and Second-Year Projects could increase the persuasiveness of his or her digital portfolio by regrouping them under the heading Engineering Expertise and Communication Abilities. The headings for your groups become *persuasive claims* ("I have engineering expertise.") and your samples become the *evidence* that supports them, as explained in Chapter 5's Guideline 4 (page 127).

To develop a truly effective site map, you will need to remain willing to refine it, perhaps radically, throughout the site's development. Remain alert and responsive to new insights that you gain about your readers' needs, desires, and intuitive ways of looking for information. Usability testing has proven to be an especially effective way to gain these insights. Unless your site is very small, your site map is likely to go through at least as many drafts as will the content of any page in your site.

Learn More

What would be a good site map for your digital portfolio? Try sketching one out, based on your list of its objectives (see Try This on page 507).

Learn More

Chapter 16 provides detailed advice for designing and conducting usability tests. Turn to page 428. Also, Guideline 10 in this chapter discusses special considerations for testing websites (page 521).

FIGURE 20.2
Website Map with Hierarchical Organization and Intuitive Categories

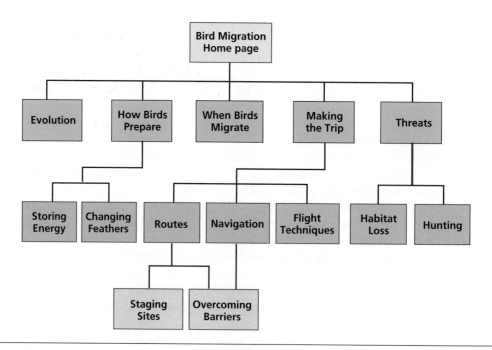

GUIDELINES FOR CONDUCTING RESEARCH

Guideline 3 | ## Gather the information your readers need

Gathering the information needed to create your website's text and graphics will always require research, even if only searching your memory. Chapter 6 and the "Writer's Reference Guide for Using Five Reader-Centered Research Methods" provide advice.

There is one additional point to emphasize about research for websites: Just because you are conducting research for a website does not make websites your best source. Books, journals, and interviews may be far superior. Focus on where you can get the best information for your readers, not on the method used.

For a digital portfolio, your primary content would be samples of your work. Pick ones that best demonstrate that you possess the knowledge and skills wanted by employers in your field. Remember that photos, videos, and similar digital items can be the best way to represent certain kinds of accomplishments. By including links to the websites of companies for which you've worked, organizations in which you've participated, and your academic program, you can fill out an employer's appreciation of your qualifications and accomplishments.

Learn More

For detailed advice for conducting research, turn to Chapter 6 (page 147) and the Writer's Reference Guide that follows it (page 165).

Guideline 4 | ## Ethics Guideline: Respect intellectual property and provide valid information

When you are creating a website, you encounter two sorts of ethical issues: those involved with what you borrow from other people when making your site and those concerning what you offer visitors to your site.

Ethics of Using Images and Text You've Downloaded

It's very easy to download images and text from someone else's website for incorporation into your own site. Some items available on the web fall into the public domain, meaning that no one has a copyright on them. You can use them freely without permission.

However, much of what's on the web is copyrighted. This includes not only text but also technical drawings, photographs, and frames from your favorite comic strip. Many sound files, animations, and videos available on the Web are also copyrighted. When you are thinking of using images from another site, check to see whether the site includes a copyright notice.

Whenever you are unsure about whether permission is needed to use something, check with your instructor or other knowledgeable person—or simply request the permission.

Even when permission isn't needed, acknowledge your source when you use something that clearly represents someone else's intellectual or creative effort. Appendix A tells how to cite web sources (pages 682 and 688).

WWW

For links to websites that provide detailed advice about copyright, intellectual property, and the web, go to Chapter 20 at **www.cengage.com/ english/anderson7e.**

Ethics of Posting

When you post something on the web, keep in mind that your readers rely on your expertise and integrity. Be sure that you have checked your facts and that your information is up to date. Remember also that all the other ethical considerations discussed in this book apply to web communications just as much as to print communications. Don't post anything that will harm other people.

Ethics and Digital Portfolios

In a digital portfolio, as in your résumé and job application letter, represent your work ethically by following the advice given in Chapter 2's discussion of ethical issues in the job search (page 58). If you are including group projects, indicate clearly what you contributed. Ability to work effectively in groups is, by itself, an important qualification to employers.

GUIDELINES FOR DRAFTING

Guideline 5 | Design pages that are easy to use and attractive

When designing web pages, your goal is the same as when designing printed pages: to create a usable and attractive communication (see Chapter 14).

Creating Easy-to-Use Pages: Signaling a Page's Organization

To be highly usable, a page must be designed so that readers can immediately see how it is organized. As with printed pages, you can accomplish this goal by using a grid of horizontal and vertical lines to establish visually the connections among related elements.

In fact, almost all web pages are constructed as tables that form grids—quite literally. The most frequently used letter in the source code for most web pages is *t*, for *table*. This letter is in the code for the beginning and end of every table row, every table cell in a row, and so on. The only pages not constructed as tables are the ones that use the entire browser window, side to side, top to bottom, as a single large area.

Shown here are two widely used grids for web pages.

Learn More

You can put this guideline's advice to immediate use by employing features of Microsoft Word or another word processor. Follow the step-by-step directions for creating a digital portfolio on page 503.

Learn More

To learn more about grid design, turn to Guideline 2 in Chapter 14, "Designing Reader-Centered Pages and Documents," page 381.

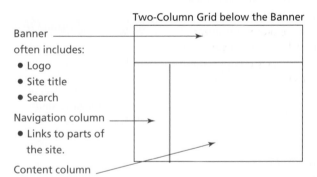

Two-Column Grid below the Banner

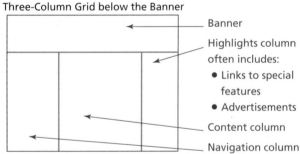

Three-Column Grid below the Banner

Banner often includes:
- Logo
- Site title
- Search

Navigation column
- Links to parts of the site.

Content column

Banner

Highlights column often includes:
- Links to special features
- Advertisements

Content column

Navigation column

By turning to pages 512–513, you can see how these two patterns form the major structural components of sample websites. In the samples, notice that the designers align related items against the same vertical or horizontal grid line to provide a visual cue that the items are related. By keeping text and graphics away from the edges of their cells, the designers also create white space (often in color in web pages) to separate unrelated items. Often, designers fill related cells with a distinctive color to indicate their connection. Sometimes, designers add borders to one or more sides of a group of cells as another way of keeping one group of related items separate from another group.

In addition to helping readers see which elements on a page are related, the design should also signal the *hierarchy* among the elements. By contrasting the appearance of two items or sets of items, you can make the more prominent set appear higher in the page's hierarchy. The techniques for giving greater visual prominence to some items is the same for web pages as for printed ones:

- Use larger type for the more major items.
- Put the larger items farther to the left by indenting less major ones.
- Use bold for the more major items.
- Use a more prominent color for the more major items.

You can see these techniques put to work in the websites on pages 514–515.

Making the Text Easy to Read

A page is usable only if readers can easily read its contents. People find it to be more difficult to read communications displayed on screen rather than on paper. Among other things, the resolution of the text on a computer screen is poor and the mechanics of scrolling make it more difficult for a reader's eyes to keep their place in a text. The following strategies will help you address the special challenges of creating legible web pages. Figure 20.3 (page 514) shows how these strategies were used to create effective web pages.

Learn More

To learn more about the use of alignment and visual grouping, turn to Guidelines 2 and 3 in Chapter 14, "Designing Reader-Centered Pages and Documents," pages 381–386.

Try This

Is the web page design used in the Writer's Tutorial on pages 504–506 the best possible one? Following the advice for this guideline, sketch a different design that works as well or better.

WWW

To view additional well-designed web pages, turn to pages 13, 107, and 136. Go also to Chapter 20 at **www.cengage.com/english/anderson7e**.

MAKING WEB PAGES EASY TO READ

- **Use typefaces that are easy to read on screen.**
- **Make sure your text is legible.**
  - **Use dark letters against a light background (or vice versa) to produce a strong contrast between the two.** White is always a good choice for the background.
  - **Use large enough text for legibility.** Test your type sizes to be sure they will be easy to read at higher resolutions, which make text smaller.
  - **Avoid intricate backgrounds that make it difficult to pick out the letters of the text.**
  - **Avoid italics.** These ornate letters can be very difficult to read on screen.
- **Present your information in small, visually distinct chunks.**
  - **Keep paragraphs short.** Break longer paragraphs into smaller ones.
  - **Use white space (blank lines) to separate paragraphs and other blocks of text,** keeping in mind that the "white space" on a web page may actually be any background color. Some sources recommend that only about 30 percent of the page be text.
  - **Edit your text for conciseness.** The fewer words, the better.

Writer's Tutorial

Designing Grid Patterns for Web Pages

BASIC DESIGNS

NATIONAL PARK SERVICE

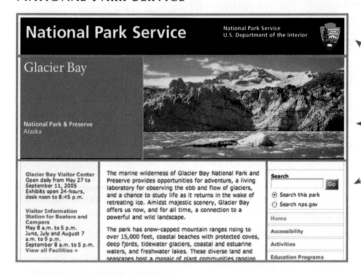

The National Park Service designed a grid with three horizontal areas.

Top: Undivided

Middle: Divided into two sides

Bottom: Divided into three parts

The design includes a very large area for the photograph to emphasize the beauty of Glacier Bay National Park.

NATIONAL ACADEMY OF ENGINEERING

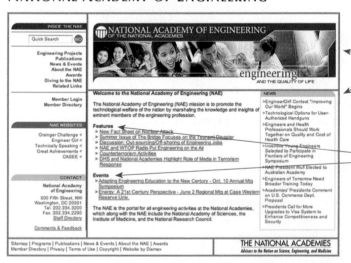

By devoting a small area for the photographic collage, the National Academy of Engineering created a grid that provides a great deal of space for links to the information readers want.

The headings that group related links also help readers find information quickly.

Learn More at the Website

For links to these websites and to see more grid patterns for web pages, go to www.cengage.com /english /anderson7e and choose "Writer's Reference Guide: Creating Eleven Types of Reader-Centered Graphics."

IBM JAPAN

For its Japanese website, IBM created a design whose central area (white) has five columns. Two of the grid lines from that area continue into the bottom area to divide its three cells.

ELABORATE DESIGNS

NATIONAL AQUARIUM IN BALTIMORE

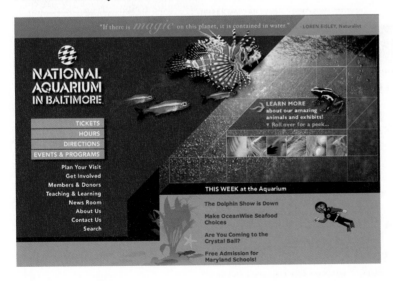

Where Do Angles and Curves Come From?

If all web pages are constructed from rectangular table cells (see page 503), how do designers make pages with angles and curves? Using specialized graphics programs, they create the overall image, and then cut it into rectangles. The three images above are among the twenty rectangles used to build the upper right-hand part of the National Aquarium in Baltimore website.

- **Use plenty of headings.** They not only help readers locate information but also break the screen into meaningful chunks.
- **Limit scrolling (it greatly increases the reading difficulty).**
  - **Avoid vertical scrolling** except on pages where users will be looking for in-depth information.
  - **Avoid horizontal scrolling altogether.**

Making Web Pages Attractive

Three design principles you can use to make your pages attractive are order, harmony, and proportion.

You can create *order* by following the advice already given for signaling the organization of a page: If it appears organized to a reader, it will also appear orderly.

FIGURE 20.3
Readable Web Pages

In-text links are clear and informative.
- Key words highlighted
- Underlining only for links

Text is chunked.
- Short paragraphs
- Many headings
- Use of a list
- White space separating blocks of text

Text is legible.
- Dark letters against a light background
- Text large enough to be readable
- No italics

Writing is clear and concise.

No scrolling required.

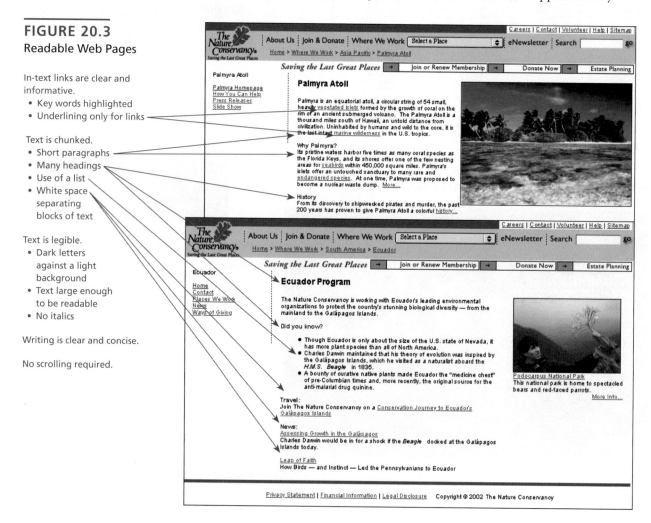

To achieve *harmony*, focus on the colors you use. Avoid colors that clash. Some word processing programs help you do this by providing palettes of harmonious colors. In some versions of Microsoft Word, choose Theme from the Format menu. Scroll through the Color Scheme option possibilities.

To review a web page for *proportion*, maximize it in a browser window, step back from your screen and determine whether some element is so much larger than others that it dominates the page. Most often, images and headings are the culprits. Headings don't need to be very large to do their job, and images don't need to be very big to communicate their information.

Guideline 6 | Provide navigational aids that help your readers move quickly through your website to the information they want

After you have carefully organized your site in a reader-centered way (Guideline 2), you have done half the work needed to help your readers locate the information they want. The other half is to design pages that include navigational aids that enable readers to move quickly from the home page to their destinations. The following strategies, which are illustrated in Figure 20.4, will help you guide your readers rapidly to any part of your site.

Try This

The Writer's Tutorial places links at the bottom of the section pages (see Figure 20.1). Where else could the links be placed that would be handy and attractive?

HELPING READERS NAVIGATE YOUR SITE

- **Include the main menu on every page.** Enable your readers to start a new "drill down" from whatever page they are on.

- **On every page, provide a link to the home page.** If a link to your home page isn't in your main menu, be sure to provide it elsewhere on every page.

- **On each page, provide a menu for the category in which that page is located.** Once readers are in a category, they may want to explore the various types of information you provide in it. By providing the category menu, you enable them to do that without backtracking to a higher-level page.

- **Place navigational aids in the same location on every page.** Once readers have learned where your navigational aids are, let them use that knowledge throughout the time they spend at your site. Placing your navigational aids in the same place on every page also helps you create a visually unified site, as explained on page 516.

- **Use clear, intuitive labels for all menu items.** The words you use for your links make a great deal of difference to your readers. Links are helpful only if readers can accurately predict what they will find if they click on them.

- **Make clear, informative in-text links.** When creating links within your text, signal clearly what your readers will find there.

    - **Highlight only the major words.** If you were to highlight all the words in the following sentence, your readers wouldn't know whether the link will lead to information on whales, the Bering Sea, or krill. Highlighting only major words makes clear what the topic of the link is.

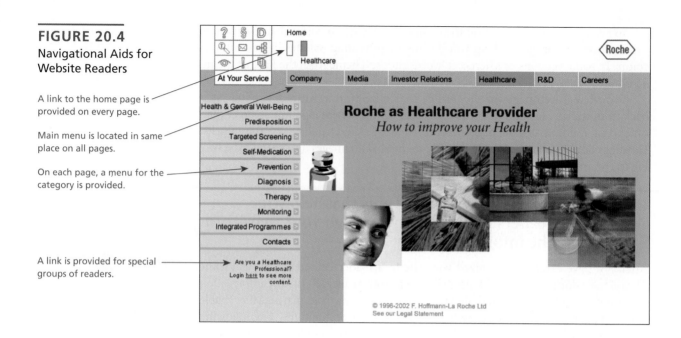

FIGURE 20.4
Navigational Aids for Website Readers

A link to the home page is provided on every page.

Main menu is located in same place on all pages.

On each page, a menu for the category is provided.

A link is provided for special groups of readers.

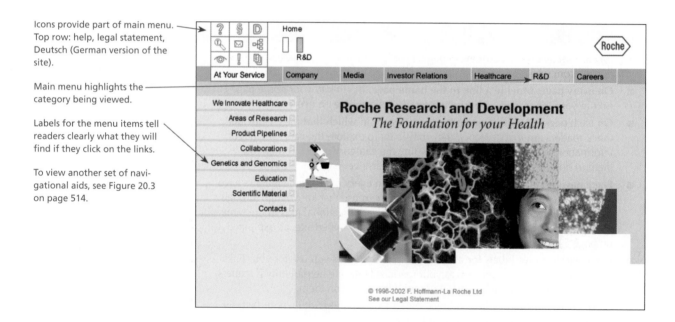

Icons provide part of main menu. Top row: help, legal statement, Deutsch (German version of the site).

Main menu highlights the category being viewed.

Labels for the menu items tell readers clearly what they will find if they click on the links.

To view another set of navigational aids, see Figure 20.3 on page 514.

When whales reach the northern Bering Sea, they feed on the plentiful krill.

Blue type indicates the link.

- **Supply explanations of the links.** These are especially helpful for links included in a list.

  International projects—Read about our current projects in Europe and Asia.

  - **Use precise, specific wording.** Choose words that tell exactly what the linked page will show.
- **Use underlining only for links.** Readers are accustomed to the convention that underlining signals links within text. To avoid confusion, don't use underlining for anything but links. Similarly, do *not* use the feature of some web page editors that would allow you to remove the underlining from your links.s
- **Include a site map.** Many readers use site maps as a navigation shortcut or as an aid to finding information they haven't otherwise been able to locate. See Figure 20.5 for an example. Each word in the lists is a clickable link to a page that has the information named. Usually, the site maps located in websites are presented as outlines; every item in the outline is a link directly to the topic it describes.

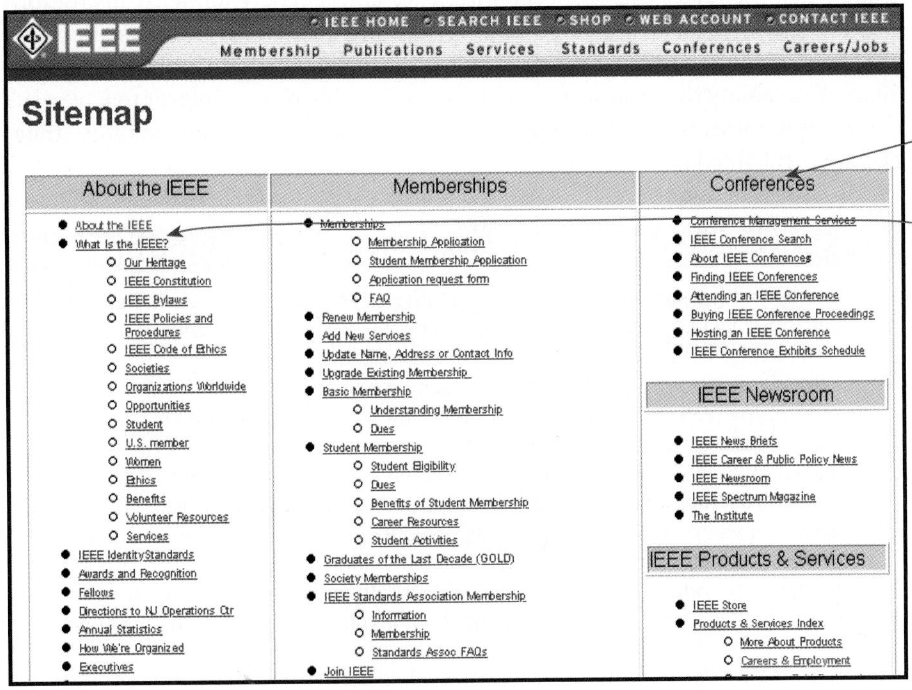

FIGURE 20.5
Site Map

Headings and subheadings show the structure of the site and help readers find specific information.

Links that have been visited are displayed in red.

The menu is longer than shown in this figure. All items in the main menu (top of page) are included, plus others.

To see other ways of designing site maps, go to Chapter 20 at www.cengage.com/english/anderson7e.

WWW

For more information on Cascading Style Sheets, go to Chapter 20 at **www.cengage.com/english/anderson7e**.

Try This

In the Writer's Tutorial for creating digital portfolios, the directions for creating section pages (page 505) shows one possible design for the section pages that use the home page's grid pattern. Can you create another design for the section pages that uses the home page's grid pattern?

Guideline 7 | Unify your website verbally and visually

Consistency in the verbal and visual style of a site increases its usability and attractiveness for readers.

Because you create each page in a website separately, it is easy to change the writing style unintentionally as you move from topic to topic. The change might appear in the way information is presented or in the phrasing of links and headings. You can avoid such shifts by planning the style you want to achieve before you begin drafting. You can check changes that have slipped in by making consistency of style one of the features to be examined when you and others review your site.

To create visual unity, apply the design principle of repetition. Use the same grid for all pages in your site that present the same type of information. Even for pages that present different kinds of information, retain some element of the grid, such as the banner across the top of the page. Also, apply the same palette of colors and use the same typefaces throughout. Figure 20.6 illustrates the application of these strategies. An advanced feature of web page design called Cascading Style Sheets saves time by letting you control the appearance and layout of many pages at once.

Guideline 8 | Ethics Guideline: Construct your website so readers with disabilities can use it

A truly reader-centered website can be used by every reader who would like to use it, including people with disabilities. In 2001, the U.S. government instituted a regulation that mandates that all federal agencies and federally affiliated institutions (including almost all U.S. colleges and universities) make their websites and other electronic and information technology accessible to persons with disabilities (Roach, 2002). Your ability to create an accessible site will enable your employer to have a site that complies with this regulation. Some of the most important accessibility strategies are highlighted below.

Some persons with a visual impairment supplement a regular keyboard for typing with another that enables them to rapidly navigate the web page as the computer interprets into Braille the headings, text, links, and image tags.

©Christina Micek

HELPING READERS WITH DISABILITIES USE YOUR SITE

- **Readers with visual impairment.** Use relative sizes, not point sizes, to designate how big your text will be. If you use relative sizes, readers with poor vision can use their browser settings to enlarge the text on your pages.

  Many people who can't see well use screen readers, which convert text to speech. To create a reader-centered site for these persons, include an <alt> tag with all images. This tag enables you to provide a text description of the image that can be synthesized into speech by screen readers. Also, when you create in-text links, highlight significant, descriptive words, not such phrases as "Click here." For example, instead of writing, "To obtain more information about our newest products, click here," write, "You can obtain more information about our newest products." Screen readers aid visually impaired readers by pulling out all the links on a page and placing them in a list. A list in which every link

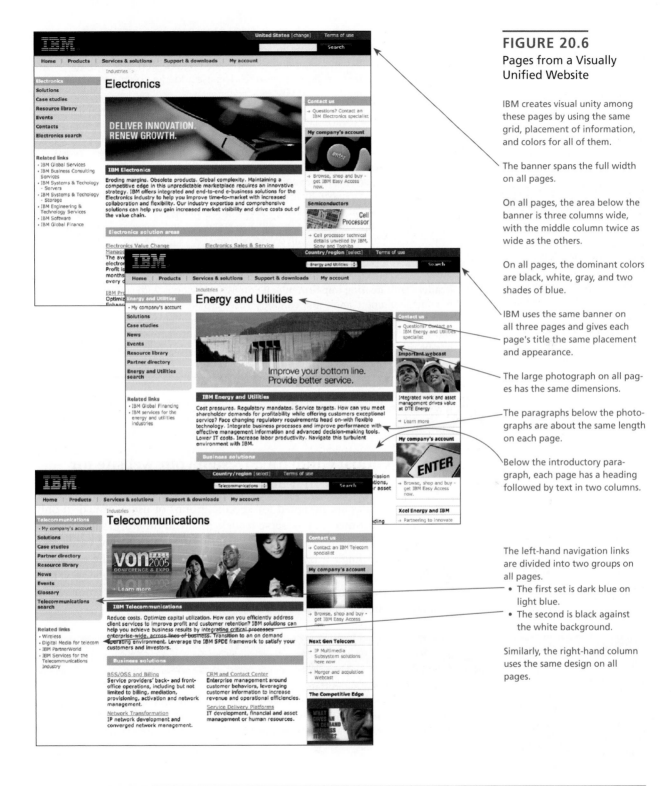

FIGURE 20.6
Pages from a Visually Unified Website

IBM creates visual unity among these pages by using the same grid, placement of information, and colors for all of them.

The banner spans the full width on all pages.

On all pages, the area below the banner is three columns wide, with the middle column twice as wide as the others.

On all pages, the dominant colors are black, white, gray, and two shades of blue.

IBM uses the same banner on all three pages and gives each page's title the same placement and appearance.

The large photograph on all pages has the same dimensions.

The paragraphs below the photographs are about the same length on each page.

Below the introductory paragraph, each page has a heading followed by text in two columns.

The left-hand navigation links are divided into two groups on all pages.
- The first set is dark blue on light blue.
- The second is black against the white background.

Similarly, the right-hand column uses the same design on all pages.

says "Click here" will be useless. In addition, use each HTML tag only for its intended purpose. For instance, don't use the heading tags <H1>, <H2>, and so on to change type size. Some screen readers use a louder volume to signal that text is a heading.

For readers with color blindness, supplement all information conveyed in color with that same information in text. For example, when you present a warning in bold red text, also include the word "Warning." In addition, because most color blindness is red-green, avoid using this combination of colors to distinguish content.

When you include movies or animations, use a program such as QuickTime Pro to add captions for visually impaired readers.

If these and similar strategies don't succeed in making your site fully accessible to visually impaired readers, create a separate text-only version. Update it whenever you update the visual version of your site.

- **Readers with limited mobility.** Some readers have difficulty controlling a mouse precisely. Make all clickable areas as large as possible.
- **Readers with hearing impairments.** If you include videos with sound, provide a volume control and captions. Also, provide concurrent captions with movies and other items that have sound. Supplement sound that gives verbal feedback by providing visual information of the same sort. For instance, if your site provides a verbal "Thank you," also include a pop-up window with that message.

WWW

For more on creating websites for readers with disabilities and for links to websites where you can test the usability of your web pages for readers with disabilities, go to Chapter 20 at **www .cengage.com/english/ anderson7e.**

Several websites will test the accessibility of your web pages for free. For links, go to Chapter 20 at www.cengage.com /english /anderson7e.

Guideline 9 | Global Guideline: Design your website for international and multicultural readers

Although some websites are designed for specific groups of users located in only one nation or culture, others aim to serve readers in any of the world's nations and cultures. The following strategies will help you create a site that will be easily read by persons for whom English is a second or third language and by the computer programs that international readers can use to automatically translate your text into their language. If your employer decides to make a second version of your site in another language, these strategies will also ease that work.

WWW

For links to websites that provide additional advice about creating websites for international and multicultural readers, go to Chapter 20 at **www.cengage.com/english/ anderson7e.**

DESIGNING PAGES FOR INTERNATIONAL AND MULTICULTURAL READERS

- **Text.** Write simply. Avoid slang, colloquial expressions, and metaphors that refer to sports, occupations, or other things specific to your nation. Also, as advised in Chapters 5, 8, and 9, consider your readers' cultural background when drafting your text. Check your draft with persons knowledgeable about the cultures of your likely readers.

- **Images.** As Chapter 12 explains, images that are meaningful to persons in your culture may be incomprehensible—even offensive—to readers in other cultures. Follow that chapter's advice when selecting your photographs, drawings, icons, and other images. Also, avoid including words in your images. Computerized translators cannot read images. And don't forget that persons creating a version of your site in another language will need to recreate all images, including buttons in menus that have words.
- **Page design.** Many English words are translated into much longer words in other languages. Similarly, the same statement may take as many as 30 percent more words in other languages. If you expect your site to be translated into other languages, design the size of your buttons, menus, and text areas to accommodate these differences.

REVISING GUIDELINE

The final step in creating a reader-centered communication is revising it.

Guideline 10 | Test your website before launching it

The way your website appears in your browser is not necessarily the way it will appear to all of your readers even if they are using only the default settings. For example, a PC and a Mac can process the same web page file so differently that a page that looks fine in one can be seriously flawed in the other. Likewise, different browsers, such as Firefox and Internet Explorer, can handle web page files in very different ways. For these reasons, you should test your site on computers with different operating systems and browsers before launching it. While testing, be sure to set the screen resolutions at the various settings that members of your target audience are likely to have. A web page that looks good at a resolution of 1024 × 768 may be too wide to fit on a screen whose highest resolution is 800 × 600.

Once you've tested your site, you are ready to show it to others, whether by posting it on a server or copying it onto a CD. As a student, you would probably post it on a server at your school. As an employee, you would probably post it on a server owned by your employer, although some employers pay companies called Internet providers to host their sites. Often posting is done by means of FTP (file transfer protocol) programs, but posting can be accomplished in other ways as well. After you've posted your site, be sure to test it once again using several browsers.

Learn More

Chapter 16 provides detailed advice for designing and conducting usability tests. Turn to page 428.

Writer's Guide and Other Resources

Figure 20.7 presents a Writer's Guide for Websites that you can use in your course and on the job. At the website for the book, you can obtain a copy of the planning guide that you can download.

FIGURE 20.7   **Writer's Guide for Creating Websites**

 To download a copy of this writer's guide, go to Chapter 20 at
www.cengage.com/english/anderson7e.

Writer's Guide
CREATING WEBSITES

Defining Your Site's Objectives

1. Define your site's usability objectives.

2. Define its persuasive objectives.

3. Consider your readers' Internet connections, hardware, and browser.

Providing Quick, Easy Access

1. Build a reader-centered, three-click hierarchy.

2. Use intuitive categories when building your hierarchy.

3. Include the main menu and link to the home page on every page.

4. On each page, include a menu for the page's category.

5. Put navigational aids in the same location on every page.

6. Use clear, intuitive, informative links.

7. Use underlining for links only.

8. Include a site map.

Designing Easy-To-Read, Attractive Pages

1. Follow Chapter 13's guidelines for page design.

2. Use legible text.

3. Present your information in small, visually distinct chunks.

4. Use plenty of headings.

5. Limit scrolling.

Assuring Your Site Will Work for Your Readers

1. Test your site on multiple platforms and browsers.

2. Create your text, images, and page design for international and multicultural readers.

3. Include features that aid readers with disabilities.

Designing Ethically

1. Observe laws concerning copyright and fair use.

USE WHAT YOU'VE LEARNED

Note to the Instructor. For a website assignment, consider the Informational Website assignment provided in Appendix B. Other projects that involve websites are included at the book's website.

EXERCISE YOUR EXPERTISE

Visit a website, perhaps one related to your major or to a project you are preparing in this course. Begin by identifying the site's target readers and its usability and persuasive objectives. Next, evaluate one of the site's pages in terms of this chapter's guidelines for designing effective web pages. Finally, evaluate the site's overall design in terms of this chapter's guidelines for building effective websites. Report the results of your analysis in a memo to your instructor.

EXPLORE ONLINE

1. Using a search engine, locate four websites that provide advice for designing websites. Compare their advice to the advice given in this chapter. Report your results in a memo to your instructor.

2. At the website of an international corporation (e.g., Sony), link to the organization's sites in three other languages. Discuss four ways in which these sites differ from the corporation's site for your country.

3. Evaluate the website for your university (or another site identified by your instructor) in light of this chapter's strategies for creating sites that can be used by people with disabilities.

4. Create your own website, following the guidelines given in the website assignment in Appendix B.

COLLABORATE WITH YOUR CLASSMATES

Working with two or three other students, evaluate the usability and persuasiveness of your school's website from the perspective of a high school student who wants to learn about your school. Together, brainstorm the top six questions you think the student would have, and then look for the answers. Report your results in a memo to your instructor.

APPLY YOUR ETHICS

Create a web page for use by students in your major at your school that provides links to information on ethical issues that are relevant to them. Include links to your school's policies concerning academic integrity, plagiarism, and related issues as well as information from professional organizations. Make the page visually appealing, informative, and easy to use.

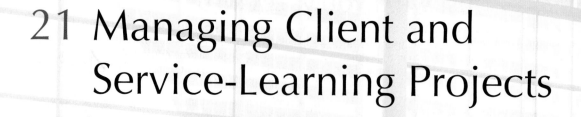

21 Managing Client and Service-Learning Projects

GUIDELINES

1. Determine exactly what your client wants and why *526*
2. Develop your own assessment of the situation *527*
3. Create a project management plan *528*
4. Submit a written proposal to your client—and ask for a written agreement *530*
5. Communicate with your client often—especially at all major decisions *531*
6. Advocate and educate, but defer to your client *536*
7. Hand off the project in a helpful way *537*

In this chapter, you will learn a valuable set of skills that will enable you to succeed when preparing a communication project for a client. Client projects are quite common: In almost all specialized fields, employees of consulting companies, think tanks, design firms, and similar organizations work under contract with corporate and government clients to prepare research studies, feasibility reports, and other information products. Similarly, numerous technical communication companies specialize in preparing instruction manuals, training programs, marketing materials, websites, and other communications for their clients. Within many organizations, some departments routinely prepare communication projects at the request of other departments who are, in essence, their clients.

Work on a communication project for a client differs significantly from work on a project you prepare without a client. With a client project, your final product must not only be usable and persuasive in your target readers' eyes, but it must also please your client. Consequently, you must thoroughly understand your client's goals for the communication, and you must obtain your client's approval for all the communication strategies and design features you employ. In some situations, your client may readily agree to whatever you suggest. In others, your client may dictate approaches that are very different from those you would choose.

You may also need much more than approvals from your client in order to complete your task. For instance, you may need your client to provide the information to be presented in your communication, information about the target audience, and access to readers to test your drafts. Because of these and many other factors, your success on a client project depends as much on your ability to build and maintain a productive relationship with the client as it does on your ability to create a communication that is usable and persuasive in its target readers' eyes.

To help you develop expertise at working with clients, your instructor may ask you to prepare a communication on behalf of a real company, university office, or similar organization. Working as a consultant, you will create a communication your client will actually use, such as a procedure manual for its customer service representatives, an addition to its website, or a feasibility report about a new venture it is considering. In a variation on this type of assignment, your instructor may ask you to help a nonprofit, volunteer, or community-service organization, such as United Way or a neighborhood health clinic. Projects of this type are often called service-learning projects. For a service-learning assignment, your instructor may ask you to learn about the needs and aspirations of the people served by the organization,

WWW

For additional information and links related to this chapter's guidelines, go to Chapter 21 at **www.cengage.com/english/anderson7e**.

Client projects give students an opportunity to research practical problems and challenges and then develop effective solutions.

© Dmitry Yashkin 2009/ Shutterstock.com

© Gabe Palmer/CORBIS

© Tony Freeman/PhotoEdit

© Masterfile/Royalty Free

understand the motivation and responsibilities of its staff, and reflect on ways you can use your talents to assist others in your community.

Whatever the details, a client or service-learning project can be one of the most satisfying assignments of your college years. You get to produce a communication that has real, practical results. You get to work with people who greatly appreciate your assistance. And you get to develop a highly valuable set of skills in project management.

OVERALL PROJECT MANAGEMENT STRATEGY

The project management strategies presented in this chapter have two overall goals. The first is to enable you to develop a close, cooperative, productive partnership with your clients. The second is to enable you to plan and conduct your work in an efficient manner that produces an excellent communication that pleases your client as well as its target readers. Underlying all of this chapter's specific advice are three key principles, which lead you through the three major stages of work on client and service-learning projects.

PRINCIPLES OF EFFECTIVE CLIENT PROJECT MANAGEMENT

- **Begin your partnership with your client by reaching a detailed, mutual understanding of all important aspects of the project.** As you start your work together, you and your client should establish a shared understanding of the project's goals, how you will pursue these goals, and what you will each contribute to the project's success. Guidelines 1 through 4 will help you develop this mutual understanding.

- **As the project proceeds, communicate continuously with your client.** Continuous communication ensures that your client understands how you are proceeding and agrees with you on all major decisions. Guidelines 5 and 6 highlight the two most important strategies for effective communication as your work on the project advances.

- **As you finish the project, help your client take the next step.** In addition to presenting your completed communication, explain what the client must do to use it as effectively as possible. Guideline 7 suggests ways to do this.

WWW

To download a copy of the Worksheet for Defining Objectives that you can modify in preparation for a meeting with your client, go to Chapter 3 at **www.cengage.com/ english/anderson7e**.

Guideline 1 | ## Determine exactly what your client wants and why

The first step in forging your partnership is to fully understand the project from your client's perspective. Ask the client to meet with you to discuss the project in depth, if possible. To the meeting, bring a list of questions you need to have answered before you begin work. Let your client speak first, however. Then you ask the questions from your list that remain unanswered. The Writer's Guide for Defining Your Communication's Objectives, Figure 3.1, page 70 can provide a helpful starting point for your list. Be sure to include questions on the following topics.

- **Client's organization.** Be sure you learn about the organization's products or services. Especially in service-learning projects, learn also its goals and values. In addition, find out about the events that led the client to request your assistance. The story of these events can help you understand what the client wants your communication to achieve. It can also reveal a wealth of information about factors you need to consider as you proceed.

- **Readers.** Determine whom the client sees as the target audience. Are there subgroups? How will members of the target audience use the communication? What are their current attitudes about its subject matter, and what are their important characteristics that will influence the way they respond to the communication?

- **Usability and persuasive objectives.** Learn what the client wants the communication to enable the readers to do. How does the client want it to affect the readers' attitudes?

- **Stakeholders.** Learn who besides the target readers will be affected by the communication. This information will help you understand the ethical dimensions of the project.

- **Deadlines.** Determine when the client wants you to complete the project. Are there interim deadlines for completing certain parts of it?

- **Preferences and requirements.** Find out what characteristics your client would like or will require the communication to have. Must it be written in a certain style? If it will be printed, is there a page limit? If it's an online communication, must it work on all platforms or only PC or Mac? If it's a website, which browsers must it work with?

- **Resources.** Learn which members of the client's staff will answer your questions. What hardware, software, or other equipment will the client provide? When will these people and other resources be available to you?

- **Budget.** If you will be paid or if production expenses are involved, ask how much the client has allocated for the project.

Learn More

Review Chapter 3 for advice about describing audience and purpose.

By interviewing your client, you gain information needed to develop a detailed, effective project plan.

Guideline 2 | Develop your own assessment of the situation

The next step in preparing your proposal is to deepen your understanding of your communication's goals by seeking your own answers to each of the questions you asked your client. When describing projects, clients sometimes forget to mention important facts. Sometimes they don't realize the importance of certain kinds of information that you know to be critical. As you size up the situation yourself, focus particularly on gathering information about the readers, their tasks, and their attitudes. Talk with some of them, if possible, or with other people knowledgeable about them.

As you assess the situation, you may reach some conclusions that differ from your client's. For example, you may reach a different understanding of what the readers want, what will inspire them to change their attitudes in the desired way, and even who the readers will be. Discuss these differences with your client as soon as possible.

Consider also the work you will be doing and the resources you will have. In their initial descriptions of a project, clients sometimes underestimate how much work a project involves, expecting more than can be accomplished with the time and money allotted. You and your client will both benefit from reaching a clear, mutual understanding of what you are to produce and what your client will provide.

Guideline 3 | Create a project management plan

After you've developed a full understanding of the project, create a project management plan. Once it's completed, you can share the plan as part of a written proposal that you submit to your client for approval or for discussion and refinement (see Guideline 4). In the plan, describe in detail each of the following key elements of the project: *the deliverable* (the finished communication you will deliver to your client at the end of the project), the resources you'll need from the client to complete the project, and the schedule you and your client will follow as you work together.

Your Deliverable

Describe your deliverable in as much detail as you can without prematurely making decisions that are best made as the project unfolds. Here are three major topics to include.

- **Size.** You and your client will both be unhappy with the results of your project if one of you is expecting the final product to be twice the size the other is expecting. When describing size, use numbers rather than adjectives. Instead of saying that the report you are producing will be "short," decide whether you are going to deliver 10, 20, or 50 pages. Rather than saying the website you are creating will "provide all the information visitors might desire," say whether it will have approximately 12 screens or 120.
- **Overall content and major features.** Although you should remain open to new ideas and insights as you work on the project, your proposal should describe the general nature of the communication you will deliver. Be careful not to promise more than you can prepare in the time available and (if you are being paid) for the price you are quoting. Don't agree to create features that require you to acquire new skills unless you have the time necessary to learn the skills.
- **Technical aspects.** If you are creating an online communication, describe its functionality. Will it be created only for a PC or Mac platform, or for both? If it is a website, which versions of which browsers will it display in? If the communication will be printed, will it be in black and white or full color? In part,

you may be repeating requirements your client has given you. By repeating, you will ensure that you have understood correctly, and you and your client will be protected against later changes of mind.

Your Client's Contributions

In addition to describing the deliverable you will give your client, identify all the things you will need from your client. In most cases, these will include information as well as access to people who can answer the questions you encounter along the way. Perhaps you will also need access to certain equipment or facilities or the use of certain software. If you are going to do user testing for usability and persuasiveness, you may need your client to provide you with a certain number of persons from the target audience to serve as test users. Certainly you'll need responses to outlines, drafts, and other materials you provide for review. If you are working on a service-learning project, specify the ways the client will help you learn about the organization and community.

Your Schedule

Finally, create a detailed project schedule. It will serve you in many ways. For instance, the schedule will enable you to determine how many hours, days, and weeks it will take you to complete the project. Placed in your proposal, the schedule assures your client that you have a workable plan and tells your client when to expect drafts, progress reports, and other communications from you. As you work on the project, your schedule guides your activities, telling you what you need to do, when you need to complete each task, and whether you are progressing rapidly enough to complete your deliverable by the deadline.

As the first step in creating your schedule, list the major activities you will perform. Begin with the framework provided in this book: defining objectives (including learning in detail about the communication's readers and purpose), planning (including conducting necessary research), drafting, and revising (including testing for usability and persuasiveness, if appropriate). Next, identify the specific tasks you will need to perform as part of each of these major activities. Will your research include interviewing subject matter experts as well as reading additional information? While drafting, will you create charts, drawings, and other graphics from scratch as well as acquiring existing ones?

Learn More

For advice about schedule charts, see page 377.

Once you've identified the activities you'll perform, create a detailed schedule. Work backwards from the final deadline, allotting a reasonable proportion of the available time to each activity. Establish *milestones*, or minideadlines, within the overall schedule. Establish milestones at the dates when each task needs to be completed so that there is adequate time for the final project to be completed. For instance, indicate when research needs to be done so that you can begin drafting.

In most projects, several tasks can be performed simultaneously. For example, research can be conducted for one section of a website while another section is being built. In these cases, identify the sequence of tasks that determines the quickest time the project can be completed. This is the sequence along which the addition of one hour or day to any task postpones completion of the entire project. It is called the *critical path*. For large projects, you may find it helpful to use project management software such as Microsoft Project.

In your schedule, identify the dates when you and your client will communicate with one another. Include the dates for each meeting and for the deadlines for each draft and progress report you will submit. Be sure to schedule an interaction at the points where you need to check your results with your client. For example, if your work includes audience analysis, set a time to go over the results to be sure that your client agrees with your analysis. Disagreement here could lead to very different expectations about what you'll produce. Other times to check with your client are after you've created a detailed outline of the communication, after you've completed your usability test, and when you have completed an early or partial draft that shows how you will implement your communication strategies.

Finally, be sure your schedule clearly tells your client when you will need things. Specify not only the dates you will submit drafts or other material but also the dates you need to hear back from the client. In terms of your ability to complete your project on time, it makes all the difference in the world whether you must wait two days or two weeks for the approval or comments you need before you can proceed to the next task.

Guideline 4 | Submit a written proposal to your client— and ask for a written agreement

A written proposal is an excellent tool for reaching a common understanding with your client about your project. It is also a valuable guide to your ongoing work together on the project. It can serve two important purposes.

- **To build your client's confidence in your ability to do an excellent job.** A well-written proposal persuades the client that you thoroughly comprehend the client's goals, that you understand the kind of product needed to achieve the goals, and that you will follow a process that will effectively create that product. In business situations, where individuals and organizations often compete for work from clients, these persuasive dimensions of a proposal can be extremely important. However, establishing the client's confidence in your knowledge and professional approach is also important in the client and service-learning projects you might do in school.

- **To ensure that you and your client reach substantial agreement about the project before you begin work on it.** In a well-written proposal, you provide explicit details about the final product you will deliver to your client, the things your client will provide you, and the schedule and other arrangements for your work together. It provides an excellent opportunity for you and your client to be sure that you both have the same vision of the project. This will spare you both frustration and lost time as the project advances. Also, the proposal can provide a framework that helps you both plan your time and other commitments. Finally, it can protect you both from the problems that might arise if you had only an oral agreement and later discovered that you recollected some significant aspect of that agreement differently. In a school setting, such situations can lead to pressure on you to do additional work. In a business setting, it can lead to disagreements about whether you have completed the job and can be paid for it.

To highlight for your client the importance of studying the details of your proposal, ask for a written response to the proposal. If the client agrees to the proposal as you've written it, then a one-sentence memo will do. If not, then your proposal provides the perfect basis for detailed, specific negotiations. In either event, it provides you—and your client—with an invaluable measure of protection.

When preparing your proposal, be cautious about appearing to promise more than you can deliver. If there are some things you think you might be able to provide but aren't sure, say that they will be included if it is feasible within the time and budget. Also, be sure that the proposal includes all the significant things you want to be able to rely on your client to provide so that your client isn't surprised by unexpected requests once the project begins.

Even though the proposal is a formal agreement that is designed to protect you and your client, its overall goal is to create a cordial, cooperative, mutually satisfying partnership. Be careful to avoid taking on a demanding tone, even when describing what your client will give you. Write in a cordial and businesslike manner. Convey enthusiasm for working with the client on the project.

Figure 21.1 shows a sample client proposal.

Learn More

Chapter 23 provides detailed advice for writing proposals.

WWW

For additional sample client proposals and links to advice about writing them, go to Chapter 21 at **www.cengage .com/english/anderson7e.**

Guideline 5 | Communicate with your client often— especially at all major decisions

Many of your contacts with your client will occur according to the project schedule you include in your proposal. These will be times, for instance, when you agreed to provide a progress report or submit something for your client to review. Be prepared, in addition, to contact your client promptly at any other time when the need arises, as when you require some information or have encountered a problem the client should know about. Two of the most common problems involve a project's schedule and scope. Schedule problems can arise either because it's taking you longer than anticipated to complete some part of the project or because your client is taking longer than you expected to provide needed information or feedback on a draft or other material. Scope problems can arise when either you or the client begins to enlarge the project beyond what you both agreed to in your proposal. Sometimes the size of a project increases in small increments, a situation called *scope creep*. In other situations, an entirely new task is contemplated. The key to completing your project successfully is to bring all problems to your client's attention so that the two of you can determine how best to address them.

FIGURE 21.1
Client Proposal Written by
a Student Team

October 7, 2011

Dr. G. F. Hargis, Curator
Wright Zoology Museum
Crandell University
Iowa City, IA 52240

Dear Dr. Hargis:

The student team opens by continuing to build a positive relationship with their client, Dr. Hargis.

We are very excited about the opportunity to work with you to develop a website for the Wright Zoology Museum. As you know, our work on the site will serve as a project for our technical communications course with Professor Dellapiana.

The students tell what they want Dr. Hargis to do with the letter: determine whether he agrees with what they say.

In this letter, we explain our understanding of your objectives for the site and our proposed strategy for creating it. If you approve of what we say, please send us a memo authorizing us to begin work. If you want to discuss the project further, please contact Tim at T_Banner@Crandell.edu.

Background

The students briefly summarize information they've heard from Dr. Hargis to demonstrate that they've understood him accurately.

We understand that you wish to increase greatly the number of visits the museum receives from elementary school classes in grades three through six. You have recently developed new programs suitable for these. As you've also explained, very few teachers in the region know that the museum exists because it has been used primarily as a resource for students here at the university. Also, you have reported that some teachers who have brought their classes to the museum have been disappointed because its programs have been designed for older students.

Objectives

The students state their understanding of Dr. Hargis's objectives.

Through talking with you, we have established the following objectives for the website.

1. To introduce the museum's programs in a way that encourages teachers to visit.

2. To enable teachers to pick the programs best suited to their classes.

3. To enable teachers to request additional information or arrange a class trip online.

FIGURE 21.1
(continued)

We also have these additional objectives.

1. To provide a site that you or your graduate students can easily manage and update.

2. To help you develop a plan for publicizing the site.

Final Product

We propose to achieve these objectives by creating a site that has the following features.

1. An attractive home page that provides easy navigation to the information teachers most want.

2. Detailed information about each of your programs.

3. Lesson plans teachers can use to prepare their students for a visit.

4. Information about arranging visits and the downloadable forms and email links that will enable teachers to make their arrangements online.

5. Photographs, drawings, and other images that make the museum and its programs seem attractive.

We will build the site in Dreamweaver for use on the museum's server. The site will include approximately 60 pages and about 45 images. Each page will download in fewer than 20 seconds on a 56k modem.

Process for Developing the Website

While creating the site, we will focus on five types of activities.

Research. We will study the museum's new programs, interview teachers from your target audience, and visit websites for small scientific museums that offer similar programs for elementary schools. We will learn what types of information teachers will want from the site and what type of presentation they would find usable and persuasive. We will report our research results to you in a memo, asking you to add your own comments and insight.

Planning. We will develop three alternative screen designs for the home page and three designs for the other pages in the site. After you select the elements you like from the alternatives, we will combine them into a final design. We will also develop a site map, which we will submit for your approval.

Building. We will build the site on developmental server space arranged by Professor Dellapiana.

By adding two objectives that Dr. Hargis had not mentioned but that will benefit the museum, the students demonstrate their knowledge of web design and emphasize that their goal is to help the museum.

The students begin their description of what they will deliver to Dr. Hargis by describing their proposed site's major features. Each feature addresses one or more of the major goals Dr. Hargis has for the site.

The students provide specific details to prevent the client from expecting more than the students will be able to do.

The students describe their process in ways that establish confidence in their ability to successfully complete the project.

The students highlight the times they will submit their ideas and results to the client for review and approval.

FIGURE 21.1
(continued)

By describing the user testing they will perform, the students assure Dr. Hargis that the site they propose will succeed in achieving the museum's objectives.

User testing. When planning the site, we will ask selected teachers to respond to paper copies of our alternative designs. Besides asking how well each design appeals to the teachers, we will ask them what they would expect to see if they clicked on the links we plan to include. Through these and similar questions, we will learn how to increase the site's appeal and usability. After we've built the site, we will ask users to test it online. We will send the user test results to you, along with a list of our proposed revisions for your approval.

Technical testing. We will test the site's performance when displayed on a variety of computers using a variety of web browsers.

Schedule

The following schedule is our plan to ensure timely and accurate completion of the project. Significant deadlines for you are in bold.

The schedule identifies key interim deadlines for the student team and the client.

The students use bold type to help Dr. Hargis spot the deadlines for steps the museum will perform.

Activity	Date of Completion
Conduct research	October 15
Deliver report	October 19
Receive your response	**October 21**
Create and test site map	October 21
Create and test screen designs	October 21
Submit alternatives	October 26
Receive your decisions	**October 28**
Build site	October 29–November 26
Show the site to you	November 27
Receive your comments	**November 29**
Revise site in light of your comments	November 30–December 5
Conduct user test online	December 3
Submit user test report	December 8
Receive your response	**December 10**
Test site on several platforms and browsers	December 9
Submit a report to you	December 11
Receive your response	**December 13**
Revise the site	December 13–16
Deliver the final site	December 17

FIGURE 21.1
(continued)

Qualifications ◄

We possess the skills needed to create an effective website for the museum. Sheila, a computer science major, has created several websites using Adobe Dreamweaver, a professional-level design program. Knowledge Tim gained in his marketing minor will help us design a website that will entice teachers to bring their classes to the museum. Using information management skills learned during an internship at the Houston Environmental Office, Vijah will organize our research efforts.

The students cite specific knowledge and skills they have that are required to create a website that will achieve Dr. Hargis' objectives.

Museum's Contributions ◄

To meet all of your deadlines, we will need your assistance. First, we will need you to provide program descriptions and materials, as well as photographs and other images for the site. To help us meet the deadlines above, we will need in three days the responses to the questions we ask you. We will also need to use your digital camera to take pictures needed by the site.

The students detail the things the client must contribute.

Conclusion ◄

We are eager to begin working with you on the Wright Museum's website. Please let us know as soon as convenient whether the arrangements we described in this letter are acceptable to you.

The students conclude with enthusiasm. By saying that they want to "work with" the client, they reinforce the point that they want to be partners with the client.

Sincerely,

Tim Banner
Tim Banner
T_Banner@Crandell.edu

Sheila Esterbook
Sheila Esterbook
E_Esterbook@Crandell.edu

Vijah Singe
Vijah Singe
V_Singe@Crandell.edu

Whenever you communicate with your client, whether at a prearranged time or because need has arisen, follow these practices.

COMMUNICATING WITH YOUR CLIENT

- **Be candid.** Don't hide or minimize problems. Clients need to know about difficulties so they can work with you to solve them and so that they can make contingency plans, if necessary.
- **Be specific.** The more precisely you describe the progress you've made, the information you need, the problems you've encountered, or the strategy you are recommending, the better you are preparing your client to help you complete the project successfully. Similarly, when submitting materials for review, let your client know the specific issues you'd like comments on— overall strategy? phrasing? color choice?—while also inviting comments on any topic.
- **Communicate constructively.** Even if you are discussing a problem caused by your client, remember that your goal is to prepare a communication that your client approves. You stand the best chance of achieving this goal if you always maintain a positive relationship.
- **Respect your client's time.** If you can do so without delaying your work, collect questions so you can ask many of them at once, or incorporate them in a regularly scheduled progress report. Find out when it would be most convenient for you to call. Ask whether your client prefers e-mail or paper mail.

Guideline 6 | Advocate and educate, but defer to your client

If you disagree with your client, explain why.

As you share your strategies, outlines, drafts, and other materials with your client, you may discover that you and your client disagree about some large or small aspect of the communication you are preparing. Your differences may concern content, writing style, page design—any aspect of the communication whatever.

If this happens, explain to your client the reader-centered reasons for the choices you think best. Using your knowledge of what makes a communication usable and persuasive, educate your client and advocate for your positions.

Remember that the client is always the client.

In the end, however, remember that the communication you are preparing is not your communication but the client's. As communication specialist Tony Marsico (1997) says, "The client may not always be right, but the client is always the client." The only exception is a situation where you believe it would be unethical to proceed. Even in this extreme case, however, you can't force the client to give your version of the communication to the target audience. Your option is to resign from the project.

When educating and advocating, take a strategic approach. Distinguish large matters—those that have the greatest impact on the communication's effectiveness—from small ones. Then focus first on the large issues, postponing discussion of small ones or letting them go entirely. Also, know when to quit. As long as your client is listening, help the client make good decisions. But once the client's decision has been made and you have no new perspectives to offer, stop. If you continue, the time you take from the client and the irritation you cause will only harm your relationship and diminish or completely eliminate your ability to persuade about future issues and to get future work with the client.

Guideline 7 | Hand off the project in a helpful way

When you have completed your deliverable, you are not necessarily finished with the project. In the workplace, almost all client projects are accompanied by a transmittal letter that reviews the assignment, describes important features of the deliverable, and closes in a cordial fashion. In some cases, you can best serve your client by writing a longer transmittal letter or preparing a separate document to provide additional information, advice, and assistance. For example, if you have created a website, you may be able to accompany the site with instructions for updating the content, recommendations for drawing people to the site, and information about sources for future assistance with it. You may also have generated ideas about ways the site could be expanded—ways you didn't pursue because the work required to do so would have exceeded the scope that you and your client originally defined for the project. Your client would also benefit from information about the software you used to create the site and the logic you used in creating it because this information could help others working on the site in the future. By carefully planning how you will hand off your project, you provide added assistance and gain added praise for your work.

Learn More

For advice about writing transmittal letters, see page 314.

© James Marshall/The Image Works

At the end of client projects, students often meet with their clients to summarize their results, present their findings, and answer questions.

CONCLUSION

When working for a client, you must be a manager as well as a communicator. This chapter's guidelines have focused on the most important strategies for managing both the project and your relations with your client: Reach a written agreement about the project before you begin, work carefully with your client throughout the process, and hand off your deliverable in a helpful way. Figure 21.2 provides a Writer's Guide for Managing Client and Service-Learning Projects.

FIGURE 21.2 Writer's Guide for Managing Client and Service-Learning Projects

 To download a copy of this Writer's Guide, go to Chapter 21 at www.cengage.com/english/anderson7e.

Writer's Guide
MANAGING CLIENT AND SERVICE-LEARNING PROJECTS

Determining What's Needed

1. Learn about your client's organization.
2. Learn the client's perception of the readers, communication objectives, and stakeholders.
3. Learn the client's deadline, requirements for the communication, and budget.
4. Learn what information and other resources the client will provide to you.
5. Develop your own assessment of the situation.

Creating a Project Plan

1. Determine the size, overall content, major features, and technical aspects of your deliverable.
2. Create a project schedule that includes project milestones, dates for meeting with or submitting material to your client, and dates for your client's responses.
3. Include your client's contributions.

Writing a Proposal

1. Be as specific as possible to protect your client and yourself.
2. Ask for a written agreement from your client.

Conducting the Project

1. Monitor your progress continuously by comparing it against your project schedule.
2. Talk with your client about changes to your original agreement.
3. Communicate with your client in ways that are candid, specific, and respectful of your client's time.
4. Advocate for your positions, but ultimately defer to your client when necessary.
5. Hand off your project with a transmittal letter and helpful information and advice.

USE WHAT YOU'VE LEARNED

For additional exercises, go to **www.cengage.com/english/anderson7e**. *Instructors:* The book's website includes suggestions for teaching the exercises.

EXERCISE YOUR EXPERTISE

1. If your instructor has assigned a client or service-learning project, list the specific questions you will need to ask your client in order to learn about the organization, the communication, and the communication's objectives. Report the results in a memo to your instructor.

2. Draft a project schedule for your project.

EXPLORE ONLINE

If your instructor has assigned a client or service-learning project, use the Internet to learn about the organization's mission and goals and its structure. Based on what you learn, speculate about the style and features you believe the client would want used in the communication you will prepare. Identify other information you believe will enable you to plan successful strategies for establishing a successful partnership with this organization. If you can't find information about this specific organization, study ones that are similar to it. Report the results in a memo to your instructor.

COLLABORATE WITH YOUR CLASSMATES

If you are part of a team that is working on a client or service-learning project, decide how you will assign responsibilities within your team. Consult Chapter 18's guidelines for creating communications with a team as well as this chapter's advice for establishing productive relations with clients. Report your results in a memo to your instructor.

APPLY YOUR ETHICS

As a service-learning project for their course, Celia and Stephen have been working together for the last six weeks to create a new website for a social services agency in their community. At the beginning of the project, they used a thorough proposal to develop mutually agreeable commitments with the director of the agency. They have worked very hard, and they have had good cooperation from the agency, which has devoted many hours to working with them. However, the project has turned out to be larger than they anticipated, so they have been unable to complete the project by the end of the term. Their instructor understands the circumstances and will allow the work they have completed to serve as their project so that they will be able to finish the course. However, the agency will not have the new website it had counted on launching. For each of the following circumstances, discuss the ethical obligations Celia and Stephen would have to complete the project during the upcoming vacation or during next semester, when they will both have a full load of courses.

- If they were unable to complete the project this term because they promised features they didn't know how to implement

- If Celia and Stephen's inability to complete the project resulted because the director thought they had promised to do a larger site than they thought they had agreed to provide

- If the delay were caused by the director's wanting very substantial revisions when Celia and Stephen submitted drafts of the site map and sample sections to the director for review

- If the delay were caused by the director's illness, which delayed the agency's response to drafts

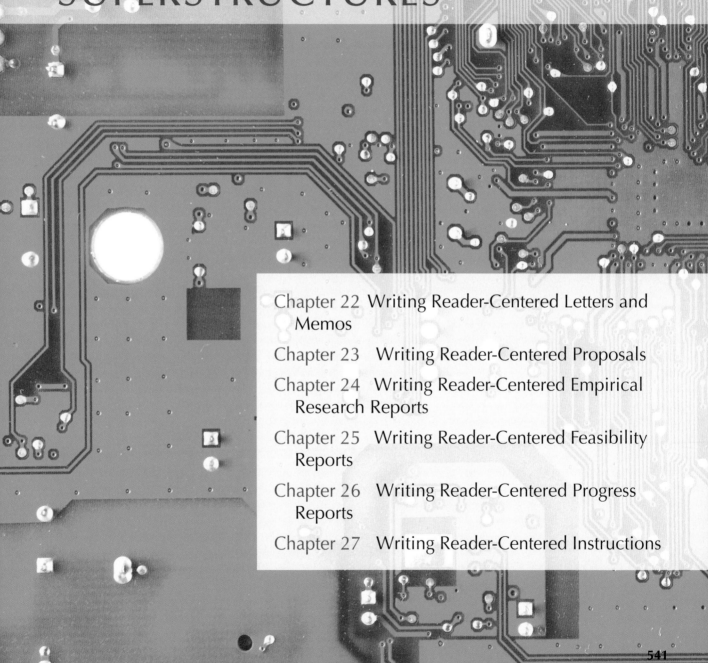

PART IX
SUPERSTRUCTURES

22 | Writing Reader-Centered Letters and Memos

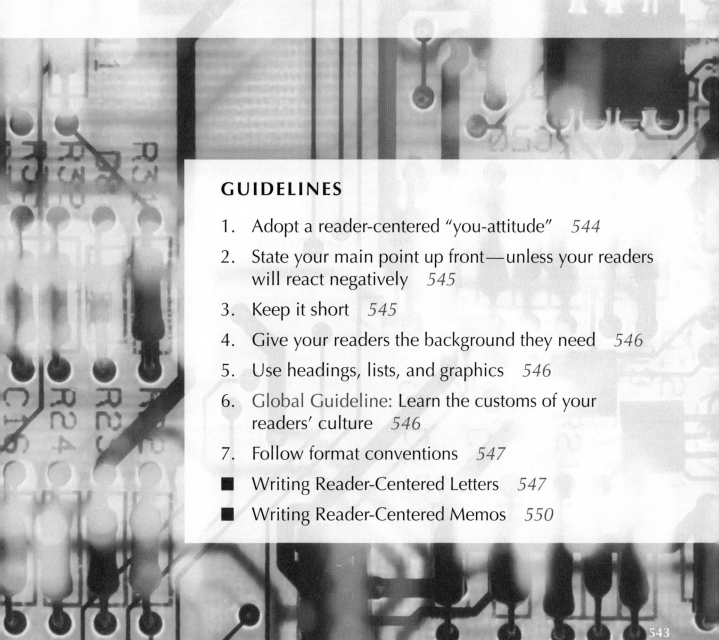

GUIDELINES

543

In the workplace, people transact much of their business through letters and memos. The two forms of communication share many features. Both are usually short, two pages or fewer, though they may be longer in certain circumstances. Both are used for a wide variety of purposes, such as making requests, providing information, giving instructions, offering suggestions, reporting data, and so on. And both are usually printed on paper, although it is becoming increasingly common to send them as e-mail attachments.

To help you learn how to write workplace letters and memos, this chapter begins with guidelines that apply to both, followed by advice specific to each form. All of this chapter's suggestions reinforce, adapt, or elaborate reader-centered guidelines you've read elsewhere in this book.

Letters and memos are usually written from one individual or small group to another. Usually they are two pages or less, though they may be longer in certain circumstances. Sometimes they are only a few sentences long. Both are used for a wide variety of purposes, such as making requests, providing information, offering suggestions, reporting data, and so on.

Letters and memos are most likely to succeed in achieving their objectives if the writer applies the reader-centered approach explored throughout this book. The following guidelines repeat, adapt, and elaborate selected guidelines found elsewhere in this book in order to apply them to the special features of these printed types of correspondence.

Guideline 1 | Adopt a reader-centered "you-attitude"

To your readers, the letters and memos you write at work will feel very much like direct, personal statements from you. You can significantly increase their positive emotional response to your messages by adopting what specialists in business correspondence call the "you attitude." This strategy involves focusing your sentences on your readers' needs, desires, and situation rather than on you or your subject matter. By crafting sentences in this reader-centered way, you reinforce all of the other reader-centered work you perform when planning and drafting your letter or memo. The following examples show how to rephrase sentences to convey the "you attitude," not the self-centered "me attitude."

- Focus sentences on the reader, not the writer. Sentences that use I as the grammatical subject appear to show that the writer is more concerned with himself or herself than with the reader.

Lacks you-attitude

I have arranged to have equipment needed for your fieldwork delivered to Boise office.

Has you-attitude

You can pick up the equipment needed for your fieldwork at the Boise office.

- Phrase sentences so they create positive feelings in the reader. For instance, where the message is positive, use you.

| We will reimburse you fully for the expenses resulting from this error. | Lacks you-attitude

| You will receive full reimbursement for your expenses resulting from this error. | Has you-attitude

- Phrase sentences so they avoid creating negative feelings in the reader. For instance, avoid you when conveying criticism and other negative messages.

| Your conclusions are inaccurate because you failed to account for all variables. | Lacks you-attitude

| Because all variables were not accounted for, the conclusions are inaccurate. | Has you-attitude

Guideline 2 | State your main point up front—unless your readers will react negatively

Readers usually read most efficiently when the writer announces his or her main point at the beginning of a communication. With this advance knowledge, readers can see more easily how the other parts of the communication fit together to form a coherent message.

When they lead off with their main point, writers are using the direct pattern of organization. Knowing how much this pattern contributes to usability and persuasiveness, some writers even include the heart of their message in the subject line of their memos: "Recommendation for Replacing the Afterburner," "Phoenix Site Is the Best Location for Building Our New Factory."

At times, however, you increase your chances of writing successfully by employing a different pattern of organization, for example, when you expect the reader to have an immediate negative response to your main point, as might happen when you are refusing a request or communicating some other decision or action that will not please the reader. In the vast majority of working situations, you can serve your readers best by stating your main point at the beginning of your letters and memos.

Learn More

For more on the way that the direct pattern enhances usability and persuasiveness, go to pages 103–105 and 130–133.

Guideline 3 | Keep it short

Few things help readers of correspondence as much as brevity. Without cutting substance, shorten your message in the following ways.

STRATEGIES FOR KEEPING CORRESPONDENCE SHORT

- **Delete unnecessary words.** You will find several strategies for finding and deleting unnecessary words in Chapter 9's first guideline on constructing sentences (page 270).
- **Stick to the point.** No matter how interesting a fact is to you or how proud you are of some accomplishment, don't mention it unless your readers will find the information useful or persuasive.
- **Provide only as much detail as your readers need.** Offer to give them more information if they want it, but in your correspondence include only what readers really need.

Guideline 4 | Give your readers the background they need

While it may be obvious to you, your readers may not always know how your message relates to them, their goals, and their areas of responsibility. If you think they may need an explanation, tell them. Also, readers who don't work closely with you may not fully understand the terms and concepts at the heart of your message.

Learn More

To learn how to identify background information your readers may need, see Guideline 5 in Chapter 10, page 295.

Guideline 5 | Use headings, lists, and graphics

Headings and bulleted lists, which enable readers to quickly grasp the organization and main points of long documents, are equally helpful in correspondence. Similarly, tables, graphs, diagrams, and other graphics are just as effective at presenting and interpreting data in letters, memos, and e-mail attachments as they are in lengthy reports and proposals.

Learn More

For advice on using headings and bullets, see Guideline 6 in Chapter 8, pages 216–225. For graphics, see Chapter 13, page 329.

Guideline 6 | Global Guideline: Learn the customs of your readers' culture

Customs concerning correspondence vary from culture to culture—including not only regional and ethnic cultures but also organizational cultures. With respect to organizations, the differences can be relatively minor, though you should still learn and follow your employer's conventions. For example, in most companies, employees initial memos by their names in the "From" line, while in others they sign at the bottom of the memo.

Some regional and ethnic differences are much more substantial. In Japan, letters customarily begin with a reference to the season. In some Spanish-speaking cultures, letters begin with statements that seem florid by the more straightforward conventions—for instance, in the United States.

Even paper sizes for letters differ from region to region. Word processing programs let you choose between preparing documents for printing on $8\frac{1}{2}'' \times 11''$ paper or the longer and narrower A4 paper. With the exceptions of the ted States and Canada, A4 is the standard for letters around the world.

Some customs, though known, are better avoided because they hinder good communication. For example, at some workplaces writers are in the habit of peppering their correspondence with a variety of wordy and inelegant expressions that they do not use in conversation or other kinds of writing. Here are some examples of these correspondence clichés.

WWW

For resources you can use to learn about correspondence customs in other cultures, see page 79 and also go to Chapter 22 at **www.cengage.com/english/anderson7e.**

Replacing correspondence clichés with plain language

CORRESPONDENCE CLICHÉ	PLAIN LANGUAGE
As per your request	As you requested
Prior to completion	Before completing
Contingent on	Depending on
Pursuant to	According to
In lieu of	Instead of

It has come to our attention	We learned
Enclosed please find	We've enclosed; Here is
Please be aware that	Please note that

Guideline 7 | **Follow format conventions**

One element of reader-centered writing is that it conforms to the conventions with which your readers are familiar. You gain credibility by doing so because you demonstrate to your readers that you know "how things are done here," that you belong. By following the conventions for letters and memos you also benefit from the many writers and readers in the past who have found that these conventions promote effective communication from the perspective of both groups.

Here, for instance, are three conventions that letters and memos share. All three also promote easy reading.

- Use single spacing.
- Separate paragraphs by a blank line.
- Use short paragraphs.

There are, of course, additional conventions specific to each of these forms of correspondence. Some concern format and some concern conventional ways to present your message within the formats. The rest of this chapter describes these conventions and highlights special considerations for each type of correspondence.

WRITING READER-CENTERED LETTERS

At work, you will usually use letters when writing relatively short messages to readers outside of your employer's organization. These readers may include customers, clients, suppliers, and government agencies, among others.

Conventional Format for Letters

Conventional formats for letters are described on pages 548–549.

Special Considerations for Writing Letters

The following considerations supplement the general guidelines for writing reader-centered correspondence given earlier in this chapter.

- Follow the conventions for opening and closing a letter. As noted on page 300, conventions about the opening of letters vary considerably among cultures. In the United States and Canada, letters are generally more formal than memos and

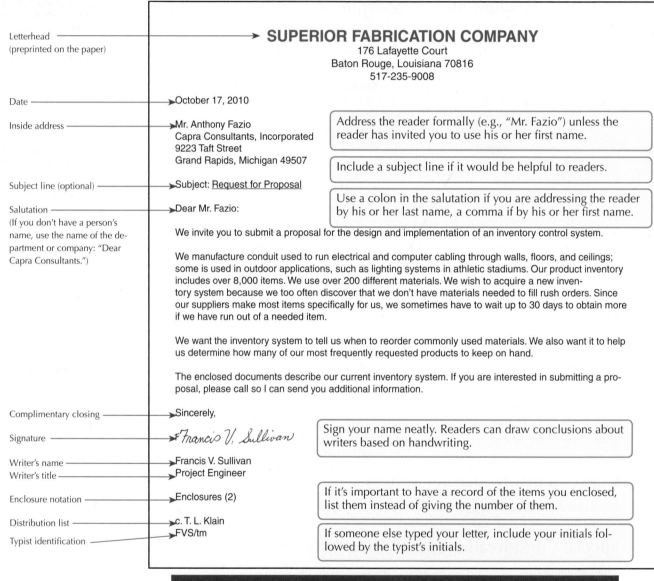

Letterhead (preprinted on the paper)

SUPERIOR FABRICATION COMPANY
176 Lafayette Court
Baton Rouge, Louisiana 70816
517-235-9008

Date

October 17, 2010

Inside address

Mr. Anthony Fazio
Capra Consultants, Incorporated
9223 Taft Street
Grand Rapids, Michigan 49507

Address the reader formally (e.g., "Mr. Fazio") unless the reader has invited you to use his or her first name.

Subject line (optional)

Subject: Request for Proposal

Include a subject line if it would be helpful to readers.

Salutation
(If you don't have a person's name, use the name of the department or company: "Dear Capra Consultants.")

Dear Mr. Fazio:

Use a colon in the salutation if you are addressing the reader by his or her last name, a comma if by his or her first name.

We invite you to submit a proposal for the design and implementation of an inventory control system.

We manufacture conduit used to run electrical and computer cabling through walls, floors, and ceilings; some is used in outdoor applications, such as lighting systems in athletic stadiums. Our product inventory includes over 8,000 items. We use over 200 different materials. We wish to acquire a new inventory system because we too often discover that we don't have materials needed to fill rush orders. Since our suppliers make most items specifically for us, we sometimes have to wait up to 30 days to obtain more if we have run out of a needed item.

We want the inventory system to tell us when to reorder commonly used materials. We also want it to help us determine how many of our most frequently requested products to keep on hand.

The enclosed documents describe our current inventory system. If you are interested in submitting a proposal, please call so I can send you additional information.

Complimentary closing

Sincerely,

Signature

Francis V. Sullivan

Sign your name neatly. Readers can draw conclusions about writers based on handwriting.

Writer's name
Writer's title

Francis V. Sullivan
Project Engineer

Enclosure notation

Enclosures (2)

If it's important to have a record of the items you enclosed, list them instead of giving the number of them.

Distribution list
Typist identification

c. T. L. Klain
FVS/tm

If someone else typed your letter, include your initials followed by the typist's initials.

TIPS

- For letters that don't fill the page, place the content so its middle is slightly above the middle of the page.
- On letterhead paper, type your message with the same margins that the letterhead uses. Keep the same margins on subsequent pages.

LETTER PRINTED ON PLAIN PAPER (NOT LETTERHEAD)

Place the return address 1 inch below the top on either the right or left of the page.

Type the date one return below the inside address.

Start the inside address at least two returns below the date.

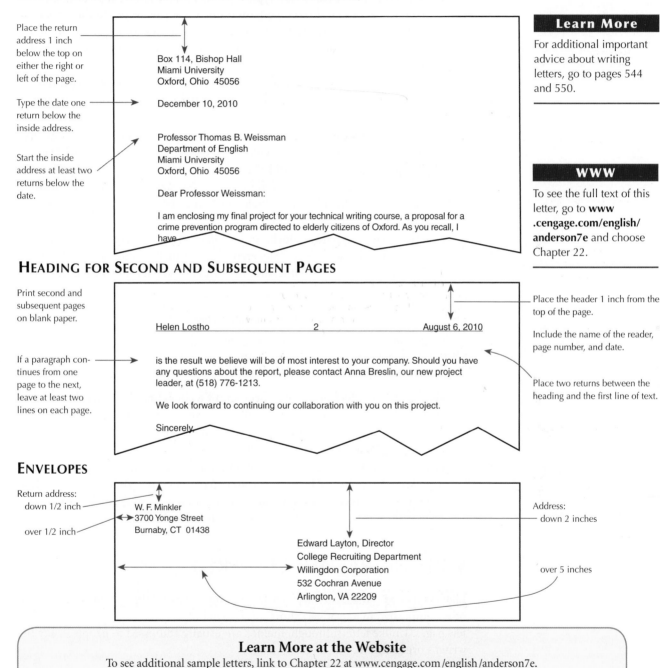

Box 114, Bishop Hall
Miami University
Oxford, Ohio 45056

December 10, 2010

Professor Thomas B. Weissman
Department of English
Miami University
Oxford, Ohio 45056

Dear Professor Weissman:

I am enclosing my final project for your technical writing course, a proposal for a crime prevention program directed to elderly citizens of Oxford. As you recall, I have

Learn More

For additional important advice about writing letters, go to pages 544 and 550.

WWW

To see the full text of this letter, go to **www .cengage.com/english/ anderson7e** and choose Chapter 22.

HEADING FOR SECOND AND SUBSEQUENT PAGES

Print second and subsequent pages on blank paper.

If a paragraph continues from one page to the next, leave at least two lines on each page.

Helen Lostho   2   August 6, 2010

is the result we believe will be of most interest to your company. Should you have any questions about the report, please contact Anna Breslin, our new project leader, at (518) 776-1213.

We look forward to continuing our collaboration with you on this project.

Sincerely,

Place the header 1 inch from the top of the page.

Include the name of the reader, page number, and date.

Place two returns between the heading and the first line of text.

ENVELOPES

Return address:
down 1/2 inch
over 1/2 inch

W. F. Minkler
3700 Yonge Street
Burnaby, CT 01438

Edward Layton, Director
College Recruiting Department
Willingdon Corporation
532 Cochran Avenue
Arlington, VA 22209

Address:
down 2 inches

over 5 inches

Learn More at the Website

To see additional sample letters, link to Chapter 22 at www.cengage.com /english /anderson7e.
Also, go to pages 53, 88, 126, and 135.

e-mail. Avoid slang and jargon. Choose a formal vocabulary—but avoid the correspondence clichés and bureaucratese described on page 265.

- The *introductory* paragraph typically indicates the letter's topic, explains its purpose, and indicates, perhaps implicitly, its relevance to the reader. Often an introductory paragraph expresses gratitude or appreciation of the reader—for instance by thanking the reader for a letter to which the writer is responding.
- *Closing paragraphs* often make a social gesture—for example by expressing pleasure in working with the reader, offering to assist the reader, and the like.

- **Use a formal greeting in the salutation.** Consistent with the relative formality of letters, it is customary to use the reader's last name in the salutation: "Dear Mr. Takjeki:" "Dear Dr. Reynolds:" However, if you call the person by his or her first name in conversation or if the reader has signed a letter to you with only a first name, then it is acceptable (and expected) that you will use the person's first name in the salutation.

- **Create a formal signature block.** At the end of the letter, writers usually end with a formal phrase, such as "Sincerely" or "Cordially," and identify themselves by typing their full names and their title. Between these formalities, they sign their name in ink.

Sincerely,	Complimentary close
Ibrahim Ettouney	Writer's signature
Ibrahim Ettouney	Writer's name
Process Engineer	Writer's title

- **Sign neatly. Readers draw inferences about writers based on their handwriting.** Usually writers sign their full name when first addressing a reader. Afterwards, they may sign only their first name to signal that they would be comfortable being addressed by their name in the future.

- **List the people to whom you are sending copies.** As a courtesy to your readers, let them know who else will be reading your letter. Sometimes the list assures readers that the writer has conveyed the letter's information to all who should have it. It also lets the reader know who might ask them about the letter's contents.

WRITING READER-CENTERED MEMOS

Like letters and e-mail, memos are used for almost any workplace purpose that can be accomplished in a relatively short space, generally between a few sentences and a few pages. Unlike letters, though, memos are usually addressed to people inside the writer's organization.

Memo Format

The conventional format for memos is described on page 552. The memo's distinguishing feature is the heading, which looks like a form with slots for the writer's name, reader's name, date, and (perhaps) the memo's subject. Memos don't include return addresses because none is needed for correspondence sent within the same company or agency.

Special Considerations for Writing Memos

The following considerations supplement the general guidelines for writing reader-centered correspondence given earlier in this chapter.

- **Provide a specific, informative subject line.** A heading that says "Results" doesn't tell the reader what the results are from. One that says "Gallagher Project" doesn't communicate whether the memo asks questions, provides answers, makes a proposal, or reports progress. Much more informative for current readers and future readers looking for information about the completed project is "Results of Turbulence Tests for Gallagher Project."

- **State the purpose and main point in the opening sentences.** Memos usually get off to a faster start than letters. Use the direct pattern of organization, as described in this chapter's second guideline on page 545—unless you have a very strong reason for using the indirect pattern.

- **Initial the memo.** In most organizations a writer's initials substitute for a full signature. Put your initials by your name in the "From" line.

- **List the people to whom you are sending copies.** As in a letter, you are treating your readers courteously when you let them know who else will be reading your letter. You assure your readers that you have given the letter's information to all who should have it. You have also alerted them to possibility of receiving questions about your memo from the individuals listed.

Learn More

For more on the direct and indirect patterns of organization, see pages 104–106 and 131–133.

The heading and these words are printed on the company's memo paper.

Rentscheller Company
Interoffice Memorandum

If your employer's memo paper doesn't include a space for a subject line, help your readers by typing in the subject yourself.

For additional important advice about writing memos, go to pages 550–551.

To   Dwight Levy
From   Natalie Sebastian
Date   June 16, 2010

Natalie provides a very specific subject line.

Subject   Results of Radon Testing at Worthington Warehouse

Writers initial memos in most organizations.

Her opening states the major finding of the tests.

I've completed the radon testing in our Worthington warehouse. Results show slightly elevated but not dangerous radon concentrations.

Test Results

Natalie provides background information she knows her reader will desire.

Radon is a naturally occurring radioactive gas that is present through-out the environment. It can accumulate, sometimes to dangerous concentrations, in enclosed places, such as the basements of houses and other underground structures. The normal, or "background," level of radon in our region is 0.40 ± 0.13 pCi/L (picocuries per liter of air). On all four above-ground floors of the warehouse, radon concentrations were within this range. However, our tests showed somewhat elevated concentrations on the two underground floors. On Floor A, the readings were 0.63 pCi/L, and on Floor B, they were 0.70 pCi/L.

Natalie provides specific data from the tests, together with other information needed to interpret the results.

Memos always begin one or two returns below the subject line, regardless of how much blank space is left at the bottom of the page.

Recommendation

Headings help readers even in a short memo.

Natalie answers the major question she anticipates her reader will have after reading the test results: "What should we do?"

Although the levels on Floors A and B are well below the levels considered dangerous, it does seem advisable to reduce them, especially since radon concentrations can vary and may be higher at other times of the year. The easiest way to reduce radon concentrations on these floors is to increase the volume of outside air that circulates through them. We can detach the ventilation ducts for these two floors from the ventilation system used for the rest of the building. We can then re-route the ducts to two new, high-capacity blowers.

I can prepare cost estimates, if you wish.

Natalie's closing suggests a next step but leaves the decision to her reader.

Second and Subsequent Pages

Place the header 1 inch from the top of the page

Hasim K. Lederer   2   November 16, 2010

Include the name of the reader, page number, and date.

with many benefits for our company and our customers. These improvements can be made with minimal disruption to our employees, suppliers, and technical sales force if we schedule the transition carefully. The most important factors will be notifying our

Place two returns between the heading and the first line of text.

Print second and subsequent pages on blank paper.

If a paragraph continues from one page to the next, leave at least two lines on each page.

Learn More at the Website

To see additional sample memos, link to Chapter 22 at www.cengage.com /english /anderson7e.
Also, go to pages 14, 138, 219, 316, 571 and 621.

23 | Writing Reader-Centered Proposals

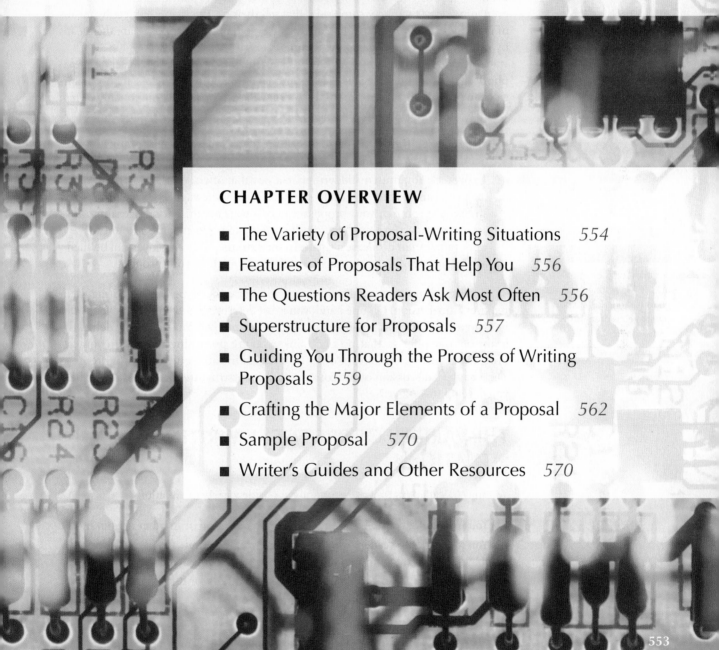

CHAPTER OVERVIEW

553

In a proposal, you make an offer and try to persuade your readers to accept it. You say that, in exchange for money or time or some other consideration, you will give your readers something they want, create something they desire, or do something they wish to have done.

Throughout your career, you will have many occasions to make such offers. You may think up a new product you could develop—if your employer will give you the time and funds to do so. Or you may devise a plan for increasing your employer's profits—if your employer will authorize you to put the plan into effect. You may even join one of the many companies that sell their products and services by means of proposals (rather than through advertising) so that *your* proposals are the means by which your employer generates income.

When one of your proposals has been accepted by your readers, it serves metaphorically—and, often, legally—as a contract. You and your readers are both obligated to provide what your proposal says you will each provide. Consequently, when writing a proposal, you have two goals:

- **To persuade.** If you don't persuade your readers, you won't get the money, time, or other things you want.
- **To protect.** You must avoid writing in a way that offers, or seems to offer, more than you can provide or wish to provide. If you write vaguely or make overly ambitious promises, your proposal can cost your organization huge sums of money because your employer may be willing to perform large amounts of work rather than displease a client. In the extreme, your employer can be sued for failing to fulfill the contractual obligations made in your proposal.

Because proposals must both protect and persuade, expertise in writing them requires a dual vision. On one hand, you must determine how to present your offer in as appealing a way as possible. On the other hand, you must carefully define the limits of your offer so that no one thinks you are promising more than you can provide for the money, time, or other compensation you request from your readers. This chapter's discussion of the conventional superstructure for proposals will help you develop this expertise.

WWW

For additional examples and advice for writing propvosals, go to Chapter 23 at **www.cengage.com/english/anderson7e**.

THE VARIETY OF PROPOSAL-WRITING SITUATIONS

You may be called on to write proposals in a wide variety of situations:

- You may write to employees in your own organization or in other organizations.
- You may write to readers who have invited you to submit a proposal, or you may submit it on your own initiative.
- Your proposal may be in competition against others, or it may stand or fall on its own merits.

- You may have to obtain approval from people in your organization before you submit it to your readers, or you may be authorized to submit it directly yourself.
- You may have to follow regulations governing the content, structure, and format of your proposal, or you may be free to write your proposal as you think best.
- After you have delivered your proposal, your readers may follow any of a wide variety of methods for evaluating it.

To illustrate these variables in actual proposal-writing situations, the following paragraphs describe the circumstances in which two successful proposals were written. The information provided here will be useful to you later in this chapter, where several pieces of advice are explained through the examples of these proposals.

Example Situation 1

Helen wanted permission to develop a customized computer program that employees at her company could use to reserve conference rooms. On several occasions, she had arrived at a conference room she had reserved only to find that someone else had reserved it also. Because she is employed to write computer programs, she is well qualified to write this one. However, she cannot work on this special project without permission from her manager and her manager's boss. She wrote her proposal to gain that permission.

Helen's proposal-writing situation was uncomplicated. She had only one reader. Because her employer had no guidelines for the way internal proposals should be written, she could use the content, structure, and format she thought most effective. She did not need anyone's approval to submit her proposal to her managers, and she did not have to worry about competition from other proposals. Hers would be evaluated strictly on its own merits.

Example Situation 2

The second proposal was written under very different circumstances. To begin with, it was written by three people, not just one. The writers were a producer, a scriptwriter, and a business manager at a public television station that seeks funds from nonprofit corporations and from the federal government to produce television programs. The writers had learned that the U.S. Department of Education was interested in sponsoring programs about environmental concerns. To learn more about what the department wanted, the writers obtained copies of its "request for proposals" (RFP). After studying the RFP, they decided to propose to develop a program that high school teachers and community leaders could use to teach about hazardous wastes.

In their proposal, the writers addressed an audience very different from Helen's. The Department of Education receives about four proposals for every one it can

Proposals for many kinds of projects are evaluated by experts in the area of research or other activity for which funds are available. Sometimes the reviewers from various parts of a country are flown to a central location to read and discuss the proposals. They determine which proposals will be funded.

fund. To evaluate these proposals, the department follows a procedure widely used by government agencies and other organizations. It sends the proposals to experts around the country. These experts, called *reviewers,* rate and comment on each proposal. Every proposal is seen by several reviewers. Then the reviews for each proposal are gathered and interpreted by staff members at the department. The proposals that have received the best reviews are funded. To secure funding, the writers of the environmental proposal needed to persuade their reviewers that their proposed project came closer to meeting the objectives of the Department of Education than did at least three-quarters of the other proposed projects.

Before the writers could even mail their proposal to the department, they had to obtain approval for it from several administrators at the television station. That's because the proposal, if accepted, would become a contract between the station and the Department of Education.

FEATURES OF PROPOSALS THAT HELP YOU

Two features of proposals can be very helpful to you. First, when decision makers read proposals on almost any topic, they tend to ask the same set of questions. Second, proposals are prepared so often in the workplace that a set of conventions has developed concerning their content and organization. These conventions have proven successful in helping writers provide the information their readers want in a structure the readers find easy to use. The conventions constitute the *superstructure* or *genre* of the proposal. As you research, plan, and write proposals in college and during your career, you will be helped immensely by your knowledge of the readers' typical questions and the superstructure for answering them. The next two sections describe the questions and superstructure.

Learn More

For a detailed explanation of superstructures, see Guideline 3 in Chapter 4, page 105.

THE QUESTIONS READERS ASK MOST OFTEN

Proposal readers ask the same set of questions because they all approach proposals the same way: as investors. In a proposal, you ask your readers to invest some resource, such as time or money, so you can work on the project you're proposing. No matter how much they like the project, your readers have limited resources, which they must invest carefully. Supporting any project leaves them with fewer resources for other activities and initiatives. To grant Helen two weeks to create a new scheduling

Try This

Proposal readers are investors.

program, her readers had to agree that she would not spend that time on other work. To support the television team's environmental project, the Department of Education had to turn down other proposals seeking the same dollars.

When deciding whether to invest resources in a project, decision makers ask questions on the following topics:

- **Problem.** Your readers will want to know why you are making your proposal and why they should be interested in it. What problem, need, or goal does your proposal address—and why is that problem, need, or goal important to them?
- **Solution.** Your readers will want to know exactly what you propose to make or do and how it relates to the problem you describe. They will ask, "What kinds of things will a successful solution to this problem have to do?" and "How do you propose to do those things?" They will examine your responses carefully, trying to determine whether your overall strategy and your specific plans are likely to work.
- **Costs.** Your readers will want to know what it will cost to implement your proposal and whether the cost will be worth it to them.

In addition, if you are proposing to perform some work (rather than supplying a product), your readers will want an answer to this question:

- **Capability.** If your readers pay or authorize you to perform the work, how will they know whether they can depend on you to deliver what you promise?

SUPERSTRUCTURE FOR PROPOSALS

The conventional superstructure for proposals provides a framework for answering those questions and includes ten topics. In some proposals you may need to include information on all ten, but in others you will need to cover only some of them. Even in the briefest proposals, however, you will probably need to treat four: introduction, problem, solution, and costs.

SUPERSTRUCTURE FOR PROPOSALS		
TOPICS	**READERS' QUESTIONS**	**YOUR PERSUASIVE POINTS**
*Introduction	What is this communication about?	Briefly, I propose to do the following.
*Problem	Why is the proposed project needed? What will we gain that is important to us if we invest in it?	The proposed project addresses a problem, need, or goal that is important to you.

Objectives	What features will a solution to the problem need in order to be successful?	A successful solution can be achieved if it has these features.
*Solution	What will your proposed solution look like?	Here's what I plan to produce, and it has the features necessary for success.
Method Resources Schedule Qualifications Management	Are you going to be able to deliver what you describe here?	Yes, because I have a good plan of action (method); the necessary facilities, equipment, and other resources; a workable schedule; appropriate qualifications; and a sound management plan.
*Costs	What will it cost?	The cost is reasonable.

*Topics marked with an asterisk are important in almost every proposal, whereas the others are needed only in certain ones.

Whichever question you are answering, remember the twin qualities of all workplace writing: usability and persuasiveness. To be usable, your proposal must enable your readers to find quickly and understand readily the answer to each of their questions. To be persuasive, your proposal must present each answer in a way that leads your readers to conclude that you will produce an outstanding result if your proposal is accepted. Consequently, to write a successful proposal, you must not only employ the proposal superstructure when determining your proposal's overall content and organization but also follow this book's other reader-centered strategies when selecting, organizing, and drafting each section's prose and graphics. Your goal is to lead your readers through the following sequence of thought:

Ideal sequence of readers' thoughts while reading a proposal

1. The readers learn generally what you want to do. (Introduction)
2. The readers are persuaded that there is a problem, need, or goal that is important to them. (Problem)
3. The readers are persuaded that the proposed action will be effective in solving the problem, meeting the need, or achieving the goal that they now agree is important. (Objectives, Solution)
4. The readers are persuaded that you are capable of planning and managing the proposed solution. (Method, Resources, Schedule, Qualifications, Management)
5. The readers are persuaded that the cost of the proposed action is reasonable in light of the benefits the action will bring. (Costs)

There is no guarantee, of course, that your readers will read your proposal from front to back or concentrate on every word. Long proposals usually include a summary or abstract at the beginning. Instead of reading the proposal straight through, many readers will read the summary, perhaps the first few pages of the body, and then skim through the other sections.

In fact, in some competitive situations, readers are prohibited from reading the entire proposal. For instance, companies competing for huge contracts to build satellites for the National Aeronautics and Space Administration (NASA) submit their proposals in three volumes: one explaining the problem and their proposed solution, one detailing their management plan, and one analyzing their costs. Each volume (sometimes thousands of pages long) is evaluated by a separate set of experts: technical experts for the first volume, management experts for the second, and budget experts for the third.

Even when readers do not read your proposal straight through—and even when your proposal is only a page or two long—the account just given of the relationships among the parts will help you write a tightly focused proposal in which all the parts support one another.

GUIDING YOU THROUGH THE PROCESS OF WRITING PROPOSALS

When you write a proposal, you engage in the same process as when preparing any other workplace communication. All the guidelines in Chapters 3 through 16 can help you as you define the proposal's objectives, conduct research, draft, and revise. Yet, proposal writing also involves special challenges of its own. The following suggestions will help you address these challenges efficiently and effectively.

Learn More

Remember that a superstructure is not an outline. For general advice about using superstructures, see page 106.

Defining Your Proposal's Objectives

People read proposals in order to make a decision: Should we support the proposed project or not? From the readers' desire to decide flow the objectives that all proposal share. The usability objective is to present all the information the reader believes to be relevant to the decision and to do so in a way that enable the reader to easily find, understand, and evaluate that information. The persuasive objective is to persuade readers that they should support the proposed project.

The Writer's Guide for Defining Your Communication's Objectives (page 70) will help you identify the additional information about your proposal's purpose, readers, and context that is needed to fully define its objectives.

Defining Your Proposed Project's Objectives

When preparing proposals, you have a second set of objectives to define, the objectives of the project you are proposing. From your perspective, the project's objective will be obvious: to achieve some goal that is important to you. As explained on page 556, however, proposal readers are investors. You cannot hope to win support for

For many proposals, a crucial objective is persuading readers that the project you are proposing addresses a problem, need, or goal that is important to them.

Learn More

To learn more about RFPs, go to Chapter 23 at **www .cengage.com/english/ anderson7e.**

In some situations, you will have to persuade your readers that your communication addresses a problem, need, or goal of theirs.

Try This

Think of an actual or fanciful project that would produce an outcome you would like. Whose support would you need in order to conduct that project? How would you argue that the project would address a goal, need, or desire of that person or group?

a project unless you persuade the readers that it addresses a problem, need, or goal that is important to them.

Sometimes readers will make it very easy for you to persuade them that your project addresses their needs and desires. They will tell you what they want and ask for a proposal from you. They may share this information informally, for instance in a meeting or e-mail. Or they may convey it in a Request for Proposal (RFP), such as the one to which the educational television group responded with its proposal.

In other situations, you will need to employ your creativity to persuade readers that your proposed project will help them achieve one of their goals or solve one of their problems. Helen wrote her proposal on her own, without receiving an invitation for it. She developed the idea of writing a program for scheduling conference rooms in response to the personal frustration she experienced repeatedly when arriving at rooms she had reserved, only to find that they had been double-booked. No matter how intense her frustration, her managers were not likely to approve her project simply to make her feel better. For her proposal to win their approval, Helen had to persuade them that the program would satisfy a need or goal of their own.

The educational television group also had to use creativity—plus research—to persuade readers that their project would meet the readers' goals. The Department of Education had issued an RFP, but it gave only a general sense of what the department wanted: "educational resources that could be used to teach about the relation of the natural and artificial environments to the total human environment." The team's proposal had to persuade the Department of Education's reviewers that its educational program was the kind of resource the department wanted.

Later in this chapter, you will find suggestions for ways to persuade readers that your projects will produce something they want (page 567). The key point here is that when you are describing your project's objectives, you must focus on what it has to offer them.

Conducting Research

It may seem that you can skip research when you are preparing a proposal. After all, you know what you want to propose. But be cautious. Knowing what you want to do does not necessarily mean that you are prepared to answer all of your readers' questions. Conduct any research needed to assure that you are prepared to write a proposal in which you do the following.

- **Demonstrate that you fully understand your readers' goals.** As individuals, proposal readers often have more goals and concerns than are evident at first. In addition, many proposals are reviewed by readers from a variety of backgrounds who will ask questions from a diversity of perspectives. Unless you are confident that you are fully prepared to satisfy all of these readers, talk with as many readers as possible.

- **Describe in detail the outcomes of your project.** The public television team had expertise in producing high-quality educational programs. They didn't need to conduct research related to that aspect of their proposed project. However, they did need to learn how to develop effective instructors' guides that would help teachers use the programs effectively in their classes. Your ability to describe in detail what you will give your readers and how you will produce these items will greatly increase your proposal's persuasiveness.

- **Explain persuasively how your project is superior to the alternatives.** In competitive proposal-writing situations, your readers will want to know why your project is better than those of your competitors. Conduct any research required so you can provide persuasive evidence on this point. Even when you are writing noncompetitive proposals, readers may still wonder why they can't reach their goal in some other, better way than you are proposing. For example, Helen realized her managers would want to know whether they could achieve the same goal by buying a commercially available scheduling program. To address this question in her proposal, Helen had to do research into the capabilities of commercial programs.

Planning

The conventions concerning the content and organization of proposals are strong. For most situations, you can organize your proposal effectively by picking the elements from the superstructure shown on page 562 that are relevant to your situation and presenting them in the order shown there. However, combining some element and reordering them will sometimes make your proposal more persuasive and easier for your readers to use. Take the reader-centered approach of adapting the superstructure to suit your readers and situation.

Sometimes the order of elements in the superstructure provides an effective organizational pattern.

Drafting and Revising

When you are drafting and revising a proposal, three pieces of advice from Chapter 5's discussion of persuasion deserve special attention.

- **Create a tight fit among the parts.** There's a tight fit between the proposal's problem and solution when the solution described addresses every aspect of the problem. Nothing is missed. Similarly, the solution and costs have a tight fit if there is a cost for every component of the solution—and no costs that aren't attached to the solution. Create such obvious, direct connections among all the elements you include from the superstructure. As described earlier in this chapter, being careful investors, proposal readers check for this kind of correspondence among parts to assure themselves that the proposed project is carefully planned.

- **Provide sufficient, reliable evidence to support all of your persuasive claims.** As discussed in Chapter 5's Guideline 4, the amount and kinds of information readers find to be sufficient and reliable vary from one reader to another and one situation to another. The insights you gather about your specific readers

Learn More

For more on creating a tight fit among the parts of a proposal see Guideline 5 in Chapter 5, page 130.

Learn More

For more on providing sufficient, reliable evidence, see Guideline 4 in Chapter 5, page 127.

while defining your proposal's objectives will help you determine what they will want. At the very least, give specific details rather than general statements. Chapter 9's discussion of using concrete, specific words will also help you persuade your readers that there is good reason to trust your proposal.

Learn More

For more on addressing counterarguments see Guideline 3 in Chapter 5, page 125.

- **Address likely counterarguments.** Being careful investors, proposal readers also look for counterarguments to the various claims you make in your proposal, such as the claim that your method will succeed in producing the solution you've described. Conduct research that will help you predict what these counterarguments might be. Even if you are proposing a project for which research won't help, consider your claims and assumptions with the same critical eye your readers will cast on them. Then address the counterarguments your readers are likely to raise.

CRAFTING THE MAJOR ELEMENTS OF A PROPOSAL

All of the detailed advice that Chapters 8 through 14 provide about drafting workplace communications applies to proposals. The following sections explain ways you can use some of this advice most productively when writing proposals. As you read, keep in mind that the conventional superstructure is only a general plan. You must use your imagination and creativity to adapt it to particular situations.

In addition, as you plan and write your proposal, remember that the ten topics identify kinds of information you need to provide, not necessarily the titles of the sections you include. In brief proposals, some parts may take only a sentence or a paragraph, or several sections may be grouped together. For instance, writers often combine their introduction, their discussion of the problem, and their explanation of their objectives under a single heading, which might be "Introduction," "Problem," or "Need."

Introduction

At the beginning of a proposal, you want to do the same thing that you do at the beginning of anything else you write on the job: Tell your readers what you are writing about. In a proposal, this means announcing what you are proposing.

How long and detailed should the introduction be? In proposals, introductions are almost always rather brief. By custom, writers postpone the full description of what they are proposing until later, after they have discussed the problem their proposal will help solve.

You may be able to introduce your proposal in a single sentence, as Helen did:

Helen's introduction | I request permission to spend two weeks writing, testing, and implementing a program for scheduling conference rooms in the plant.

When you are proposing something more complex, your introduction will probably be longer. You may want to provide background information to help your readers understand what you have in mind. Here, for example, is the introduction from the proposal written by the employees of the public television station to the Department of Education:

> Chemicals are used to protect, prolong, and enhance our lives in numerous ways. Recently, however, society has discovered that some chemicals also present serious hazards to human health and the environment. In the coming years, citizens will have to make many difficult decisions to solve the problems created by these hazardous substances and to prevent future problems from occurring.
>
> To provide citizens with the information and skills they will need to decide wisely, WPET Television proposes to develop an educational package entitled "Hazardous Substances: Handle with Care!" This package, designed for high school students, will include five 15-minute videotape programs and a thorough teacher's guide.

Introduction to the television writers' proposal

Learn More

For more ideas on writing introductions, see Chapter 10, "Beginning a Communication."

Problem

Once you've announced the project you are proposing, you must persuade your readers that it addresses some problem, need, or goal that is significant to them. As explained earlier in this chapter (page 556), you cannot hope to win approval for your project unless you persuade your readers they want the outcome your project will produce.

Often this requires considerable creativity and research. The following paragraphs offer advice that applies to each of three situations you are likely to encounter on the job: when your readers define the problem for you, when your readers provide you with a general statement of the problem, and when you must define the problem yourself.

When Your Readers Define the Problem for You

When your readers have told you in detail, perhaps in an RFP, what they want, your task is to persuade them that you understand their desire in the same way they do. In your problem or goal section, describe the problem in a way that corresponds with and builds on the readers' description but doesn't merely repeat it. You can also build confidence in your understanding of the problem by using, with precision, the technical or specialized terms the readers use.

When Your Readers Provide a General Statement of the Problem

At times, your readers may describe the problem only vaguely. This happened to the television team that submitted the environmental education proposal. In its RFP, the Department of Education provided only the general statement that it perceived a need for "educational resources that could be used to teach about the relation of the natural and artificial environments to the total human environment." Each group that wished to submit a proposal had to identify a more specific problem that the Department would find important.

Finding a problem that someone else thinks is important can be difficult. Often, it will require you to hunt for clues to your readers' values, attitudes, concerns, and opinions. You may have to interview people (including your target readers, where possible), visit the library, or search the web. The television team found its major clue in the following statement, which appeared elsewhere in the RFP: "Thus, environmental education should be multifaceted, multidisciplinary, and issue-oriented." This statement suggested to the writers that they should define the problem as the need experienced by some group of people to gain the multifaceted, multidisciplinary knowledge that would enable them to make good decisions about a practical environmental issue.

But who would that group be? And what is the environmental issue this group needs to address? First, the television writers selected an issue faced by many communities: how to handle and regulate hazardous substances that are produced or stored within their boundaries. This issue seemed ideal because citizens can make good decisions about hazardous substances only if they possess "multifaceted, multidisciplinary" knowledge from such fields as health, economics, and technology. Next, the writers also decided to target their project to high school students. These students seemed like a perfect group to address because they soon would be eligible to vote on questions involving hazardous substances in their communities. Finally, through research, the writers determined that, in fact, no such materials existed. They really were needed.

Thus, by working creatively from the Department of Education's vague statement, the television writers identified a specific problem that their proposed project would help to solve: the lack of educational materials that high school teachers can use to provide their students with the multifaceted, multidisciplinary knowledge the students need in order to make wise decisions, as voters, on the hazardous substance issues facing their communities.

WWW

For more information on RFPs, go to Chapter 23 at **www.cengage.com/ english/anderson7e.**

When You Must Define the Problem Yourself

In other situations, you may not have the aid of explicit statements from your readers to help you formulate the problem. This is most likely to happen when you are preparing a proposal on your own initiative, without being asked by someone else to submit it. Describing the problem in such situations can be particularly challenging because the arguments that will be persuasive to your readers might be entirely distinct from your own reasons for writing the proposal.

The reasons you offer your readers for supporting your proposal may be different from your reasons for writing it.

Try this approach: Learn about or review your readers' overall goals, looking for ones you can link to your project. What goals or responsibilities do your readers have that your proposal will help them with? What concerns do they typically express that your proposal could help address?

You can also gain ideas about ways to link your proposed project to your readers' needs and desires by speaking with the readers. This conversation can have two advantages. First, it will let you know whether your proposal has at least some chance of succeeding. If it doesn't, it's best to find out before you invest time writing it. Second, by talking with people who will be readers of your proposal, you can find out how they view the problem you are thinking of describing in your proposal.

When Helen spoke to her boss, she learned something she hadn't thought of before: Sometimes the rooms are used for meetings with customers. When customers see confusion arising over something as simple as meeting rooms, they may wonder whether the company has similar problems that would affect its products and services.

Objectives

When using the conventional superstructure for proposals, writers usually state the objectives of their proposed solution after describing the problem they are proposing.

Your statement of objectives plays a crucial role in the logical development of your proposal: It links your proposed action to the problem by telling how the action will solve the problem. To make that link tight, you must formulate each of your objectives so that it grows directly out of some aspect of the problem you describe. Here, for example, are three of the objectives that the writers devised for their proposed environmental education program. To help you see how these objectives grew out of the writers' statement of the problem, each objective is followed by the point from their problem statement that serves as the basis for it. (The writers did not include the bracketed sentences in the objectives section of their proposal.)

1. To teach high school students the definitions, facts, and concepts necessary to understand both the benefits and risks of society's heavy reliance upon hazardous substances. [This objective is based on the evidence presented in the problem section that high school students do not have that knowledge.]

2. To employ an interdisciplinary approach that will allow students to use the information, concepts, and skills from their science, economics, government, and other courses to understand the complex issues involved with our use of hazardous substances. [This objective is based on the writers' argument in the problem section that people need to understand the use of hazardous substances from many points of view to be able to make wise decisions about them.]

3. To teach a technique for identifying and weighing the risks and benefits of various uses of hazardous substances. [This objective is based on the writers' argument in the problem section that people must be able to weigh the risks and benefits of the use of hazardous substances if they are to make sound decisions about that use.]

Objectives of the environmental education proposal

The writers created similar lists of objectives for the other parts of their proposed project, each based on a specific point made in their discussion of the problem.

Like the writers at the television station, Helen built her objectives squarely on her description of the problem. She also took into account something she learned through her research: Even with a computerized system, managers wanted a single person in each department to coordinate room reservations. They didn't want each individual employee to be able to enter changes into the system.

Similarly, you will need to consider the concerns of your readers when framing the objectives for your proposals. Here are two of Helen's objectives:

Helen's objectives

1. To provide a single, up-to-date room reservation schedule that can be viewed by the entire company. [This objective corresponds to her argument that the problem arises partly because there is no convenient way for people to find out what reservations have been made.]
2. To allow a designated person in each department to add, change, or cancel reservations with a minimum of effort. [This objective relates to Helen's discovery that the managers want a single person to handle room reservations in each department.]

In proposals, writers usually describe the objectives of their proposed solution without describing the solution itself at all. The television writers, for example, wrote their objectives so that their readers could imagine achieving them through the creation of a textbook, an educational movie, a series of lectures, or a website—as well as through the creation of the videotape programs that the instructors proposed. Similarly, Helen described objectives that might be achieved by many kinds of computer programs. She withheld her ideas about the design of her program until the next section of her proposal.

The purpose of separating the objectives from the solution is not to keep readers in suspense. Rather, this separation enables readers to evaluate the aims of the project separately from the writers' particular strategies for achieving those aims.

Proposal writers usually present their objectives in a list or state them very briefly. For instance, Helen used only one paragraph to present her objectives. The members of the television group presented all their objectives in three pages of their 98-page proposal.

Solution

When you describe your solution, you describe your plan for achieving the objectives you have listed. For example, Helen described the various parts of the computer program she would write, explaining how they would be created and used. The television writers described each of the four components of their environmental educational package. Scientists seeking money for cancer research would describe the experiments they wish to conduct.

When describing your solution, you must persuade your readers of two things:

- **That your solution will successfully address each of the objectives.** For instance, to be sure that the description of their proposed educational program matched their objectives, the writers included detailed descriptions of the following: the definitions, facts, and concepts that their videotape program would teach (Objective 1); the strategy they would use to help students take an interdisciplinary approach to the question of hazardous waste (Objective 2); and the technique for identifying and weighing risks and benefits that they would help students learn (Objective 3).

- **That your solution offers a particularly desirable way of achieving the objectives.** For example, the television writers planned to use a case study approach in their materials, and in their proposal offered this explanation of the advantages of the case study approach they planned to use:

> Because case studies represent real-world problems and solutions, they are effective tools for illustrating the way that politics, economics, and social and environmental interests all play parts in hazardous substance problems. In addition, they can show the outcome of the ways that various problems have been solved—successfully or unsuccessfully—in the past. In this way, case studies provide students and communities with an opportunity to learn from past mistakes and successes.

Television writers' argument for the special desirability of their solution

Of course, you should include such statements only where they won't be perfectly obvious to your readers. In her proposal, Helen did not include any because she planned to use standard practices whose advantages would be perfectly evident to her readers, both of whom had several years' experience doing her kind of work.

The description of your proposed solution is one of the places where you must be very careful to protect yourself and your organization from seeming to promise more than you can deliver. The surest protection is to be very precise. For instance, Helen included specific details about the capabilities of her program. It would support the scheduling of six conference rooms (not more), it would provide reservation capabilities up to twelve months in advance (not longer), and it would operate on an NT server (not necessarily other servers). If you want to identify things you might be able to do, explicitly identify them as possibilities, time and opportunity permitting, not as promises.

Learn More

When describing your solution, you may find it helpful to use the strategies for describing an object or process (pages 246–251).

Method

Readers of proposals sometimes need to be assured that you can, in fact, produce the results that you promise. That happens especially in situations where you are proposing to do something that takes special expertise.

To assure themselves that you can deliver what you promise, your readers will look for information about several aspects of your project: your method or plan of action for producing the result; the facilities, equipment, and other resources you plan to use; your schedule; your qualifications; and your plan for managing the project. This section is about method; the other topics are discussed in the sections that follow.

To determine how to explain your proposed method, imagine that your readers have asked you, "How will you bring about the result you have described?"

In some cases, you will not need to answer that question. For example, Helen did not talk at all about the programming techniques she planned to use because her readers were already familiar with them. On the other hand, they did not know how she planned to train people to use her program. Therefore, in her proposal she explained her plans for training.

In contrast, the television writers needed to describe their method in detail in order to persuade their readers that they would produce effective educational materials. An important part of their method, for instance, was to use three review teams: a team of scientists to assess the accuracy of the materials they drafted, a team of filmmakers to advise about plans for making the videotapes, and a team of specialists in high school education to advise about the effectiveness of the script. In their proposal, they described these review teams in great detail, emphasizing the way each would enhance the effectiveness of the programs the writers were proposing to produce.

In addition, the writers described each phase of their project to show that they would conduct all phases in a way that would lead to success. These phases include research, scripting, review, revision, production, field testing, revision, final production, and distribution. As they described each step, they explained how their plans for conducting it would contribute to a successful outcome.

Resources

By describing the facilities, equipment, and other resources to be used for your proposed project, you assure your readers that you will use whatever special equipment is required to do the job properly. If part of your proposal is to request that equipment, tell your readers what you need to acquire and why.

If no special resources are needed, you do not need to include a section on resources. Helen did not include one. In contrast, the television writers needed many kinds of resources. In their proposal, they described the excellent library facilities that were available for their research. Similarly, to persuade their readers that they could produce high-quality programs, they described the videotaping facilities they would use.

Schedule

Proposal readers have several reasons for wanting to know what your schedule will be. First, they want to know when they can enjoy the final result. Second, they want to know how you will structure the work so they can be sure that the schedule is reasonable and sound. In addition, they may want to plan other work around the project: When will your project have to coordinate with others? When will it take people's attention from other work? When will other work be disrupted and for how long? Finally, proposal readers want a schedule so they can determine if the project is proceeding according to plan.

The most common way to provide a schedule is to use a schedule chart, which is sometimes accompanied by a prose explanation of its important points.

In many projects, your ability to complete your work on time depends on your getting timely responses and cooperation from other people, including the client or

Learn More

When describing your method, you may find it helpful to use the strategies for describing a process (pages 248–251).

Learn More

For information on creating a schedule chart, see page 377.

other person or organization who approved your proposal. When discussing your schedule, be sure to include the dates when you will need something from your client, such as approvals of your drafts and information you must have in order to proceed with your work.

Qualifications

When they are thinking about investing in a project, proposal readers want to be sure that the proposers have the experience and capabilities required to carry it out successfully. For that reason, a discussion of the qualifications of the personnel involved with a project is a standard part of most proposals. For example, the television writers discussed their qualifications in two places. First, in a section entitled "Qualifications," they presented the qualifications of each of the eight key people who would be working on the project. In addition, they included a detailed résumé for each participant in an appendix.

In other situations, much less information might be needed. For instance, Helen's qualifications as a programmer were known to her readers because they were employing her as one. If that experience alone were enough to persuade her readers that she could carry out the project successfully, Helen would not have needed to include any section on qualifications. However, her readers might have wondered whether she was qualified to undertake the particular program she proposed because different kinds of programs require specialized knowledge and skills. Therefore, Helen wrote the following:

> As you know, I am thoroughly familiar with the kind of database design that would be used in the reservation system. I recently completed an evening course in interactive media where I learned how to program dynamic web pages that can be used to request a database search and display the results. In addition, as an undergraduate I took a course in scheduling and transportation problems, which will help me here.

Helen's statement of her qualifications

In some situations, your readers will want to know not only the qualifications of the people who will work on the proposed project, but also the qualifications of the organization for which they work.

To protect yourself and your organization, guard against implying that you and those working with you have more qualifications than you really have. Time spent learning new programs, gaining new knowledge, or otherwise gaining background needed to complete a project adds to the time (and money) required for completion.

Management

When you propose a project that will involve more than about four people, you can make your proposal more persuasive by describing the management structure of your group. Proposal readers realize that on complex projects even the most highly qualified people can coordinate their work effectively only if a well-designed project management plan is in place. In projects with relatively few people, you can describe the management structure by first identifying the person or persons

Learn More

For information on creating an organizational chart, see page 376.

who will have management responsibilities and then telling what their duties will be. In larger projects, you might need to provide an organizational chart for the project and a detailed description of the management techniques and tools that will be used.

Because her project involved only one person, Helen did not establish or describe any special management structure. However, the television writers did. Because they had a complex project involving many parts and several workers, they set up a project planning and development committee to oversee the activities of the principal workers. In their proposal they described the makeup and functions of this committee, and in the section on qualifications they described the credentials of the committee members.

Costs

As emphasized throughout this chapter, when you propose something, you are asking your readers to invest resources—usually money and time. Naturally, then, you need to tell them how much your proposed project will cost.

Learn More

To create a budget, follow the guidelines for creating a table. See page 358.

One way to discuss costs is to include a budget statement. Sometimes, a budget statement needs to be accompanied by a prose explanation of any unusual expenses and the method used to calculate costs.

In proposals where dollars are not involved, information about the costs of required resources may be provided elsewhere. For instance, in her discussion of the schedule for her project, Helen explained the number of hours she would spend, the time that others would spend, and so on.

In some proposals, you may demonstrate that the costs are reasonable by also calculating the savings that will result from your project.

SAMPLE PROPOSAL

Figure 23.1 shows the proposal written by Helen. For additional examples, see the book's website at www.cengage.com /english /anderson7e.

WRITER'S GUIDES AND OTHER RESOURCES

WWW

To download Writer's Guides for planning and revising proposals, go to Chapter 23 at **www.cengage.com/ english/anderson7e.**

Figure 23.2 (pages 576–578) presents a Writer's Guide for Revising Proposals that you can use in your course and on the job. At this book's website, you can download a copy of this Writer's Guide as well as one for planning proposals. The website also includes other resources that will help you write effective proposals.

Note to the Instructor For a proposal-writing assignment, consider the "Formal Report or Proposal" project provided in Appendix B. Other projects and cases that involve report writing are included at the book's website.

FIGURE 23.1
Sample Proposal in the
Memo Format

PARKER MANUFACTURING COMPANY

Memorandum

TO: Floyd Mohr and Marcia Valdez
FROM: Helen Constantino
DATE: July 14, 2010
RE: Proposal to Write a Program for Scheduling Conference Rooms

I request permission to spend two weeks writing, testing, and implementing a program for scheduling the buliding's conference rooms in the plant. This program will eliminate several problems with conference room schedules that have become acute in the last three months.

◄ In the introduction, Helen tells what she is proposing and what it will cost.

Present System

At present, the chief means of coordinating room reservations is the monthly "Reservations Calendar" distributed by Peter Svenson of the Personnel Department. Throughout each month, Peter collects notes and phone messages from people who plan to use one of the conference rooms sometime in the next month. He stores these notes in a folder until the fourth week of the month, when he takes them out to create the next month's calendar. If he notices two meetings scheduled for the same room, he contacts the people who made the reservations so they can decide which of them will use one of the other seven conference rooms in the new and old buildings. He then e-mails the calendar to the heads of all seventeen departments. The department heads usually give the calendars to their administrative assistants.

◄ Because her introduction includes much background information, Helen makes a separate section for it; she labels the section "Present System" rather than "Background" to provide her readers with a more precise heading.

Someone who wants to schedule a meeting during the current month usually checks with the department's administrative assistant to see whether a particular room has been reserved on the monthly calendar. If not, the person asks the assistant to note his or her reservation on the department's copy of the calendar. The assistant is supposed to call the reservation in to Peter, who will see whether anyone else has called about using that room at that time.

(continued)

FIGURE 23.1
(continued)

For her problem section, Helen
also uses a more precise heading.

Problems with the Present System

The present system worked adequately until about four months ago, when two
important changes occurred:

- The new building was opened, bringing nine departments here from
  the old Knoll Boulevard plant.
- The Marketing Department began using a new sales strategy of bringing
  major customers here to the plant.

Helen pinpoints the problem.

These two changes have greatly complicated the work of scheduling rooms. With
the greatly increased use of the conference rooms, employees often schedule
more than one meeting for the same time in the same room. If one of the
meetings involves customers brought here by the Marketing Department, we end
up giving a bad impression of our ability to manage our business.

Helen describes the consequences
of the problem that are important to her readers.

Objectives

To solve our room scheduling problems, we need a reservation system that will
do the following:

- Provide a single, up-to-date room reservation schedule that can be
  viewed by the entire company, thereby ending the confusion caused
  by having a combination of departmental and central calendars
- Allow a designated person in each department to add, change, or cancel
  reservations with a minimum of effort
- Show the reservation priority of each meeting so that people scheduling
  meetings with higher priority (such as those with potential customers)
  will be able to see which scheduled meetings they can ask to move to
  free up a room

Helen ties the objectives of her
proposed project directly to the
points she raised when describing
the problem.

2

FIGURE 23.1
(continued)

Proposed Solution

I propose to solve our room-scheduling problems by creating a web-based system for reserving our six conference rooms up to six months in advance. A designated person in each department will be able to enter, alter, or cancel reservations quickly and easily through his or her web browser. Located on an NT server, the program will automatically revise the reservation list every time someone makes an addition or other change so people anywhere in the company can view a completely updated schedule at any time. When they enter a reservation, people will include information about the priority of their meeting, so that other individuals will be able to view this information if they are having trouble finding a place for their own meeting. The program will also be able to handle reservations for projection equipment, thereby ending problems in that area. This web-based program will have the added advantage of freeing Peter Svenson from the time-consuming task of manually maintaining the monthly reservations calendar.

After stating the general nature of her proposed project, Helen links the features of her product to the objectives she identified earlier.

Helen protects herself from misinterpretations of what she will produce by precisely describing the capabilities of her proposed program.

Method for Developing the Program

I propose to create and implement the program in three steps: writing it, testing it, and training people in its use.

Helen provides an overview of her method, then explains each step in a way that shows how each will contribute to a successful outcome.

Writing the Program

The program will have three features. The first will display the reservations that have been made. When users access the program, the system will prompt them to tell which day's schedule they want to see and whether they want to see the schedule organized by room or by the hour. The calendar will display a name for the meeting, the name of the person responsible for organizing it, and the projection equipment needed.

The second feature will handle entries and modifications to the schedule. When users call up this program, they will be asked for their company identification number. To prevent tampering with the calendar, only people whose identification number is on a list given to the computer will be able to proceed. To make, change, or cancel reservations, users will simply follow prompts given by the system.

3

(continued)

FIGURE 23.1
(continued)

Once a user completes his or her request, the system will instantly update the calendar that everyone can view. In this way, the calendar will always be absolutely up to date.

Third, when a higher priority meeting displaces one already scheduled, the program will automatically e-mail all persons listed as planning to attend the lower-priority meeting. This is among the features of my program that are not available on commercially available, free-standing scheduling programs.

Testing the Program

I will test the program by having the administrative assistants in four departments use it to create an imaginary schedule for one month. The assistants will be told to schedule more meetings than they usually do to be sure that conflicts arise. They will then be asked to reschedule some meetings and cancel some others.

Training

Training in the use of the program will involve preparing a brief user's guide and conducting training sessions. I will write the user's guide, and I will work with Joseph Raab in the Personnel Department to design and conduct the first training session. After that, he will conduct the remaining training sessions on his own.

Resources Needed

To write this program, I will need the cooperation of several people. I have already contacted four people to test the program, and Vicki Truman, head of the Personnel Department, has said that Joseph Raab can work on it because that department is so eager to see Peter relieved of the work he is now having to do under the current system. I will also need your authorization to use a total of about four hours of our server administrator's time.

4

Helen has already planned the testing.

Helen tells what resources she already has (people's time, in this case).

She protects herself by identifying the resources her readers must provide (again, people's time).

FIGURE 23.1
(continued)

Schedule

I can write, test, and train in eight 8-hour days, beginning August 15.

Task	Hours
Designing Program	12
Coding	24
Testing	8
Writing User's Manual	12
Training First Group of Users	8
Total	64

In her schedule, Helen also tells the cost (in hours of work). She protects herself by identifying all the costs.

The eight hours estimated for training include the time needed both to prepare the session and to conduct it one time.

Qualifications

As you know, I am thoroughly familiar with the kind of database design that would be used in the reservation system. I recently completed an evening course in interactive media where I learned how to program dynamic web pages that can be used to request a database search and display the results. In addition, as an undergraduate I took a course in scheduling and transportation problems, which will help me here.

Helen focuses on her qualifications that relate directly to the project she is proposing.

Conclusion

I am enthusiastic about the possibility of creating this much-needed program for scheduling conference rooms and hope that you are able to let me work on it.

5

FIGURE 23.2 **Writer's Guide for Revising Proposals**
To download a copy of this Writer's Guide as well as another guide for Planning Proposals, go to Chapter 23 at www.cengage.com/english/anderson7e.

Writer's Guide
REVISING PROPOSALS

Does your draft include each of the elements needed to create a proposal that your readers will find to be usable and persuasive? Remember that some elements of the superstructure may be unnecessary for your specific readers and purpose and that the elements may be organized in various ways.

Introduction

☐ Tells clearly what you propose to do

☐ Provides background information the readers will need or want

☐ Forecasts the rest of your proposal, if this would help your readers

Problem

☐ Explains the problem, need, or goal of your proposed action

☐ Persuades your readers that the problem, need, or goal is important to them

Objectives

☐ Relates your objectives directly to the problem, need, or goal you described

☐ Presents your objectives *without* naming your solution

Solution (often the longest section of a proposal)

☐ Describes your solution in a way that assures your readers can understand it

☐ Persuades that your solution will achieve each of the objectives you described

☐ Persuades that your solution offers an especially desirable way of achieving the objectives

☐ Protects you and your employer by clearly promising only what you and your employer want to deliver to your readers

Method

☐ Describes clearly the steps you will follow in preparing the solution

☐ Persuades that the method you plan to use for creating the solution will work

Resources

☐ Persuades that you have or can obtain the needed resources

☐ Protects you and your employer by clearly identifying any resources your readers must supply

Schedule

☐ Tells when your project will be completed

☐ Persuades that you have scheduled your work reasonably and soundly

---

FIGURE 23.2 *(continued)*

Writer's Guide
REVISING PROPOSALS
(continued)

☐ Protects yourself and your employer by clearly stating what your readers must do in order for you to be able to meet your deadlines

☐ Includes a schedule chart, if one would make your proposal more usable and persuasive

Qualifications

☐ If necessary, persuades that you have the ability to complete the project successfully

Management

☐ If your project is large, persuades that you will organize the people working on it effectively

☐ Includes an organizational chart, if one would make your proposal more usable and persuasive

Costs

☐ Persuades that you have presented all the costs

☐ Persuades that the costs are reasonable

☐ Protects you and your employer by including all your costs in your budget

☐ Includes a budget table, if one would make your proposal more usable and persuasive

Conclusion

☐ Summarizes your key points

☐ Concludes the proposal on a positive note that builds confidence in your ability to do a good job

Reasoning (See Chapter 5)

☐ States your claims and conclusions clearly

☐ Provides sufficient evidence, from the readers' viewpoint

☐ Provides evidence your readers will find to be reliable

☐ Explains, if necessary, the line of reasoning that links your facts and your claims

☐ Addresses any counterarguments or objections that your readers are likely to raise at any point in your report

☐ Avoids making false assumptions and overgeneralizing

☐ Achieves a tight fit among its parts

Prose (See Chapters 4, 5, 8, and 9)

☐ Presents information in a clear, usable, and persuasive manner

☐ Uses a variety of sentence structures and lengths

☐ Flows in a way that is interesting and easy to follow

☐ Uses correct spelling, grammar, and punctuation

(continued)

FIGURE 23.2 *(continued)*

Writer's Guide
REVISING PROPOSALS
(continued)

Graphics (See Chapter 13)

☐ Included wherever readers would find them helpful or persuasive

☐ Looks neat, attractive, and easy to read

☐ Referred to at the appropriate points in the prose

☐ Located where your readers can find them easily

Page Design (See Chapter 14)

☐ Looks neat and attractive

☐ Helps readers find specific information quickly

Ethics

☐ Treats all the report's stakeholders ethically

☐ Presents all information accurately and fairly

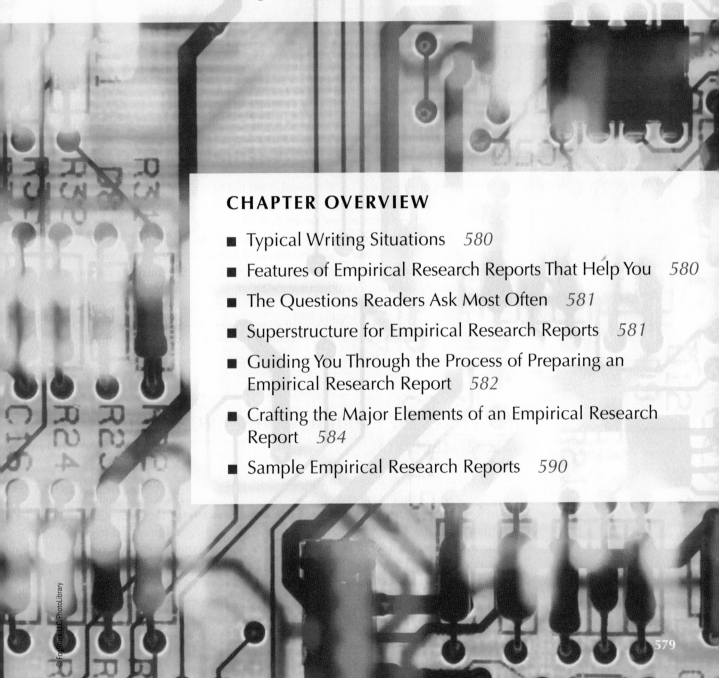

24 | Writing Reader-Centered Empirical Research Reports

CHAPTER OVERVIEW

In empirical research, investigators gather information through carefully planned, systematic observations or measurements. When scientists send a satellite to investigate the atmosphere of a distant planet, when engineers test jet engine parts made of various alloys, and when pollsters ask older citizens what kinds of outdoor recreation they participate in, they all are conducting empirical research. In your career, you will almost certainly perform empirical research—and report on it in writing. This chapter guides you through the process of writing an empirical research report.

TYPICAL WRITING SITUATIONS

Empirical research has two distinct purposes. Most aims to help people make practical decisions. The engineers who test jet engine parts are trying to help designers determine which alloy to use in a new engine. Similarly, the researchers who study older persons' recreational activities are trying to help decision makers in the state park system determine what sorts of services and facilities to provide for senior citizens.

A smaller portion of empirical research aims not to support practical decisions but rather to extend human knowledge. Here, researchers set out to learn how fish remember, what the molten core of the earth is like, or why people fall in love. Such research is usually reported in scholarly journals, such as the *Journal of Chemical Thermodynamics,* the *Journal of Cell Biology,* and the *Journal of Social Psychology,* whose readers are concerned not so much with making practical business decisions as with extending the frontiers of human understanding.

These two aims of research sometimes overlap. Some organizations sponsor basic research in the hope that what is learned can later have practical applications. Likewise, some practice-based research produces results that help explain something basic about the world in general.

Empirical research reports propel advances in engineering, science, and medicine.

FEATURES OF EMPIRICAL RESEARCH REPORTS THAT HELP YOU

Two features of empirical research reports can be very helpful to you. First, whether the reports aim to support practical decisions, extend human knowledge, or achieve some combination of these two purposes, their readers want the same kinds of information from them. Second, empirical research reports are prepared so often in the workplace that a set of conventions has developed concerning their content and organization. These conventions have proven successful in helping writers provide the information their readers want in a structure the readers find easy to use.

The conventions constitute the *superstructure* or *genre* of the empirical research report. As you research, plan, and write empirical research reports in college and in your career, you will be helped immensely by your knowledge of the readers' typical questions and the superstructure for answering them. The next two sections describe the questions and superstructure.

Learn More

For a detailed explanation of superstructures, see Guideline 3 in Chapter 4 (page 105).

THE QUESTIONS READERS ASK MOST OFTEN

The following seven questions describe the kinds of information readers seek when reading empirical research reports:

- **Why is your research important to us?** Readers concerned with solving specific practical problems want to know what problems your research will help them address. Readers concerned with extending human knowledge want to know what your research contributes to that pursuit.

- **What were you trying to find out?** A well-designed empirical research project is based on carefully formulated research questions that the project will try to answer. Readers want to know what those questions are so they can determine whether they are significant.

- **Was your research method sound?** Unless your method is appropriate to your research questions and unless it is intellectually sound, your readers will not place any faith in your results or in your conclusions and recommendations.

- **What results did your research produce?** Naturally, your readers want to learn what results you obtained.

- **How do you interpret those results?** Your readers want you to interpret your results in ways that are meaningful to them.

- **What is the significance of those results?** How do your results relate to the problems your research was to help solve or to the area of knowledge it was meant to expand?

- **What do you think we should do?** Readers concerned with practical problems want to know what you advise them to do. Readers concerned with extending human knowledge want to know what you think your results imply for future research.

SUPERSTRUCTURE FOR EMPIRICAL RESEARCH REPORTS

The superstructure for empirical research reports contains seven elements. Each element responds to one of the decision makers' seven basic questions.

SUPERSTRUCTURE FOR EMPIRICAL RESEARCH REPORTS	
REPORT ELEMENT	**READERS' QUESTION**
Introduction	Why is your research important to us?
Objectives of the research	What were you trying to find out?
Method	Was your research method sound?
Results	What results did your research produce?
Discussion	How do you interpret those results?
Conclusions	What is the significance of those results?
Recommendations	What do you think we should do?

GUIDING YOU THROUGH THE PROCESS OF PREPARING EMPIRICAL RESEARCH REPORTS

When creating empirical research reports, writers perform the same activities as when they are preparing any workplace communication: defining the communication's objectives, conducting research, planning, drafting, and revising. The following sections tell how you can perform each activity in ways especially suited to this special kind of communication.

Defining Your Report's Objectives

Readers read empirical research reports to learn something of value or interest to them. If the research being reported on is intended to help them make a practical decision, they want help with the decision. If the report is intended to extend human knowledge, they want to learn something related to their professional interests. From the readers' desires flow the objectives that all empirical research reports share. The usability objective is to present the information in a way that makes it easy for the reader to find and understand all the information related to the research that interests them. The persuasive objective arises from the cautious nature of workplace readers, who want to feel confident that the information is valid. Thus the persuasive objective is to persuade readers that the information presented is accurate, relevant, and reliable.

The Writer's Guide for Defining Your Communication's Objectives (page 70) will help you identify the additional information about your report's purpose, readers, and context that is needed to fully define its objectives.

Conducting Research

The key to writing a report your readers will find persuasive is to conduct research in a way they will consider to be credible. No skill with words will make up for a research design that readers don't respect. Think of it this way: Writing is not the last step in research; research is one of the first steps in writing. Here are three ways to increase the credibility of your empirical research and, therefore, your reports about it.

- **Understand the context in which your readers will view your research.** Your report will seem interesting and helpful to your readers only if it directly addresses their goals and concerns. If you are trying to answer a practical question, gain a full sense of the problem, situation, or desire that prompted your readers to ask you for your help. If your goal is to extend knowledge, understand what is already known about your topic so you can determine what will make your research interesting or significant in your readers' eyes.
- **Define your research questions precisely, linking them to your reader's interests.** The more sharply you define the question you are trying to answer, the better able you will be to produce results that will be interesting and useful to your readers.
- **Use methods your reader will find credible to gather information, facts, and data.** Whether you are conducting a survey, running a scientific experiment, or testing an engineering product, some methods of data gathering are considered more credible than others. Pick methods your readers respect.
- **Analyze the information, facts, and data using techniques your readers will find credible.** Picking the wrong statistical tool or using faulty logic undermines otherwise good research.

Planning

You can plan an effective organization for many empirical research reports by adopting the superstructure for these reports as your outline. Readers are accustomed to reading empirical research reports that take up its seven topics in the order in which the superstructure lists them, though some topics may be combined. For example the introduction and research question are often combined, as are the results and discussion.

However, don't use the superstructure for your pattern if another organization will be more helpful to your readers. For example, if your research uses several distinct research methods, each producing its own results, you can probably create a more usable report by organizing around the research question each method is designed to address.

Drafting and Revising

Some people mistakenly think that drafting a report could or should wait until all the research has been completed. There are advantages, however, to drafting parts as soon as you are able. By drafting your criteria as you begin your research, you sharpen your sense of what you need to find. Likewise, drafting your methods section as you are designing your research enables you to step back to review and possibly refine your research plan. Drafting as you go can also help you avoid the all-too-common situation in which writers leave so much drafting until right before the deadline that they don't have time to revise and polish their reports.

Learn More

For detailed advice about defining research questions and methods of information gathering, see Chapter 6 and the Writer's Reference Guide on "Using Five Reader-Centered Research Methods."

Try This

Think of a topic you might like to research. It could concern your classes, community, or another aspect of your life. What would be your research question? What methods would you use to gather relevant information or data? Would you need to change your methods if you also needed to persuade others that your results were reliable?

Learn More

For advice on analyzing research results, see Chapter 7.

Learn More

Remember that a superstructure is not an outline; you may combine the elements of a superstructure in many ways (see page 557).

CRAFTING THE MAJOR ELEMENTS OF AN EMPIRICAL RESEARCH REPORT

All of the detailed advice that Chapters 8 through 14 provide about drafting workplace communications applies to empirical research reports. The following sections tell how you can apply some of this advice most productively when writing this type of report. Much of the advice is illustrated through the use of two sample reports. The first is presented in full on pages 591–607. Its aim is practical. It was written by engineers who were developing a satellite communication system that will permit companies with large fleets of trucks to communicate directly with their drivers at any time. In their report, the writers tell decision makers and other engineers in their organization about the first operational test of the system, in which they sought to answer several practical engineering questions.

The aim of the second sample report is to extend human knowledge: a famous study by Robert B. Hays (1985) of the ways people develop friendships. Selected passages from this report, which appeared in the *Journal of Personality and Social Psychology*, are quoted throughout this section.

Introduction

Learn More

For additional advice on writing an introduction, see Chapter 10, "Beginning a Communication."

In the introduction to an empirical research report, you should seek to answer the readers' question, "Why is this research important to us?" Typically, writers answer that question in two steps: They announce the topic of their research and then explain the importance of the topic to their readers.

Announcing the Topic

You can often announce the topic of your research simply by including it as the key phrase in your opening sentence. For example, here is the first sentence of the report on the satellite communication system:

Topic of report

> For the past eighteen months, the Satellite Products Laboratory has been developing a system that will permit companies with large, nationwide fleets of trucks to communicate directly to their drivers through a satellite link.

Here is the first sentence of the report on the way people develop friendships:

Topic of report

> Social psychologists know very little about the way real friendships develop in their natural settings.

Explaining the Importance of the Research

To explain the importance of your research to your readers, you can use either or both of the following methods: tell how your research is relevant to your organization's goals or tell how it expands existing knowledge on the subject.

- **Relevance to organizational goals.** In reports written to readers in organizations (whether your own or a client's), you can explain the relevance of your research by relating it to some organizational goal or problem. Sometimes, in fact, the importance of your research to the organization's needs will be so obvious to your readers that merely naming your topic will be sufficient. At other times, you will need to discuss at length the relevance of your research to the organization. In the first paragraph of the satellite report, for instance, the writers mention the potential market for the satellite communication system they are developing. That is, they explain the importance of their research by saying that it can lead to a profit. For detailed advice about how to explain the importance of your research to readers in organizations, see Guideline 1 of Chapter 10, "Beginning a Communication."

- **Expansion of existing knowledge.** A second way to establish the importance of your research is to show the gap in current knowledge that it will fill. The following passage from the opening of the report on the friendship study illustrates this strategy.

> A great deal of research in social psychology has focused on variables influencing an individual's attraction to another at an initial encounter, usually in laboratory settings (Bergscheid & Walster, 1978; Bryne, 1971; Huston & Levinger, 1978), yet very little data exists on the processes by which individuals in the real world move beyond initial attraction to develop a friendship; even less is known about the way developing friendships are maintained and how they evolve over time (Huston & Burgess, 1979; Levinger, 1980).

The writer tells what is known about his topic.

The writer identifies the gaps in knowledge that his research will fill.

The writer continues this discussion of published research for three paragraphs. Each follows the same pattern: It identifies an area of research, tells what is known about that area, and identifies gaps in the knowledge—gaps that will be filled by the research the writer has conducted. These paragraphs serve an important additional function performed by many literature reviews: They introduce the established facts and theories that are relevant to the writer's work and necessary to an understanding of the report.

Objectives of the Research

Every empirical research project has carefully constructed objectives. These objectives define the focus of your project, influence the choice of research method, and shape the way you interpret your results. Readers of empirical research reports want and need to know what the objectives are.

The following example from the satellite report shows one way you can inform your readers about your objectives.

> In particular, we wanted to test whether we could achieve accurate data transmissions and good-quality voice transmissions in the variety of terrains typically encountered in long-haul trucking. We wanted also to see what factors might affect the quality of transmissions.

Statement of research objectives

When reporting research that employs statistics, you can usually state your objectives by stating the hypotheses you tested. Where appropriate, you can explain these hypotheses in terms of existing theory, again citing previous publications on the subject. In the following passage, the writer who studied friendship explains some of his hypotheses. Notice that he begins by stating the research's overall goal.

<table>
<tr><td>Overall goal</td><td rowspan="4">The goal of the study was to identify characteristic behavioral and attitudinal changes that occurred within interpersonal relationships as they progressed from initial acquaintance to close friendship. With regard to relationship benefits and costs, it was predicted that both benefits and costs would increase as the friendship developed, and that the ratings of both costs and benefits would be positively correlated with ratings of friendship intensity. In addition, the types of benefits listed by the subjects were expected to change as the friendships developed. In accord with Levinger and Snoek's (1972) model of dyadic relatedness, benefits listed at initial stages of friendship were hypothesized to be more activity centered and to reflect individual self-interest (e.g., companionship, information) than benefits at later stages, which were expected to be more personal and reciprocal (e.g., emotional support, self-esteem).</td></tr>
<tr><td>First objective (hypothesis)</td></tr>
<tr><td>Second objective (hypothesis)</td></tr>
<tr><td>Third objective (hypothesis)</td></tr>
</table>

Method

Readers of your empirical research reports will look for precise details concerning your method. Those details serve three purposes. First, they let your readers assess the soundness of your research design and its appropriateness for the problems you are investigating. Second, the details enable your readers to determine the limitations that your method might place on the conclusions you draw. Third, they provide information that will help your readers repeat your experiment if they wish to verify your results or conduct similar research projects of their own.

The nature of the information you should provide about your method depends on the nature of your research. The writers of the satellite report provided three paragraphs and two tables explaining their equipment (truck radios and satellite), two paragraphs and one map describing the eleven-state region covered by the trucks, and two paragraphs describing their data analysis.

The friendship study's writer began his method discussion this way.

At the beginning of their first term at the university, first-year students selected two individuals whom they had just met and completed a series of questionnaires regarding their relationships with those individuals at 3-week intervals through the school term.

Learn More

Often a description of empirical research methods uses the pattern for organizing a description of a process (see pages 248–251).

In the rest of that paragraph, the writer explains that the questionnaires asked the students to describe such matters as their attitudes toward each of the other two individuals and the specific things they did with each of them. However, that paragraph is just a small part of the researcher's account of his method. He then provides a 1,200-word discussion of the students he studied and procedures he used.

How can you decide which details to include? The most obvious way is to follow the general reporting practices in your field. Find some research reports that use a method similar to yours and see what they report. Depending on the needs of your readers, you may need to include some or all of the following:

- Every procedural decision you made while planning your research
- Every aspect of your method that your readers might ask about
- Any aspect of your method that might limit the conclusions you can draw from your results
- Every procedure that other researchers would need to understand in order to design a similar study

Results

The results of empirical research are the data you obtain. Although the results are the heart of any empirical research report, they may take up a very small portion of it. Generally, results are presented in one of two ways:

- **Tables and graphs.** The satellite report, for instance, uses two tables. The report on friendship uses four tables and eleven graphs.
- **Sentences.** When placed in sentences, results are often woven into a discussion that combines data and interpretation, as the next paragraphs explain.

Learn More

For information on using tables and other graphics for reporting numerical data, see pages 359–367.

Discussion

Sometimes writers briefly present all their results in one section and then discuss them in a separate section. Sometimes they combine the two in a single, integrated section. Whichever method you use, your discussion must link your interpretative comments with the specific results you are interpreting.

One way of making that link is to refer to the key results shown in a table or other graphic and then comment on them, as appropriate. The following passage shows how the writers of the satellite report did that with some of the results they presented in one of their tables:

> As Table 3 shows, 91% of the data transmissions were successful. These data are reported according to the region in which the trucks were driving at the time of transmission. The most important difference to note is the one between the rate of successful transmissions in the Southern Piedmont region and the rates in all the other regions. In the Southern Piedmont area, we had the truck drive slightly outside the ATS-6 footprint so that we could see if successful transmissions could be made there. When the truck left the footprint, the percentage of successful data transmissions dropped abruptly to 43%.

Writers emphasize a key result shown in a table.

Writers draw attention to other important results.

Writers interpret those results.

When you present your results in prose only (rather than in tables and graphs), you can weave them into your discussion by beginning your paragraphs with general statements that serve as interpretations of your data. Then you can cite the relevant results as evidence in support of the interpretations. Here is an example from the friendship report.

General interpretation

Specific results presented
as support for the
interpretation

Intercorrelations among the subjects' friendship intensity ratings at the various assessment points showed that friendship attitudes became increasingly stable over time. For
example, the correlation between friendship intensity ratings at 3 weeks and 6 weeks was
.55; between 6 weeks and 9 weeks, .78; between 9 weeks and 12 weeks, .88 (all p • .001).

In a single report, you may use both of these methods of combining the presentation
and discussion of your results.

Conclusions

Besides interpreting the results of your research, you need to explain what those
results mean in terms of the original research questions and the general problem you
set out to investigate. Your explanations of these matters are your conclusions.

For example, if your research project is focused on a single hypothesis, your conclusion can be brief, perhaps only a restatement of your chief results. However, if your
research is more complex, your conclusion should draw all the strands together.

In either case, the presentation of your conclusions should correspond very
closely to the objectives you identified toward the beginning of your report. Consider,
for instance, the correspondence between objectives and conclusions in the satellite
study. The first objective was to determine whether accurate data transmissions and
good-quality voice transmissions could be obtained in the variety of terrains typically encountered in long-haul trucking. The first of the conclusions addresses that
objective.

The Satellite Products Laboratory's system produces good-quality data and voice transmissions throughout the eleven-state region covered by the satellite's broadcast footprint.

The second objective was to determine what factors affect the quality of transmissions, and the second and third conclusions relate to it.

The most important factor limiting the success of transmissions is movement outside
the satellite's broadcast footprint, which accurately defines the satellite's area of effective
coverage.

The system is sensitive to interference from certain kinds of objects in the line of sight
between the satellite and the truck. These include trees, mountains and hills, overpasses, and buildings.

The satellite research concerns a practical question. Hence its objectives and
conclusions address practical concerns of particular individuals—in this case, the
engineers and managers in the company that is developing the satellite system. In
contrast, research that aims primarily to extend human knowledge often has objectives and conclusions that focus on theoretical issues.

For example, at the beginning of the friendship report, the researcher identifies
several questions that his research investigated, and he states what answers he
predicted his research would produce. In his conclusion, then, he systematically
addresses those same questions in terms of the results his research produced. Here

is a summary of some of his objectives and conclusions. (Notice how he uses the technical terminology commonly employed by his readers.)

OBJECTIVE

As they develop friendships, do people follow the kind of pattern theorized by Guttman, in which initial contacts are relatively superficial and later contacts are more intimate?

Do *both* the costs (or unpleasant aspects) and the benefits of personal relationships increase as friendships develop?

Are there substantial differences between friendships women develop with one another and the kind men develop with one another?

CONCLUSION

Yes. "The initial interactions of friends . . . correspond to a Guttman-like progression from superficial interaction to increasingly intimate levels of behavior."

Yes. "The findings show that personal dissatisfactions are inescapable aspects of personal relationships and so, to some degree, may become immaterial. The critical factor in friendships appears to be the amount of benefits received. If a relationship offers enough desirable benefits, individuals seem willing to put up with the accompanying costs."

Apparently not. "These findings suggest that—at least for this sample of friendships—the sex differences were more stylistic than substantial. The bonds of male friendship and female friendship may be equally strong, but the sexes may differ in their manner of expressing that bond. Females may be more inclined to express close friendships through physical or verbal affection; males may express their closeness through the types of companionate activities they share with their friends."

Conclusions from the friendship report

Typically, in presenting the conclusions of an empirical research project directed at extending human knowledge, writers discuss the relationship of their findings to the findings of other researchers and to various theories that have been advanced concerning their subject. The writer of the article on friendship did that. The table above presents only a few snippets from his overall discussion, which is several thousand words long and is full of thoughts about the relationship of his results to the results and theories of others.

In the discussion section of their empirical research reports, writers sometimes discuss any flaws in their research method or limitations on the generalizability of their conclusions. For example, the writer on friendship points out that his subjects all were college students and that most lived in dormitories. It is possible, he cautions, that what he found while studying this group may not be true for other groups.

Recommendations

The readers of some empirical research reports, especially those directed at solving a practical problem, want to know what, based on the research, the writer thinks should be done. Consequently, such reports usually include recommendations.

For example, the satellite report contains three. The first is the general recommendation that work on the project be continued. The other two involve specific actions that the writers think should be taken: design a special antenna for the trucks, and develop a plan that tells what satellites would be needed to provide coverage throughout the forty-eight contiguous states, Alaska, and southern Canada. As is common in research addressed to readers in organizations, these recommendations concern practical business and engineering decisions.

Even in reports designed to extend human knowledge, writers sometimes include recommendations. These usually convey ideas about future studies that should be made, adjustments in methodology that seem to be called for, and the like. At the end of the friendship report, for instance, the writer suggests that researchers study different groups (not just students) in different settings (not just college) to establish a more comprehensive understanding of how friendships develop.

SAMPLE EMPIRICAL RESEARCH REPORTS

Figure 24.1 (pages 591–607) shows the full report on the truck-and-satellite communication system, which is addressed to readers within the writers' own organization. For additional examples, see the book's website. To see examples of empirical research reports presented as journal articles, consult journals in your field.

Writer's Guides and Other Resources

Figure 24.2 (pages 608–609) presents a Writer's Guide for Revising Empirical Research Reports that you can use in your course and on the job. At this book's website you can download a copy of this Writer's Guide as well as one for planning empirical research reports. The website also includes other resources to help you write effective reports.

WWW

To download Writer's Guides for planning and revising empirical research reports, go to Chapter 24 at **www.cengage .com/english/anderson7e**.

Note to the Instructor   For an assignment involving an empirical research report, you can ask students to write on research they are conducting in their majors, or you can adapt the "Formal Report or Proposal" project given in Appendix B. You could also use Appendix B's User Test and Report project. Other projects and cases that involve report writing are included at the book's website. To tailor the Writer's Guide to your course, download them from the website.

FIGURE 24.1
Empirical Research Report

ELECTRONICS CORPORATION OF AMERICA

Truck-to-Satellite Communication System:
First Operational Test

September 29, 2010

Internal Technical Report

Number TR-SPL-0931

This report was written by four engineers who conducted a research project. On the cover, the team placed the items their employer requires on the covers of all research reports. The company also specifies the layout of the covers. Each requirement is intended to help readers work efficiently.

The company uses colored type for report titles so that readers can easily spot the titles when searching through company files.

The company requires that a report's date be placed on the cover to help future readers know how old the report is.

The company places report numbers on the cover to make finding copies easy. Reports are filed and cited by this number.

The cover does not have a page number.

Learn More

For more information on creating reader-centered report covers, see page 315.

(continued)

FIGURE 24.1
(continued)

Like the cover, the title page contains exactly the information specified by the company for all research reports.

As is typical in many organizations, the title page includes more information than the cover does.

This company permits the researchers to be named. In some companies, a report's writers are not named.

The company requires that all research reports be reviewed and approved by the laboratory director. She examines drafts carefully and usually requires changes before giving approval.

Learn More

For more information on review processes, see page 415.

ELECTRONICS CORPORATION OF AMERICA

**Truck-to-Satellite Communication System:
First Operational Test**

September 29, 2010

Research Team

Margaret C. Barnett

Erin Sanderson

L. Victor Sorrentino

Raymond E. Wu

Internal Technical Report
Number TR-SPL-0931

Read and Approved:

Beverly Fisher   September 29, 2010

                                      Date

FIGURE 24.1
(continued)

EXECUTIVE SUMMARY

For the past eighteen months, the Satellite Products Laboratory has been developing a system that will permit companies with large, nationwide fleets of trucks to communicate directly to their drivers at any time through a satellite link. Costs would be lower than for cell phones, and a GPS system mounted in each truck would enable a company to determine the exact location of all its vehicles at one time. During the week of May 18, we tested our concepts for the first time, using the ATS-6 satellite and five trucks that were driven over an eleven-state region with our prototype mobile radios.

More than 91% of the 2500 data transmissions were successful and more than 91% of the voice transmissions were judged to be of commercial quality. The most important factor limiting the success of transmissions was movement outside the satellite's broadcast footprint. Other factors include the obstruction of the line of sight between the truck and the satellite by highway overpasses, mountains and hills, trees, and buildings.

Overall, the test demonstrated the soundness of the prototype design. Work on it should continue as rapidly as possible. We recommend the following actions:

- Develop a new antenna designed specifically for use in communications between satellites and mobile radios.

- Explore the configuration of satellites needed to provide thorough footprint coverage for the 48 contiguous states, Alaska, and Southern Canada at an elevation of 25° or more.

The researchers provide a brief executive summary in order to help their readers quickly understand the report's main points. The summary also enables readers to determine whether they could meet their practical needs by reading the entire report.

Every piece of information the researchers include in the summary is also in the body of the report.

In the first sentence, the researchers identify the topic of their report.

In one sentence, the researchers summarize their research method, providing specific details about the data, the satellite used, the number of trucks, and region covered.

In precise, succinct statements, the researchers detail the results that are most relevant to their readers' major question: "Does the prototype design work?"

Their conclusion is the answer to their readers' question: "Yes, the design does work."

Learn More

For more information on executive summaries, go to page 320.

The researchers use a bulleted list to highlight their two major recommendations for continued work on the satellite project.

(continued)

FIGURE 24.1
(continued)

In addition to telling readers where they can find particular kinds of information, the table of contents also provides them with an outline of the report.

The researchers include the major headings within their sections.

They align the page numbers on the right-hand side.

They number the front matter with roman numerals. They begin using Arabic numerals on the first page of the report's body.

Learn More

For more information on writing the front matter of a report, go to Chapter 12, page 315.

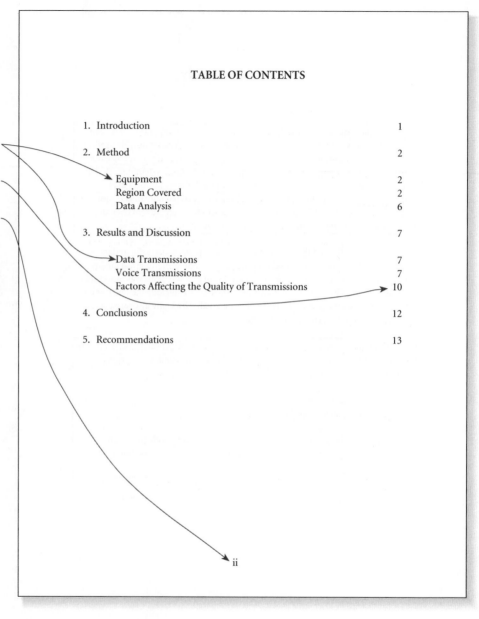

TABLE OF CONTENTS

ii

FIGURE 24.1
(continued)

Section 1

INTRODUCTION

For the past eighteen months, the Satellite Products Laboratory has been developing a system that will permit companies with large, nationwide fleets of trucks to communicate directly to their drivers at any time through a satellite link. Costs would be lower than for cell phones, and a GPS system mounted in each truck would enable a company to determine the exact location of all its vehicles at one time. Several trucking lines and supermarket chains have expressed an interest in such a service. At present, they can communicate with their long-distance drivers only when the drivers pull off the road to phone in, meaning that all contacts are originated by the drivers, not the central dispatching service. The potential market for such a satellite service also includes many other companies and government agencies (such as the National Forest Service) that desire to communicate with trucks, cars, boats, or trains that regularly operate outside the very limited range of urban cellular telephone systems.

This report describes the first operational test of the system we have developed. Such tests were particularly important to conduct before continuing further with the development of this system because our system is much different from those currently being used with commercial satellites. Specifically, our system will transmit to mobile ground stations by using the short antennas and the low power provided by conventional terrestrial broadcasting systems.

In particular, we wanted to test whether we could achieve accurate data transmissions and good-quality voice transmissions in the variety of terrains typically encountered in long-haul trucking. We wanted also to see what factors might affect the quality of transmissions.

The test results indicate that our design is basically sound, although a new mobile antenna needs to be designed and the satellite configuration needs to be examined.

1

The researchers begin the body of their report by stating its topic.

They immediately tell their readers the significance of their research to the company.

The researchers describe the overall objectives of their test.

Learn More

For advice about presenting the objectives of empirical research, see page 585.

They tell their readers their main finding ("the bottom line") at the very beginning of their report.

Learn More

For advice about writing the introduction to an empirical research report, see page 584. In addition, see Chapter 10's guidelines for writing the beginning of a communication.

(continued)

FIGURE 24.1
(continued)

The researchers' major goal in the methods section is to persuade their readers that they used sound procedures. They know that their readers will not trust the researchers' results unless they are persuaded that the research method was solid.

Their second goal in the methods section is to provide details that will enable the readers to interpret the results.

The researchers begin with a one-sentence overview of their method.

In their next sentence, they forecast the organization of their methods section.

The researchers organize this paragraph (and many others) so that it flows from general ideas to particular ones.

They provide precise details about the equipment they used.

In this sentence the researchers refer readers to the technical data in Table 1.

Anticipating a possible question from their readers, the researchers explain their justification for using the large ground station at King of Prussia.

Section 2

METHOD

In this experiment, we tested a full-scale system in which five trailer trucks communicated with an earth ground station via a satellite in geostationary orbit 23,200 miles above the earth. This section describes the equipment we used, the area covered by the test, and the data analyses we performed.

Equipment

The five trucks, operated by Smithson Moving Company, were each equipped with a prototype of our newly developed 806 megahertz (MHz) two-way mobile radio equipment. Each radio had a speaker and microphone for voice communications, along with a ten-key keyboard and digital display for data communications. Equipped with dipole antennas, the radios broadcast at 1650 MHz with 12 to 15 watts of power. They received signals at 1550 MHz and had an equivalent antenna temperature of 800°K, including feedline losses. Technical specifications for the receivers and transmitters of these radios are given in Table 1.

The satellite used for this test was the ATS-6, which has a larger antenna than most commercial communication satellites, making it more sensitive to the low-power signals sent from the mobile stations. Technical specifications for its receiver and transmitter are given in Table 2.

Through the ATS-6 satellite, the five trucks communicated with the Earth Ground Station in King of Prussia, Pennsylvania. This facility is a relatively large station, but not larger than is planned for a fully operational commercial system.

Region Covered

The five trucks drove throughout the region covered by the "footprint" of ATS-6. The footprint is shown as the area within the oval in Figure 1. It is defined as the area in which the broadcast signals received are within at least 3 dB of the signal received at the center of the beam. In all, the trucks covered eleven states: Georgia, South Carolina, North Carolina, Tennessee, Virginia, West Virginia, Ohio, Indiana, Illinois, Iowa, and Nebraska.

2

Learn More

For advice about writing the methods section of an empirical research report, see page 586.

Learn More

For advice about integrating graphics and text, see Guideline 6 in Chapter 13, page 348.

Learn More

For guidelines on organizing sections and paragraphs in a reader-centered way, see Chapter 8, page 214.

FIGURE 24.1
(continued)

Table 1

Specifications for Satellite-Aided Mobile Radio

Transmitter

Frequency	1655.050 MHz
Power Output	16 watts nominal
	12 watts minimum
Frequency Stability	±0.0002% (− 30° to +60°C)
Modulation	$16F_3$ Adjustable from 0 to ±5 kHz swing FM with instantaneous modulation limiting
Audio Frequency Response	Within +1 dB and − 3 dB of a 6 dB/octave pre-emphasis from 300 to 3000 HZ per EIA standards
Duty Cycle	EIA 20% Intermittent
Maximum Frequency Spread	±6 MHz with center tuning
RF Output Impedance	50 ohms

Receiver

Frequency	1552.000 MHz
Frequency Stability	±0.0002% (−30° to +60°C)
Noise Figure	2.6 dB referenced to transceiver antenna jack
Equivalent Receiver Noise Temperature	238° Kelvin
Selectivity	−75 dB by EIA Two-Signal Method
Audio Output	5 watts at less than 5% distortion
Frequency Response	Within +1 and −8 dB of a standard 6 dB per octave deemphasis curve from 300 to 3000 Hz
Modulation Acceptance	±7 kHz
RF Input Impedance	50 ohms

3

With this table, the writers provide detailed data in a highly usable manner.

The researchers give their table a title that briefly but precisely tells their readers what the table shows.

Because the table contains text rather than numbers, the researchers have aligned the contents of their columns on the left.

Learn More

Pages 358–359 provide guidelines for creating reader-centered tables.

(continued)

FIGURE 24.1
(continued)

Table 2

Performance of ATS-6 Spacecraft L-Band Frequency Translation Mode

Receive

Receiver Noise Figure (dB)	6.5
Equivalent Receiver Noise Temperature (°K)	1005
Antenna Temperature Pointed at Earth (°K)	290
Receiver System Temperature (°K)	1295
Antenna Gain, peak (dB)	38.4
Spacecraft G/T, peak (dB/°K)	7.3
Half Power Beamwidth (degrees)	1.3
Gain over Field of View (dB)	35.4
Spacecraft G/T over Field of View (dB/°K)	4.3

Transmit

Transmit Power (dBw)	15.3
Antenna Gain, peak (dB)	37.7
Effective Radiated Power, peak (dBw)	53.0
Half Power Beamwidth (degrees)	1.4
Gain over Field of View (dB)	34.8
Effective Radiated Power over Field of View (dBw)	50.1

4

FIGURE 24.1
(continued)

With this map, the researchers
describe the area covered by the
satellite beam much more clearly
than they could with sentences.

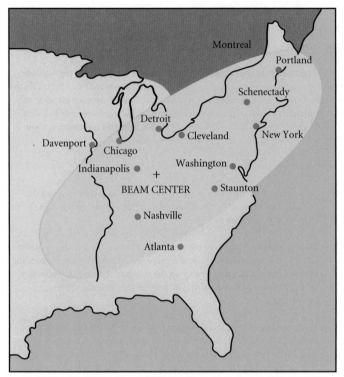

Figure 1.   ATS-6P Footprint. (Oval is −3dB Contour Relative to Beam
            Center.)

5

(continued)

FIGURE 24.1
(continued)

The researchers describe their method of analysis to persuade their readers of the soundness of their study. They know that their readers will want this information because the study results would be incorrect if a wrong or flawed method of analysis had been used.

The research team used three methods for analyzing data. Each method corresponds to one of their research objectives, which they describe in the report's introduction.

To help their readers see the correspondence between their analysis methods and their research objectives, the team describes their three analysis methods in the same order they used when naming their research objectives in the introduction.

Within this region, the trucks drove through the kinds of terrain usually encountered in long-haul trucking, including both urban and rural areas in open plains, foothills, and mountains.

Data Analysis

All test transmissions were recorded at the earth station on a high-quality reel-to-reel tape recorder. The strength of the signals received from the trucks via the satellite was recorded on a chart recorder that had a frequency response of approximately 100 hertz. In addition, observers in the trucks recorded all data signals received and all data codes sent. They also recorded information about the terrain during all data and voice transmissions.

We analyzed the data collected in several ways:

- To determine the accuracy of the data transmissions, we compared the information recorded by the observers with the signals recorded on the tapes of all transmissions.

- To determine the quality of the voice transmissions, we had an evaluator listen to the tape using high-quality earphones. For each transmission, the evaluator rated the signal quality on the standard scale for the subjective evaluation of broadcast quality. On it, Q5 is excellent and Q1 is unintelligible.

- To determine what factors influenced the quality of the transmissions, we examined the descriptions of the terrain that the observers recorded for all data transmissions that were inaccurate and all the voice communications that were rated 3 or less by the evaluator. We also looked for relationships between the accuracy and quality of the transmissions and the distance of the trucks from the edge of the broadcast footprint of the ATS-6 satellite.

6

FIGURE 24.1
(continued)

Section 3

RESULTS AND DISCUSSION

In total, the test transmissions took 603 hours of satellite time. They included 690 data communications and 5351 seconds of voice communications. This section reports and discusses the results of these transmissions and examines the factors that influence transmission quality.

Data Transmissions

As Table 3 shows, 91% of the data transmissions were successful. These data are reported according to the region in which the trucks were driving at the time of transmission. The most important difference to note is the one between the rate of successful transmissions in the Southern Piedmont region and the rates in all the other regions. In the Southern Piedmont area, we had the truck drive slightly outside the ATS-6 footprint so that we could see if successful transmissions could be made there. When the truck left the footprint, the percentage of successful data transmissions dropped abruptly to 43%. If transmissions in that area are eliminated from the calculations, 95% of all data transmissions were successful.

Voice Transmissions

The success rate for the voice transmissions was also very high. As the two left-hand columns of Table 4 show, 91% of the communications were rated as Q4 or Q5. That is a very good rating because Q3 is usually considered adequate to provide useful commercial communications. As with data communications, the rate of successful transmissions drops off sharply in the Southern Piedmont area, where the truck involved left the ATS-6 footprint. If the transmissions made in the South Piedmont area are eliminated from the calculations, the system produced voice communications rated a Q4 or Q5 96% of the time.

Thus, within the footprint of the satellite, 95% or more of the transmissions of both data and voice were of commercial quality. That success rate is very good: the specifications for the mobile radio systems used by police and fire departments usually require only 90% effectiveness for the area covered.

7

The research team begins this section with general information.

The researchers tell their readers the key result first.

They then present other important results.

Immediately after presenting their results, the researchers interpret the results. This interpretation is one of the discussion parts of their results and discussion section.

Again, the researchers begin with the key result. In this sentence they also tell their readers the main point to draw from Table 4.

As soon as they have presented the results, the researchers interpret the results from the point of view of their readers' main question: Does the system work enough for commercial application?

The team summarizes the data and voice transmission results, emphasizing the main points of interest to their readers.

Learn More

For advice on writing the results and discussion sections of empirical research reports, see page 587. For advice about integrating graphics and text, see Guideline 6 in Chapter 13, page 348.

(continued)

FIGURE 24.1
(continued)

Table 3

Success in Decoding DTMF Automatic Transmitter

REGION	STATES	ATS TRANSMISSIONS		ELEVATION
		Sent by Vehicle	Received and Decoded Correctly	Angle to Satellite (°)
Open Plains	Indiana, Ohio, Nebraska, Illinois, Iowa	284	283 (100%)	17–26
Western Appalachian Foothills	Ohio, Tennessee	55	53 (96%)	15–19
Appalachian Mountains	West Virginia	112	93 (83%)	15–17
Piedmont	Virginia, North Carolina	190	178 (97%)	11–16
Southern Piedmont	Georgia, South Carolina	49	21 (43%)	17–18
TOTAL		690	628 (91%)	

8

Table 4
Quality of Voice Communication Signal[1]

Area	Transmission Time (Secs)	No Blockage Time (Secs)		Trees			Mountains and Hills			Overpasses (Momentary Dropouts)			Buildings		
		Q5	Q4	Q3	Q2	Q1	Q3	Q2	Q1	Q3	Q2	Q1	Q3	Q2	Q1
Open Plains	2481	2334	73	1	1	0	4	1	3	13	2	20	15	10	4
Western Foothills	344	322	2	0	0	0	6	12	13	0	0	0	0	0	0
Appalachian Mountains	1037	614	267	42	17	0	20	31	37	0	2	7	0	0	0
Piedmont	1219	481	623	5	15	19	25	16	10	4	1	4	8	4	1
Southern Piedmont	270	0	149	109	3	12	0	0	0	0	0	2	0	0	0
TOTAL TIME (seconds)	5351	3751	1114	157	36	31	55	60	63	17	5	33	23	14	5

[1]Total times in seconds for each quality of received signals. Q5 is excellent; Q1 is unintelligible.

9

FIGURE 24.1
(continued)

To help their readers use this table without confusion, the research team placed vertical lines between the columns, which are very close together. Such lines are not needed in the other tables, where the space between the columns is greater.

(continued)

FIGURE 24.1
(continued)

In this section, too, the researchers begin with the most important point.

The team refers to Table 4 again. This time, however, they want their readers to notice something different—and they tell the readers what that is.

The researchers continue to present the information that is most important to their readers before the information that is less important to them. In this section, they start with the factor that causes the most disruptions: trees.

The researchers devote this paragraph to answering the question they imagine their readers will ask: "What can be done about the trees?"

Factors Affecting the Quality of Transmissions

The factor having the largest effect on transmission quality is the location of the truck within the footprint of ATS-6. The quality of transmissions is even and uniformly good throughout the footprint, but almost immediately outside of it the quality drops well below acceptable levels.

Several factors were found to disrupt transmissions even when the trucks were in the satellite's footprint. Table 4 shows what these factors are for the 4% of the transmissions in the footprint that were not of commercial quality. In all cases the cause is some object passing in the line of sight between the satellite and the truck.

Trees caused 45% of the disruptions, more than any other source. At the frequencies used for broadcasting in this system, trees and other foliage have a high and very sharp absorption. Of course, the tree must be immediately beside the road and also tall enough to intrude between the truck and the satellite, which is at an average elevation of 17° above the horizon. And the disruption caused by a single tree will create only a very brief and usually insignificant dropout of one second or less. Only driving past a group of trees will cause a significant loss of signal. Yet this happened often in the terrain of the Appalachian Mountains and the South Piedmont.

We believe we could eliminate many of the disruptions caused by trees if we developed an antenna specifically for use in communications between satellites and mobile radios. In the test, we used a standard dipole antenna. Instead, we might devise a Wheeler-type antenna that is omnidirectional in azimuth and with gain in the vertical direction to minimize ground reflections.

Mountains and hills caused 36% of the disruptions. That happened mostly in areas where the satellite's elevation above the horizon was very low. Otherwise a hill or mountain would have to be very steep to block out a signal. For example, if a satellite were only 17° above the horizon, the hill would have to rise over 1500 feet per mile to interfere with a transmission—and the elevation would have to be precisely in a line between the truck and the satellite.

10

FIGURE 24.1
(continued)

Most of the disruptions caused by mountains and hills can be eliminated by using satellites that have an elevation of at least 25°. That would place them above all but the very steepest slopes.

Highway overpasses accounted for 11% of the disruptions, but these disruptions had little effect on the overall quality of the broadcasts. As one of the test trucks drove on an open stretch of interstate highway, the signal was strong and steady, with fading less than 2 decibels peak-to-peak. About two seconds before the truck entered the overpass, there was detectable but not severe multipath interference. The only serious disruption was a one-second dropout while the truck was directly under the overpass. This one-second dropout was so brief that it did not cause a significant loss of intelligibility in voice communications. Only a series of overpasses, such as those found where interstate highways pass through some cities, cause a significant problem.

Finally, buildings and similar structures accounted for about 8% of the disruptions. These were experienced mainly in large cities, and isolated buildings usually caused only brief disruptions. However, when the trucks were driving down city streets lined with tall buildings, they were unable to obtain satisfactory communications until they were driven to other streets.

In the format used for this report, each section begins on its own page. Therefore, the last page of a section will have blank space if there isn't enough text to fill it.

11

(continued)

FIGURE 24.1
(continued)

The researchers present their conclusions in an easy-to-read bullet list.

Each conclusion corresponds to one of the research objectives that the team named in the introduction.

Learn More

For advice on writing the conclusions section of an empirical research report, see page 588. In addition, see Chapter 11's guidelines for writing the ending of a communication.

Section 4

CONCLUSIONS

This test supports three important conclusions:

- The Satellite Products Laboratory's system produces good-quality data and voice transmissions through the eleven-state region covered by the satellite's broadcast footprint.

- The most important factor limiting the success of transmissions is movement outside the satellite's broadcast footprint, which accurately defines the satellite's area of effective coverage.

- The system is sensitive to interference from certain kinds of objects in the line of sight between the satellite and the truck. These include trees, mountains and hills, overpasses, and buildings.

12

FIGURE 24.1
(continued)

Section 5

RECOMMENDATIONS

Based on this test, we believe that work should proceed as rapidly as possible to complete an operational system. In that work, the Satellite Products Laboratory should do the following things:

1. **Develop a new antenna designed specifically for use in communications between satellites and mobile radios.** Such an antenna would probably eliminate many of the disruptions caused by trees, the most common cause of poor transmissions.

2. **Define the configuration of satellites needed to provide service throughout our planned service area.** We are now ready to determine the number and placement in orbit of the satellites we will need to launch in order to provide service to our planned service area (48 contiguous states, Alaska, Southern Canada). Because locations outside of the broadcast footprint of a satellite probably cannot be given satisfactory service, our satellites will have to provide thorough footprint coverage throughout all of this territory. Also, we should plan the satellites so that each will be at least 25° above the horizon throughout the area it serves; in that way we can almost entirely eliminate poor transmissions due to interference from mountains and hills.

13

The researchers present their overall recommendation first: Work on the project should continue.

They use bold type to highlight each of their other recommendations.

They then add the details that answer their readers' question: What would be involved in the work you recommend?

After making their last recommendation, the research team ends its report.

Learn More

For advice on writing the recommendations in an empirical research report, see page 590. In addition, see Chapter 11's guidelines for various ways to end a communication plus ways to select among these alternatives.

FIGURE 24.2 **Writer's Guide for Revising Empirical Research Reports**

To download a copy of this Writer's Guide as well a Writer's Guide for Planning Empirical Research Reports, go to Chapter 24 at www.cengage.com/english/anderson7e.

Writer's Guide
REVISING EMPIRICAL RESEARCH REPORTS

Does your draft include each of the elements needed to create a report that your readers will find to be usable and persuasive? Remember that some elements of the superstructure may be unnecessary for your specific readers and purpose and that the elements may be organized in various ways.

Introduction

☐ Announces the topic of research presented in the report

☐ Persuades readers that this research is important to them

☐ Explains the relevance of the research to the organization's goals and, if appropriate, existing knowledge on the topic

☐ States briefly your main conclusions and recommendations, if the readers would welcome or expect them at the beginning of the report

☐ Provides background information the readers will need or want

☐ Forecasts the rest of your report, if this would help your readers

Objectives of Research

☐ Describes precisely what you were trying to find out through your research

Method

☐ Tells the things your readers want to know about the way you obtained the facts and ideas presented in the report

☐ Persuades the readers that this method would produce reliable results

Results

☐ Presents in clear and specific terms the things you found out

☐ Includes material that is relevant to your readers, and excludes material that isn't

Discussion (often presented along with the results)

☐ Interprets your results in a way that is useful to your readers

Conclusions (often presented along with the discussion)

☐ Explains the significance—from your readers' viewpoint—of your results and generalizations about them

☐ States the conclusions plainly

FIGURE 24.2   *(continued)*

Writer's Guide
REVISING EMPIRICAL RESEARCH REPORTS
(continued)

Recommendations

☐ Tells what you think the readers should do

☐ Makes the recommendations stand out prominently (for instance, by presenting them in a numbered list)

☐ Indicates how your recommendations are related to your conclusions and to your readers' goals

☐ Suggests some specific steps your readers might take to act on each of your recommendations, unless the steps will be obvious to the readers

Reasoning (See Chapter 5)

☐ States your claims and conclusions clearly

☐ Provides sufficient evidence, from the readers' viewpoint

☐ Explains, if necessary, the line of reasoning that links your facts and your claims

☐ Addresses any counterarguments or objections that your readers are likely to raise at any point in your report

☐ Avoids making false assumptions and overgeneralizing

Prose (See Chapters 4, 5, 8, and 9)

☐ Presents information in a clear, usable, and persuasive manner

☐ Uses a variety of sentence structures and lengths

☐ Flows in a way that is interesting and easy to follow

☐ Uses correct spelling, grammar, and punctuation

Graphics (See Chapter 13)

☐ Included wherever readers would find them helpful or persuasive

☐ Look neat, attractive, and easy to read

☐ Referred to at the appropriate points in the prose

☐ Located where your readers can find them easily

Page Design (See Chapter 14)

☐ Looks neat and attractive

☐ Helps readers find specific information quickly

Ethics

☐ Treats all the report's stakeholders ethically

☐ Presents all information accurately and fairly

25 | Writing Reader-Centered Feasibility Reports

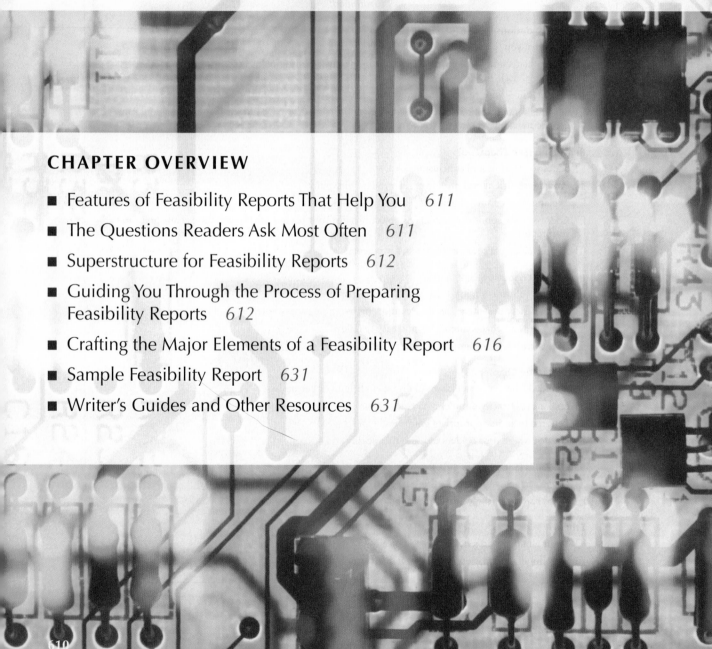

CHAPTER OVERVIEW

The workplace buzzes with decisions about future action. Among the more common are decisions about whether to do something new or in a new way. Although some of these decisions are easily made, others involve substantial changes. A company that manufactures wind turbines might want to know whether it should build a second factory several states away from its original location. A hospital might want to know whether it should replace its conventional telephone system with the Internet-based technology known as Voice Over Internet Protocol (VOIP) that works through computers and computer lines rather than telephone switchboards and wiring.

To determine whether such large changes are practical and desirable, decision makers usually ask consultants or some of their own employees to conduct extensive research. This research is typically called a *feasibility study*. The study's results are presented in a *feasibility report*. Whether on a large or small scale, you are likely to be asked to prepare a feasibility report many times in your career. This chapter guides you through the process.

FEATURES OF FEASIBILITY REPORTS THAT HELP YOU

Two features of feasibility reports can be very helpful to you. First, as decision makers consider the choices being considered, they tend to ask the same set of general questions. These questions can guide your research and writing. Second, feasibility reports are prepared so often in the workplace that a set of conventions have developed concerning their content and organization. These conventions have proven successful in helping writers provide the information their readers want in a structure the readers find easy to use. The conventions have proven to help both writers and decision-making readers answer the readers' questions. These conventions constitute the *superstructure*, or *genre*, of the feasibility report. As you research, plan, and write feasibility reports in college and in your career, you will be helped immensely by your knowledge of the readers' typical questions and the superstructure for answering them. The next two sections describe the questions and superstructure.

> **Learn More**
>
> For a detailed explanation of superstructures, see Guideline 3 in Chapter 4 (page 105).

THE QUESTIONS READERS ASK MOST OFTEN

As decision makers think about the choices they must make, they ask many questions. From situation to situation, their basic questions remain the same, and the superstructure for feasibility reports is a widely used way of providing the answers. Here are the readers' typical questions:

- **Why is it important for us to consider these alternatives?** In some cases, the key decision makers need to be told why they have to make a choice in the first place. They may need a detailed explanation of the problem before they can appreciate the urgency of considering alternative courses of action. If they are already familiar with the problem, they may need to be reassured that you understand it in the same way they do.

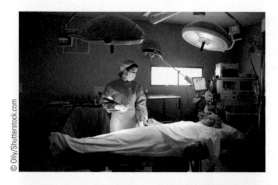

Decisions and actions based on feasibility studies affect every aspect of our lives and environment.

- **What are the important features of our alternatives?** So that they can understand your detailed discussion of the alternatives, readers want you to highlight the key features of each one.
- **Are your criteria reasonable and appropriate?** To help your readers choose between alternative courses of action, you must evaluate the alternatives in terms of specific criteria. At work, people want these criteria to reflect the needs and aims of their organization.
- **Are your facts reliable?** Evaluating alternatives involves comparing facts about them against the criteria for judging them. Your readers want to know that they can rely on the accuracy of the facts you present.
- **How do the alternatives stack up against your criteria?** The heart of a feasibility study is your evaluation of the alternatives in terms of your criteria. Your readers want to know the results.
- **What overall conclusions do you draw about the alternatives?** Based on your detailed evaluation of the alternatives, you will reach general conclusions about the merits of each. Your readers need to know what your conclusions are because these overall judgments are the basis on which they will make their decision.
- **What do you think we should do?** Because of your expertise on the subject, your readers want you to tell what you recommend. In addition, they may desire suggestions about how to proceed if they decide to follow your recommendation.

SUPERSTRUCTURE FOR FEASIBILITY REPORTS

The superstructure for feasibility reports contains seven elements. The superstructure is helpful to writers and readers because each element responds to one of the decision makers' seven basic questions.

GUIDING YOU THROUGH THE PROCESS OF PREPARING FEASIBILITY REPORTS

When creating feasibility reports, writers perform the same activities as when they are preparing any workplace communication: defining the communication's objectives, conducting research, planning, drafting, and revising. The following sections

SUPERSTRUCTURE FOR FEASIBILITY REPORTS	
REPORT ELEMENT	**READERS' QUESTION**
Introduction	Why is it important for us to consider these alternatives?
Overview of alternatives	What are the important features of the alternatives?
Criteria	Are your criteria reasonable and appropriate?
Method	Are your facts reliable?
Evaluation	How do the alternatives stack up against your criteria?
Conclusions	What overall conclusions do you draw about the alternatives?
Recommendations	What do you think we should do?

tell how you can perform each activity in ways especially suited to this special kind of communication.

Defining Your Communication's Objectives

A feasibility report's purpose flows directly from its goal of helping decision-making readers choose between two or more courses of action, one of which is usually to continue doing things as they are now done. The report's usability purpose is to provide in an easy-to-use form the information and advice its readers need to make this decision. By identifying the specific way your readers will ask the last four questions above, you can define exactly what your report should provide.

The report's persuasive purpose is to assure its decision-making readers that it provides a solid basis for decisions they must make. As explained in Chapter 5, decision makers take a cautious, professionally skeptical approach when reading. Your answers to the readers' questions about the research methods and criteria for evaluating the alternatives address their major concerns about the accuracy, completeness, and reliability of the information your report provides.

The Writer's Guide for Defining Your Communication's Objectives (page 70) will help you identify the additional information about your communication's purpose, readers, and context that is needed to fully define its objectives.

Learn More

For detailed advice about defining your report's objectives, see Chapter 3.

Conducting Research

When you define the objectives of your report, you also define the objectives of the research, which has only one purpose: to provide information and advice you need in order to write a usable and persuasive report.

Chapters 6 and 7 provide detailed advice about conducting this research. Here are a few additional points that deserve emphasis when you are preparing a feasibility report.

- **Gain a full understanding of your readers' criteria.** Begin by researching your readers. Learn how they will evaluate the alternatives. This knowledge can guide the rest of your research efforts, assuring that you gather the information your readers want. It also saves you from spending time gathering information that is irrelevant to your readers.

- **Investigate all implications of the alternatives being considered.** In particular, consider the actions that would be required to implement the alternatives. The hospital thinking of replacing its conventional phone system with Internet technology needs to consider many factors other than the features and cost of the two systems. Can its central computer server support the Internet system? What training in using that system would the doctors, nurses, and other staff members need? Who would provide it? Because the entire hospital couldn't be converted in one day, what needs to happen on the days when some departments have the new system and others have the old system?

- **Consult several kinds of sources.** Different kinds of sources can provide different perspectives and types of information. Chapter 6 suggests that you consult each of these four types (page 153):
  - Persons who would be affected by changing to each of the alternatives
  - Persons who would help to implement the alternatives if they were chosen
  - Other organizations or groups using the alternatives
  - Professional publications that may report on or evaluate the alternatives

- **Avoid bias in your information gathering.** Employees conducting feasibility studies sometimes start out favoring a particular alternative. They may then tend to seek out information that supports that choice, rather than gathering the full range of relevant information.

- **Consider creating tables for analyzing the information you've gathered.** Tables or matrixes can help you examine the alternatives the way your readers want to: point-by-point against the criteria. This approach matches the way your readers will use your report.

Planning

When planning a feasibility report, your most important decision is how to organize content between your introduction and your recommendations.

Sometimes the most effective way to organize is to create five middle sections, one for each of the five middle elements of the superstructure. Figure 25.1 shows how Magnus used this organization when reporting on the feasibility of installing a heat-pump system to replace the conventional heating and air-conditioning systems at his employer's building. This organization works best when the readers have only one decision to make and the writer used only a single method to gather the information about the alternatives.

A second organization can be more effective for readers who have two or more decisions to make. Megan and Rajiv addressed such readers in their report on the feasibility of switching to an Internet telephone system. Their readers had to consider not only the system's features but also who might install it, how training could be handled, and so

Try This

Think of a change you would like to see at your school or in your community. Who could decide to bring about that change? If this person or group were to request a feasibility study related to the change, what criteria would the person or group use to evaluate the alternatives?

Learn More

For detailed advice about conducting research, see Chapters 6 and 7.

Sometimes the order of elements in the superstructure provides an effective organizational pattern.

If your readers have several decisions to make, they will probably find it easiest to use a report organized around those decisions.

FIGURE 25.1
Feasibility Report Organized by Devoting One Section to Each of the Seven Elements in the Superstructure for Feasibility Reports

Replacing the Building's Heating and Air-Conditioning System with a Heat-Pump System: A Feasibility Study

I. Introduction

A. The heating and air-conditioning system in our building needs to be replaced
B. I was asked to study the feasibility of changing to a heat-pump system

II. Types of heat pumps for use in office buildings

A. Air source (exchanges heat with air outside the building)
B. Geothermal (exchanges heat with ground)
C. Heat absorption (runs on gas to exchange heat with outside air)

III. Criteria

A. Efficiency in our climate (different systems work better in different climates)
B. Amount of damage installation would cause to our wooded property
C. Cost (installation and operation)

IV. Research methods

A. Consulted credible sources in print and online
   i. U.S. government resources on green technologies
   ii. Engineering journals
B. Interviewed knowledgeable people
   i. Representatives of companies that sell heat-pump systems in our area
   ii. Owners of buildings in the area that use heat pumps

V. Evaluation

A. Effectiveness in our area's climate
B. Amount of damage installation would cause to our property
C. Cost (installation and operation)

VI. Conclusions

A. Over 6 years, total cost for all heat pumps is lower than standard heat and air
B. Geothermal heat pumps are most effective here, though more costly and most potentially damaging to our property

VII. Recommendations

A. Select a geothermal system
B. Use the vertical rather than horizontal installation of the pipes to minimize damage to landscape around the building
C. Next step: Invite proposals from companies that install heat-pump systems

The organization that Magnus created for this report matches the order of the elements in the superstructure for feasibility reports.

Introduction
Magnus created a brief, one-paragraph introduction focused on the background for his feasibility study and report.

Overview of alternatives
To assure that his readers understood each of the alternatives, Magnus provided detailed explanations including schematic diagrams of each alternative.

Criteria
Magnus used only a few sentences to describe the criteria.

Method
He used the same methods for gathering information related to all of the criteria.

Evaluation
In the section for each criterion, Magnus presented the relevant information about all three types of systems.

Conclusions
He stated each of his major conclusions and cited the specific facts discussed in the Evaluation section that supported the conclusion.

Recommendations
Magnus highlighted the main actions he believed his employer should take.

Learn More

For information on patterns for organizing comparisons, go to pages 242–245.

on. To help readers make these separate decisions, you can treat each decision in its own section or chapter that discusses the criteria, methods, results, and evaluation relevant to that decision. Figure 25.2 shows the organization of Megan and Rajiv's report.

In other cases, organizing around the criteria helps readers.

A third organization works well when your readers will make a single decision but you used a different research method for gathering information relevant to each of the criteria. In such cases, readers find it helpful to have a separate chapter or section devoted to each criterion. Each of the chapters would include the research method, results, and evaluation related to the specific criterion the chapter addresses. Figure 25.3 shows how Phil used this pattern when reporting on the feasibility of using a new ingredient to make paper at one of his employer's paper mills.

To organize in these reader-centered ways, you need to remember that the superstructure for feasibility reports represents the elements of the report, not their outline. The elements may be combined and rearranged in many ways to meet your readers' needs.

Learn More

Remember that a superstructure is not an outline; you may combine the elements of a superstructure in many ways to serve your readers' needs (see page 557).

Drafting and Revising

Some people mistakenly think that drafting a report could or should wait until all the research has been completed. There are advantages, however, to drafting parts as soon as you are able. By drafting your criteria as you begin your research, you sharpen your sense of what you need to find. Likewise, drafting your methods section as you are designing your research enables you to step back to review and possibly refine your research plan. Drafting as you go also helps you avoid the all-too-common situation in which writers leave so much drafting until right before the deadline that they don't have time to revise and polish their reports.

CRAFTING THE MAJOR ELEMENTS OF A FEASIBILITY REPORT

Learn More

For additional advice on writing an introduction, see Chapter 10, "Beginning a Communication."

All of the detailed advice that Chapters 8 through 14 provide about drafting workplace communications applies to feasibility reports. The following sections tell how you can apply some of this advice most productively when writing this type of report.

Introduction

In the introduction to a feasibility report, you should answer your readers' question, "Why is it important for us to consider these alternatives?" The most persuasive way to answer this question is to identify the problem your feasibility report will help your readers solve or the goal it will help them achieve: to reduce the number of rejected parts, to increase productivity, and so on. Beyond that, your introduction should announce the alternative courses of action you studied and tell generally how you investigated them.

Consider, for example, the way Phil, a process engineer in a paper mill, wrote the introduction of a feasibility report he prepared. (Phil's entire report appears in Figure 25.5, pages 621–626.) Phil was asked to study the feasibility of substituting one ingredient for another in the furnish for one of the papers the mill produces (*furnish* is the combination of ingredients used to make the pulp for paper):

FIGURE 25.2

Feasibility of Switching to an Internet Telephone System at Whitney Hospital

I. Introduction

   A. Current system provides familiar services satisfactorily
   B. Internet system has features that increase efficiency

II. Infrastructure

   A. New phone on every desk (alternatives, criteria, method, evaluation)
   B. Server to support the system (alternatives, criteria, method, evaluation)
   C. Ethernet wiring (alternatives, criteria, method, evaluation)

III. Installation

   A. What needs to be done (criteria, method)
   B. Alternative 1: Use our IT department (description of this alternative)
   C. Alternative 2: Contract with an outside firm (description of this alternative)
   D. Evaluation of alternatives

IV. Training

   A. What needs to be done (criteria, method)
   B. Alternative 1: Use our IT department (description of this alternative)
   C. Alternative 2: Contract with an outside firm (description of this alternative)
   D. Evaluation of alternatives

V. Management of the Changeover

   A. Overview of changeover
   B. Alternatives for minimizing disruption to doctors, etc. (criteria, method, evaluation)
   C. Alternatives for maintaining full communication while changing (criteria, method, evaluation)

VI. Conclusions and Recommendations

FIGURE 25.2

Feasibility Report Whose Middle Sections Are Organized Around the Decisions the Readers Will Make

Megan and Rajiv investigated the feasibility of installing a new phone system. In the middle of their report, they devote one chapter to each of four major decisions their readers would need to make.

Introduction
They explain the reasons for considering a new phone system and describe the decisions to be made.

Overview of alternatives

Chapters II through V
If the hospital switches, it will need to make decisions in four areas: infrastructure, installation, training, and management of the changeover. Megan and Rajiv write a separate chapter for each of these decisions. In each of these chapters, they include the relevant criteria, methods, alternatives, and evaluation.

Conclusions and recommendations
They combined these two elements in a final chapter draws together the evaluations from the earlier chapters and recommends specific actions to the readers.

Learn More

For information on patterns for organizing comparisons, go to pages 242–245.

At present we rely on the titanium dioxide (TiO2) in our furnish to provide the high brightness and opacity we desire in our paper. However, the price of TiO2 has been rising steadily and rapidly for several years. We now pay roughly $1400 per ton for TiO2, or about 70¢ per pound.

Problem

FIGURE 25.3

Feasibility Report Whose Middle Sections Are Organized Around the Criteria the Readers Will Use to Evaluate the Alternatives

In this feasibility report, Phil organizes the central chapters around the criteria his readers will use to decide whether to change the extender in the paper mill where he works.

Introduction

Overview of alternatives

First three criteria
For each of the first three criteria, Phil describes the relevant criterion, method, results, and evaluation.

Fourth and fifth criteria
For the fourth and fifth criteria, the methods of gathering information are obvious, so Phil doesn't report them.

Sixth criterion

Conclusions and recommendation
Because these two elements are so brief, Phil combined them in one chapter.

Feasibility of Using Silicate Extenders for 36-Pound Book Paper

I. Introduction

    A. The cost of TiO2 is rising; some paper mills are using silicate extenders

    B. I evaluated two silicate extenders: Zenolux and Tri-Sil

II. Physical Properties

    A. Retention
        i. Criterion (named only: the readers need no explanation)
        ii. Method
        iii. Evaluation
    B. Opacity
        i. Criterion (named only; the readers need no explanation)
        ii. Method
        iii. Evaluation
    C. Brightness
        i. Criterion (named only; the readers need no explanation)
        ii. Method
        iii. Evaluation

III. Material Handling

    A. Storage
        i. Criterion
        ii. Evaluation
    B. Safety
        i. Criterion
        ii. Evaluation

IV. Cost

    A. Criteria
    B. Evaluation

V. Conclusions and Recommendation

Methods and Evaluation
Because he used different methods to gather information related to each of the criteria, Phil described the relevant method in the section on that criterion. Likewise, he presented his evaluations in terms of each criterion in the chapter or section that discussed it.

Some mills are now replacing some of the TiO2 in their furnish with silicate extenders. Because the average price for silicate extenders is only $500 per ton, well under half the cost of TiO2, the savings are very great.

To determine whether we could enjoy a similar savings for our 30-pound book paper, I have studied the physical properties, material handling requirements, and cost of two silicate extenders, Tri-Sil 606 and Zenolux 26 T.

Possible solution

What Phil did to investigate the possible solution

Generally, the introduction to a long feasibility report (and most short ones) should also include a preview of the main conclusions and, perhaps, the major recommendations. Phil included his major conclusion:

I conclude that one of the silicate extenders, Zenolux 26 T, looks promising enough to be tested in a mill run.

Phil's main point

As another example, consider the way Ellen wrote the introduction of a feasibility report she prepared for the board of directors of the bank where she works (see Figure 25.4). Ellen was asked to evaluate the feasibility of opening a new branch in a particular suburban community. She began by announcing the topic of her report:

This report discusses the feasibility of opening a branch office of Orchard Bank in Rolling Knolls, Tennessee.

Ellen's introduction

Then, after giving a sentence of background information about the source of the bank's interest in exploring this possibility, Ellen emphasized the importance of conducting a feasibility study:

In the past, Orchard Bank has approached the opening of new branches with great care, which is undoubtedly a major reason that in the twelve years since its founding it has become one of Tennessee's most successful small, privately owned financial institutions.

Ellen also included her major conclusion:

Overall, the Rolling Knolls location offers an enticing opportunity but would present Orchard Bank with some challenges it has not faced before.

She ended her introduction with a brief summary of her major recommendation:

We should proceed carefully.

The introduction of a feasibility report is often combined with one or more of the other six elements, such as a description of the criteria, a discussion of the method of obtaining facts, or an overview of the alternatives. It may also include the various kinds of background, explanatory, and forecasting information that may be found in the beginning of any technical communication (see Chapter 10).

Criteria

Criteria are the standards that you apply in a feasibility study to evaluate the alternative courses of action you are considering. For instance, to assess the feasibility of

FIGURE 25.4
How Ellen Used the
Superstructure for
Feasibility Reports

In her report on the feasibility
of opening a new branch of her
bank, Ellen mixes in several ways
the elements of the superstructure for feasibility reports.

Introduction
Ellen explains the importance
of considering the feasibility of
opening a new branch.

Overview of alternatives
For Ellen's study there are two
alternatives: Open a new branch
or don't open it. Because her
readers already know what it is
like to have no new branch, she
describes only what it would be
like to open one.

Evaluation of the alternatives
Ellen divides her evaluation of
the alternative of opening a new
branch into three sections, one
for each of her criteria:
• Marketability
• Ability to make a profit
• Costs

Criteria
Ellen identifies each of her criteria at the beginning of the appropriate section. For instance, at
the beginning of her section on
market analysis, she discusses her
criteria related to marketability.

**Feasibility of Opening a Branch of Orchard Bank
in Rolling Knolls, Tennessee**

I. Introduction

    A. Orchard Bank has built its success on carefully selecting locations
       for its offices
    B. A new branch office in Rolling Knolls has been suggested
    C. This report evaluates the proposed location

II. Proposed Location

    A. The branch would serve Rolling Knolls and Pickett
    B. The site selected is Lot P1-C in Rolling Knolls

III. Market Analysis

    A. This is a prosperous, growing area
    B. The population has two distinct groups
       1. Long-time residents of the area
       2. Newcomers, mainly young commuters
    C. The competition
       1. Local offices of banks and savings and loans
       2. Nashville banks used by many commuters
    D. Potential marketing strategies
       1. A two-pronged strategy for appealing to this two-part market
       2. The strategy would require Orchard to do some things it hasn't
          done before

IV. Financial Analysis

    A. The branch could become profitable after three years if a good
       marketing plan is established and followed
    B. Estimated deposits
    C. Estimated income and expenses

V. Fixed Asset Expenditures

VI. Conclusions and Recommendations

    A. Rolling Knolls offers opportunity but also risk
    B. Recommendations
       1. Hire marketing firm to help assess situation more precisely
       2. Take an option on the lot in Rolling Knolls

Method
Ellen explains each of her methods of gathering facts
at the beginning of each analysis. For instance, she
describes how she learned about the prosperity and
growth of the Rolling Knolls area at the beginning of
Section II.A.

Conclusions
Recommendations
Ellen combines these two elements of the
superstructure in her final section.

REGENCY INTERNATIONAL PAPER COMPANY

MEMORANDUM

FROM: Phil Hines, Process Engineer
TO: Jim Shulmann, Senior Engineer
DATE: December 13, 2010

**SUBJECT: FEASIBILITY OF USING SILICATE EXTENDERS
FOR 30-POUND BOOK PAPER**

Summary

I have investigated the feasibility of using a silicate extender to replace some of the TiO_2 in the furnish for our 30-pound book paper. Because the cost of the extenders is less than half the cost of TiO_2, we could enjoy a considerable savings through such a substitution.

The tests show that either one of the two extenders tested can save us money. In terms of retention, opacity, and brightness, Zenolux is more effective than Tri-Sil. Consequently, it can be used in smaller amounts to achieve a given opacity or brightness. Furthermore, because of its better retention, it will place less of a burden on our water system. With respect to handling and cost, the two extenders are roughly the same.

I recommend a trial run with Zenolux.

Introduction

At present we rely on the titanium dioxide (TiO_2) in our furnish to provide the high brightness and opacity we desire in our paper. However, the price of TiO_2 has been rising steadily and rapidly for several years. We now pay roughly $1400 per ton for TiO_2, or about 70¢ per pound.

FIGURE 25.5
Feasibility Report
(*Courtesy of Allen J. Hines*)

In this report, Phil reports on the feasibility of replacing one chemical with another in the "furnish" his employer uses to manufacture paper.

Phil includes a precise and informative subject line so his readers know exactly what his report is about.

Phil uses a 125-word summary of the entire report to enable his readers to understand the main points immediately.

In his summary, Phil emphasizes the content that will most interest his readers: his conclusion and recommendation.

In the first sentence of his introduction, Phil presents the background information his readers will need in order to understand why his report is important to them.

Phil identifies the specific problem his report addresses.

(continued)

FIGURE 25.5

Phil tells his readers about a possible solution that other paper mills have used: Replace TiO_2 with a silicate extender.

He describes what he did to investigate the feasibility of the possible solution.

Phil states his major conclusion and recommendation.

He identifies his criteria for evaluating the physical properties of the silicate extenders.

Phil helps his readers find key information by briefly stating the overall result of his tests of physical properties before going into the details of the tests.

Phil explains the importance of testing retention.

To build his readers' confidence in his results, Phil gives precise details about his method for testing retention.

He combines his results with his discussion of them. In this paragraph, he gives exact percentages.

Phil introduces his method for testing opacity.

Some mills are now replacing some of the TiO_2 in their furnish with silicate extenders. Because the average price for silicate extenders is only $500 per ton, well under half the cost of TiO_2, the savings are very great.

To determine whether we could enjoy a similar savings for our 30-pound book paper, I have studied the physical properties, material handling requirements, and cost of two silicate extenders, Tri-Sil 606 and Zenolux 26 T. I conclude that one of the silicate extenders, Zenolux 26 T, looks promising enough to be tested in a mill run.

Tests of Physical Properties

The three physical properties I tested are retention, opacity, and brightness. In all three areas, Zenolux is superior.

Retention

As with any ingredient in our furnish, we must be concerned with the proportion of a silicate extender that will be retained in the paper and the proportion that will be left in the water, where it is wasted and may cause problems in our water system. To test retention of the two silicate extenders, I made two dozen handsheets, each containing the equivalent of 3 grams of oven-dried pulp and 2 grams of oven-dried extender. By weighing the finished handsheets, I determined how much silicate extender had been lost from each.

The results showed that the average retention for Zenolux was 75%, whereas the average retention for Tri-Sil was 51%. Higher retention should result in higher opacity and brightness because more particles remain in the furnish to prevent clumping of the TiO_2.

Opacity

To determine the effectiveness of each extender in preventing light from passing through the paper, I conducted a two-stage test of opacity. First, I investigated the opacity of TiO_2, Tri-Sil, and Zenolux when each is used alone. To do that, I made the following sets of handsheets:

FIGURE 25.5
(continued)

Silicate Extenders   Page 3

> - 8 handsheets containing each of the following loadings of TiO_2: 2%, 4%, 6%, 8%, 10%, 12%, 14%, and 16%.
> - 8 handsheets containing each of the same loadings of Tri-Sil.
> - 8 handsheets containing each of the same loadings of Zenolux.
>
> I stored the handsheets at the standard conditions of 50% relative humidity and 23°C for 24 hours. Then, I found the TAPPI opacity of each handsheet. The results, given for the average opacity at each loading, are shown in Figure 1. Again, Zenolux is superior to Tri-Sil.

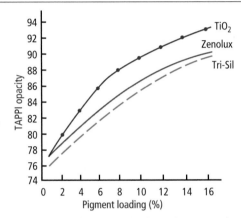

Figure 1. TAPPI Opacity for Various Loadings of TIO_2, Zenolux, and Tri-Sil.

In the second stage of the opacity test, I made two additional sets of handsheets:

- 5 handsheets with each of the following pigment loadings: 100% TiO_2 and 0% Tri-Sil; 75% TiO_2 and 25% Tri-Sil; 50% TiO_2 and 50% Tri-Sil; 25% TiO_2 and 75% Tri-Sil; and 0% TiO_2 and 100% Tri-Sil.
- 5 handsheets with each of the same proportions of TiO_2 and Zenolux.

As in his section on testing physical properties, Phil includes precise information about his testing method in order to persuade his readers that he used a sound method. To making reading easy, he presents some details in a bulleted list.

Phil refers his readers to a graph that presents the results of his test in a more understandable way than sentences would. However, he also states the main point he wants his readers to derive from the graph: Zenolux is superior.

He places the graph at the point in the report when his reader will want to look at it.

Again, Phil describes his method in detail, and again he uses a graph to present the results in a reader-centered way (next page). He also states the main point he wants his readers to see when they look at the graph.

(continued)

FIGURE 25.5
(continued)

Silicate Extenders   Page 4

As Figure 2 shows, Zenolux is again superior.

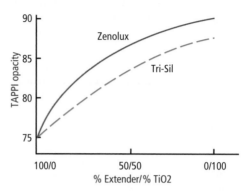

Figure 2. TAPPI Opacity for Various Ratios of Zenolux and Tri-Sil to TiO$_2$.

Brightness

Using the three sets of handsheets employed in the first-stage opacity tests, I calculated the GE brightness achieved by each of the three pigments. Figure 3 shows the results. Although not as bright as TiO$_2$, Zenolux is brighter than Tri-Sil.

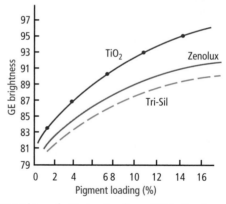

Figure 3. GE Brightness for Various Loadings of TiO$_2$, Zenolux, and Tri-Sil.

When describing his tests of brightness, Phil uses the same reader-centered strategies he employed in the section on opacity:

- He briefly but precisely describes his method.
- He presents his test results in graphs.
- He states in a sentence the main point readers should take away from each graph.

Learn More

To learn the design guidelines Phil followed to create these reader-centered graphs, see page 350.

FIGURE 25.5

Silicate Extenders   Page 5

Using the two sets of handsheets employed in the second-stage opacity tests, I examined the brightness of each of the ratios of TiO_2 to the extenders. The results, shown in Figure 4, indicate that, as expected, TiO_2 with Zenolux is brighter than TiO_2 with Tri-Sil.

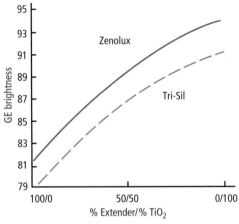

Figure 4. GE Brightness for Various Ratios of Zenolux and Tri-Sil to TiO_2.

Material Handling

With respect to material handling, I found no basis for choosing between Zenolux and Tri-Sil. Both are available from suppliers in Chicago. Both are available in dry form in bags and as a slurry in bulk hopper cars. Zenolux slurry is also available in 20,000-gallon tank cars, which provide a small savings, but we do not have the storage facility needed to receive it in this way.

Similarly, both silicate extenders are quite safe. Both are 100% pigment and both are chemically stable. Neither poses a hazard of fire or explosion and neither is hazardous when mixed with any other substance, whether liquid, solid, or gas. Both create the same effects in the event of overexposure: dehydration of the respiratory tract, eyes, and skin. These effects can be avoided by purchasing the extenders in slurry rather than dry form.

In the first sentence of his section on material handling, Phil states his main finding: There are no significant differences among the three extenders he tested.

Phil then provides details about each aspect of material handling that he investigated. In this way, he assures his readers that he didn't overlook any important consideration.

(continued)

FIGURE 25.5
(continued)

Phil realized that his readers will want to know how much a conversion to one of the silicate extenders will cost. Therefore, he provides not only the cost per ton of each item, but also considers how much of each the paper mill would need to buy in order to produce good-quality paper. In reaching this conclusion, Phil is building on the results of his tests of opacity and brightness.

Phil ends his report with the following items:
- His conclusion
- The results and explanation that support his conclusion
- His recommendation

Silicate Extenders   Page 6

Cost

Either silicate extender could save us money, and they both cost about the same: $421 per ton for Zenolux and $391 per ton for Tri-Sil. The $30 difference is only 7% of the cost of Zenolux. This savings may be offset by the larger amounts of Tri-Sil needed to achieve the same brightness and opacity. Thus, based on the information available at this time, it seems that cost is not a factor that would help us choose between the two extenders.

Conclusion and Recommendation

Although the two extenders are comparable in many ways, Zenolux appears to be the better choice for us. Both extenders could save us money and both are about the same in terms of handling and cost. However, Zenolux can be used in smaller amounts to obtain a given brightness and opacity. Furthermore, its higher retention rate will place less of a burden on our aging water system.

I recommend that we test Zenolux on our 30-pound paper machine.

opening the new branch office, Ellen used many criteria, including the existence of a large enough market, the possibility of attracting depositors away from competitors, the likelihood that profits on deposits at the branch would exceed the expenses of operating it, and the reasonableness of the financial outlay required to open the office. If she had found that the proposed branch failed to meet any of those criteria, she would have concluded that opening it was not feasible. Likewise, Phil evaluated the two silicate extenders by applying several specific criteria.

Two Ways of Presenting Criteria

There are two common ways of telling your readers what your criteria are:

- **Devote a separate section to identifying and explaining them.** Writers often do this in long reports or in reports in which the criteria themselves require extended explanation.
- **Integrate your presentation of them into other elements of the report.** Phil did this in the following sentence from the third paragraph of his introduction:

> To determine whether we could enjoy a similar savings for our 30-pound book paper, I have studied the physical properties, material handling requirements, and cost of two silicate extenders, Tri-Sil 606 and Zenolux 26 T.

Phil names his three criteria.

For each of the general criteria named in this sentence, Phil had some more specific criteria, which he described when he discussed his methods and results. For instance, at the beginning of his discussion of the physical properties of the two extenders, he named the three properties he evaluated.

Importance of Presenting Criteria Early

Whether you present your criteria in a separate section or integrate them into other sections, you should introduce them early in your report. There are three good reasons for doing this. First, because your readers know that the validity of your conclusion depends on the criteria you use to evaluate the alternatives, they will want to evaluate the criteria themselves. They will ask, "Did you take into account all the considerations relevant to this decision?" and "Are the standards you are applying reasonable in these circumstances?"

Second, your discussion of the criteria tells readers a great deal about the scope of your report. Did you restrict yourself to technical questions, for instance, or did you also consider relevant organizational issues such as profitability and management strategies?

The third reason for presenting your criteria early is that your discussion of the alternative courses of action will make much more sense to your readers if they know in advance the criteria by which you evaluated the alternatives.

Sources of Your Criteria

You may wonder how to come up with the criteria you will use in your study and report. Often, the person who asks you to undertake a study will tell you what criteria to apply. In other situations, particularly when you are conducting a feasibility study that requires technical knowledge that you have but your readers don't, your readers may expect you to identify the relevant criteria for them.

In either case, you are likely to refine your criteria as you conduct your study. The writing process itself can help you refine your criteria because as you compose you must think in detail about the information you have obtained and decide how best to evaluate it.

You may refine your criteria while you are writing your report.

Four Common Types of Criteria

As you develop your criteria, you may find it helpful to know that, at work, criteria often address one or more of the following questions:

- **Will this course of action really do what's wanted?** This question is especially common when the problem is a technical one: Will this reorganization of the department really improve the speed with which we can process loan applications? Will the new type of programming really reduce computer time?
- **Can we implement this course of action?** Even though a particular course of action may work technically, it may not be practical. For example, it may require overly extensive changes in operations, equipment, or materials that are not readily available, or special skills that employees do not possess.
- **Can we afford it?** Cost can be treated in several ways. You may seek an alternative that costs less than some fixed amount or one that will save enough to pay for itself in a fixed period (for example, two years). Or you may simply be asked to determine whether the costs are "reasonable."
- **Is it desirable?** Sometimes a solution must be more than effective, implementable, and affordable. Many otherwise feasible courses of action are rejected because they create undesirable side effects. For example, a company might reject a plan for increasing productivity because it would impair employee morale.

Ultimately, your selection of criteria for a particular feasibility study will depend on the problem at hand and on the professional responsibilities, goals, and values of the people who will use your report. In some instances, you will need to deal only with criteria related to the question, "Does it work?" At other times, you might need to deal with all the criteria mentioned above, plus others. No matter what your criteria, however, announce them to your readers before you discuss your evaluation.

Method

By explaining how you obtained your facts, you answer your readers' question, "Are your facts reliable?" That is, by showing that you used reliable methods, you assure your readers that your facts form a sound basis for decision making.

The source of your facts will depend on the nature of your study—library research, calls to manufacturers, interviews, meetings with other experts in your organization, surveys, laboratory research, and the like.

How much detail should you provide about your methods? That depends on your readers and the situation, but in every case your goal is to say enough to satisfy your readers that your information is trustworthy. For example, Ellen used some fairly technical procedures to estimate the amount of deposits that Orchard Bank could expect from a new branch in Rolling Knolls, Tennessee. However, because those procedures are standard in the banking industry and well-known to her readers, she did not need to explain them in detail.

In contrast, Phil provided very specific details about his methods of testing the extenders. Even a small mistake could produce inaccurate results that, in turn, could lead the paper mill to make a very expensive error. He knew, therefore, that his readers would want to review for themselves each step in his test procedure.

The best place for describing your methods depends partly on how many techniques you used. If you used only one or two—say, library research and interviews—you might describe each in a separate paragraph or section near the beginning of your report, perhaps in the introduction. On the other hand, if you used several different techniques, each pertaining to a different part of your analysis, you might describe each of them at the point where you present and discuss the results you obtained.

Of course, if your methods are obvious, you may not need to describe them at all. You must always be sure, however, that your readers know enough about your methods to accept your facts as reliable.

Overview of Alternatives

To understand your detailed evaluation of the alternatives, your readers must first understand what the alternatives are. Often, you can foster this understanding with a short explanation, perhaps only a sentence long. For instance, as a consultant to a chain of convenience stores you would need only a few words to enable your readers to understand whether increasing starting pay would help your company attract and retain skilled store managers.

However, you may sometimes need to explain the alternatives in detail. Although the hospital executives were familiar with what conventional phones could do, they didn't know anything about the capabilities of an Internet telephone system, such as their ability to forward voicemail messages to the e-mail account of the person called. Consequently, Megan and Rajiv described the Internet system in detail, highlighting features conventional telephones don't have that could help doctors, nurses, and other hospital personnel.

Evaluation of the Alternatives

The heart of a feasibility report is your evaluation of the alternatives you examined. As explained earlier in this chapter (page 616), a primary consideration is organizing your evaluation in the way that makes it easiest for your readers to use your evaluation as they make their decisions. Two additional strategies can also assist your readers.

Put Your Most Important Points First

Chapter 8's advice to begin each segment of your communication with the most important information applies to the segment in which you evaluate the alternatives, as well as to its parts. By presenting the most important information first, you save your readers the trouble of trying to figure out what generalizations they should draw from the details you are presenting. For instance, in her report to the bank, Ellen begins one part of her evaluation of the Rolling Knolls location by saying:

> The proposed Rolling Knolls branch office would be profitable after three years if Orchard Bank successfully develops the type of two-pronged marketing strategy outlined in the preceding section.

Learn More

See Guideline 3 in Chapter 8, page 214.

Beginning of Ellen's evaluation

Ellen then spends two pages discussing the estimates of deposits, income, and expenses that support her overall assessment.

Similarly, Phil begins one part of his evaluation of the silicate extenders in this way:

> With respect to material handling, I found no basis for choosing between Zenolux and Tri-Sil.

Beginning of Phil's evaluation

Phil then spends two paragraphs reporting the facts he has gathered about the physical handling of the two extenders.

Dismiss Obviously Unsuitable Alternatives

Lengthy discussions of clearly unsuitable alternatives provide no benefit to your readers. They only increase the time required to read your report.

If you discover that an alternative fails to meet one or more of the critical criteria, explain this fact briefly, providing only the details needed to persuade your readers that you were right to drop the alternative from consideration.

Provide this explanation in the introduction (when you are talking about your report's scope) or in the overview of the alternatives. If you delay the explanation to a later section, the mistaken thought that you have missed an important alternative may distract your readers as they read the rest of your report.

Conclusions

Your conclusions constitute your overall assessment of the feasibility of the alternative courses of action you studied. You might present your conclusions in two or three places in your report. You should certainly mention them in summary form near the beginning. If your report is long (say, several pages), you might also remind your readers of your conclusions at the beginning of the evaluation segment. Finally, you should provide detailed discussion of your conclusions in a separate section following your evaluation of the alternatives.

Recommendations

It is customary to end a feasibility report by answering the decision makers' question, "What do you think we should do?" Because you have investigated and thought about the alternatives so thoroughly, your readers will place special value on your recommendations. Depending on the situation, you might need to take only a single sentence or many pages to present your recommendations.

Sometimes your recommendations will pertain directly to the course of action you studied: "Do this," or "Don't do this." At other times you may perform a preliminary feasibility study to determine whether a certain course of action is promising enough to warrant a more thorough investigation. Ellen's report about opening a new bank office is of that type. She determined that there was a substantial possibility of making a profit with the new branch, but she felt the need for expert advice before making a final decision. Consequently, she recommended that the bank hire a marketing agency to evaluate the prospects.

Sometimes you may discover that you were unable to gather all the information you needed to make a firm recommendation. Perhaps your deadline was too short or your funds too limited. Perhaps you uncovered an unexpected question that needs further investigation. In such situations, you should point out the limitations of your report and let your readers know what else they should find out so they can make a well-informed decision.

SAMPLE FEASIBILITY REPORT

Figure 25.5 (pages 621–626) shows the feasibility report Phil wrote to help his employer decide whether to use a silicate extender at one of its paper mills. Phil's outline for this report is shown in Figure 25.3 (page 618).

WRITER'S GUIDES AND OTHER RESOURCES

Figure 25.6 (pages 631–632) presents a Writer's Guide for Revising Feasibility Reports that you can use in your course and on the job. At the website for this book, you can download a copy of this Writer's Guide as well as one for planning feasibility reports.

Note to the Instructor For an assignment involving an empirical research report, you can ask students to write on research they are conducting in their majors, or you can adapt the "Formal Report or Proposal" project given in Appendix B. Other projects and cases that involve report writing are included at the book's website. To tailor the writer's guides to your course, download them from the website.

WWW

To download Writer's Guides for planning and revising feasibility reports, go to Chapter 25 at **www.cengage.com/ english/anderson7e.**

FIGURE 25.6   **Writer's Guide for Revising Feasibility Reports**
To download a copy of this Writer's Guide as well as a Writer's Guide for Planning Feasibility Reports, go to Chapter 25 at www.cengage.com/english/anderson7e.

Writer's Guide
REVISING FEASIBILITY REPORTS

Does your draft include each of the elements needed to create a report that your readers will find to be usable and persuasive? Remember that some elements of the superstructure may be unnecessary for your specific readers and purpose and that the elements may be organized in various ways.

Introduction

☐ Identifies the action or alternatives you investigated

☐ Tells (or reminds) your readers why you conducted this study

☐ Persuades readers that the study is important to them

☐ States briefly your main conclusions and recommendations, if the readers would welcome or expect them at the beginning of the report

☐ Provides background information the readers will need or want

☐ Forecasts the rest of your report, if this would help your readers

Overview of Alternatives

☐ Presents a general description of each alternative

Criteria (often included in the introduction)

☐ Identifies the standards by which the action or alternatives were evaluated

☐ Focuses on criteria that are important to the reader

Method

☐ Tells the things your readers want to know about the way you obtained the facts and ideas presented in the report

(Continued)

FIGURE 25.6 *(Continued)*

Writer's Guide
REVISING FEASIBILITY REPORTS
(continued)

☐ Persuades the readers that this method would produce reliable results

Evaluation (usually the longest part of a feasibility report)

☐ Evaluates the action or alternatives in terms of the criteria

☐ Presents the facts and evidence that support each evaluative statement

Conclusions (often presented along with the discussion)

☐ Explains the significance—from your readers' viewpoint—of your facts and generalizations about them

☐ States the conclusions plainly

Recommendations

☐ Tells which course of action or alternative you recommend

☐ Makes the recommendations stand out prominently (for instance, by presenting them in a numbered list)

☐ Indicates how your recommendations are related to your conclusions and to your readers' goals

☐ Suggests some specific steps your readers might take to act on each of your recommendations, unless the steps will be obvious to the readers

Reasoning (See Chapter 5)

☐ States your claims and conclusions clearly

☐ Provides sufficient evidence, from the readers' viewpoint

☐ Explains, if necessary, the line of reasoning that links your facts and your claims

☐ Addresses any counterarguments or objections that your readers are likely to raise at any point in your report

☐ Avoids making false assumptions and overgeneralizing

Prose (See Chapters 4, 5, 8, and 9)

☐ Presents information in a clear, usable, and persuasive manner

☐ Uses a variety of sentence structures and lengths

☐ Flows in a way that is interesting and easy to follow

☐ Uses correct spelling, grammar, and punctuation

Graphics (See Chapter 13)

☐ Included wherever readers would find them helpful or persuasive

☐ Looks neat, attractive, and easy to read

☐ Referred to at the appropriate points in the prose

☐ Located where your readers can find them easily

Page Design (See Chapter 14)

☐ Looks neat and attractive

☐ Helps readers find specific information

Ethics

☐ Treats all the report's stakeholders ethically

☐ Presents all information accurately and fairly

26 | Writing Reader-Centered Progress Reports

CHAPTER OVERVIEW

A progress report is a report on work you have begun but not yet completed. The typical progress report is one of a series submitted at regular intervals, such as every week or month.

TYPICAL WRITING SITUATIONS

There are two basic types of progress reports. In the first, you tell your readers about your progress on one project. Lee is a geologist assigned to evaluate the site where a city would like to build a civic center but worries that the land may not be geologically suited for this construction. Every two weeks, Lee submits a progress report to his supervisor and the city engineer. His supervisor reads the report to make sure Lee is working in a technically sound manner. The city engineer uses the report to see that Lee is on schedule. She also looks for indications of the likely outcome of Lee's study. She could accelerate or halt other work as a result of Lee's preliminary findings.

Some progress reports concern a single project.

In the second type of progress report, you describe your work on all your projects. A chemist, Jacqueline manages a department that develops improved formulas for the laundry detergents her company manufactures—making them clean better, smell better, more economical to produce, and safer for the environment. Every month, she prepares a report that summarizes her department's progress on approximately 12 projects. Her readers include her immediate superiors, who want to be sure that her department's work is proceeding satisfactorily; researchers in other departments, who want to see whether her staff has made discoveries that can be used in other products, such as dishwashing detergents; and corporate planners, who want to anticipate changes in formulas that will require alterations in production lines or advertising.

Some progress reports concern all projects worked on during a single period.

As the examples of Lee and Jacqueline indicate, progress reports may differ in many ways: They may cover one project or many; they may be addressed to people inside the writer's organization or outside it; and they may be used by people who have a variety of reasons for reading them.

READERS' CONCERN WITH THE FUTURE

Despite their diversity, however, almost all progress reports have this in common: Their readers are primarily concerned with the *future*. That is, even though most progress reports talk primarily about what has happened in the past, their readers usually want that information so that they can plan for the future.

Although progress reports talk about the past, they are used to make decisions about the future.

For example, from your report they may be trying to learn the things they need to know in order to manage *your* project. They will want to know, for instance, what they should do (if anything) to keep your project going smoothly or to get it back on track. The progress reports written by Lee and Jacqueline are used for this purpose by some of their readers.

Other readers may be reading your progress reports to learn what they need to know in order to manage *other* projects. Almost all projects in an organization are interdependent with other projects. For instance, suppose you are conducting a marketing survey whose results will be used by another group as it designs an advertising campaign. If you have fallen behind your schedule, the other group's schedule may need to be adjusted.

Your readers may also be interested in the preliminary results of your work. Suppose, for instance, that you complete one part of a research project before you complete the others. Your readers may very well be able to use the results of that part immediately. The city engineer who reads Lee's reports about the possible building site wishes to use each of Lee's results as soon as it is available.

FEATURES OF PROGRESS REPORTS THAT HELP YOU

Two features of progress reports can be very helpful to you. First, whatever specific use readers want to make of the information in these reports, they are looking for the same basic kinds of information. Second, progress reports are prepared so often in the workplace that a set of conventions has developed concerning their content and organization. These conventions have proven successful in helping writers provide the information their readers want in a structure the readers find easy to use. The conventions constitute the *superstructure* or *genre* of the progress report. As you research, plan, and write progress reports in college and your career, you will be helped immensely by your knowledge of the readers' typical questions and the superstructure for answering them. The next two sections describe the questions and superstructure.

Learn More

For a detailed explanation of superstructures, see Guideline 3 in Chapter 4, page 105.

THE QUESTIONS READERS ASK MOST OFTEN

Your readers' concern with the implications of your progress for their future work and decisions leads them to want your progress report to answer the following questions.

- **What work does your report cover?** To be able to understand anything else in a progress report, readers must know what project or projects and what time period the report covers.
- **What is the purpose of the work?** Readers need to know the purpose of your work to see how it relates to their own responsibilities and to the other work, present and future, of the organization.
- **Is your work progressing as planned or expected?** Your readers will want to determine whether adjustments are needed in the schedule, budget, or number of people assigned to the project or projects on which you are working.

- **What results have you produced?** The results you produce in one reporting period may influence the shape of work in future periods. Also, even when you are still in the midst of a project, readers will want to know about any results they can use in other projects now, before you finish your overall work.
- **What progress do you expect during the next reporting period?** Again, your readers' interests will focus on such management concerns as schedule and budget and on the kinds of results they can expect.
- **How do things stand overall?** This question arises especially in long reports. Readers want to know what the overall status of your work is, something they may not be able to tell readily from all the details you provide.
- **What do you think we should do?** If you are experiencing or expecting problems, your readers will want your recommendations about what should be done. Your ideas about how to improve the project will also be welcome.

SUPERSTRUCTURE FOR PROGRESS REPORTS

Learn More

Remember that a superstructure is not an outline; you may combine the elements of a superstructure in many ways (see page 557).

The conventional superstructure for progress reports provides a very effective framework for answering your readers' questions about your projects:

SUPERSTRUCTURE FOR PROGRESS REPORTS	
REPORT ELEMENT	READERS' QUESTION
Introduction	What work does your report cover? What is the purpose of the work?
Facts and Discussion	
Past Work	Is your work progressing as planned or expected? What results have you produced?
Future work	What progress do you expect during the next reporting period?
Conclusions Recommendations	In long reports—How do things stand overall? What do you think we should do?

GUIDING YOU THROUGH THE PROCESS OF PREPARING PROGRESS REPORTS

When creating progress reports, writers perform the same activities as when they are preparing any workplace communication: defining the communication's objectives, conducting research, planning, drafting, and revising. The following sections suggest ways of performing these activities that are especially suited to writing progress reports.

Defining Your Report's Objectives

When defining the objectives of a progress report, focus on the specific ways your readers will use the information you provide. Are they principally interested in whether your projects are on schedule? Then let them know while trimming other information. Do they want to use your latest findings or results in their own work? Then describe your outcomes and accomplishments in enough detail to enable them to build their own work on yours.

Many organizations have strong conventions, even printed or online forms, for writing progress reports. When defining your progress report's objectives, learn your readers' expectations.

The Writer's Guide for Defining Your Communication's Objectives (page 70) will help you identify the additional information about your report's purpose, readers, and context that is needed to fully define its objectives.

Conducting Research

Employees rarely need to conduct research when writing progress reports. However, you may find it helpful to "research" your memory by making a few notes about what you want to say before you draft. These notes could help you focus on making your most important points directly and succinctly. Brevity is a virtue in progress reports.

Planning

One goal when writing progress reports is to finish your writing quickly (and well) so you can return to the work on which you are reporting. A key to writing rapidly is to plan a simple organization for providing the information your readers need. Here are three of the many organizational plans you might use.

If you are reporting on a single project, you might organize in this way.

I. What are the major events that happened during the most recent time period
II. What major events do I expect to occur during the next time period

Erin used this organization when reporting every two weeks on her assignment to increase efficiency at her employer's steel mills. Figure 26.1 shows the outline of her report for the two weeks in which she discovered employees at one plant were taking too long to change the blade on a hot saw, which is a machine that cuts the white-hot ingots of metal.

If you are working on two or more projects simultaneously, you can expand the organization Erin used.

FIGURE 26.1
Outline for Erin's Progress
Report on Increasing
Efficiency

This outline shows how Erin
organized a report in which she
described her progress on a project aimed at reducing the time
needed to change a saw blade
used in her employer's manufacturing process.

She devotes one section of her
report to each of the elements of
the superstructure for progress
reports.

Her full report is shown in
Figure 26.3 on page 643.

Introduction ——————
Erin describes the work covered in
her progress report, including the
+work's purpose. She identifies
the time period covered.

Past work ——————
Erin emphasizes her accomplishments, including her discovery
of a problem that is causing
inefficiency.

Future work ——————
She tells specifically what she
will do.

Conclusions ——————
Erin indicates that she's behind
schedule and asks for assistance.

Recommendations ——————
She describes actions that can be
taken immediately.

Progress on Efficiency Project

I. Introduction
    A. I am reporting on the efficiency project.
    B. This report covers my work for the past two weeks.
    C. I focused on the hot saw.

II. Past Work
    A. I discovered a problem with the location of the tools.
    B. I discovered a problem with the blades our supplier provides.

III. Future Work
    A. During the next two weeks, I will visit Winnipeg to check on
       their hot-saw process.
    B. I will also check on their loading dock procedures.

IV. Conclusions
    A. I'm behind schedule; please let Larry help me for a few days.
    B. I have two recommendations.
        1. Build a tool stand next to the hot saw.
        2. Look for a new supplier or get ours to meet our specifications.

I. What happened during the most recent time period
   A. Project A
   B. Project B
II. What's expected to happen during the next time period
   A. Project A
   B. Project B

A third option is to organize around your projects rather than time periods.

I. Work on Project A
   A. What happened during the last time period
   B. What I expect to happen during the next time period
II. Work on Project B
   A. What happened during the last time period
   B. What I expect to happen during the next time period

This organization works very well in reports where you have more than a few sentences to say about one or more of your projects because it keeps all the information on each project together, making the report easy for readers to follow. Lloyd organized this way when reporting on the progress of a large project he directed, which was to introduce a new line of high-fashion women's clothes (Figure 26.2).

Drafting and Revising

As mentioned, in routine progress reports, writing efficiently is a major goal. You can save time at drafting and revising by including only the most important points and by stating your information in a straightforward manner. No flourishes. No spin. This suggestion does not mean you should write carelessly. Review your draft carefully before submitting it.

CRAFTING THE MAJOR ELEMENTS OF A PROGRESS REPORT

All of the advice that Chapters 8 through 14 provide about drafting workplace communications applies to progress reports. The following sections tell how you can apply some of this advice most productively when writing this type of report.

Introduction

In the introduction to a progress report, you can address the readers' first two questions. You can usually answer the question, "What work does your report cover?" by opening with a sentence that identifies the project or projects your report concerns and the time period it covers.

Sometimes you will not need to answer the second question—"What is the purpose of the work?"—because all your readers will already be quite familiar with its purpose. At other times, however, it will be crucial for you to tell your work's purpose

Try This

Imagine that you are going to write a progress report about what you've done in the past week or few weeks on a personal effort or course project. To whom would you write? How would your reader use the information in your report? What would your main points be? Alternatively, imagine that you are writing the progress report to yourself.

Learn More

For additional advice about writing an introduction, see Chapter 10, "Beginning a Communication."

FIGURE 26.2
Outline of Lloyd's Progress
Report on Introducing a
New Line of Clothes

In this report, Lloyd tells his
readers about progress in several
related projects, devoting a separate section to each one.

Introduction
Lloyd describes the work covered
in his progress report, including
the work's purpose. He identifies
the time period covered.

Progress on Manufacturing
- **Past work**
  Lloyd focuses on a problem.
- **Conclusion**
  Lloyd explains the possible
  consequences of the problem.
- **Recommendation**
  He recommends actions to be
  taken in response to the manufacturing problem he has
  described. If his recommendations are accepted, they will
  be his future work.

Progress on three other projects
Lloyd briefly describes past work
on three other projects, emphasizing accomplishments. Because
there are no problems, he has no
need to make recommendations.
Similarly, because his readers
already know what he will be
doing, he does not need to describe his future work.

Conclusion
Lloyd summarizes by telling how
things stand overall.

Lloyd ends with a reference to his
recommendations in Section II.B.

Progress on Development of New Line of Clothes

I. Introduction
   A. I am reporting on progress during the past month on the High Look project.
   B. The report covers the activities of all the groups working under me to introduce our line of high-fashion clothes for women.

II. Manufacturing
   A. Huge problem
      1. Last week, garment workers at the Saucon factory went on a wildcat strike.
      2. Saucon is making 65% of our dresses and 45% of our blouses.
      3. Garments are going to arrive late at retail stores if the strike continues another two days; that would decrease sales by very large amounts.
      4. This morning, Saucon management predicted no quick settlement.
   B. Recommendations
      1. Contract with Webster Corporation to begin manufacturing the garments if we can get the fabrics.
      2. Immediately begin picking alternative fabrics.

III. Marketing
   A. Orders continue to arrive.
   B. We will probably make waves, if we can get our goods to market.

IV. Designs for Next Year
   A. Progress is on schedule.
   B. Designers are playing with ideas for High Look.

V. All Other Areas: Proceeding Well

VI. Conclusion
   A. Although things are going well in most areas, we face a major crisis if our initial offering fails to sell because our goods don't make it to retailers on time.
   B. Let's do something.

---

because your readers will include people who don't know or may have forgotten it. You are especially likely to have such readers when your progress reports are widely circulated in your own organization or when you are reporting to another organization that has hired your employer to do the work you describe. You can usually explain its purpose most helpfully by describing the problem that your work will help your readers solve.

> This report covers the work done on the Focus Project from July 1 through September 1. Sponsored by the U.S. Department of Energy, the Focus Project aims to overcome the technical difficulties encountered in constructing photovoltaic cells that can be used to generate commercial amounts of electricity.

Project and period covered

Purpose of project

Of course, your introduction should also provide any background information your readers will need in order to understand the rest of your report.

Facts and Discussion

In the discussion section of your progress report, you should answer these readers' questions: "Is your work progressing as planned or as expected?" "What results have you produced?" and "What progress do you expect during the next reporting period?"

Because the work to be accomplished during each reporting period is usually planned in advance, you can indicate your progress by comparing what happened with what was planned. Where there are significant discrepancies between the two, your readers will want to know why. The information you provide about the causes of problems will help your readers decide how to remedy them. It will also help you to explain any recommendations you make later in your report.

How much information should you include in your progress reports? Generally, readers prefer brief reports. Although you need to provide your readers with specific information about your work, don't include details unless they will help readers decide how to manage your project or unless you believe readers will be able to make immediate use of them. As you work on a project, many minor events will occur, and you will have lots of small setbacks and triumphs along the way. Avoid discussing such matters. No matter how important these details may be to you, they are not likely to be interesting to your readers. Stick to the information your readers can use.

Conclusions

Your conclusions are your overall views on the progress of your work. In short progress reports, there may be no need to include any conclusions, but if your report covers several projects or tasks, conclusions may help your readers understand the general state of your progress.

Often, you can help your readers by stating your conclusions near the beginning of your progress report, for instance in the introduction. This location alerts readers right away that the project is on schedule or that work has fallen behind.

Recommendations

Your readers will want your thoughts about how to improve the project or increase the value of its results. Your recommendations might be directed at overcoming a difficulty you have experienced or anticipate encountering in the future. Or they might be directed at refocusing or otherwise altering your project.

Place your recommendation where they will be most helpful to your readers. Often that location will be the end. Erin used this placement in her report on her efficiency project for the steel mill (Figure 26.3, page 643). If you are reporting on a project with parts, you may be able to help your readers by placing your recommendation in the section on the part that has the problem. For example, when reporting on progress at introducing a new clothing line, Lloyd cited a problem with manufacturing and placed his recommendation for manufacturing at that same location (Figure 26.2, page 640).

TONE IN PROGRESS REPORTS

You may wonder what tone to use in the progress reports you prepare. Generally, you want to persuade your readers that you are doing a good job. That is especially likely when you are new on the job and when your readers might discontinue a project if they feel that it isn't progressing satisfactorily.

Because of this strong persuasive element, some people adopt an inflated or highly optimistic tone. However, this sort of tone can lead to difficulties. It might lead you to make statements that sound more like an advertising claim than a professional communication. Such a tone is more likely to make your readers suspicious than agreeable. Also, if you present overly optimistic accounts of what can be expected, you risk creating an unnecessary disappointment if things don't turn out the way you seem to be promising. And if you consistently turn in overly optimistic progress reports, your credibility with your readers will quickly vanish.

In progress reports, it's best to be straightforward about problems so that your readers can take appropriate measures to overcome them and so they can adjust their expectations realistically. You can sound pleased and proud of your accomplishments without exaggerating them.

Sample Progress Report

Figure 26.3 shows the progress report in which Erin reported on her investigation of the excessive time taken to change the hot saw at one of her employer's steel mills. The outline for Erin's report is shown in Figure 26.1 (page 638).

WWW

To download Writer's Guides for planning and revising progress reports, go to Chapter 26 at **www.cengage.com/english/anderson7e.**

WRITER'S GUIDES AND OTHER RESOURCES

Figure 26.4 (pages 644–645) presents a Writer's Guide for Revising Progress Reports that you can use in your course and on the job. At the website for the book, you can download a copy of this Writer's Guide as well as one for planning progress reports. The website has additional resources to help you write effective reports.

FIGURE 26.3
Erin's Progress Report on
Increasing Efficiency

Memorandum

To	Amad Rajani
From	Erin Malene
Date	July 10, 2010
Subject	Progress Report on Efficiency Project

In my efficiency project, I've concentrated during the past two weeks on studying the amount of time required to change the hot-saw blade. The Winnipeg plant shuts down its production for an average of 5 minutes each time it makes this change. In contrast, we stop production an average of 14 minutes.

Results from Last Two Weeks
By watching several blade changes and by talking with the workers, I've identified two problems. First, the tools are stored in a workroom 30 yards from the hot saw. When an extra tool is needed, minutes are lost while someone retrieves it. Second, the contractor who supplies the blades occasionally makes the mounting holes too large, so extra time is needed to insert shims.

Plans for Next Two Weeks
During the next two weeks, I will travel to Winnipeg, where I will observe their blade changes and also examine their procedures on the loading dock. I was scheduled to make that trip last week but was unable to do so because I had to help solve the boil-over problem in Building 3.

Conclusion
I'm a week behind schedule, so it would be helpful if Larry could assist me in the lab for a few days.

In the meantime, here are two actions we can take right now.

1. Build a tool stand next to the hot saw with all the equipment.
2. Begin looking for a new supplier or gain our current supplier's commitment to meet our specifications.

In this report, Erin describes her progress on a project aimed at reducing the time needed to change a saw blade used in her employer's manufacturing process.

An outline of her report is shown in Figure 26.1 on page 638.

Introduction
Erin reminds her reader of the purpose of her work.

She tells the time period covered in this report, and she describes the work she has done in this period.

Past work
Erin emphasizes her accomplishments, including her discovery of a problem that is causing inefficiency. By sharing this information, she enables her reader to begin considering solutions immediately, even though Erin has more work to do on her overall project.

Future work
She tells specifically what she will do, and she provides information that explains why she is behind schedule.

Conclusions
Erin indicates that she's behind schedule and asks for assistance.

Recommendations
She describes actions that can be taken immediately.

Note to the Instructor   For an assignment involving a progress report, see the Progress Report project given in Appendix B. Other projects and cases that involve report writing are included at the book's website. To tailor the Writer's Guides to your course, download them from the website.

FIGURE 26.4 **Writer's Guide for Revising Progress Reports**

 To download a copy of this Writer's Guide as well as a Writer's Guide for Planning Progress Reports, go to Chapter 26 at www.cengage.com/english/anderson7e.

Writer's Guide
REVISING PROGRESS REPORTS

Does your draft include each of the elements needed to create a report that your readers will find to be usable and persuasive? Remember that some elements of the superstructure may be unnecessary for your specific readers and purpose and that the elements may be organized in various ways.

Introduction

☐ Identifies the work your report covers

☐ Indicates the purpose of the work, if the readers need to be reminded

☐ Identifies the period covered

☐ Forecasts the rest of your report, if doing so would help your readers

Body of Report

☐ States whether your work on each of your tasks is progressing as planned with respect to schedule, budget, or other concerns of your readers

☐ Tells what you accomplished on each of your major tasks during the period covered by the report

☐ Identifies planned work that is not complete

☐ Reports results or accomplishments that your readers would like to know about immediately

☐ Identifies any significant problems your readers would want to know about

☐ Identifies specific tasks you will be performing during the next reporting period

☐ Describes the progress you expect to achieve during the next period

☐ Identifies any upcoming problems that your readers should know about

☐ Identifies any help you feel you need

Conclusion

☐ Includes final statement that lets your readers know how things stand overall

Recommendations

☐ Indicates any actions you think your readers should take

Prose (See Chapters 4, 5, 8, and 9)

☐ Presents information in a clear, usable, and persuasive manner

☐ Uses a variety of sentence structures and lengths

FIGURE 26.4   *(continued)*

Writer's Guide
REVISING PROGRESS REPORTS
(continued)

☐   Flows in a way that is interesting and easy to follow
☐   Uses correct spelling, grammar, and punctuation

Graphics (See Chapter 13)
☐   Included wherever readers would find them helpful or persuasive
☐   Looks neat, attractive, and easy to read
☐   Referred to at the appropriate points in the prose
☐   Located where your readers can find them easily

Page Design (See Chapter 14)
☐   Looks neat and attractive
☐   Helps readers find specific information

Ethics
☐   Treats all the report's stakeholders ethically
☐   Presents all information accurately and fairly

27 | Writing Reader-Centered Instructions

Instructions come in many lengths, shapes, and levels of complexity. They range from the terse directions on a shampoo bottle ("Lather. Rinse. Repeat.") to the huge manuals that are hundreds or thousands of pages long for servicing airplane engines, managing large computer systems, and performing biomedical procedures.

Although some instructions are prepared by professional writers and editors, most other employees also need to prepare instructions. Whether you are developing a new procedure, training a new coworker, or preparing to leave for vacation, you may need to provide written directions to someone else. You may write for people who will read your instructions on paper, a computer monitor, a telephone screen, or other electronic device.

This chapter provides advice that you will find valuable regardless of the subject or size of the instructions you write. The advice given in the first part of this chapter applies equally to instructions written for paper and screen. A special section at the end of the chapter provides additional suggestions for instructions that will be delivered digitally as a website or as a digital movie.

WWW

For additional examples and advice for writing instructions, go to Chapter 27 at **www .cengage.com/english/ anderson7e.**

FEATURES OF INSTRUCTIONS THAT HELP YOU

Two features of instructions can be very helpful to you. First, no matter what procedure they want help with, readers want the same basic kinds of information. Second, instructions are prepared so often in the workplace that a set of conventions has developed concerning their content, organization, and other key elements. These conventions have proven successful in helping writers provide the information their readers want in a structure the readers find easy to use. The conventions constitute the *superstructure* or *genre* of instructions. As you research, plan, and write instructions in college and your career, you will be helped immensely by your knowledge of the readers' typical questions and the superstructure for answering them. The next two sections describe the questions and superstructure.

Learn More

For a detailed explanation of superstructures, see Guideline 3 in Chapter 4, page 105.

THE QUESTIONS READERS ASK MOST OFTEN

Readers read instructions in many different ways. Some follow the directions meticulously, concentrating on every word. Others look at the directions only if they get stumped while relying solely on their experience and intuition. Whether they read every word or look only occasionally at instructions, the questions readers ask are almost always versions of the following six.

Instructions must meet the needs of readers performing vastly different tasks in a wide range of settings and circumstances. To keep both hands free, astronaut Kathryn Thornton has her instructions strapped to her arm.

- **What will these instructions help me do?** Some readers will ask this question exactly as it reads. When others use these or similar words, they are asking, "Do I really have to read these?"
- **Is there anything special I need to know to be able to use these instructions effectively?**
- **If I'm working with equipment, where are the parts I need to use?**
- **What materials, equipment, and tools do I need?**
- **Once I'm ready to start, what—exactly—do I do?**
- **Something isn't working correctly. How do I fix it?**

SUPERSTRUCTURE FOR INSTRUCTIONS

The superstructure for instructions includes five elements that answer the six questions readers ask most often.

The simplest instructions contain only the directions. More complex instructions contain some or all of the other four elements listed below. And some instructions also include such additional elements as covers, title pages, tables of contents, appendixes, lists of references, glossaries, lists of symbols, and indexes.

To determine which elements to include in any instructions you write, follow this familiar advice: Consider your readers' aims and needs as well as their characteristics that will shape the way they read and respond to your communication.

SUPERSTRUCTURE FOR INSTRUCTIONS

TOPIC	READERS' QUESTIONS
Introduction	What will these instructions help me do? Is there anything special I need to know to be able to use these instructions effectively?
Description of the equipment	If I'm working with equipment, where are the parts I need to use?
List of materials and equipment needed	What materials, equipment, or tools do I need?
Directions	Once I'm ready to start, what—exactly—do I do?
Troubleshooting	Something isn't working correctly. How do I fix it?

GUIDING YOU THROUGH THE PROCESS OF PREPARING INSTRUCTIONS

When creating instructions, writers perform the same activities as when they are preparing any workplace communication: defining the communication's objectives, conducting research, planning, drafting, and revising. The following sections describe ways to perform each activity in ways especially suited to this special kind of communication.

Defining Your Communication's Objectives

The overall usability objective for instructions is obvious: to enable readers to perform a procedure correctly and efficiently. By thinking about what people do while reading instructions, you can define the instructions' objectives more specifically. People read a step, do a step, read the next step, and do the step. Your instructions' objectives should include helping readers perform each step quickly and accurately. Each time they complete a step, they must find their place again in the instructions. Your instructions' objectives should include helping readers find that place quickly. People also use instructions as reference sources, looking for a certain part without wanting to use the other parts. Another usability objective for your instructions is helping them locate the part they want very rapidly.

Instructions' persuasive objectives are much less obvious than their usability objectives, but they are equally important. Many people dislike reading instructions. They want to start right in on the task without taking time to look at the instructions. Thus, one persuasive objective for instructions is to persuade people to read them at all. When people do read instructions, they can be impatient and easily frustrated. Another persuasive objective is to entice readers to look back at the page after they've looked away to perform a step. If your instructions are for a product made by your employer, they will have a third persuasive objective: to persuade readers to feel so good about the product that they will buy from your employer again and recommend that others do likewise.

The Writer's Guide for Defining Your Communication's Objectives on page 70 will help you identify the additional information about your instructions' purpose, readers, and context that is needed to fully define their objectives.

> **Learn More**
>
> For detailed advice about defining your instructions' objectives, see Chapter 3.

Conducting Research

The amount of research you'll have to do when writing instructions can vary widely. Sometimes, you'll know the procedure and readers so well that no research will be required at all.

At other times, you may be asked to write instructions for a procedure you know little or nothing about. This is often the case for professional technical communicators but also happens to engineers, scientists, and specialists in other fields. Chapter 6 provides detailed advice about conducting research. When you are relying solely

> **Learn More**
>
> For detailed advice on conducting research, see Chapter 6. For instructions, interviewing is very commonly the main research method used.

on your own memory, you might conduct a simple form of research that checks on your memory. Write down all the steps in the process and then perform the process by following the steps you listed. You may be surprised at how many steps you didn't remember to write down even though you haven't forgotten to perform them.

Planning

For the instructions to achieve their usability and persuasive objectives, three features of instructions must work together harmoniously. For all three, planning overlaps with drafting.

Learn More

For advice about how to group the steps in your process, see the discussion of segmenting, page 248.

- **Organization of the directions.** By organizing the directions hierarchically, you can help readers find the next step as they look back at the instructions after completing the previous step. This organization can also help them find particular information when consulting the instructions as a reference document.

  To create a hierarchical organization, begin by listing all the steps in the process. Next, check the list for thoroughness as described in the previous paragraph. Then, group related steps under headings, such as "preparing the equipment," "using the equipment," and "cleaning up." If the instructions are long, shorter groups can be gathered into larger ones.

Learn More

For advice about creating graphics, see Chapter 13.

- **Graphics.** For many purposes, well-designed graphics are even more effective than words. Words cannot show readers where the parts of a machine are located, how to grasp a tool, or what the result of a procedure should look like. Graphics are especially helpful in instructions for readers who speak languages other than your own. Sometimes graphics alone can convey all the information your readers need (see Figure 13.1 on page 332). Look actively for places where adding a drawing, diagram, photo, or other graphic would make your directions easier for your readers to understand. Chapter 13 provides suggestions for designing graphics that your readers will welcome.

Learn More

For advice about designing pages and screens, see Chapter 14. Chapter 20 provides additional advice about screen design.

- **Page design.** Strategically designed pages can help you and your readers in several ways. Page design can help readers find their places as they bounce back and forth between reading steps and performing them. Good page design helps readers see the connections between related blocks of information, such as a written direction and the drawing that accompanies it. An attractive design can entice readers' attention back to instructions they would otherwise choose to ignore. Chapters 14 and 20 can guide you through the process of designing effective printed and on-screen pages.

Learn More

For advice about writing clearly and succinctly, see Chapter 9.

Drafting and Revising

Later in this chapter, you will find suggestions for drafting each element of your instructions. The following advice applies to *all* elements: Write clearly and succinctly. Choose words that convey your meaning clearly. Construct sentences your readers will comprehend effortlessly. Use as few words as possible. More words make more work for your readers and increase the chances your readers will stop reading what you've written.

When you want to find ways to revise a draft of instructions, nothing beats watching members of your target audience using the draft to perform the procedure. Where you see them succeed with some steps, you know that part of your draft is effective. Where they have problems, you have an opportunity to improve. Chapter 16 guides you through the process of planning, conducting, and interpreting the results of user tests. When testing, remember to evaluate your draft's ability to achieve its persuasive objectives as well as its usability objectives.

Learn More

To learn how to conduct user tests, see Chapter 16.

CRAFTING THE MAJOR ELEMENTS OF INSTRUCTIONS

All the advice about drafting provided in Chapters 8 through 14 can help you write effective instructions. The following sections supplement that advice with suggestions for writing the five elements of the superstructure for instructions.

Introduction

Introductions should be as short as possible—or nonexistent. Many instructions don't need one. The title alone provides all the introductory information readers require. See Figure 27.4 (page 660). On the other hand, readers sometimes do need information up front. The following sections describe the eight elements most commonly included in introductions, together with suggestions for deciding whether your readers need each of them.

Subject

As mentioned above, often the title will fully convey the subject of your instructions. Especially in longer instructions, however, you may need to announce the subject in an introduction. Here is the first sentence from the 50-page operator's manual for a 10-ton machine used in the manufacture of automobile and truck tires:

This manual tells you how to operate the Tire Uniformity Optimizer (TUO).

Opening sentence that announces the subject

The first page of a manual for a popular desktop publishing program reads:

The Microsoft Word User's Guide contains detailed information about using Microsoft® Word 2007 for Windows™ and Microsoft Word 2008 for Mac.

Purpose of the Procedure

If the purpose of the procedure your instructions describe isn't obvious from the title, announce it in the introduction. You may be able to convey your instructions' aim by listing the major steps in the procedure or the capabilities of the equipment whose operation you are describing. Here is the second sentence of the manual for the Tire Uniformity Optimizer:

Depending upon the options on your machine, it may do any or all of the following jobs:

- Test tires
- Find irregularities in tires
- Grind to correct the irregularities, if possible
- Grade tires
- Mark tires according to grade
- Sort tires by grade

Intended Readers

When they pick up instructions, people often want to know whether the instructions are directed to them or to people who differ from them in interests, responsibilities, level of knowledge, or some other variable.

Sometimes, they can tell merely by reading the instructions' title. For instance, the operator's manual for the Tire Uniformity Optimizer is obviously addressed to people hired to operate that machine.

In contrast, people who consult instructions for a computer program may wonder whether the instructions assume that they know more (or less) about computers than they actually do. In such situations, answer their question in your introduction. Readers who don't already possess the required knowledge can then seek help or acquire the necessary background.

Scope

By stating the scope of your instructions, you help readers know whether the instructions contain directions for the specific tasks they want to perform. The manual for the Tire Uniformity Optimizer describes the scope of its instructions in the third and fourth sentences:

This manual explains all the tasks you are likely to perform in a normal shift. It covers all of the options your machine might have.

A manual for Microsoft Windows NT, an operating system, describes its scope in this way:

In Part I, you'll learn the basic features of Windows NT 4.0, including the new Windows 2000 interface. Designed to get you up-to-speed quickly and easily, Part I provides the step-by-step procedures you'll need to get started. Part II lists the system requirements for running this new version, and then guides you through installing this new operating system.

Organization

By explaining how the instructions are organized, an introduction can help readers understand the overall structure of the tasks they will perform and locate specific pieces of information without having to read the entire set of instructions.

Often introductions explain scope and organization together. If you look back at the statement of scope from the Microsoft Windows NT manual, you will see that it also describes the manual's organization: It announces that the manual is organized into two parts, each with two types of content.

Similarly, the introduction to the Tire Uniformity Optimizer devotes several sentences to explaining that manual's organization, and this information also fills out the readers' understanding of the manual's scope.

> The rest of this chapter introduces you to the major parts of the TUO and its basic operation. Chapter 2 tells you step by step how to prepare the TUO when you change the type or size of tire you are testing. Chapter 3 tells you how to perform routine servicing, and Chapter 4 tells you how to troubleshoot problems you can probably handle without needing to ask for help from someone else. Chapter 5 contains a convenient checklist of the tasks described in Chapters 3 and 4.

Paragraph describing a manual's organization

Conventions

If your instructions use abbreviations or conventions that the reader needs to know in order to interpret the directions correctly, explain them in the introduction. For instance, the introduction for a manual for operating a machine for harvesting corn says, "Right-hand and left-hand sides are determined by facing the direction of forward travel."

Motivation

As pointed out above (and as you may know from your own experience), some people are tempted to toss instructions aside and rely on their common sense. A major purpose of many introductions is to persuade readers to read the instructions. You can accomplish this goal by employing an inviting and supportive tone and by creating an attractive design. You can also include statements that tell readers directly why it is important to pay attention to the instructions. The following example is from instructions for a ceiling fan that purchasers install themselves.

> We're certain that your Hampton Bay fan will provide you with many years of comfort, energy savings, and satisfaction. To ensure your personal safety and to maximize the performance of your fan, please read this manual.

Statement of scope

Motivation to read the instructions

Safety

Your readers depend on you to prevent them from taking actions that could spoil their results, damage their equipment, or cause them injury. Moreover, product liability laws require companies to pay for damages or injuries that result from inadequate warnings in their instructions.

To satisfy your ethical and legal obligations, you must provide prominent, easy-to-understand, and persuasive warnings. If a warning concerns a general issue that covers the entire set of instructions (e.g., "Don't use this electrical tool while standing on wet ground"), place it in your introduction. If it pertains to a certain step, place it before that step. The following principles apply to warnings in either location:

- **Make your warnings stand out visually.** Try printing them in large, bold type and surrounding them with a box. Sometimes, writers use the following international hazard alert symbol to draw attention to the warning.

You may also include an icon to convey the nature of the danger. Here are some icons developed by Westinghouse.

Electrical Shock   Fire   Eye Protection

- **Place your warnings so that your readers will read them before performing the action the warnings refer to.** It won't help your readers to discover the warning after they've performed the step and the damage has been done.
- **State the nature of the hazard and the consequences of ignoring the warning.** If readers don't know what could happen, they may think that it's not important to take the necessary precautions.
- **Tell your readers what steps to take to protect themselves or avoid damage.**

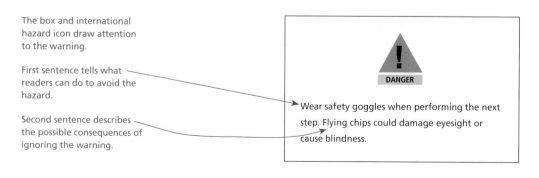

The box and international hazard icon draw attention to the warning.

First sentence tells what readers can do to avoid the hazard.

Second sentence describes the possible consequences of ignoring the warning.

DANGER

Wear safety goggles when performing the next step. Flying chips could damage eyesight or cause blindness.

Sample Introductions

Figure 27.1 shows the introduction to the instruction manual for the Tire Uniformity Optimizer. The introduction to another manual appears in Figure 10.2 (page 299).

Notice that the manual for the Tire Uniformity Optimizer uses the word *Introduction* and the introduction to the Detroit Diesel Engine Series 53 manual (Figure 10.2) is headed "General Information." The material that this chapter refers

Chapter 1—Introduction

This manual tells you how to operate the Tire Uniformity Optimizer (TUO) and its controller, the Tire Quality Computer (TQC). The TUO has many options. Depending upon the options on your machine, it may do any or all of the following jobs:

- Test tires
- Find irregularities in tires
- Grind to correct the irregularities, if possible
- Grade tires
- Mark tires according to grade
- Sort tires by grade

This manual explains all the tasks you are likely to perform in a normal shift. It covers all of the options your machine might have.

The rest of this chapter introduces you to the major parts of the TUO and its basic operation. Chapter 2 tells you step-by-step how to prepare the TUO when you change the type or size of tire you are testing. Chapter 3 tells you how to perform routine servicing, and Chapter 4 tells you how to troubleshoot problems with the TUO. Chapter 5 contains a convenient checklist of the tasks described in Chapter 3.

Major Parts of the Tire Uniformity Optimizer

You can find the major parts of the TUO by looking at Figure 1-1. To operate the TUO, you will use the Operator's Control Panel and the Computer Panel.

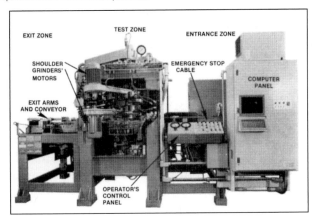

Figure 1-1. *Overview of the TUO and TQC.*

Page 1

The first sentence identifies the **subject** of the manual.

The second sentence and list identify the **purposes of the procedures** that can be performed by following the instructions.

This sentence describes the **scope** of the manual: all of the procedures the reader is likely to perform during a normal shift.

This paragraph describes the **organization** of the manual.

The photograph and labels provide readers with a **description of the equipment** that enables them to locate all the major parts they will have to find while following the instructions. More detailed photos are provided later in the manual to guide the reader when using the Operator's Control Panel and other parts.

The manual presents **safety** warnings and information to **motivate** readers to follow certain parts of the procedures at the appropriate places later in the manual. This manual does not use any **conventions** that need to be explained in the introduction.

to as the *Introduction* is called many other names in other instructions. Sometimes, it is given no title at all.

Description of the Equipment

Learn More

To describe equipment, use the pattern for describing an object (pages 246–247) or use a photograph or drawing (pages 368–371).

To be able to operate or repair a piece of equipment, readers need to know the location of its parts. Sometimes, they need to know their functions as well. Instructions often include a description of the equipment to be used, usually by including a labeled photograph or drawing of it. For example, the first page of the manual for the Tire Uniformity Optimizer displays a labeled photograph of the machine. In some instructions, such illustrations are accompanied by written explanations of the equipment and its parts.

List of Materials and Equipment Needed

Some procedures require materials or equipment that readers wouldn't normally have at hand. If yours do, include a list of these items. Present the list *before* giving your step-by-step instructions. This will save your readers from the unpleasant surprise of discovering that they cannot go on to the next step until they have gone to the shop, supply room, or store to obtain an item that they didn't realize they would need.

Directions

At the heart of a set of instructions are the step-by-step directions that tell readers what to do. The following sections describe strategies for writing directions your readers will find easy to understand and use. Figure 27.2 illustrates much of this advice.

Write Each Direction for Rapid Comprehension and Immediate Use

Readers want to understand as quickly as possible what they should do next.

1. **In each direction give readers only enough information to perform the next step.** If you give more, they may forget some or become confused.
2. **Present the steps in a list.** A list format helps readers see each step as a separate action.
3. **Use the active voice and the imperative mood.** Active, imperative verbs give commands: "*Stop* the engine." (This statement is much simpler than "The operator should then stop the engine.")
4. **Highlight key words.** In some instructions, a direction may contain a single word that conveys the critical information. You can speed the readers' task by using bold, all-capital letters or a different typeface to make this word pop off the page. Example: Press the **RETURN** key.

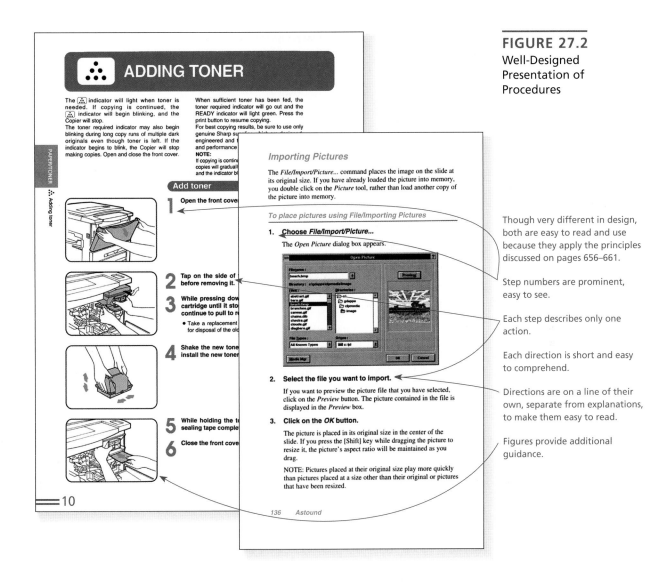

FIGURE 27.2
Well-Designed
Presentation of
Procedures

Though very different in design, both are easy to read and use because they apply the principles discussed on pages 656–661.

Step numbers are prominent, easy to see.

Each step describes only one action.

Each direction is short and easy to comprehend.

Directions are on a line of their own, separate from explanations, to make them easy to read.

Figures provide additional guidance.

Help Readers Locate the Next Step Quickly

There are many ways you can help your readers as they turn their eyes away from the task and back to your text:

1. **Number the steps.** With the aid of numbers, readers will not have to reread earlier directions to figure out which one they last read.

2. **Put blank lines between steps.** This white space helps readers pick out a particular step from among its neighbors.

3. **Give one action per step.** It's easy for readers to overlook a direction that is tucked in with another direction rather than having its own number.

4. Put step numbers in their own column. Instead of aligning the second line of a direction under the step number, align it with the text of the first line. Not this:

Step number is obscured.

> **2.** To quit the program, click the CLOSE button in the upper right-hand corner of the window.

But this:

Step number is in its own column.

> **2.** To quit the program, click the CLOSE button in the upper right-hand corner of the window.

Within Steps, Distinguish Actions from Supporting Information

When actions don't stand out from supporting information, readers can make errors.

1. Present actions before responses. As the following example shows, you make reading unnecessarily difficult if you put the response to one step at the beginning of the next step.

The computer response obscures the action to be performed.

> **4.** Press the RETURN key.
> **5.** The Customer Order Screen will appear. Click on the TABS button.

Instead, put the response after the step that causes it.

Improved placement of the computer reaction lets the actions stand out.

> **4.** Press the RETURN key.
>   The Customer Order Screen will appear.
> **5.** Click on the TABS button.

2. Make actions stand out visually from other material.

In the following example, bold is used to signal to the readers that the first part of step 4 is an action and the second part is the response.

Use bold and layout to make actions stand out.

> **4. Press the RETURN key.** The Customer Order Screen will appear.

You can also use layout to make such distinctions.

> **4. Press the RETURN key.**
>   • The Customer Order Screen will appear.

And you can use similar techniques when explaining steps.

> **7. Enter ANALYZE.** This command prompts the computer to perform seven analytical computations.

Group Related Steps Under Action-Oriented Headings

By arranging the steps into groups, you divide your procedure into chunks that readers are likely to find manageable. You also help them *learn* the procedure so that they will be able to perform it without instructions in the future. Moreover, if you use action-oriented headings and subheadings for the groups of steps, you aid readers who need directions for only one part of the procedure. The headings enable them to locate quickly the information they require.

To create action-oriented headings, use participles, not nouns, to describe the task. For example, use *Installing* rather than *Installation* and use *Converting* rather than *Conversion*. Here are some of the action-oriented headings and subheadings from Chapter 4 of the Microsoft Windows NT manual.

> Setting Up Your Computers on Your Network
>> Connecting to Computers on Your Network
>> Sharing Your Printer
>>> Viewing Network Drives
>>> Using Dial-Up Networking
>> Using Peer Web Services
>>> Installing Peer Web Services
>>> Configuring and Administering Peer Web Services

The first word in each heading and subheading is a participle.

Use Many Graphics

Drawings, photographs, and similar illustrations often provide the clearest and simplest means of telling your readers such important things as:

1. **Where things are.** For instance, Figure 27.3 shows the readers of an instruction manual where to find four control switches.

2. **How to perform steps.** For instance, by showing someone's hands performing a step, you provide your readers with a model to follow as they attempt to follow your directions (see Figure 27.4).

3. **What should result.** By showing readers what should result from performing a step, you help them understand what they are trying to accomplish and help them determine whether they have performed the step correctly (see Figure 27.5).

Learn More

Chapter 13 tells how to design effective graphics for instructions.

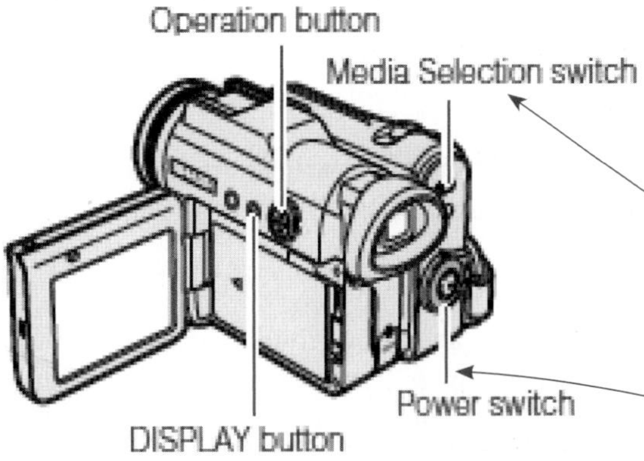

Operation button
Media Selection switch
DISPLAY button
Power switch

FIGURE 27.3
Drawing That Shows Readers Where to Locate Parts of a Camcorder

Writers at Sharp Electronics used this diagram to show new owners of a camcorder the location of buttons and switches they will use.

The writers placed the labels far enough from the drawing to stand out.

To avoid ambiguity, they drew the arrows directly to the labeled part.

FIGURE 27.4
Drawings That Show How
to Do Something

Writers for One Touch wrote
these instructions to tell people
with diabetes how to obtain the
drop of blood they need in order
to test their insulin levels.

Their title provides all the
information readers need to
understand and use these
instructions. No separate
introduction is needed.

The Lancet is a sharp needle used
to prick the skin.

The writers designed each
drawing to show exactly how to
hold the PENLET.

In the drawing for step 4,
the writers highlighted the
placement of the PENLET against
the side of a finger.

In the drawing for step 5, the
writers emphasized that the drop
of blood must hang from the
finger so that it may be applied
to a test strip (in the next part of
the procedure).

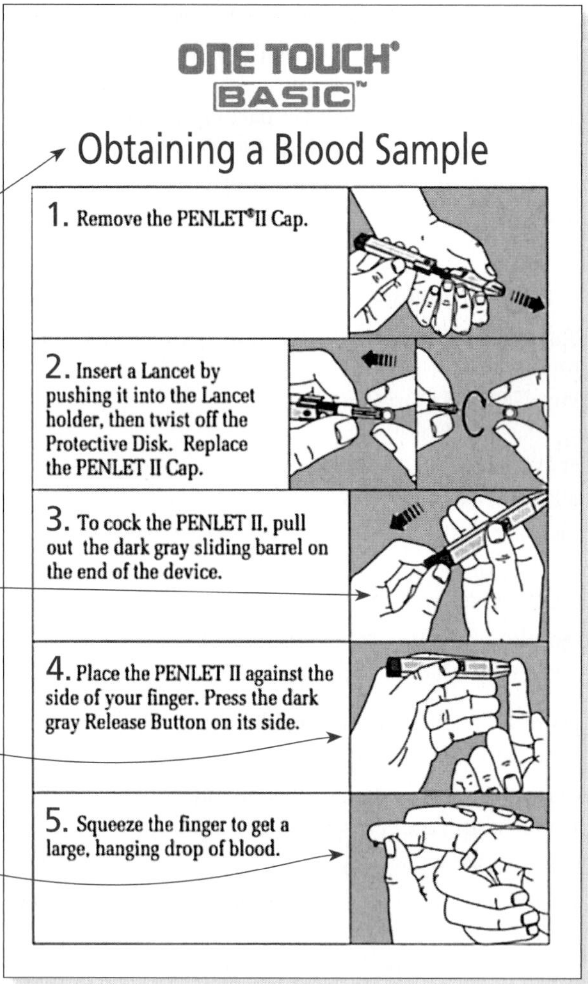

ONE TOUCH®
BASIC™

Obtaining a Blood Sample

1. Remove the PENLET®II Cap.

2. Insert a Lancet by pushing it into the Lancet holder, then twist off the Protective Disk. Replace the PENLET II Cap.

3. To cock the PENLET II, pull out the dark gray sliding barrel on the end of the device.

4. Place the PENLET II against the side of your finger. Press the dark gray Release Button on its side.

5. Squeeze the finger to get a large, hanging drop of blood.

Chapter 13 provides detailed advice for using page design to help readers see which figure goes with which text.

Present Branching Steps Clearly

Sometimes instructions include alternative courses of action. For example, a chemical analysis might require one procedure if the acidity of a solution is at a normal level and another if the acidity is high. In such a situation, avoid listing only one of the alternatives.

Possibly confusing direction | **6.** If the acidity is high, follow the procedure described on page 20. |

4. Adjust safety harness across your hips to provide maximum protection for internal organs.

FIGURE 27.5
Photographs That Show a Successful Outcome

Correct   Incorrect

 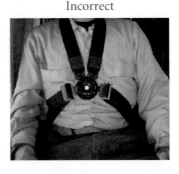

Instead, describe the step that enables readers to determine which alternative to choose (in the example, checking the acidity is that step) and then format the alternatives clearly:

6. Check the acidity.
   • If it is high, follow the procedure described on page 20.
   • If it is normal, proceed to Step 7.

Revised direction

Follow the same logic with other places where your instructions branch into two or more directions. The following example is from instructions for a computer program.

9. Determine which method you will use to connect to the Internet:
   • If you will use PPP (Point to Point Protocol), see Chapter 3.
   • If you will use SLIP (Serial Line Internet Protocol), see Chapter 4.

Tell What to Do in Case of a Mistake or Unexpected Result

Try to anticipate the places where readers might make mistakes or obtain an unexpected result. Unless the remedy is obvious, tell your readers what to do.

5. Depress and release the RUN switches on the operator's panel.
   NOTE: If the machine stops immediately and the FAULT light illuminates, reposition the second reel and repeat Step 5.

Troubleshooting

In various circumstances, readers find it easier to have information about correcting mistakes or unexpected results gathered into a single section. Often, a table format works best. Figure 27.6 shows the chapter of the manual for the Tire Uniformity Optimizer that tells how to troubleshoot the TUO's Tire Quality Computer (TQC).

Chapter 4—Troubleshooting

This chapter tells you what to check when troubleshooting the TQC. It lists the problems that may occur, the probable causes, and the remedies.

The first list in this chapter consists of the error messages that appear on the CRT when a problem occurs. Next to the error messages are the causes of the problem and the possible remedies. A list of all the error messages can be found in Appendix B. The second list consists of observable phenomena that are listed in order of normal TQC operation.

One easily solved problem is caused by entering entries too quickly to the TQC through the keyboard. If the operator does not wait for the TQC to respond to one request before entering another, errors and inaccurate data will result. Make sure you allow sufficient time for the TQC to respond to your input before you press another key.

The writers used color to highlight a warning.

Warning

EXTERNAL TEST EQUIPMENT CAN DAMAGE THE TQC. If you use external equipment to troubleshoot the TQC, make sure that it does not introduce undesired ground currents or AC leakage currents.

Troubleshooting with Error Messages

Power-up Error Messages

The writers created the three-column table to enable readers to locate quickly the error message given by the TQC and read across for the relevant information and remedy.

Error Message	Probable Cause	Remedy
BACKUP BATTERY IS LOW	1. Battery on Processor Support PCB	1. Replace the battery on the Processor Support PCB.
CONTROLLER ERROR	1. PC interface PCB 2. Processor Support PCB	1. Swap the PC Interface PCB. 2. Swap the Processor Support PCB.
EPROM CHECKSUM ERROR	1. Configuration tables 2. Analog Processor PCB	1. Check the configuration tables. 2. Swap the Analog Processor PCB 88/40.
KEYBOARD MALFUNCTION: PORT	1. Keyboard or keyboard cable 2. Processor Support PCB	1. Check the keyboard and cable. 2. Swap the Processor Support PCB.
RAM FAILURE AT 0000:	1. Main Processor 86/30	1. Swap the 86/30.
RAM FAILURE AT 1000:	1. Main Processor 86/30	1. Swap the 86/30.
TIGRE PROGRAM CHECKSUM ERROR	1. TIGRE program	1. Reenter the TIGRE program or debug the program.

Table 4-1. *Power-up error messages.*

Page 59

Physical Construction of Instructions

The physical construction of instructions is an important element of their design. Computer manuals are often printed in a small format because readers use them on crowded desktops. Cookbooks are sometimes printed on glossy paper to withstand kitchen spills. Be sure to adapt your instructions to the environment in which they will be used.

Sample Printed Instructions

Figure 27.7 explains how a student followed the advice you have just read while creating instructions for a lab procedure used in paper mills. For additional examples, see the book's website.

FIGURE 27.7
Instructions Written by a Student

**Determining the Percentages of
Hardwood and Softwood Fiber
in a Paper Sample**

These instructions tell you how to analyze a paper sample to determine what percentage of its fibers is from hardwood and what percentage from softwood. This information is important because the ratio of hardwood to softwood affects the paper's physical properties. The long softwood fibers provide strength but bunch up into flocks that give the paper an uneven formation. The short hardwood fibers provide an even formation but little strength. Consequently, two kinds of fibers are needed in most papers, the exact ratio depending on the type of paper being made.

Herman explains the importance of the procedure.

To determine the percentages of hardwood and softwood fiber, you perform the following major steps: preparing the slide, preparing the sample slurry, placing the slurry on the slide, staining the fibers, placing the slide cover, counting the fibers, and calculating the percentages. The procedure described in these instructions is an alternative to the test approved by the Technical Association of the Pulp and Paper Industry (TAPPI). The TAPPI test involves counting fibers in only one area of the sample slide. Because the fibers can be distributed unevenly on the slide, that procedure can give inaccurate results. The procedure given here produces more accurate results because it involves counting all the fibers on the slide.

He provides an overview of the procedure and indicates why the reader should follow it.

EQUIPMENT

He lists all of the equipment needed so readers can assemble it before starting the procedure.

Microscope	Hot plate
Microscope slide	Paper sample
Microscope slide	Blender
cover	Beaker
Microscope slide	Eyedropper
marking pen	Graff "C" stain
Acetone solvent	Pointing needle
Clean cloth	

(continued)

FIGURE 27.7
(continued)

2

Herman creates small groups of related steps. He places them under headings that help readers understand the overall procedure and quickly locate the directions they need when referring to the instructions in the future.

Herman explains the reason for the caution as a way of motivating readers to avoid making a mistake.

He uses bold for the action taken in each step, thereby making the action stand out from explanatory information.

He places the graphic immediately after the step it helps to explain. He keeps the graphic within the gridlines for the directions and out of the grid column for the step numbers.

PREPARING THE SLIDE

1. **Clean slide.** Using acetone solvent and a clean cloth, remove all dirt and fingerprints. NOTE: Do not use a paper towel because it will deposit fibers on the slide.

2. **Mark slide.** With a marking pen, draw two lines approximately 1.5 inches apart across the width of the slide.

3. **Label slide.** At one end, label the slide with an identifying number. Your slide should now look like the one shown in Figure A.

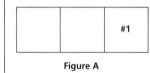

Figure A

4. **Turn on hot plate.** Set the temperature at warm. NOTE: Higher temperatures will "boil" off the softwood fibers that you will later place on the slide.

5. **Place slide on hot plate.** Leave the slide there until it dries completely, which will take approximately 5 minutes.

6. **Remove slide from hot plate.** Leave the hot plate on. You will use it again shortly.

PREPARING THE SAMPLE SLURRY

1. **Pour 2 cups of water in blender.** This measurement can be approximate.

2. **Obtain paper sample.** The sample should be about the size of a dime.

3. **Tear sample into fine pieces.**

4. **Place sample into blender.**

5. **Turn blender on.** Set blender on high and run it for about 1 minute.

6. **Check slurry.** After turning the blender off, see if any paper clumps remain. If so, turn the blender on for another 30 seconds. Repeat until no clumps remain.

7. **Pour slurry into beaker.**

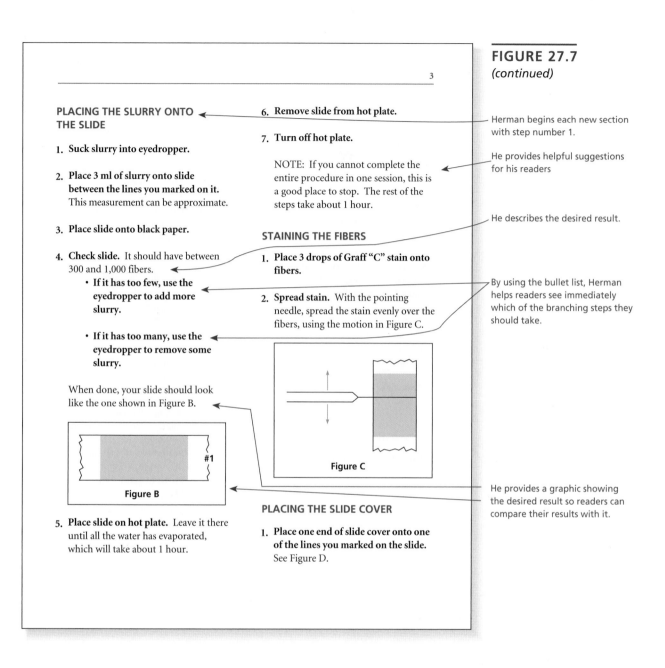

FIGURE 27.7
(continued)

3

PLACING THE SLURRY ONTO THE SLIDE

1. Suck slurry into eyedropper.

2. Place 3 ml of slurry onto slide between the lines you marked on it. This measurement can be approximate.

3. Place slide onto black paper.

4. Check slide. It should have between 300 and 1,000 fibers.
   • If it has too few, use the eyedropper to add more slurry.
   • If it has too many, use the eyedropper to remove some slurry.

   When done, your slide should look like the one shown in Figure B.

Figure B

#1

5. Place slide on hot plate. Leave it there until all the water has evaporated, which will take about 1 hour.

6. Remove slide from hot plate.

7. Turn off hot plate.

   NOTE: If you cannot complete the entire procedure in one session, this is a good place to stop. The rest of the steps take about 1 hour.

STAINING THE FIBERS

1. Place 3 drops of Graff "C" stain onto fibers.

2. Spread stain. With the pointing needle, spread the stain evenly over the fibers, using the motion in Figure C.

Figure C

PLACING THE SLIDE COVER

1. Place one end of slide cover onto one of the lines you marked on the slide. See Figure D.

Herman begins each new section with step number 1.

He provides helpful suggestions for his readers

He describes the desired result.

By using the bullet list, Herman helps readers see immediately which of the branching steps they should take.

He provides a graphic showing the desired result so readers can compare their results with it.

(continued)

Crafting the Major Elements of Instructions ■   665

FIGURE 27.7
(continued)

Herman labels items in his graphic to help readers understand what it illustrates.

He uses a figure to explain a procedure that would be difficult to understand if presented in words alone.

4

Cover slide

Stain

Slide

Figure D

2. **Adjust magnification.** You should be able to distinguish black fibers from dark purple ones.

3. **Move slide to show upper left-hand corner of area with fibers.**

4. **Count whole fibers.** Move the slide so that your view of it changes in the manner shown in Figure E, counting

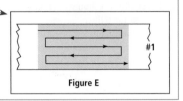

Figure E

the whole softwood and hardwood fibers you see. Ignore fragments of fibers, which you will count later.

* **Recognizing softwood fibers.** Softwood fibers are long and flat. They have blunt ends. The stain dyes the fibers colors that range from slightly purple (almost translucent) to a dark purple. See Figure F.

2. **Slowly lower the other side of the slide cover.** Be sure that no air gets trapped under the slide cover.

3. **Drain excess stain.** With a cloth underneath, turn the slide onto one of its longest edges so that the excess stain will run off.

4. **Clean slide.** Use acetone solvent to remove residue and fingerprints.

COUNTING THE FIBERS

1. **Place slide onto microscope.**

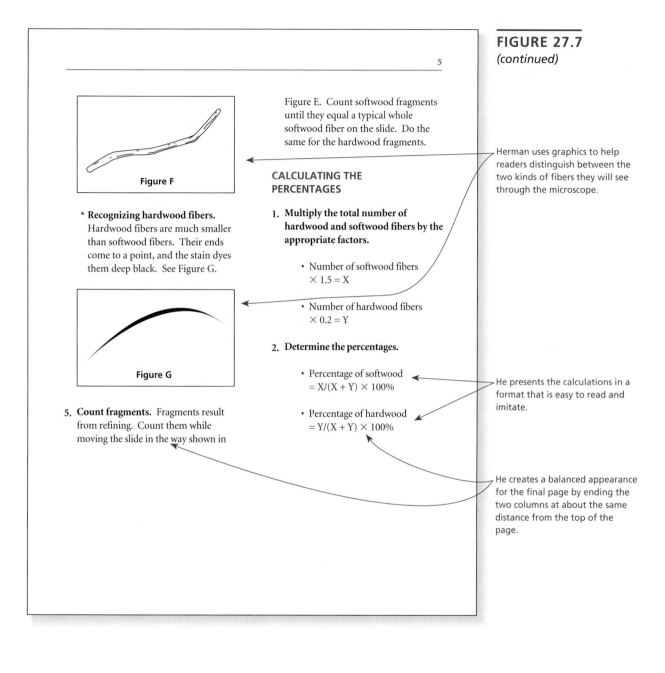

FIGURE 27.7
(continued)

5

Figure F

Figure E. Count softwood fragments until they equal a typical whole softwood fiber on the slide. Do the same for the hardwood fragments.

Herman uses graphics to help readers distinguish between the two kinds of fibers they will see through the microscope.

* **Recognizing hardwood fibers.** Hardwood fibers are much smaller than softwood fibers. Their ends come to a point, and the stain dyes them deep black. See Figure G.

Figure G

5. **Count fragments.** Fragments result from refining. Count them while moving the slide in the way shown in

CALCULATING THE PERCENTAGES

1. **Multiply the total number of hardwood and softwood fibers by the appropriate factors.**

   - Number of softwood fibers $\times 1.5 = X$

   - Number of hardwood fibers $\times 0.2 = Y$

2. **Determine the percentages.**

   - Percentage of softwood $= X/(X + Y) \times 100\%$

   - Percentage of hardwood $= Y/(X + Y) \times 100\%$

He presents the calculations in a format that is easy to read and imitate.

He creates a balanced appearance for the final page by ending the two columns at about the same distance from the top of the page.

WEB PAGE INSTRUCTIONS

WWW

To view other examples of online instructions, go to Chapter 27 at **www.cengage.com/english/anderson7e.**

As explained at the beginning of this chapter, all the advice you have read so far applies not only to printed instructions but also to web page instructions, which are becoming increasingly common. Often they are included as the online Help in computer programs. Many sites on the Internet provide instructions meant for on-screen use. Although there are special computer programs designed specifically for creating web page instructions, you can also use the feature of your desktop publishing program that enables you to convert text files into website pages. The following guidelines highlight some of the advice given earlier in the chapter that you will find most helpful when creating web page instructions. Figure 27.8 shows sample pages from professionally developed online instructions. Figure 27.9 shows pages from online instructions created by a student.

Learn More

For detailed advice on creating websites and web pages from which they are constructed, see Chapter 20.

GUIDELINES FOR WEB PAGE INSTRUCTIONS

1. **Organize the steps into small, related groups arranged hierarchically.** This organizational strategy facilitates quick access to the specific directions that a reader wants. If you devote a separate web page to each of the smallest groups, you also save your readers from the inconvenience and, sometimes, confusion caused when people need to scroll through directions.

2. **Give readers quick access to a complete list of the topics.** When using online instructions, many readers want to skip around among tasks. Help them by creating a complete list of items in your hierarchy so they don't have to move through layers of pages to locate what they want. Make this list available on every page, for instance, by putting a link to it in a navigation bar at the top or side of each page.

3. **Provide links to helpful information.** If your directions use a term that some of your readers may not know, provide a link to an explanation. If readers would find it helpful to learn about an alternative or related procedure, include a link to it.

4. **Write succinct, precise directions.** Even more than on paper, succinctness is valued in online instructions. Extra words not only make extra reading but also extra scrolling.

5. **Include clear, well-labeled graphics.** Tables, drawings, screenshots, and other graphics are just as helpful online as in print.

6. **Use a consistent design on all pages.** As in a printed document, a consistent design makes online instructions easier for readers to use.

7. **Use all the page design strategies you would use on paper.** Despite the many differences between a computer screen and a printed page, online instructions are pages. All of the design features that make paper pages usable and persuasive do the same for digital pages.

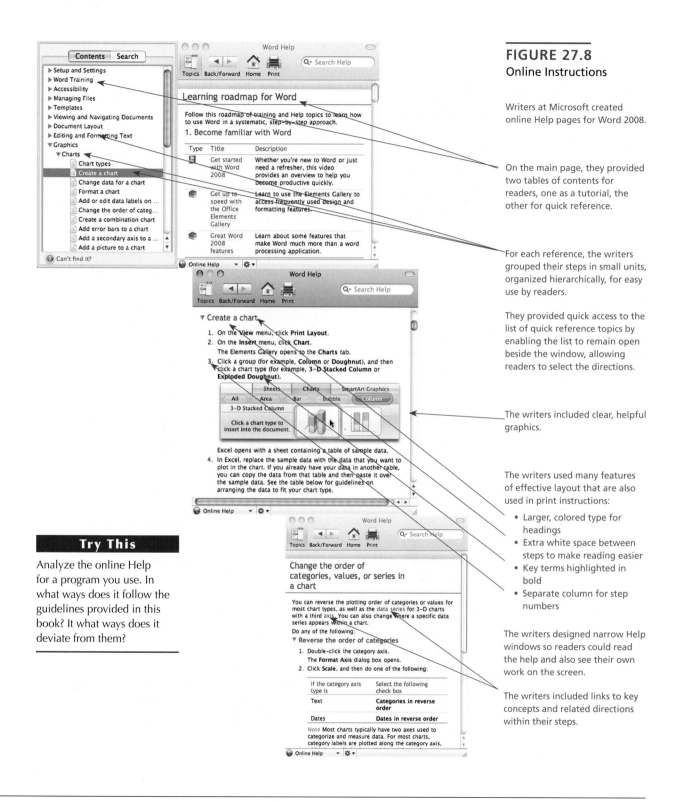

FIGURE 27.8
Online Instructions

Writers at Microsoft created online Help pages for Word 2008.

On the main page, they provided two tables of contents for readers, one as a tutorial, the other for quick reference.

For each reference, the writers grouped their steps in small units, organized hierarchically, for easy use by readers.

They provided quick access to the list of quick reference topics by enabling the list to remain open beside the window, allowing readers to select the directions.

The writers included clear, helpful graphics.

The writers used many features of effective layout that are also used in print instructions:

- Larger, colored type for headings
- Extra white space between steps to make reading easier
- Key terms highlighted in bold
- Separate column for step numbers

The writers designed narrow Help windows so readers could read the help and also see their own work on the screen.

The writers included links to key concepts and related directions within their steps.

Try This

Analyze the online Help for a program you use. In what ways does it follow the guidelines provided in this book? It what ways does it deviate from them?

FIGURE 27.9
Online Instructions
Created by a Student

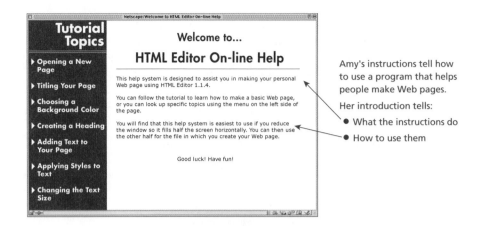

Amy's instructions tell how to use a program that helps people make Web pages.

Her introduction tells:

- What the instructions do
- How to use them

Amy uses the same basic design for all pages.

Most of her screens require no scrolling (though the menu on the left must be scrolled).

Note her use of figures.

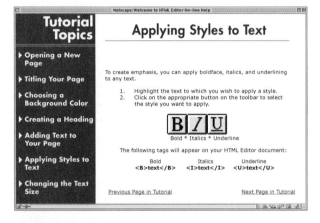

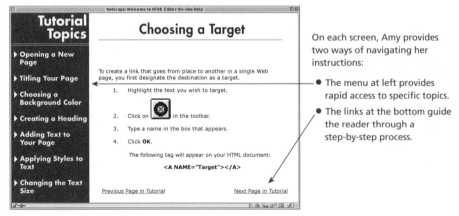

On each screen, Amy provides two ways of navigating her instructions:

- The menu at left provides rapid access to specific topics.
- The links at the bottom guide the reader through a step-by-step process.

8. **If the instructions are for a computer program, enable readers to see your instructions and their work simultaneously.** When people are working on a computer, nothing is more frustrating to them than having to switch between windows to follow directions. Design your web pages and their contents so that readers can keep their work open in a window beside yours.

9. **Conduct user testing.** All instructions, whether in print or online, have to be tested by members of the target audience, who try to use them under conditions identical to those of the anticipated actual use.

DIGITAL MOVIE INSTRUCTIONS

Increasingly, instructions are prepared as digital movies, with the text being spoken and the graphics being either a series of still images or videocam clips. New cars now come with DVDs that provide instructions to new owners. Company websites include movies that tell people how to use the company's products. The Help feature of many computer programs, such as Microsoft Word, include digital movies that explain how to use some of the program's features. YouTube features an ever-growing collection of movie instructions.

Although web page instructions closely resemble printed instructions, digital movie instructions are quite different. You couldn't create an effective script simply by reading directions prepared for print or a web page. In movie instructions, the relationship between images and script are coordinated through timing rather than placement on the page. In the movie version, you need to have images continuously. You can't leave the visual portion of the movie blank until you come to a place where you would use the next graphic in printed or web page instructions. Tone of voice and pace of reading make a great difference in the effectiveness of digital movie instructions.

Movie-making programs are available for free on almost every computer. All recent versions of Windows include an easy-to-use program called Windows Movie Maker. Apple computers have iMovie. Some computers are equipped with cameras and microphones.

At this book's website, you can read a bonus chapter about making digital movie instructions, link to examples made by professionals, and view samples made by students. Go to www.cengage.com/english/anderson7e and click on Chapter 27.

Try This

In the Help feature of a program on your computer, find a video that includes step-by-step instructions. Most desktop publishing programs, such as Microsoft Word, have them. Imagine that you are making a set of print instructions for the same procedure. Quickly draft the written instructions for the first few steps, including text and graphics. How are your print instructions like the movie instructions? How do they differ? Alternatively, write the video script, including graphics, for a simple, step-by-step procedure.

WRITER'S GUIDES AND OTHER RESOURCES

WWW

To download Writer's Guides for planning and revising instructions, go to Chapter 27 at **www.cengage.com/ english/anderson7e.**

Figure 27.10 presents a Writer's Guide for Revising Instructions that you can use in your course and on the job. At this book's website, you can download a copy of this Writer's Guide as well as one for planning instructions.

Note to the Instructor   To tailor the Writer's Guides to your course, download them from the website.

FIGURE 27.10   **Writer's Guide for Revising Instructions**

To download a copy of this Writer's Guide as well as a Writer's Guide for Planning Instructions, go to Chapter 27 at www.cengage.com/english/anderson7e.

**Writer's Guide
REVISING INSTRUCTIONS**

The following headings reflect the elements of the conventional superstructure for instructions. In many situations, some of the elements are unnecessary. Include only the elements that will make your instructions usable and persuasive to your readers.

Introduction

☐ Describes the purpose, aim, or desired outcome of the procedure, if it is not clear from the title

☐ Identifies the intended readers (their knowledge level, job descriptions, etc.)

☐ Tells the scope and organization, either separately or together

☐ Provides motivation (reasons for following the instructions rather than using another procedure)

☐ Explains the conventions used

☐ Includes safety information

Description of the Equipment

☐ Shows the location of the key parts

☐ Explains the function of the key parts

Materials and Equipment

☐ Tells readers what materials and equipment to gather before starting the procedure

Directions

☐ Support rapid comprehension and immediate use by providing only the information your readers need, listing the steps, using the active voice and imperative mood ("Do this."), and highlighting key words

☐ Help readers locate the next step quickly by numbering the steps and putting blank lines between them, giving one action per step, and putting step numbers in their own column

☐ Distinguish action from supporting information by putting actions before responses, and making actions stand out visually

---

FIGURE 27.10 *(continued)*

**Writer's Guide
REVISING INSTRUCTIONS
*(continued)***

☐ Group related steps under action-oriented headings
☐ Use many graphics
☐ Present branching steps clearly

Troubleshooting
☐ Tells what to do if something fails or there is an unexpected result

Graphics (See Chapter 13)
☐ Included wherever readers would find them helpful or persuasive
☐ Look neat, attractive, and easy to read
☐ Referred to them at the appropriate points in the prose
☐ Located where your readers can find them easily

Page Design (See Chapter 14)
☐ Looks neat and attractive
☐ Helps readers find specific information quickly

Proofreading (See Chapter 15)
☐ Uses correct spelling, grammar, and punctuation

USE WHAT YOU'VE LEARNED

EXERCISE YOUR EXPERTISE

1. Find and photocopy a short set of instructions (five pages or less). Analyze the instructions, noting how the writers have handled each element of the superstructure for instructions. If they have omitted certain elements, explain why you think they did so. Comment on the page design and graphics (if any). Then evaluate the instructions. Tell what you think works best about them and identify ways you think they can be improved.

2. Complete the Instructions project in Appendix B.

EXPLORE ONLINE

Choose one:

a. Find a set of instructions designed to be read on the World Wide Web and discuss ways they might be changed to be effective in print.

b. Find a set of print instructions and discuss ways they might be changed to make them as effective as possible if viewed on the World Wide Web.

COLLABORATE WITH YOUR CLASSMATES

Working with another student, conduct an informal user test of a set of online instructions for a program on your computer. First, evaluate the screen design together. Next, while one of you uses part of the instructions, the other should record observations about parts of the text and graphics that work well and parts that could be improved. Present your results in a memo to your instructor. Consider the guidelines given on pages 656–661 and your own experience as you prepare your memo.

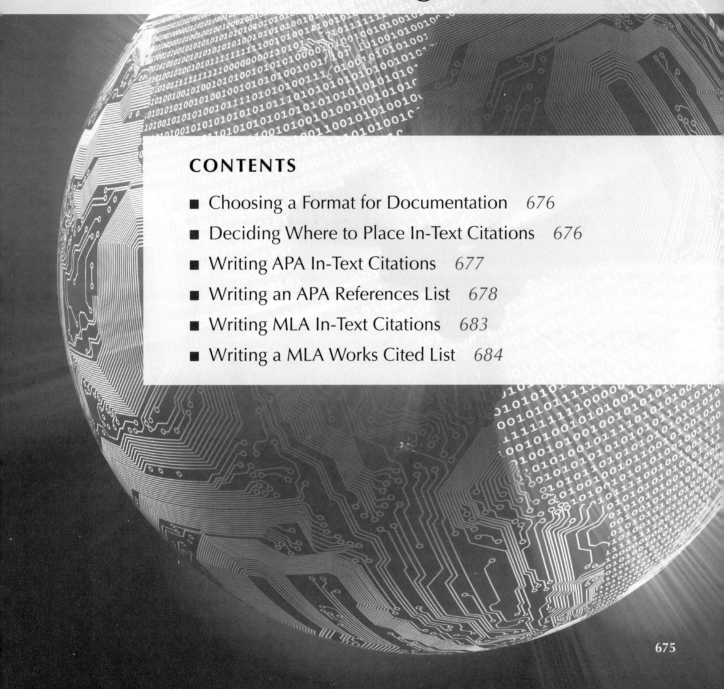

Appendix A
Documenting Your Sources

CONTENTS

In many of the communications you will write at work, you will want to tell your readers about other sources of information concerning your subject. You may have any of the following reasons for wanting to do so:

- **To acknowledge the people and sources that have provided you with ideas and information.** For a discussion of this reason for citing your sources, see Chapter 6's Guideline 7 on page 158.
- **To help your readers find additional information about something you have discussed.**
- **To persuade your readers to pay serious attention to a particular idea.** By showing that an idea was expressed by a respected person or in a respected publication, you are arguing that the idea merits acceptance.
- **To explain how your research contributes to the development of new knowledge in your field.** In research proposals and in research reports published in professional journals, writers often include a literature review to demonstrate how their own research advances knowledge. For more information on using references in this way, see the discussion of literature reviews in the discussion of empirical research reports, page 585.

CHOOSING A FORMAT FOR DOCUMENTATION

There are numerous formats for documentation, some very distinct from one another and some differing only in small details. No one format is the best for all situations. To document correctly, you must use the format required or most appropriate for the particular circumstances in which you are writing. Most organizations specify the documentation style they want their employees to use. Also, when you are writing to people in another organization, you may find that they have their own preferences or requirements, which you should follow.

To describe the documentation format they prefer, many organizations issue style guides that detail their rules and provide sample citations. Other organizations ask writers to follow a style guide published by a professional organization such as the American Psychological Association (APA) or the Modern Language Association (MLA). The rest of this appendix explains the APA and MLA styles. Most other documentation styles resemble one of these two.

WWW

For additional information about documenting your sources, including links to other documentation systems, go to Appendix A at **www.cengage.com/english/anderson7e.**

DECIDING WHERE TO PLACE IN-TEXT CITATIONS

In both styles, you cite a source by putting the name of the author in parentheses at the appropriate place in the body of your communication (the APA style includes the year of publication also). This citation refers your readers to the full bibliographic information that you provide in an alphabetical list of sources at the end of your communication. Between the two styles there are many differences in the way you write the in-text citations and the entries in the reference list. However, the rules for positioning the in-text citations are the same for both.

Your primary objective when placing citations in your communications is to make clear to your readers what part of your text each citation is referring to. This is easy with citations that pertain to a single fact, sentence, or quotation. You simply place your citation immediately after the appropriate material (note that the APA style for formatting these citations is used in the following examples):

> According to F. C. Orley (2010, p. 37), "We cannot tell how to interpret these data without conducting further tests."

> Researchers have shown that a person's self-esteem is based upon performance (Dore, 2009), age (Latice, 1998), and weight (Swallen & Ditka, 2010).

If your citation refers to material that appears in several sentences, place the citations in a topic sentence that introduces the material. Your readers will then understand that the citation covers all the material that relates to that topic sentence. As a further aid to your readers, you may use the author's name (or a pronoun) in successive sentences:

> A much different account of the origin of oil in the earth's crust has been advanced by Thomas Gold (2008). He argues that . . . To critics of his views, Gold responds . . .

WRITING APA IN-TEXT CITATIONS

To write an APA in-text citation, enclose the author's last name and the year of publication in parentheses inside your normal sentence punctuation. Place a comma between the author's name and the date. Use *p.* if you are citing a specific page, and use *pp.* if citing more than one page.

> The first crab caught in the trap attracts others to it (Tanner, 2007, pp. 33–34).

If you incorporate the author's name in the sentence itself, give only the year and pages (if any) in parentheses:

> According to Tanner (2007, pp. 33–34), the first crab caught in the trap attracts others to it.

Here are some other types of citations:

(Hoeflin & Bolsen, 2004)	Two authors
(Wilton, Nelson, & Dutta, 1932)	First citation for three, four, or five authors
(Wilton et al., 1932)	Second and subsequent citations for three to five authors. Omit year in subsequent citations in the same paragraph.

WWW

For additional information and details about the APA documentation style, including links to other documentation systems, go to Appendix A at **www.cengage.com/english/ anderson7e.**

(Norton et al., 2004)	First and subsequent citation for six or more authors (*et al.* is an abbreviation for the Latin phrase *et alii,* which means "and others")
(Angstrom, 2010, p. 34)	Reference to a particular page
(U.S. Department of Energy, 1997)	Government or corporate author
("Geologists Discover," 2009)	No author listed (use the first few words of the title). In the example, the words from the title are in quotation marks because the citation is to an article; if a book is cited, the words would be in italics with no quotation marks.
(Justin, 1998; Skol, 1972; Weiss, 1966)	Two or more sources cited together (arrange them in alphabetical order)

In some communications, you might cite two or more sources by the same author. If they were published in *different* years, your readers will have no trouble telling which work you are referring to. If they were published in the same year, you can distinguish between them by placing lowercase letters after the publication dates in your citations and in your reference list:

(Burkehardt, 1998a)

(Burkehardt, 1998b)

WRITING AN APA REFERENCES LIST

Illustrated and explained here are APA References list entries for the most common types of print, electronic, and other sources. To create entries that are not listed here, follow the logic of these examples or else consult the *Publication Manual of the American Psychological Association*, Fifth Edition, available in the reference section of most libraries. Figure A.1 shows how to arrange entries in your References list.

Print Sources

1. Book, One Author—APA

Lightman, A. P. (2005). *A sense of the mysterious: Science and the human spirit.* New York, NY: Pantheon Books.

- Give the author's last name followed by a comma and initials (not full first or middle names).
- Place the copyright date in parentheses, followed by a period.
- Italicize the title, and capitalize only the first word of the title, the first word of the subtitle (if any), and proper nouns.

References

Adams, D. J., Dyson, P. J., & Tavener, S. J. (2004). *Chemistry in alternative reaction media.* Hoboken, NJ: Wiley.

Gould, S. J. (1989). *Wonderful life: The burgess shale and the nature of history.* New York, NY: Norton.

Gould, S. J. (1995). *Dinosaur in a haystack: Reflections in natural history.* New York, NY: Harmony.

International Business Machines. (n.d.). What's user-centered design? Retrieved from https://www-01.ibm.com/software/ucd/ucd.html#whatisucd.

McLaurin, J., & Chakrabartty, A. (1997). Characterization of the interactions of Alzheimer ß-amyloid peptides with phospholipid membranes. *European Journal of Biochemistry, 245,* 355–63.

Rethinking traditional design. (1997). *Manufacturing Engineering 118(2),* 50.

Younger manufacturing workers can teach old dogs new tricks. (2007). *Manufacturing, 86(5),* 5.

Second and subsequent lines are indented.

When the list includes two or more items by the same author, the oldest appears first.

For the second and subsequent items by the same person, the author's name is repeated.

Items by corporate and government groups are alphabetized by the groups' names (spelled out).

Items without authors are alphabetized by the title.

- Follow the city of publication with a comma and the two-letter postal abbreviation for the state.
- Indent the second and subsequent lines by 5 spaces.

2. Book, Two or More Authors—APA

Budinski, K. G., & Budinski M. K. (2005). *Engineering materials: properties and selection*. Upper Saddle River, NJ: Pearson.

Adams, D. J., Dyson, P. J., & Tavener, S. J. (2004). *Chemistry in alternative reaction media.* Hoboken, NJ: Wiley.

3. Anthology or Essay Collection—APA

Bodeker, G., & Burford, G. (Eds). (2007). *Traditional, complementary and alternative medicine: policy and public health perspectives*. London: Imperial College Press.

- If there is only one editor, use this abbreviation: (Ed.).

4. Second or Subsequent Edition—APA

Simmons, L. H., & Olin, H, B. (2001). *Construction: principles, materials, and methods* (7th ed.) New York, NY: John Wiley.

5. Government Report—APA

Frankforter, J. D., & Emmons, P. J. (1997). *Potential effects of large floods on the transport of atrazine into the alluvial aquifer adjacent to the Lower Platte River, Nebraska* (U.S. Geological Survey Water-Resources Investigation Report 96–4272). Denver, CO: U.S. Geological Survey.

- If the report doesn't list an author, use the name of the agency that published it as the author. If it is a United States government agency, use the abbreviation "U.S." Example: "U.S. Geological Survey."
- If the report has an identifying number, place it immediately after the title.

6. Corporate Report—APA

Daimler-Benz AG. (2009). *Environmental report 2009.* Stuttgart, Germany: Author.

- List the names of the individual authors rather than the corporation if the names are given on the title page.
- If the names of the individual authors aren't given on the title page, list the corporation as the author. (In the example, "Daimler-Benz AG" is the name of a company.)
- When the author and publisher are the same, use the word *Author* as the name of the publisher.

7. Essay in a Book—APA

Moor, J. H. (2008). Why we need better ethics for emerging technologies.
In J. Hoven & J. Weckert (Eds.), *Information technology and moral philosophy*
(pp. 24–43). New York, NY: Cambridge University Press.

8. Encyclopedia Article—APA

Rich, E. (2003). Artificial intelligence. In *Encyclopedia Americana* (Vol. 2,
pp. 407–412). Danbury, CT: Grolier.

9. Pamphlet or Brochure—APA

Ohio Department of Natural Resources. (2010). *Ring-necked pheasant management
in Ohio.* [Pamphlet]. Columbus, OH: Author.

- When the author and publisher are the same, use the word *Author* as the name
of the publisher.

10. Article in Journal That Numbers Its Pages Continuously through Each Volume—APA

McLaurin, J., & Chakrabartty, A. (1997). Characterization of the interactions of
Alzheimer ß-amyloid peptides with phospholipid membranes. *European
Journal of Biochemistry, 245,* 355–363.

- After the journal's name (note use of capital letters), add a comma and the
volume number (in italics).
- The word *Alzheimer* is capitalized because it is a proper name.

11. Article in Journal That Numbers Its Pages Separately for Each Issue—APA

Matthews, J. N. A. (2009). Superconductors to boost wind power. *Physics Today.
62*(4), 25–26.

- After the journal title, include the volume number, followed by the issue
number (in parentheses).
- Put the volume number—but not the issue number—in italics.

12. Article in a Popular Magazine—APA

Cowley, G. (1997, April 28). Cardiac contagion. *Newsweek, 129,* 69–70.

- Give the full date of the issue, placing the year first.
- Provide the volume number (in the example: 129) but not the issue
number.

13. Newspaper Article—APA

Wade, N. (1997, May 24). Doctors record signals of brain cells linked to memory. *New York Times* [National edition], p. A8.

- In front of the page number for newspapers, write *p.* (for one page) or *pp.* (for more than one page).

14. Article with No Author Listed—APA

Younger manufacturing workers can teach old dogs new tricks. (2007). *Manufacturing 86*(5), 5.

- Begin with the article's title. This example gives the issue number in parentheses because the journal numbers its pages separately for each issue (see Example 1).

Electronic Sources

www

For information about the APA documentation style for other types of electronic sources, go to Appendix A at **www .cengage.com/english/ anderson7e.**

15. Online Report That Is Not Available in Print—APA

International Business Machines. (n.d.). What's user-centered design? Retrieved from https://www-01.ibm.com/software/ucd/ucd.html#whatisucd

- Because the site doesn't give the date this page was published on the web, the abbreviation *n.d.* is used to mean "no date."

Because placing a period at the end of an Internet address can cause confusion, final periods are omitted in APA entries for online sources.

16. Online Journal Article That Is Not Available in Print—APA

Toll, D. (1996). Artificial intelligence applications in geotechnical engineering. *Electronic Journal of Geotechnical Engineering.* Retrieved from http://geotech.civen.okstate.edu/ejge/Abs9608.htm

17. Online Journal Article That Is Also Available in Print—APA

Porfiri, M. (2008). Charge dynamics in ionic polymer metal composites. [Electronic version]. *Journal of Applied Physics, 104,* 915–25. Retrieved from http://scitation.aip.org/journals/doc/JAPIAU-ft/vol_104/iss_10/104915_1.html

Because material that is available on the World Wide Web changes continuously, APA entries do not include the date you retrieved the item.

Additional Common Sources

18. E-mail—APA

The APA style includes references to e-mails only in parentheses in the text, not in the References list. The parenthetical citation in the text includes the author's initials as well as his or her last name and an exact date:

(M. Grube, personal communication, December 4, 2010)

19. Letter—APA

The APA style treats letters the same way it treats e-mails (see Example 18).

(L. A. Cawthorne, personal communication, August 24, 2010)

20. Interview—APA

The APA style treats interviews the same way it treats e-mails (see Example 18).

WRITING MLA IN-TEXT CITATIONS

A basic MLA citation contains two kinds of information: the author's name and the specific page or pages on which the cited information is to be found. Enclose these in parentheses—with no punctuation between them—and place them inside your normal sentence punctuation. If you are citing the entire work, omit the page numbers. The following citation refers the reader to pages 33 and 34 of a work by Tanner. Note that MLA style uses a hyphen to separate page ranges. In large numbers, only the last two digits are provided unless more are necessary (252–53, but 295–305).

The first crab caught in the trap attracts others to it (Tanner 33-34).

If you incorporate the author's name in the sentence itself, give only the page numbers in parentheses:

According to Tanner, the first crab caught in the trap attracts others to it (33-34).

Here are some other types of citations:

(Hoeflin and Bolsen 16)	Two authors
(Wilton, Nelson, and Dutta 222)	Three authors
(Norton et al. 776)	Four or more authors (list all authors or list the first author followed by *et al.,* which is the abbreviation for the Latin phrase *et alii,* which means "and others")
(U.S. Department of Energy 4-7)	Government or corporate author
("Geologists Discover" 31)	No author listed (use the first few words of the title). In the example, the words from the title are in quotation marks because the citation is to an article; if a book is cited, the words would be in italics with no quotation marks. When there are two or more anonymous works by the same title, include in the parenthetical reference either a publication year for a book or a periodical name for an article.
(Justin 23; Skol 1089; Weiss 475)	Two or more sources cited together

WWW

For additional information and details about the MLA documentation style, including links to other documentation systems, go to Appendix A at **www.cengage .com/english/anderson7e.**

In some of your communications, you might cite two or more sources by the same author. If they were published in different years, your readers will have no trouble telling which work you are referring to. If they were published in the same year, distinguish between them by placing a comma after the author's name, followed by a few words from the title.

Whole work cited | (Burkehardt, "Gambling Addiction")
Specific pages cited | (Burkehardt, "Obsessive Behaviors," 81-87)

WRITING A MLA WORKS CITED LIST

The following sections describe how to write entries for the most common types of print, electronic, and other sources. To create entries that are not listed here, follow the logic of these examples or else consult those available in the reference section of most libraries. Figure A.2 shows how to arrange entries in your list of works cited.

Print Sources

1. Book, One Author—MLA

Lightman, Alan P. *A Sense of the Mysterious: Science and the Human Spirit*. New York: Pantheon Books, 2005. Print.

- Give the author's last name followed by a comma and then the first and middle names or initials—exactly as they appear on the title page.
- *Italicize* the title, and capitalize all major words.
- Follow the city of publication with a colon, the publisher's name, a comma, and the publication date.
- Indent (5 spaces) all lines after the first one.

2. Book, Two or More Authors—MLA

Budinski, Kenneth G., and Michael K. Budinski. *Engineering Materials: Properties and Selection*. Upper Saddle River, NJ: Pearson, 2005. Print.

Adams, Dave J., Paul J. Dyson, and Stewart J. Tavener. *Chemistry in Alternative Reaction Media*. Hoboken, NJ: Wiley, 2004. Print.

- Give the first author's name in reverse order (last name first).
- Give the names of additional authors in normal order (first name first).

3. Anthology or Essay Collection—MLA

Bodeker, Gerard, and Gemma Burford, eds. *Traditional, Complementary and Alternative Medicine: Policy and Public Health Perspectives*. London: Imperial College Press, 2007. Print.

- Use the abbreviation "ed." for a single editor and "eds." for multiple editors.

Works Cited

Adams, Dave J., Paul J. Dyson, & Stewart J. Tavener. *Chemistry in Alternative*
  Reaction Media. Hoboken, NJ: Wiley, 2004. Print.

Gould, Stephen Jay. *Dinosaur in a Haystack: Reflections in Natural History.* New York:
  Harmony, 1995. Print.

---. *Wonderful Life: The Burgess Shale and the Nature of History.* New York: Norton, 1989.

International Business Machines. *What's User-Centered Design?* n.d. Web. 25 Jan. 2010.
  Print. <https://www-01.ibm.com/software/ucd/ucd.html#whatisucd>.

McLaurin, JoAnne, and Avijit Chakrabartty. "Characterization of the Interactions of
  Alzheimer ß-amyloid Peptides with Phospholipid Membranes." *European Journal of*
  Biochemistry, 245 (1997): 355-63. Print.

"Rethinking Traditional Design." *Manufacturing Engineering 118.2* (1997): 50. Print.

"Younger Manufacturing Workers Can Teach Old Dogs New Tricks." (2007).
  Manufacturing 86(5), 5. Print.

Second and subsequent lines are indented.

When the list includes two or more items by the same author, they are alphabetized by title.

For the second and subsequent items by the same person, the author's name is replaced by three hyphens followed by a period.

Items by corporate and government groups are alphabetized by the groups' names (spelled out).

Items without authors are alphabetized by the title.

4. Second or Subsequent Edition—MLA

Simmons, Leslie H., and Harold Bennett Olin. *Construction: Principles, Materials, and Methods*. 7th ed. New York: John Wiley, 2001. Print.

5. Government Report—MLA

Frankforter, J. D., and P. J. Emmons. *Potential Effects of Large Floods on the Transport of Atrazine into the Alluvial Aquifer Adjacent to the Lower Platte River, Nebraska*. U.S. Geological Survey Water-Resources Investigation Report 96–4272. Denver: U.S. Geological Survey, 1997. Print.

- If the report doesn't list an author, begin the entry with the name of the government, followed by a period and the name of the agency that issued the document. Example: "United States. Geological Survey."
- If the report has an identifying number, place it immediately after the title.
- In the example, the authors' initials (not first and middle names) are given because that is how their names appear on the title page.

6. Corporate Report—MLA

Daimler-Benz AG. *Environmental Report 2009*. Stuttgart, Germany: Daimler-Benz AG, 2009. Print.

- List the names of the individual authors rather than the corporation if the names are given on the title page.
- If the names of the individual authors aren't given on the title page, list the corporation as the author. (In the example, "Daimler-Benz AG" is the name of a company.)

7. Essay in a Book—MLA

Moor, James H. "Why We Need Better Ethics for Emerging Technologies." Information Technology and Moral Philosophy. Ed. Jeroen van den Hoven and John Weckert. New York: Cambridge University Press, 2008. 24-43. Print.

- Use the abbreviation "Ed." for a single editor and for multiple editors.

8. Article in an Encyclopedia, Dictionary, or Similar Reference Work—MLA

Rich, Elaine. "Artificial Intelligence." *Encyclopedia Americana*. 2003 ed. Print.

- If no author is listed, begin with the article's title.
- If entries in the work are arranged alphabetically, do not give volume or page number.
- When citing familiar reference works, give the edition number (if provided) and year of publication, but not the publisher or city of publication.

9. Pamphlet or Brochure—MLA

Ohio State. Department of Natural Resources. *Ring-Necked Pheasant Management in Ohio*. Columbus: State of Ohio, 1996. Print.

- If the pamphlet or brochure doesn't list an author, begin the entry with the name of the government or other organization that published it, followed by a period and the name of the agency that issued the document.
- If the pamphlet or brochure lists no author and no publisher, begin with the document's title.

10. Article in Journal That Numbers Its Pages Continuously through Each Volume—MLA

McLaurin, JoAnne, and Avijit Chakrabartty. "Characterization of the Interactions of Alzheimer ß-amyloid Peptides with Phospholipid Membranes." *European Journal of Biochemistry* 245 (1997): 355-63. Print.

- Place the article's title in quotation marks and capitalize all major words.
- After the journal's name, give the volume number, followed by the year (in parentheses).
- For page numbers larger than 99, give only the last two digits unless more are needed for clarity. Examples: "355-63" and "394-405."

11. Article in Journal That Numbers Its Pages Separately for Each Issue—MLA

Bradley, John, and Gilbert Soulodre. "The Acoustics of Concert Halls." *Physics World* 10.5 (1997): 33-37. Print.

- After the volume number, add a period and the issue number. (In the example, "10.5" signifies volume 10, issue 5.)

12. Article in a Popular Magazine—MLA

Cowley, Geoffrey. "Cardiac Contagion." *Newsweek* 28 Apr. 1997: 69-70. Print.

- Give the full date of the issue, beginning with the day and abbreviating the month.
- Do not give the issue or volume number.

13. Newspaper Article—MLA

Wade, Nicholas. "Doctors Record Signals of Brain Cells Linked to Memory." *New York Times* 24 May 1997, natl. ed.: A8. Print.

- If the newspaper lists an edition (for example, "late edition") in the masthead, place a comma after the date and add the edition's name, using abbreviations where reasonable. In the example, "natl. ed." indicates "national edition."

14. Article with No Author Listed—MLA

"Younger Manufacturing Workers Can Teach Old Dogs New Tricks."
Manufacturing 86.5(2007): 5. Print.

- Begin with the article's title.
- This example gives the issue number ("5") because the journal numbers its pages separately for each issue (see Example 11).

Electronic Sources

WWW

For more about the MLA documentation style for other types of electronic sources, go to Appendix A at **www .cengage.com/english/ anderson7e.**

15. Online Report That Is Not Available in Print—MLA

International Business Machines. *What's User-Centered Design*? n.d. Web. 25 Jan. 2010 <https://www-01.ibm.com/software/ucd/ucd.html#whatisucd>.

- If the title of the overall website is distinct from the title of the work, give the website title after the title of the work.
- The title is italicized if it is independent; it is in roman type and quotation marks if it is part of a larger work.
- After the title, give the date of the publication or posting if known. If not known, use "n.d."
- After the date of publication or posting, give the medium of publication (web) followed by the date of access.

Because material that is available on the World Wide Web changes continuously, MLA entries include the date you most recently viewed the item online.

16. Online Journal Article That Is Not Available in Print—MLA

Toll, David. "Artificial Intelligence Applications in Geotechnical Engineering." *Electronic Journal of Geotechnical Engineering* 1 (1996). Web. 12 July 2010. <http://geotech.civen.okstate.edu/ejge/Abs9608.htm>.

To prevent confusion about punctuation in Internet addresses, the MLA style encloses these addresses in angled brackets: < >.

17. Online Journal Article That Is Also Available in Print—MLA

Porfiri, Maurizio. "Charge Dynamics in Ionic Polymer Metal Composites." *Journal of Applied Physics, 104* (2008): 104915-25. Web. Nov. 2010 <http://scitation.aip. org/journals/doc/JAPIAU-ft/vol_104/iss_10/104915_1.html>.

- First, give complete print publication information just as you would if you were describing the print publication only. Follow it with the medium of publication (Web) and the date you accessed the source, then the URL.

If the URL of a document is excessively long and complicated, give the URL of the search page, rather than the URL of the document. If the site does not have a search page, give the URL of the home page, followed by the word *Path*, a colon, and sequence of links (use semicolons to separate names of the links and a period after the last link).

Additional Common Sources

18. E-mail—MLA

Grube, Melvin. "Effects of Sun Spots." Message to Justin Timor. 4 Dec. 2010. E-mail.

- Place the title (taken from the subject line) in quotation marks.

19. Letter—MLA

Cawthorne, Linda A. Letter to the author. 24 Aug. 2010. TS.

- The medium of publication is typescript: (TS).

Oparka, Sabrina. Telephone interview. 17 Oct. 2009.

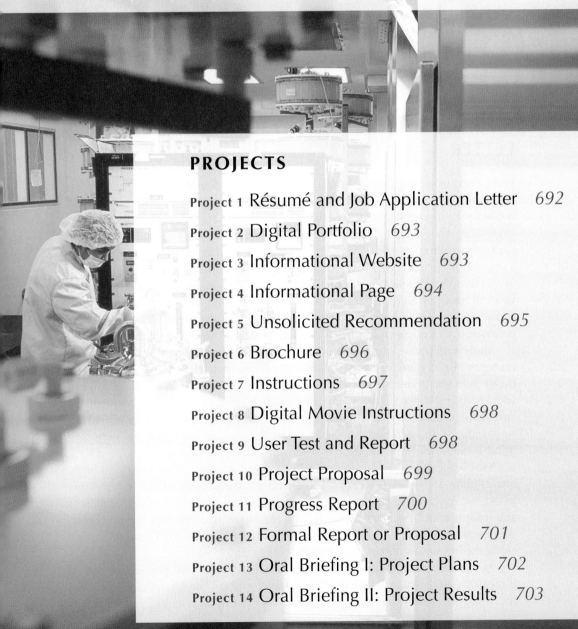

Appendix B
Projects

PROJECTS

This appendix contains writing and speaking assignments that your instructor might ask you to complete. All share this important feature: They ask you to communicate to particular people for specific purposes that closely resemble the purposes you will have for writing in your career.

Some of the assignments contain specifications about such things as length and format. Your instructor may change these specifications to tailor the assignments to your writing course.

Note to the Instructor. You will find helpful notes about these assignments in the *Instructor's Manual.* The book's website includes versions of these assignments you can download and adapt to your course. It also includes additional assignments and cases.

PROJECT 1: RÉSUMÉ AND JOB APPLICATION LETTER

Write a résumé and job application letter addressed to some real person in an organization with which you might actually seek employment. If you will graduate this year, you will probably want to write for a full-time, permanent position. If you aren't about to graduate, you may want to apply for a summer position or an internship. If you are currently working, imagine that you have decided to change jobs—perhaps to obtain a promotion, secure higher pay, or find more challenging and interesting work.

To complete this project, you may need to do some research. Among other things, you will have to find an organization that really employs people in the kind of job you want, and you will need to learn something about the organization so you can persuade your readers in the organization that you are knowledgeable about it.

Many employers publish brochures about themselves; your campus placement center or library may have copies. Many employers also post company websites. If the sources you find don't give the name of some specific person to whom you can address your letter, call the organization's switchboard to ask for the name of the employment director or the manager of the particular department in which you would like to work. While you work on this assignment, keep this real person in mind—even if you will not actually send your letter to him or her.

Your letter should be an original typed page, but your résumé may be a high-quality photocopy of a typed original. Remember that the appearance of your résumé, letter, and envelope will affect your readers, as will your attention to such details as grammar and spelling. Enclose your letter and résumé in an envelope complete with your return address and your reader's name and address. You will find information about the formats for letters and envelopes on pages 548–549.

As part of your package, include the names, addresses, and phone numbers of three or four references. These may be included within your résumé. If you choose instead to have your résumé say, "References available upon request," enclose a copy of the list of references you would send if the list were requested by an employer. Throughout your work on this project, you should carefully and creatively follow the advice given in Chapter 2 on résumés and letters of application.

PROJECT 2: DIGITAL PORTFOLIO

As Chapter 20 explains, a digital portfolio can be a powerful supplement to your résumé and job application letters. It demonstrates your abilities by showing projects you have completed. For this assignment, create a digital portfolio you could give employers in your job search.

In the portfolio, include five or more samples of your work. These may be printed documents, video clips of presentations you have made, pictures of projects you've completed, or other appropriate items. You should have at least two groups of two or more samples, such as two engineering projects and two communication projects. Include your résumé.

Provide your portfolio with a home page and with section pages for each of your groups of samples. On each section page, supply helpful and persuasive information about each sample plus a link to the sample. In the portfolio, also include at least two links to external websites.

As you plan, draft, and revise your portfolio, remember that its purpose is to persuade employers that you have the knowledge and abilities they desire in a person they would hire for the job you want.

PROJECT 3: INFORMATIONAL WEBSITE

For this project, you are to create an informational website on a topic that interests you. Begin by identifying a target audience, purpose, and subject for your site. The subject may be related to your major or to something you enjoy and know a lot about. Your site should have the following elements:

- Four or more pages, one of which is a home page (entitled "index.html").
- Links among your pages.
- One or more links to other websites related to your topic.
- At least one image.

To count as a "page" in this project, a page must fill one window on a computer screen when the window is at its maximum size. The material in the window must include some text, but may also contain one or more images.

All pages, including the home page, should be informative, not just entertaining. The text is to be text you have written, not text you have downloaded or otherwise copied from someone else (see the Ethics Guideline on pages 158–162.). You may obtain your images by downloading them from the web, provided that you are sure they are not copyrighted and that use of them is not otherwise restricted. Avoid movies, large images, and other items that consume much memory and take a long time to load.

Your site may include a link to your personal web page. However, your personal web page does not count among the four that must be about your subject.

Your site should possess the following characteristics:

- All information is presented in a way that is suited to your audience and purpose.
- The text is accurate, informative, and well written.
- The pages are easy to read and use when accessed on a computer screen.
- The pages are easy to navigate individually and as a group.
- The site is well organized.
- The pages are unified with one another in terms of visual design, writing style, and purpose.

    Turn in the following items:

- A memo to your instructor that identifies the audience and purpose of your site.
- A disk with all files for your site. Be sure your name is on the disk. If your instructor permits, you may instead post your site on a server and include the URL in your memo.

PROJECT 4: INFORMATIONAL PAGE

Project Created by Jennie Dautermann

A bus schedule, a chart comparing the features of competing computer programs, and the centerfold in a natural history magazine that uses a timeline, diagrams, and words to explain the evolution of horses—all these are "informational pages." They convey a complex set of information or ideas principally by means of visual design, rather than relying solely on sentences and paragraphs.

For this assignment, you are to create an informational page addressed to some specific set of readers. Use tables or charts. Mix diagrams with words. Or use other ways of presenting information visually to achieve your communication objectives. But use only one side of one sheet of paper.

When planning your informational page, think carefully about how your readers will use your information and about how you can use graphics to make the facts you present as accessible, understandable, and useful as possible. The advice given in Chapters 13 ("Creating Reader-Centered Graphics") and 14 ("Designing Reader-Centered Pages and Documents") can be especially helpful.

Here are informational pages you might create (note that each targets a specific group of readers):

- An explanation of a basic concept in your major, addressed to students who have just begun course work in it
- An explanation of a concept from your major that is important for members of the general public to understand as they make practical decisions
- A description of a process that would be important for clients or customers to know if they are going to purchase products or services from the kind of employer you would like to work for after graduation
- A reference card describing the computer labs on your campus, addressed to students who would want to use the labs
- A study guide that might help someone in one of your courses learn a topic that will be on an exam

PROJECT 5: UNSOLICITED RECOMMENDATION

This assignment is your chance to improve the world—or at least one small corner of it. You are to write a letter of 400 to 800 words in which you make an unsolicited recommendation for improving the operation of some organization with which you have personal contact—perhaps the company that employed you last summer, a club you belong to, or your sorority or fraternity.

There are four important restrictions on the recommendation you make:

1. Your recommendation must concern a real situation in which your letter could *really* bring about change. As you consider possible topics, focus on situations that can be improved by the modest measures that you can argue for effectively in a relatively brief letter. It is not necessary, however, that your letter aim to bring about a complete solution. In your letter, you might aim to persuade one of the key people in the organization that your recommendation will serve the organization's best interests.

2. Your recommendation must be unsolicited—that is, it must be addressed to someone who has not asked for your advice.

3. Your recommendation must concern the way an organization operates, not just the way one or more individuals think or behave.

4. Your recommendation may *not* involve a problem that would be decided in an essentially political manner. Thus, you are not to write on a problem that would be decided by elected officials (such as members of Congress or the city council), and you may not address a problem that would be raised in a political campaign.

Of course, you will have to write to an *actual* person, someone who, in fact, has the power to help make the change you recommend. You may have to investigate to learn who that person is. Try to learn also how that person feels about the situation you hope to improve. Keep in mind that most people are inclined to reject advice they haven't asked for; that's part of the challenge of this assignment. From time to time throughout your career, you will find that you want to make recommendations that your reader hasn't requested.

In the past, students have completed this assignment by writing on such matters as the following:

- A no-cost way that the student's summer employer could more efficiently handle merchandise on the loading dock
- A detailed strategy for increasing attendance at the meetings of a club the student belonged to
- A proposal that the Office of the Dean of Students establish a self-supporting legal-aid service for students

Bear in mind that one essential feature of a recommendation is that it compares two alternatives: keeping things the way they are now and changing them to the way you think they should be. You will have to make the change seem to be the better alternative *from your reader's point of view.* To do this, you will find it helpful to understand why the organization does things in the present way. By understanding the goals of the present method, you will probably gain insight into the criteria that your reader will apply when comparing the present method with the method you recommend.

When preparing this project, follow the advice about the letter or memo format on pages 548 and 552. Figure 5.11 (page 144) shows an unsolicited recommendation written by a student in response to this assignment.

PROJECT 6: BROCHURE

Create a brochure about some academic major or student service on your campus. Alternatively, create a brochure for a service organization in your community.

Begin by interviewing people at the organization to learn about their aims for such a brochure. Then follow the advice given in Chapter 3 ("Defining Your Communication's Objectives") to learn about the target audience for the brochure. Remember that to be effective the brochure must meet the needs of both the organization and the readers.

Use a folded 8½ × 14-inch sheet of paper so that there are three columns (or panels) on each side of the sheet. When the brochure is folded, the front panel should serve as a cover. In addition to any artwork you may decide to include on the cover, use at least one graphic (such as a table, flowchart, drawing, or photograph) in the text of the brochure. Note that your cover need not have any artwork; it may consist solely of attractively lettered and arranged words that identify the topic of your brochure. Along with your brochure, turn in copies of any existing brochures or other printed material you used while working on this project.

Your success in this project will depend largely on your ability to predict the questions your readers will have about your subject—and on your ability to answer those questions clearly, concisely, and usefully. Think very carefully as well about how you want your brochure to alter your readers' attitudes about your subject. Remember that along with the prose, the neatness and graphic design of your brochure will have a major effect on your readers' attitudes.

PROJECT 7: INSTRUCTIONS

Write a set of instructions that will enable your readers to operate some device or perform some process used in your major. The procedure must involve at least twenty-four steps.

With the permission of your instructor, you may also choose from topics that are not related to your major. Such topics might include the following:

- Using some special feature of a desktop publishing or spreadsheet program
- Operating a piece of equipment used in your major
- Changing a bicycle tire
- Rigging a sailboat
- Starting an aquarium
- Some other procedure of interest to you that includes at least twenty-four steps

Your instructions should guide your readers through some specific process that your classmates or instructor could actually perform. Do not write generic instructions for performing a general procedure. For instance, do not write instructions for "Operating a Microscope," but rather for "Operating the Thompson Model 200 Microscope."

Be sure to divide the overall procedure into groups of steps rather than presenting all the steps in a single list. Use headings to label the groups of steps.

When preparing your instructions, pay careful attention to the graphic design of your finished communication. You must include at least one illustration, and you may rely heavily on figures if they are the most effective way for you to achieve your objectives. In fact, your instructions need not contain a single sentence.

Finally, your instructor may require you to use a page design that has two or more columns (rather than having a single column of type that runs all the way from the left-hand margin to the right-hand margin). In a two-column design, you might put all of your steps in the left-hand column and all of your accompanying illustrations in the right-hand column. Alternatively, you might mix both text and figures in both columns. Large figures and the title for the instructions can span both columns.

You may use the format for your instructions that you believe will work best—whether it is a single sheet of 8½ × 11-inch paper, a booklet printed on smaller paper, or some other design.

Don't forget that your instructions must be accurate.

Note to the Instructor. If you wish to make this a larger project, increase the minimum length of the procedure about which your students write. In this case, you may also want to specify that they create an instruction manual rather than, for example, an instruction sheet.

PROJECT 8: DIGITAL MOVIE INSTRUCTIONS

There are many ways to provide instructions for step-by-step procedures. Changing the medium you use can significantly affect what you say and show as well as how you say and show it.

In this assignment, your challenge is to create the first two minutes of a digital movie that guides viewers through a procedure they haven't performed before. To succeed, you'll need to take a much different approach than you would use to prepare printed instructions. Instead of writing a list of numbered steps, you will prepare a script. Rather than using graphics that illustrate selected steps, you will design a visual track that will be on the screen every moment from start to finish. Rather than placing your words and graphics side by side on the page, you must create a film that weaves speech and images together. The outcome must be clear, precise, well integrated, and easy to follow.

To make your digital movie instructions, you can use such free programs as Windows MovieMaker, which is included for free with the Windows operating systems for PCs, or iMovie, which is included for free with MacIntosh operating systems. Your images may be photos, drawings, or video clips.

As you write your script, provide viewers with same understanding of the structure of the process that headings provide in printed instructions. When recording the script, control the pace of your speech to make your instructions clear and understandable upon one hearing.

Your visual track should enable viewers to see all significant parts and actions. All steps should be illustrated visually.

Your sound and visual tracks should be fully integrated and carefully coordinated so that viewers simultaneously see and hear about the same actions.

PROJECT 9: USER TEST AND REPORT

As you learned in Chapter 6, one excellent way to evaluate the usability and persuasiveness of a draft is to conduct a user test in which you give your draft to members of your target audience, asking them to use it in the same way that your target readers will use the final draft. For this assignment, you are to conduct a user test of a nearly finished draft of a project you are preparing in this course (or another course, if your instructor permits), then report the results in a memo to your instructor. Instructions make an excellent subject for a user test, but other communications can be evaluated in this way also.

For your test, use a draft that is as close as possible to what you envision for your final draft. Ask two people to serve as your test readers, and arrange for them to work independently. If you cannot recruit test readers who are from your target audience, choose people who resemble the target readers as closely as possible. Similarly, if you cannot arrange for your test readers to read in exactly the same circumstances that your target readers will, simulate those conditions in some reasonable way. Also use simulation if your communication involves a potentially dangerous step (such as jacking up a car or pouring a strong acid from one container to another) that might result in injury to your test readers if they make a mistake.

Write your report in the memo format (page 552) and use the superstructure for empirical research reports (page 581). To make your report readable and informative for your instructor, consider the following advice about the sections of your report:

- **Introduction.** Remind your instructor of the topic and target audience of the communication you are testing.
- **Objectives.** Identify the objectives of your communication (see Chapter 3) and those of your test. Remember that even instructions have to be persuasive as well as usable.
- **Method.** Describe your draft (how closely does it resemble your planned final draft?), your test readers (who were they, and why are they good representatives of your target audience?), the location of your test (how closely does it resemble the setting in which your target readers will use your communication?), and your procedure (what did you ask your test readers to do, and how did you gather information from them?).
- **Results and discussion.** Report the results your test produced (where did your test readers have difficulties, and what did they say about your communication?) and tell what these indicate about your communication. Be quite specific in this section.
- **Conclusion.** Tell what you learned overall from your test and indicate the specific revisions you will make as a result of what you learned.

PROJECT 10: PROJECT PROPOSAL

Write a proposal seeking your instructor's approval for a project you will prepare later this term.

Your work on this proposal serves three important purposes. First, it provides an occasion for you and your instructor to agree about what you will do for the later project. Second, it gives you experience at writing a proposal, a task that will be very important to you in your career. Third, it gives you a chance to demonstrate your mastery of the material in Chapters 3 ("Defining Your Communication's Objectives"), 4 ("Planning for Usability"), and 5 ("Planning Your Persuasive Strategies").

Notice that while working on this assignment, you will have to define the objectives of two different communications: (1) the *proposal* you are writing now, which is addressed to your instructor; and (2) the *project* you are seeking approval to write, whose purpose and audience you will have to describe to your instructor in the proposal.

When writing your proposal, you may think of your instructor as a person who looks forward with pleasure to working with you on your final project and wants to be sure that you choose a project from which you can learn a great deal and on which you can do a good job. Until your instructor learns from your proposal some details about your proposed project, however, his or her attitude toward it will be neutral. While reading your proposal, your instructor will seek to answer many questions, including the following:

- What kind of communication do you wish to prepare?
- Who will its readers be?
- What is its purpose?
  - What is the final result you want it to bring about?
  - What task will it enable its readers to perform?
  - How will it alter its readers' attitudes?
- Is this a kind of communication you will have to prepare at work?
- Can you write the communication effectively in the time left in the term using resources that are readily available to you?

For additional insights into the questions your instructor (like the reader of any proposal) will ask, see Chapter 23.

Your proposal should be between 400 and 800 words long. Write it in the memo format (see page 552), use headings, and include a schedule chart (page 377).

PROJECT 11: PROGRESS REPORT

Write a report of between 400 and 800 words in which you tell your instructor how you are progressing on the writing project you are currently preparing. Be sure to give your instructor a good sense not only of what you have accomplished but also of what problems you have encountered or anticipate. Use the memo format (see page 552).

PROJECT 12: FORMAL REPORT OR PROPOSAL

Write an empirical research report, feasibility report, or proposal. Whichever form of communication you write, it must be designed to help some organization—real or imaginary—solve some problem or achieve some goal, and you must write it in response to a request (again, real or imaginary) from the organization you are addressing.

A real situation is one you have actually encountered. It might involve your employer, your major department, or a service group to which you belong—to name just a few of the possibilities. Students writing on real situations have prepared projects with such titles as:

- "Feasibility of Using a Computer Database to Catalog the Art Department's Slide Library." The student wrote this feasibility report at the request of the chair of the Art Department.

- "Attitudes of Participants in Merit Hotel's R.S.V.P. Club." The student wrote this empirical research report at the request of the hotel, which wanted to find ways of improving a marketing program that rewarded administrative assistants who booked their companies' visitors at that hotel rather than at one of the hotel's competitors.

- "Expanding the Dietetic Services at the Campus Health Center: A Proposal." The student wrote this proposal to the college administration at the request of the part-time dietitian employed by the Health Center.

An imaginary situation is one that you create to simulate the kinds of situations that you will find yourself in once you begin your career. You pretend that you have begun working for an employer who has asked you to use your specialized training to solve some problems or answer some questions that face his or her organization. You may imagine that you are a regular employee or that you are a special consultant. Students writing about imaginary situations have prepared formal reports with titles such as:

- "Improving the Operations of the Gift Shop at Six Flags over Georgia." The student who wrote this proposal had worked at this shop for a summer job; she imagined that she had been hired by the manager to study its operation and recommend improvements.

- "Performance of Three Lubricants at Very Low Temperatures." The student wrote this empirical research report about an experiment he had conducted in a laboratory class. He imagined that he worked for a company that wanted to test the lubricants for use in manufacturing equipment used at temperatures below $-100°F$.

- "Upgrading the Monitoring and Communication System in the Psychology Clinic." The student who wrote this report imagined that she had been asked by the Psychology Clinic to investigate the possibility of purchasing equipment that would improve its monitoring and communication system. All of her information about the clinic and the equipment were real.

For this project, prepare a communication that has a cover, title page, and the other book-like features described in Chapter 12. Remember that your purpose is to help your readers make a practical decision or take a practical action in a real or imaginary organization. The body of your report should be between twelve and twenty pages long (not counting cover, executive summary, title page, table of contents, appendixes, and similar parts).

PROJECT 13: ORAL BRIEFING I: PROJECT PLANS

At work, you will sometimes be asked to report in brief talks about projects upon which you are working. For this assignment, you are to give an oral briefing to the class about your final project. Here are the things you should cover:

- **What kind of communication are you writing?** Who will your readers be? What role will you be playing? Identify your readers by telling what organization they are in and what positions your key readers hold. Describe your role by saying whether you are imagining that you work for the company as a regular employee or have been hired as a consultant. State whom you report to.
- **What organizational problem will your communication help your readers solve?** What need or goal will it help them satisfy or reach? Provide full background so your classmates can understand the situation from your readers' point of view.
- **What are you doing to solve the problem?**
  - *Your research activities:* What kind of information are you gathering and how, or what kind of analysis are you providing and why?
  - *Your writing activities:* How do you plan to organize and present your information? What will your communication look like?
- **What is the gist of your message to your readers?** What are the main points you are planning to make?

As you prepare and deliver your oral briefing, pretend that you are interviewing for a job (or for a new job) and that the prospective employer has asked you to give an oral briefing about a project of yours for which you are now writing a report or proposal. The members of your class can play the role of the people your employer has asked to attend your presentation. Pretend that your classmates have not heard about your project as yet, even though you may have already discussed it in class several times. This means that you will have to provide all the background information that will enable your listeners to understand the organizational situation in which you are writing.

As the name implies, a briefing is a brief presentation. Make yours between four and five minutes long—no longer. Gauge the time by making timed rehearsals. In your briefing, use at least one graphic. It might show an outline for your project, or it might be one of the graphics you will use within your project. You may present this graphic as a poster, overhead transparency, or handout, and you may use more than one graphic if doing so will increase the effectiveness of your briefing.

PROJECT 14: ORAL BRIEFING II: PROJECT RESULTS

At work, people often present the results of their major projects twice: once in a written communication and a second time in an oral briefing that covers the major points of the written document. In some ways, this briefing is like an executive summary—an overview of all the important things presented in more detail in writing.

For this assignment, you are to give an oral briefing on one of your writing projects. Address the class as if it were the same audience that you address in writing, and imagine that the audience has not yet read your communication. Limit your briefing to four or five minutes—no longer (see the last paragraph of Project 13). Use at least one graphic.

If your instructor requests, include a slide presentation made with PowerPoint or a similar program.

CREDITS

This page constitutes an extension of the copyright page. We have made every effort to trace the ownership of all copyrighted material and to secure permission from copyright holders. In the event of any question arising as to the use of any material, we will be pleased to make the necessary corrections in future printings. Thanks are due to the following authors, publishers, and agents for permission to use the material indicated.

Text and Illustrations

p. 8 (Fig. 1.1): "To clean the printer" from *HP Laser-Jet 5P and 5MP Printer User's Manual,* pp. 4–19, Boise, Idaho, Hewlett Packard. Copyright © Hewlett Packard. Reproduced with permission.

p. 110 (Fig. 4.5): Microsoft product screen shot(s). Reprinted with permission from Microsoft Corporation.

p. 122 (Fig. 5.1): Welch's brochure. Reprinted with permission from Welch's Foods, Inc.

pp. 144–145 (Fig. 5.11): From COLES/VOPAT. COLES WHAT MAKES WRITING GOOD, 1e. © 1985 Heinle/Arts & Sciences, a part of Cengage Learning, Inc. Reproduced by permission. www.cengage.com/permissions.

p. 221 (Fig. 8.4): Table of Contents from a Booklet That Uses Three Kinds of Chapter and Section Headings. (Source: National Audubon Brochure on Pest Control. From a National Audubon Society brochure. Copyright © 1993 National Audubon Society. Reprinted by permission of the National Audubon Society.)

p. 223 (Fig. 8.5): Indentation Used to Indicate Hierarchy: Managing Disk Partitions from Robert Cowart and Kenneth Gregg, Windows NT Server 4.0 Administrator's Bible (Foster City, CA: IDG Books, 1996), p. 274. From Robert Cowart and Kenneth Gregg, *Windows NT Server 4.0 Administrator's Bible* (Foster City, CA: IDG Books, 1996), p. 274.

p. 232: Case: Increasing Organ Donations (Survey results are from the Gallup Organization, Inc., The American Public's Attitudes toward Organ Donation and Transplantation, conducted for the Partnership for Organ Donations, Boston, MA, February 1993. Used with permission.)

p. 238: Detecting Coronary Disease, Passage Organized According to Formal Classification, from "Wavelet Applications in Medicine," *IEEE Spectrum, 34*(5), 50–51 (1977, May) From "Wavelet Applications in Medicine," by Metin Akay, *IEEE Spectrum, 34*(6), 50–51 (1977, May). Copyright © 1977 IEEE.

p. 249: Diagram Organized by Segmentation, from Lawrence Livermore Lab, *Science & Technology Review,* June, 2004, http://www.llnl.gov/str/Juno05/Aufderheide.html Courtesy of the University of California, Lawrence Livermore National Laboratory, and the Department of Energy under whose auspices the work was performed. Scientific Editor: Maurice B. Aufderheide.

p. 251: How Planets are Formed. (Explanation Organized by Segmenting a Process. From Cochran, William (1997, July), "Extrasolar Planets," Physics World, July, 1997, by the Institute of Physics Publishing, Bristol, England.

p. 255: Passage Organized Around Cause and Effect from "Death of Dinosaurs: A True Story?" by Boyce Rensberger, from *Science Digest, 94,* 5, May, 1986, pp. 28–32. Reprinted by permission of the Author.

p. 299 (Fig. 10.2): Beginning of a Service Manual from Detroit Diesel Engines Series 5 General Information from Detroit Diesel Engines Series 53 Service Manual. Copyright © 1990. Reprinted courtesy of Detroit Diesel Corporation. 3 Service Manual (Detroit, Mich: 1990), p. 4.

p. 327 (Fig. 12.10): A List of Symbols. Helsinki University of Technology, Laboratory of Acoustics and Audio Signal Processing, Finland.

p. 328 (Fig. 12.11): Index of Instruction Manual, "*Detroit Diesel Engines Series 53 Service Manual,*" 1990, courtesy of Detroit Diesel Allison Index of Detroit Diesel Engines Series 53 Service Manual. Copyright © 1990. Reprinted courtesy of Detroit Diesel Corporation, division, General Motors Corp.

p. 332 (Fig. 13.1): Aero Safety Graphics Inc.

p. 334 (Fig. 13.2): Drawing that Shows How to Do Something from *Power Macintosh User's Manual for Power Macintosh 6500 Series Computers,* p. 206 Copyright © 1997 by Apple Computer.

p. 334 (Fig. 13.3): Fig. 11.4: Diagram that Explains a Process. From "Life's Undersea Beginnings," by J. Cone, from *Earth,* 3, 38.

p. 335 (Fig. 13.4): Visual Aids that Shows What Something Looks Like: "Silicon Microelectronics Technology," *Bell Labs Technical Journal,* 2. 94, Autumn, 1997 Reprinted by permission of Lucent Technologies, from *Bell Labs Technical Journal,* Autumn, 1997, p. 94.

p. 335 (Fig. 13.5): Display Information in a useful way, "Developing Time in Minutes," KODAK film. Reprinted courtesy of Eastman Kodak Company.

p. 335 (Fig. 13.6): Graph that Shows Trends or other Numerical Relationships from "Larval dynamics of a riverine metapopulation: Implications for Zebra mussel recruitment, dispersal, and control in a large-river system," *Journal of the American Benthological Society,* 16, (1997):591. From J.A. Stroeckel, D.W. Schnieder, L.A. Soeken, K.D. Blodgett, and R.E. Sparks, "Larval Dynamics of a Riverine Metapopulation: Implications for Zebra Mussell Recruitment, Dispersal, and Control in a Large-River System," *Journal of the North American Benthological Society,* 16 (1997): 591. Reprinted by permission of the Journal of the North American Benthological Society.

p. 345 (Fig. 13.9): Some Ways a Color's Appearance is Affected by the Surrounding Colors based on Jan V. White, *Color for the Electronic Age* (New York: Watson-Guptill, 1990), p. 16. Courtesy of Jan V. White.

p. 345 (Fig. 13.10): Bright Colors Used to Focus Attention from "High-speed silicon-germanium electronics" from *Scientific American,* 270, 67, March, 1994. Copyright © 1994 by Ian Warpole/Scientific American, Inc. All rights reserved.

p. 346 (Fig. 13.11): Cover Illustration from the *Journal of Eukaryotic Microbiology,* supplement to Vol. 50 (2003), Society of Protozoologists, Lawrence, Kansas. Cover Illustration from "The Crystal Structure of Dihydrofolate Reudctase-Thymidylate Synthase from Cryptospondium hominis Reveals Novel Architecture for the Binfunctional Enzyme" by Robert H. O'Neil, Ryan H. Lillen, Bruce R. Donald, Robert M. Stroup & Amy C. Anderson from the *Journal of Eukaryotic Microbiology,* supplement to Vol. 50, Proceedings of the Eighth International Workshops on Opportunistic Protists, Hilo, Hawaii, July 23–29, 2003, pp. 555–556.

p. 352 (Fig. 13.12): Explanatory Text Incorporated into a Figure: USCAR powder clear coat process from "Manufacturing Solutions," *Automotive Engineering,* December, 1997. Copyright © 1997 Society of Automotive Engineers International (SAE), Warrendale, PA.

p. 366: Pie Chart: "How Products are Packaged Worldwide from "Charting a course for the future of corrugated," *Pulp & Paper International,* 39, (1997):22. Copyright © 1997 by Miller Freeman, Inc. Reprinted with permission.

p. 370: Drawing of Filling Windshield Fluid: The Reserve Tank. From 1998 *Honda Accord Owner's Manual,* Honda, p. 217. Courtesy of Honda Motor Company of America.

p. 370: Drawing that Shows Wiring a Switch: Wiring from *The Home Depot, Improvement 123: Expert Advice* from The Home Depot (Des Moines: Meredith, 1995): 169. Courtesy of The Home Depot.

p. 371: External View of a Machine: The ISEP L100 Laboratory Model. "Carousel Creates Continuous Ion Exchange," from *Chilton's Food Engineering,* Vol. 65, No. 10 : 80, June, 1993. Reprinted by permission of Advanced Separation Technologies, Inc.

p. 371: Cutaway of Audi Engine from "Getting More Direct." Cutaway of Audi Engine from "Getting More Direct," *Automotive Engineering,* 105, 84 (December, 1997), Artwork courtesy of Audi of America, Inc.

p. 371: National Institute of Health

p. 371: Exploded View of the 30-mm HRG Mechanical Components from NASA, Research and technology report: Goddard Space Flight Center. From U.S. National Aeronautics and Space Administration (1996). *Research and Technology report: Goddard Space Flight Center,* U.S. Government Printing Office, 120.

p. 375 (Fig. 13.15): The Carbon Cycle from *Environmental Science: A Global Concern* (4th ed.), by William P. Cunningham & Barbara W. Saigo, 1997. Copyright © 1997 McGraw-Hill, Inc. Reprinted by permission of The McGraw-Hill Companies.

p. 375: Flow Chart Showing Special Techniques Used in System Analysis. From *Systems Analysis and Design Methods* (4th edition), by J.L. Whitten, L.D. Bentley, & K. C. Dittman, 1998, p. 384. Copyright © 1998 McGraw-Hill, Inc. Reprinted by permission of The McGraw-Hill Companies.

p. 382: Page Layout from *Water Environment Research,* Vol. 77, No. 4, July–August, 2005, p. 372. Reprinted from *Water Environment Research,* Vol. 77, No. 4, July–August, 2005, p. 372. Copyright © 2005 Water Environment Federation: Alexandria, VA.

p. 382 (Fig. 14.1b): Page Layout from *Porting UNIX /Linus Applications to Mac OS X,* page 33. Reprinted with permission from Microsoft Corporation.

p. 384 (Fig. 14.1): Nine Thumbnail Page Designs for the Same Material. Courtesy of Professor Joseph L. Cox III.

p. 389 (Fig. 14.3): Comparison of Irregularly Shaped Figures with and without Rectangular Enclosures. From Monarch Marking, *Operating Instructions for 100 Dail-A-Pricer Printer.* Reprinted by permission of Monarch Marketing Systems, Inc., now, by change of name, Paxar Americas, Inc.

p. 390 (Fig. 14.4): Grouping to Show Organization of Information on a Page from Microsoft, (1994) *Microsoft User's Guide,* p. 110. Reprinted with permission from Microsoft Corporation.

p. 393 (Fig. 14.5): Co-ordination of Size Treatment, Color and Typeface: Protecting Your Family from Lead Brochure from EPA. Based on United States Consumer Product Safety Commission, *Protect Your Family from Lead in Your Home.* (Washington D.C.: United States Environmental Protection Agency, 1995): 33–34.

p. 394 (Fig. 14.6): Rules and Icons Used to Aid Readers in Seeing the Structure of a Page, from Toshiba (1996) *Satellite Pro 400CS/500 User's Guide,* pp. xxxvii, 134, and 210. Reprinted by permission of Toshiba America, Inc., New York.

p. 396 (Fig. 14.7): Page Design Used to Create Visual Unity in a Communication from Chrysler Corporation (1997) from *1998 Dakota Pickup Owner's Manual,* pp. 7, 88, 151, 170. Copyright © 1997 Chrysler Corporation.

pp. 400–403 (Fig. 14.8): Package Insert for a Prescription Medication. Courtesy of USV Pharmaceutical Manufacturing Corporation.

p. 404 (Fig. 14.9): Sample Page for Use with Collaboration Exercise, "Slab Roller." Courtesy of Steve Oberjon.

p. 429: Layout of a Facility for Testing Computer Manuals. Courtesy of Microsoft Corporation.

p. 620: Feasibility Report. Courtesy of Allen J. Hines.

p. 655 (Fig. 27.1): Ch. 1, Introduction (and photo) to the Akron Standard, *Operator's Manual for the Tire Uniformity Optimizer,* (1986) p. 1. Used with permission of ITN Ride Quality Products.

p. 657 (Fig. 27.2): Well Designed Presentation of Procedures, first from *Plain Paper Copier Model SF-2-22 Operation Manual:* Sharp Electronics, p. 10. Reproduced with the permission of Sharp Electronics Corporation.

p. 659 (Fig. 27.3): Drawing that Shows Readers Where to Locate Parts of a Camcorder from *RCA Pro808A Camcorder User's Guide,*

Thompson Electronics, p. 28. Courtesy of Thomson Consumer Electronics.

p. 660 (Fig. 27.4): Drawings that Show How to Do Something from *Lifescan* (1998), "Obtaining a Blood Sample" (Milpitas, CA). Courtesy of Johnson & Johnson.

p. 662 (Fig. 27.6): Trouble-shooting Section from a Manual from the Akron Standard, *Operator's Manual for the Tire Uniformity Optimizer,* (1986) p. 1. Used with permission of ITN Ride Quality Products.

Photos

p. 13 (Fig. 1.2): Courtesy of Dell, Inc.

p. 21: Courtesy of Boeing, Courtesy of International Business Machines Corporation, Courtesy of General Mills, Inc., Courtesy of Johnson & Johnson, Courtesy of Mazda Motor Corporation, Courtesy of Exxon Mobil Corporation, Courtesy of Deere & Company

p. 87 (Fig. 3.2): Courtesy of Ethicon, Inc.

p. 107 (Fig. 4.3): Courtesy of Epilepsy.com

p. 110, left: © Robin Nelson /PhotoEdit; middle: Courtesy of 3M Corporation; right: Courtesy of Miami University Communications

p. 122 (Fig 5.1): Courtesy of Welch's Foods Inc.

p. 123 (Fig 5.2): Courtesy of Johnson & Johnson

p. 124 (Fig 5.3): Courtesy of General Electric Company

p. 136 (Fig 5.7): Courtesy of International Fund for Animal Welfare

p. 139 (Fig 5.9): Courtesy of National Cancer Institute

p. 176: Courtesy of www.google.com

p. 177: © 2009 Yahoo! Inc. All Rights Reserved.

p. 177: Courtesy of Astrobiology Magazine

p. 182 (Fig. WG1.5): Courtesy of Miami University Library

p. 184: Courtesy of Miami University Library

p. 185: Courtesy of OhioLink

p. 220 (Fig. 8.3): Courtesy of The Nature Conservancy

p. 221 (Fig 8.4): Courtesy of National Audubon Society, Inc.

p. 224 (Fig 8.6): Reprinted with permission of John Wiley & Sons, Inc.

p. 244: Courtesy of The Department of Environment and Conservation (NSW)

p. 299 (Fig. 10.2): © 1990. Reprinted courtesy of Detroit Diesel Corporation

p. 327 (Fig. 12.10): Courtesy of Vesa Välimäki, Helsinki University of Technology

p. 328 (Fig. 12.11): Copyright © 1990. Reprinted courtesy of Detroit Diesel Corporation

p. 342 (Fig. 13.7): Copyright Heff D. Allred and Courtesy of Vladimir Kulyukin, Ph.D.

p. 372: Courtesy of Apple Computers

p. 382: Courtesy of Penton Media, Inc.

p. 383: Courtesy of Lawrence Livermore National Laboratory

p. 404: Courtesy of Steve Oberjon

p. 429: © ThinkStock LLC /RF/Index Stock Imagery

p. 468 (Fig. 18.2): Courtesy of Adobe Systems Incorporated

pp. 480–482, 484, 489, 492: Microsoft product screen shot(s) reprinted with permission from Microsoft Corporation.

p. 484 (Fig. 19.2): Courtesy of Sony Corporation

p. 512, top: Courtesy of National Park Service; bottom: Courtesy of the National Academy of Engineering

p. 513, top: Courtesy of IBM Corporation; bottom: Courtesy of National Aquarium in Baltimore

p. 514: Courtesy of the Nature Conservancy

p. 517 (Fig. 20.5): Courtesy of Institute of Electrical and Electronics Engineers, Inc

p. 519 (Fig. 20.6a, b, c): Courtesy of IBM Corporation

p. 655 (Fig. 27.1): Used with permission of Eagle-Picher Industries, Inc., Akron Standard

p. 669: Microsoft product screen shot(s) reprinted with permission from Microsoft Corporation

p. 670: Courtesy of Amy Beaton

REFERENCES

Adler, P. S. (1975). Beyond cultural identity: Reflections on cultural and multicultural man. In L. Samovar & R. Porter (Eds.), *Intercultural communication: A reader* (2nd ed., pp. 362–378). Belmont, CA: Wadsworth.

Anderson, J. R. (1995). Attention and sensory information processing. *Cognitive psychology and its implications* (4th ed., pp. 40–48). New York: W. H. Freeman.

Andrews, D. C., & Andrews, W. D. (1992). *Business communication* (2nd ed.). New York: Macmillan.

Anonymous. (1994). Personal interview with corporate executive who requested that the company remain anonymous.

Axtell, R. E. (2007). *Essential do's and taboos: The complete guide to international business and leisure travel.* Hoboken, NJ: Wiley.

Beaufort, A. (1999). *Writing in the real world: Making the transition from school to work.* New York: Teachers College Press.

Beebe, S. A. (1974). Eye contact: A nonverbal determinant of speaker credibility. *Speech Teacher, 23,* 21–25. Cited in M. F. Vargas, *Louder than words.* Ames, IA: Iowa State UP, 1986.

Beer, D. F., & McMurrey, D. (1997). *A guide to writing as an engineer.* New York: Wiley.

Benne, K. D., & Sheats, P. (2000). Functional roles of group members. In W. L. French, C. H. Bell, Jr., & R. A. Zawacki (Eds.), *Organization development and transformation: Managing effective change.* Boston: McGraw-Hill.

Bereiter, C., & Scardamalia, M. (1993). *Surpassing ourselves: An inquiry into the nature and implications of expertise.* Chicago, IL: Open Court.

Bolten, J. (1999). International business communication: An interactive approach. In C. R. Lovitt and D. Goswami (Eds.), *Exploring the rhetoric of international professional communication: An agenda for teachers and researchers.* (pp. 139–156). New York: Baywood.

Boren, M. T., & Ramey, J. (September 2000). Thinking aloud: Reconciling theory and practice. *IEEE Transactions on Professional Communication, 43*(3), 261–278.

Bosley, D. S. (1993). Cross-cultural collaboration: Whose culture is it, anyway? *Technical Communication Quarterly, 2,* 51–62.

Bostrom, R. N. (1981). *Persuasion.* Englewood Cliffs, NJ: Prentice-Hall.

Bowman, J. P. (April 1, 2002). Selling yourself. *Business communication: Managing information and relationships.* http://www.hcob.wmich.edu/bis/faculty/bowman/job2.html

Bransford, J. D., & Johnson, M. K. (1972). Contextual prerequisites for understanding: Some investigations of comprehension and recall. *Journal of Verbal Learning and Verbal Behavior, 11,* 717–726.

Brownell, J. (1999). Effective communication in multicultural organizations: A receiver-defined activity. In C. R. Lovitt and D. Goswami (Eds.), *Exploring the rhetoric of international professional communication: An agenda for teachers and researchers.* (pp. 171–187). New York: Baywood.

Carté, P., & Fox, C. (2004). *Bridging the culture gap: A practical guide to international business communication.* Sterling, VA: Kogan Page.

Chen, G. M., & Tan, L. (1995, April). *A theory of intercultural sensitivity.* Paper presented at the annual meeting of the Eastern Communication Association, Philadelphia.

Coleman, E. B. (1964). The comprehensibility of several grammatical transformations. *Journal of Applied Psychology, 48,* 186–190.

Connor, U. (1996). *Contrastive rhetoric: Cross-cultural aspects of second-language writing.* England: Cambridge.

Couture, B., and Rymer, J. (1993). Situational exigence: Composing processes on the job by writer's role and task value. In R. Spilka (Ed.), *Writing in the workplace: New research perspectives.* Carbondale, IL: Southern Illinois UP.

Covey, S. R. (1989). *The seven habits of highly effective people.* New York: Simon & Schuster.

Day, R. A., & Gastel, B. (2005). *How to write and publish a scientific paper* (6th ed.). Westport, CT: Greenwood.

Deakin University. (2010). *Japan career guide.* Retrieved January 4, 2010, from http://www.deakin.edu.au/current-students/services/careers/country-guides/japan.php

Dias, P., & Paré, A. (Eds.) (2000). *Transitions: Writing in academic and workplace settings.* Cresskill, NJ: Hampton.

Ding, D. D. (2003). The emergence of technical communication in China. *Journal of business and technical communication, 17,* 319–345.

Dixson, K. (November 13, 2001). Crafting an e-mail resume. *Business Week.*

Doheny-Farina, S. (1992). *Rhetoric, innovation, technology: Case studies of technical communication in technology transfers.* Cambridge, MA: MIT.

Ede, L., & Lunsford, A. (1990). *Singular texts/plural authors: Perspectives on collaborative writing.* Carbondale, IL: Southern Illinois UP.

Elashmawi, F., & Harris, P. (1998). *Multicultural Management 2000: Essential cultural insights for global management success.* Houston: Gulf.

Elbow, P. (1998). *Writing without teachers* (2nd ed.). New York: Oxford University Press.

Fisher, A. (December 1998). The high cost of living and not writing well. *Fortune,* 244.

Golen, S., Powers, C., & Titkemeyer, M. A. (1985). How to teach ethics in a basic business communication class. *Journal of Business Communication, 22,* 75–84.

Grosse, R., & Kujawa, D. (1988). *International business: Theory and managerial applications.* Burr Ridge, IL: Richard D. Irwin.

Gudykunst, W. B., & Ting-Toomey, S. (1988). *Culture and interpersonal communication.* Newbury Park, CA: Sage.

Gudykunst, W. B., Yang, S. M., & Nishida, T. (1987). Cultural differences in self-consciousness and unself-consciousness. *Communication Research, 14,* 7–36.

Hall, E. T. (1976). *Beyond culture.* Garden City, NY: Anchor.

Hays, R. B. (1985). A longitudinal study of friendship development. *Journal of Personality and Social Psychology, 48,* 909–924.

Herrington, A., & Moran, C. (2005). *Genre across the curriculum.* Logan, UT: Utah State University Press.

Herzberg, F. (1968). *Work and the nature of man.* Cleveland: World.

Hofstede, G. (2001). Culture's consequences: Comparing values, behaviors, institutions, and organizations across nations. Thousand Oaks, CA: Sage Publications.

Human Japanese. (2005). Correspondence, Japanese Style. http://www.humanjapanese.com/appendix/letters.htm. October 31, 2005.

Jewett, M., & Margolis, R. (1987). A study of the effects of the use of overhead transparencies on business meetings. Reported in B. Y. Auger, *How to run better business meetings: A reference guide for managers.* New York: McGraw-Hill.

Jobera.com. (2010). *French cover letter.* Retrieved January 4, 2010, from http://www.jobera.com/job-letters/international-cover-letters/france/french-cover-letter.htm

Kachru, Y. (1988). Writers in Hindi and English. In Alan Purves (Ed.), *Writing across languages and cultures: Issues in contrastive rhetoric* (pp. 109–137). Thousand Oaks, CA: Sage.

Kelman, H. C., & Hovland, C. I. (1953). Reinstatement of the communicator in delayed measurement of opinion change. *Journal of Abnormal and Social Psychology, 48,* 327–335.

Kleck, R. E., & Nuessle, W. (1967). Congruence between indicative and communicative functions of eye-contact in interpersonal relations. *British Journal of Social and Clinical Psychology, 6,* 256–266.

Kleinke, C. L., Bustos, A. A., Meeker, F. B., & Staneski, R. S. (1973). Effects of self-attributed gaze on interpersonal evaluations between males and females. *Journal of Experimental Psychology, 9,* 154–163.

Kostelnick, C., & Roberts, D. D. (1998). *Designing visual language: Strategies for professional communicators.* Boston: Allyn & Bacon.

Kulhavy, R. W., & Schwartz, N. H. (Winter 1981). Tone of communications and climate of perceptions. *Journal of Business Communication, 18,* 17–24.

Lauer, J. (1994). Persuasive writing on public issues. In R. Winterowd & V. Gillespie (Eds.), *Composition in context* (pp. 62–72). Carbondale, IL: Southern Illinois UP.

Lay, M. M. (1989). Interpersonal conflict in collaborative writing: What can we learn from gender studies? *Journal of Business and Technical Communication, 3*(2), 5–28.

Layton, P., & Simpson, A. J. (1975). Deep structure in sentence comprehension. *Journal of Verbal Learning and Verbal Behavior, 14,* 658–664.

Levie, W. H., & Lentz, R. (1982). Effects of text illustrations: A review of research. *Journal of Educational Psychology, 73,* 195–232.

Lustig, M. W., & Koester, J. (1993). *Intercultural competence: Interpersonal communication across cultures.* New York: HarperCollins.

Mancusi-Ungaro, H. R., Jr., & Rappaport, N. H. (April 1986). Preventing wound infections. *American Family Physician, 33,* 152.

Marsico, T. Personal communication with the author. 14 April 1997.

Maslow, A. H. (1970). *Motivation and personality.* New York: Harper & Row.

Maslow, A. H. (1970). *Motivation and personality* (2nd ed.). New York: Harper & Row.

Mathes, J. C., & Stevenson, D. W. (1991). *Designing technical reports* (2nd ed.). New York: Macmillan.

Mehrabian, A. (1972). *Nonverbal communication.* Chicago: Aldine.

Miller, C. R. (1984). Genre as social action. *Quarterly Journal of Speech, 70,* 151–167.

Munter, M. (1987). *Business communication: Strategy and skill.* Englewood Cliffs, NJ: Prentice-Hall.

Murray, R. L. (2003). Ed. Manke, K. L. *Understanding radioactive waste.* (5th ed.). Columbus, OH: Battelle.

National Commission on Writing. (2004). *Writing: a ticket to work . . . or a ticket out: A survey of business leaders. The National Commission on Writing for America's Families, Schools, and Colleges.* New York: College Board.

Neulip, J. W., & McCroskey, J. C. (1997). The development of a U.S. and generalized ethnocentrism scale. *Communication Research Reports, 14,* 385–398.

Nielsen, J. (2000). *Designing Web usability: The practice of simplicity.* Indianapolis, IN: New Riders.

Nishiyama, K. (1983). Intercultural problems in Japanese multinationals. *Communication: The Journal of the Communication Association of the Pacific, 12,* 58.

Patterson, V. (1996). Résumé talk from recruiters. *Journal of Career Planning & Employment, 56,* 33–38.

Petty, R. E., & Cacioppo, J. T. (1981). *Attitudes and persuasion: Classic and contemporary approaches.* Dubuque, IA: William C. Brown.

Petty, R. E., & Cacioppo, J. T. (1986). *Communication and persuasion: Central and peripheral routes to attitude change.* New York: Springer-Verlag.

Rabinowitz, C. J. & Carr, L. W. (2001). *Modern day vikings: A practical guide to interacting with the Swedes.* Yarmouth, ME: Intercultural Press.

Rajamäki, T., Alakomi, H., Ritvanen, T., Skyttä, E., Smolander, M., & Ahvenainen R. (2006). Application of an electronic nose for quality assessment of modified atmosphere packaged poultry meat. *Food Control, 17,* 5–13.

Randolph, G., Landis, D., & Tzeng, O. (1977). The effects of time and practice upon culture assimilator training. *International journal of intercultural relations, 1,* 105–119.

Ray, G. B. (1986). Vocally cued personality prototypes: An implicit personality theory approach. *Communication Monographs, 53,* 266–276.

Ricks, D. A. (2006). *Blunders in international business* (4th ed.). Malden, MA: Blackwell.

Roach, R. (January 31, 2002). Making Web sites accessible to the disabled. *Black Issues in Higher Education, 16,* 32.

Rogers, C. R. (1952). Communication: Its blocking and its facilitation. *Harvard Business Review, 30,* 46–50.

Ruben, B. D. (1976). Assessing communication competency for intercultural adaptation. *Group and organization studies, 2,* 470–479.

Ruch, W. V. (1984). *Corporate communications: A comparison of Japanese and American practices.* Westport, CT: Quorum.

Sauer, B. A. (2003). *The rhetoric of risk: Technical documentation in hazardous environments.* Mahwah, NJ: Erlbaum.

Schriver, K. (1997). *Dynamics of document design: Creating text for readers.* New York: Wiley.

Scollon, R. (1999). Official and unofficial discourses of national identity: Questions raised by the case of contemporary Hong Kong. In R. Wodak and C. Ludwig (eds), *Challenges in a changing world: Issues in critical discourse analysis* (pp. 21–5). Vienna: Passagen Verlag.

Scollon, R., & Scollon, C. W. (2001). *Intercultural communication* (2nd ed.). Malden, MA: Blackwell.

Scott, M., & Rothman, H. (1992). *Companies with a conscience: Intimate portraits of twelve companies that make a difference.* New York: Carol.

Simonson, M., Smaldino, S., Albright, M., & Zvacek, S. (2003). *Teaching and learning at a distance: Foundations of distance education.* Upper Saddle River, NJ: Pearson.

Smith, F. (2004). *Understanding reading* (6th ed.). Hillsdale, NJ: Erlbaum.

Smith, L. B. (28 June 1993). A user's guide to PCC (politically correct communiqué). *PCWeek, 10*(25), 204.

Spilka, R. (Ed.). (1993). *Writing in the workplace: New research perspectives.* Carbondale: Southern Illinois UP.

Stanford University. (2010). *Copyright & fair use.* http://fairuse .stanford.edu/Copyright and_Fair_Use_Overview/chapter9/ index.html. January 4, 2010.

Steier, F. (1999). A relational framework for professional communication in international organizations. In C. R. Lovitt and D. Goswami (Eds.), *Exploring the rhetoric of international professional communication: An agenda for teachers and researchers* (pp. 157–170). New York: Baywood.

Sternthal, B., Dholakia, R., & Leavitt, C. (1978). The persuasive effect of source credibility: Tests of cognitive response. *Journal of Consumer Research, 4,* 252–260.

Stevenson, D. W. (1983). Audience analysis across cultures. *Journal of Technical Writing and Communication, 13,* 319–330.

Suchan, J., & Colucci, R. (1989). An analysis of communication efficiency between high-impact and bureaucratic written communication. *Management Communication Quarterly, 2,* 464–473.

Thompson, M. A. (2000). *The global resume and cv guide.* New York: Wiley.

Thorell, L. G., & Smith, W. J. (1990). *Using computer color effectively: An illustrated reference.* Englewood Cliffs, NJ: Prentice Hall.

Tinker, M. A. (1969). *Legibility of print.* Ames: University of Iowa Press.

Toulmin, S., Rieke, R., & Janik, A. (1984). *An introduction to reasoning* (2nd ed.). New York: Macmillan.

Tversky, A., & Kahneman, D. (1974). Judgment and uncertainty: Heuristics and biases. *Science, 185,* 1124–1131.

van Dijk, T. (1980). *Macrostructures: An interdisciplinary study of global structure in discourse, interaction, and cognition.* Hillsdale, NJ: Erlbaum.

Varner, I. & Beamer, L. (2005). *Intercultural communication in the global workplace.* Boston: McGraw-Hill.

Warren, E. K., Roth, L., & Devanna, M. (Spring 1984). Motivating the computer professional. *Faculty R&D*. New York: Columbia Business School, 8.

Watson, C. E. (1991). *Managing with integrity: Insights from America's CEO's*. New York: Praeger.

Weak writers. *The Wall Street Journal* (14 June 1985), A1.

Webb, J., & Keene, M. (1999). The impact of discourse communities on international professional communication. In C. R. Lovitt, & D. Goswami (Eds.), *Exploring the rhetoric of international professional communication* (pp. 81–110). New York: Baywood.

Wells, B., Spinks, N., & Hargarve, J. (June 1981). A survey of the chief personnel officers in the 500 largest corporations in the United States to determine their preferences in job application letters and personal résumés. *ABCA Bulletin, 14*(2), 3–7.

Westinghouse Corporation. (1981). *Danger, warning, caution: Product safety label handbook*. Author.

Wheildon, C. (1995). *Type & layout: How typography and design can get your message across—or get in the way*. Berkeley, CA: Strathmoor.

White, J. V. (1990). *Color for the electronic age*. New York: Watson-Guptil.

Williams, J. M. (2005). *Style: Ten lessons in clarity and grace* (8th ed.). New York: Pearson Longman.

Williams, R. (1994). *The non-designer's design book*. Berkeley, CA: Peachpit.

Wiseman, R. L., Hammer, M. R., & Nishida, H. (1989). Predictors of intercultural communication competence. *International journal of intercultural relations, 13*, 349–370.

Wolvin, A. D., & Oakley, C. (1985). *Listening*. Dubuque, IA: Brown.

INDEX

cause and effect relationships, 252–56
  explaining, 252–53
  logical fallacies in arguments about, 254
  persuading readers that relationship
     exists, 253–56
cells, in grids, 381
chalkboards, for oral presentations, 479
chapters. *See* drafting
charts
  flowcharts, 374–75
  Gantt, 377
  organizational, 376
  pie, 366–67
  schedule, 377
chat, 449, 468
checking drafts. *See* revising
checklists. *See* writer's guides
checkpoint strategy, 90
chunking, 388–90
citations. *See* documenting sources
claim, in reasoning, 128–29
clarifiers, in teams, 462
classification, 236
  formal, 236–38
  informal, 239–41
client projects, 524–39
  advocating and educating client,
     536–37
  communicating with client about,
     531–36
  deferring to client, 536–37
  deliverable in, 528–29
  determining client's objectives for,
     526–27
  handing off/transmittal, 537
  overall management strategy, 526
  project management plan for, 528–30
  writer's guide for, 538
  written proposal and agreement for,
     530–35
  your own assessment of, 527–28
closed questions, 187, 189–90
closings, in letters, 547–50
  of transmittal, 315
cloud computing, 453
cluster sketches, 171
collaboration. *See also* teams
  computer programs to aid, 466–68
  feedback and evaluation form, 472–73
  online, 452–53

persuasion in assistance of, 118–19
  in writing, 7, 452–53
color
  in graphics, 344–47
    ethical use of, 354
    guidelines for use of, 344–47
  in visual design, 392
commenting on files, for collaboration,
    467
communication
  with clients, 531–36
  cultural differences and, 79–82
  dynamic interaction in, 12–18
  effective, characteristics of, 11–12
  ethics in, 20–21
  expertise, career success an, 4
  introduction to, 3–23
  overview of reader-centered, 25–63
  single, to satisfy diverse readers, 6
  social dimensions of, 7–8
comparison, 242–45
  divided vs alternating pattern for,
    242–43
complex audiences, 85
  letter to, 88
compromisers, in teams, 462
computerized full-text sources, 183
computer projections, 479. *See also*
    PowerPoint
computers. *See also* Internet; web pages;
    websites
  for collaboration, 466–68
  for creating graphics, 347–48, 350–51
  for creating photographs, 369
  file naming rules for team projects, 458
  for oral presentations, 479, 482, 484,
    489
  for outlining, 109, 110
  for revising, 411–12
  screen shots with, 372–73
  thumbnails for page design, 384–85
computer technology, choosing, 466–68
concerns of readers, 125–27
conclusion. *See also* closings, in letters;
    ending communications
  of job application letter, 55
  of oral presentation, 485
conclusions
  in reports
    feasibility, 630

progress, 641
  research, 200, 588–90
  from research, 195, 200
  stating in graphics, 349
conflict, handling in team meetings,
    464–65
conformity, revising for, 423–24
connotation, 280
consistency, revising for, 423
constraints on writing, 89
contact information, in résumé, 31
content, design and, 385
context, of readers, 15–16, 69
  describing, 97–99
  guideline for understanding, 86–89
contrast, in visual design, 391–94
convenience samples, 192
conventions
  for correspondence, 547
  in instructions, 643
conversational style, for oral presenta-
    tions, 486–87
copies, listing people copied on letters
    and memos, 550, 551
copyright law, 159–61
correctness, revising for, 423
correspondence. *See* e-mail; job applica-
    tion letters; letters
costs, in proposals, 570
counterarguments of readers, 125–27
  addressing, 126–27, 562
  valuing, 202
covers, 312–14
  writing, 315, 317
credibility
  conveying to readers, 133–36, 254
  establishing, 294
  of information sources, 151, 156–57
  in research reports, 583
  strategies for building, 134, 135
criteria, in feasibility reports, 619–28
critical path, 530
critical thinking, 201–2
cultural characteristics, of readers, 79–84
cultural diversity, 79–84
  adapting beginning of communication
    for, 300
  adapting graphics to, 352–53
  adapting oral presentations to, 493–95
  adapting persuasive strategies to, 140

drafting (*continued*)
  human consequences and ethics in, 228–29
  moving from most important to least important information, 214
  organizational patterns for, 235–62
    cause and effect, 252–56
    combinations of, 260
    comparison, 242–45
    cultural background and, 215
    formal classification, 236–38
    informal classification, 239–41
    partitioning, 246–47
    problem and solution, 257–59
    segmenting, 248–51
    signposts and maps for, 216–25
  organization and, 214–15, 214–29
  persuasiveness and, 209
  similarities of paragraphs, sections, and chapters, 208–9
  smoothing flow from sentence to sentence, 225–27
  topic statements, at beginning, 209–13
  transitions and, 217–18, 225–26
  usability and, 208
  visual arrangement of text, 222–25
  visual elements and, 331–404
  of websites, 510–21
drafts
  discussing in team meetings, 465–66
  editing collaboratively by teams, 467–68
  revising. *See* revising
  testing. *See* testing drafts for usability and persuasiveness
drawings, 370–71, 660
  exploded, 371
  flowcharts with, 375
  as research tool, 169–72
dryboards, for oral presentations, 479
dynamic appeal, 134

echo words, 226
editing. *See* revising
educating clients, 536–37
education, in résumé, 32–33
electronic résumés, 45–49
electronic sources
  APA style for, 682
  MLA style for, 688–89

e-mail, 448–49
  citation styles for
    APA, 682
    MLA, 689
  collaboration with, 452–53
  guidelines for writing, 450, 453
  lack of privacy of, 451–52
  lists and headings in, 450, 451
  résumés submitted by, 47
  subject lines in, 288
emotions, appealing to, 120, 137–40
empirical research reports. *See* research reports, empirical
employers. *See also* job application letters; résumés
  adapting job applications letters for, 56–57
  adapting résumés for, 43–46
  identifying, for writing résumés, 27
  information sought in job application letters, 50
  learning about, 51
  point of view of, checking drafts from, 409–10
  qualities sought in new employees, 28
encouragers, in teams, 462
ending communications, 303–10.
    See also closings, in letters; conclusion
  after making last point, 304–5
  focusing on key feeling, 307–8
  identify further study needed, 308
  referring to goals, 306–7
  repeating main point, 305
  social conventions for, 309
  summarizing key points, 305–6
  telling readers how to get assistance or more information, 308
  telling readers what to do next, 308
endnotes, 326
energizers, in teams, 462
English
  plain, 266, 546–47
  readers not fluent in, 275
environmental impact statements, 112
ethical companies, 20–21
ethics, 9–10, 20–21. *See also* stakeholders
  avoiding stereotypes, 268–69
  in business, 20–21
  code of, 20, 21

defining ethical behavior, 20–21
  for documenting sources, 161–62
  for drafting text, 228–29
  for graphics, 353–54
  human consequences, remembering, 228–29
  inclusive language and, 282–83
  informed consent from test readers and, 440, 441
  intellectual property and, 158–62
  persuasive techniques and, 140–41
  for research, 158–62
  for résumés and job search letters, 58
  stakeholder impacts and, 112–13
  stakeholders for communications, identifying, 90–93
  strategies for ethical writing, 90
  unethical practices, addressing, 300–301
  in usability planning, 112–13
  for websites, 509–10, 518–20
ethos (credibility), 120
evaluation
  of collaboration, 472–73
  in feasibility report, 629–30
  of front and back matter, 314
  of research sources, 156–57, 202
evidence
  credible, 254
  in reasoning, 128–29
  sufficient and reliable, 129, 561
examples, as evidence, 129
executive summary, 106, 320, 322
expectations
  of employers, for front and back matter, 314
  of listeners, for oral presentations, 483
  of readers
    cultural expectations, determining, 109–10
    for front and back matter, 314
    matching writing style to, 264–65
    previewing organization for, 292–93
experiential résumés, 30, 34, 35
expertise, 134
expert testimony, 129
eye contact
  cultural background and, 469–70
  in oral presentations, 490–91, 494
  in team discussions, 469–70

method *(continued)*
  feasibility, 628–29
  research, 586–87
milestones
  for client projects, 529
  in project schedules, 460–61
mining accidents, ethics example of, 228–29
misleading, avoiding, 141
MLA style
  for in-text citations, 683–84
  for works cited list, 684–89
mobility limitations, websites for readers with, 520
modular design
  of reports and proposals, 106
  of websites, 87, 107
motivation, for use of instructions, 653
movies, digital, 671
multicultural audiences, 6, 520–21

name, in résumé, 31
navigational aids, for websites, 515–17
needs, 122
nervousness, in oral presentations, 496
networks. *See* globally networked environment
newspaper indexes, 181
note taking
  in interviews, 188
  in oral presentations, 492
  for research, 157–58

objectives
  defining. *See* defining objectives
  discussing during review, 414
  professional, in résumés, 31–32
  for research, 152–53, 197
  for testing drafts, 430–31
online instructions, 668–71
on-the-job communication. *See also* writing at work
  effective, 11–12
  overview of, 3–24
on-the-job research. *See* research
open (open-ended) questions, 187, 189–90
  for testing drafts for persuasiveness, 436
openings. *See also* beginning communications; introductions
  of letters, 547–50

of oral presentations, cultural background and, 494
openness, encouraging, 293–95, 415
opinion seekers, in teams, 462
oral presentations, 475–500
  adapting to audience's cultural background, 493–95
  adapting to listeners not fluent in English, 495
  assignments/projects for, 702–3
  audience and expectations for, 483
  conversational speaking style for, 486–87
  defining objectives of, 476–77, 480
  graphics for, 487–90
  interruptions and questions during, 493
  involving audience in, 490–92
  main points, focusing on, 483–85
  nervousness, dealing with, 496
  planning verbal and visual parts as single package, 477–83
  questions, inviting and responding to, 492–93
  rehearsing, 495–96
  structure for, 485–86
  by teams, 497–98
  tutorial on writing, 480–82
  type of delivery, 477–78
  visual medium for, 478–83
    advantages and disadvantages of, 479
    integrating with verbal, 483, 484
  writer's guide for, 499–500
organization. *See also* organizational patterns
  around readers' tasks, 101–5
  cultural background and, 215
  of directions/instructions, 650
  forecasting statements for, 216–17, 485
  icons and rules to signal, 394
  of information in prose, 214–15
  of instructions, 652–53
  modular design in, 106, 107
  of oral presentations, cultural background and, 494
  outlining, 108–9
  for persuasion, 130–33

previewing at beginning, 292–93
signposts in drafting, 215–29
tight fit among parts of communication and, 132–33
for usability, 101–5
visual arrangement of text and, 222–25
organizational charts, 376
organizational culture, 9, 82
organizational goals, 121, 122
organizational hierarchies, 80–81
organizational patterns, 215–29, 235–62
  cause and effect, 252–56
  combinations of, 260
  comparison, 242–45
  direct vs. indirect, 131–32
  formal classification, 236–38
  informal classification, 239–41
  linear vs. nonlinear, 215
  partitioning, 246–47
  problem and solution, 257–59
  segmenting, 248–51
  signposts for, 216–25
organizational role, of readers, 76–77
outlined talks, 478
outlining, 108–9, 211. *See also* superstructures
  computer use for, 109, 110
  informal classification and, 241
  for planning team projects, 459
  for progress reports, 637–39, 640
overgeneralization, 254
overhead transparencies, 479

pages. *See also* design; web pages
  design elements of, 380–85, 650
  grid as visual framework for, 381–85
  informational, 694–95
  size of, 398
  website, design of, 510–15
  writer's tutorial for, 382–83
paper, 381, 398
paragraphs. *See* drafting
parallel construction, 225
partitioning, 246–47
partner, presenting yourself as, 289–91
passive voice, 272–73
patent law, 158
pathos (appeals to emotion), 120, 137
performance tests, 432–34

virtual meetings, 454
virtual teams, 471
visual arrangement of text, 222–25
visual design, 7, 8, 379. *See also* design;
    graphics
  of graphics, 331–56
  of headings, 220–22
  of pages and documents, 378–404
  of résumés, 38–43
  for usability and persuasiveness, 379
visual impairment, websites for readers
    with, 518–20
visualizing, 169–71
visual media. *See also* graphics
  for oral presentations, 478–83
    cultural background and, 495
voice of verbs, active versus passive,
    272–73
voice of writer, 264–69
  adapting to readers' cultural
    background, 268
  attitude communicated by, 267
  avoiding stereotypes, 268–69
  matching to readers' expectations,
    264–65
  roles created by, 266–67
  suppressing your own, 267–68
  using your own words, 267–68

warnings, 643–44
web page résumés, 49
web pages. *See also* websites
  attractiveness of, 514–15
  combining usability and persuasive-
    ness, 13
  designing, tutorial on, 512–14
  instructions on, 668–71
  readability of, 511
  reader-centered, 124
  signaling organization of, 510–11
  usability and attractiveness of, 510–15
websites, 501–23. *See also* web pages
  accessibility for readers with
    disabilities, 518–20
  building credibility with, 136
  copyright law for, 161
  defining objectives for, 507
  digital portfolio, 503
  drafting, 510–21
  emotional appeals with, 139

ethical issues regarding, 509–10,
    518–20
gathering information for, 509
grid patterns for, 512–14
for international and multicultural
    users, 520–21
learning about readers of, 507
map of, 507–8
modular design of, 87, 107
navigational aids for, 515–17
page design for, 510–15
planning, 507–8
reader-centered process for
    creating, 502
research for, 509–10
résumés submitted by, 49
revising, 521
site maps for, 517
testing, 521
tutorial for creating, 504–6
unifying verbally and visually, 518, 519
URLs of, 158, 175
writer's guide for creating, 522
writing assignment for, 693–94
whistleblowing, 301
white space, as design element of page, 380
wikis, 452
William's principles, for design, 386
word choice, 275–83
  accurate use of words, 278–79
  associations of (connotation and
    register), 280–81
  concrete, specific words, 276–77
  cultural background, consideration
    for, 281
  inclusive language for, 282–83
  plain vs. fancy words, 279
  specialized terms, use of, 277–78
  using your own words, 267–68
wordless instructions, 332
word processors
  for consistency, 395
  for page design, 394–98
words
  cultural differences in interpretation of, 82
  echo, 266
  for Internet searches, 174–75
  transitional, 255–56
  unfamiliar, defining, 278
  voice and, 267–68

word searches, in library, 179–80
work
  communication and, 3–24
  writing and, 5–10
work experience, in résumé, 33–36
works cited list, MLA style for, 684–89
World Wide Web. *See* Internet
writer-centered facts, 50
writer's guides
  for checking your own drafts, 413
  for creating graphics, 356, 357–77
  for defining communication
    objectives, 70, 91–93
  for designing grid patterns for print,
    382–83
  for job application letters, 57
  for managing client and service-
    learning projects, 538
  for oral presentations, 480–82, 499–500
  for planning persuasive strategies, 142
  for planning usability strategies, 114
  for reader-centered organizational
    patterns, 235–62
  for reader-centered research, 163,
    165–93
  for résumés, 44–45
  for reviewing drafts, 420
  for revising empirical research
    reports, 608–9
  for revising feasibility reports, 631–32
  for revising instructions, 672–73
  for revising progress reports, 644–45
  for revising proposals, 576–78
  for testing drafts, 442
  for website creation, 522
writing, major activities of, 22
writing assignments. *See* projects
writing at work. *See also* defining
    objectives; drafting
  as action, 10
  applying cultural knowledge to, 82–83
  characteristics of, 5–10
  collaboration in, 7, 452–53
  constraints on, 89
  legal and ethical issues in, 9–10, 90–93
writing style, 19
  for résumés, 30–31

you-attitude, adopting for
    correspondence, 544–45